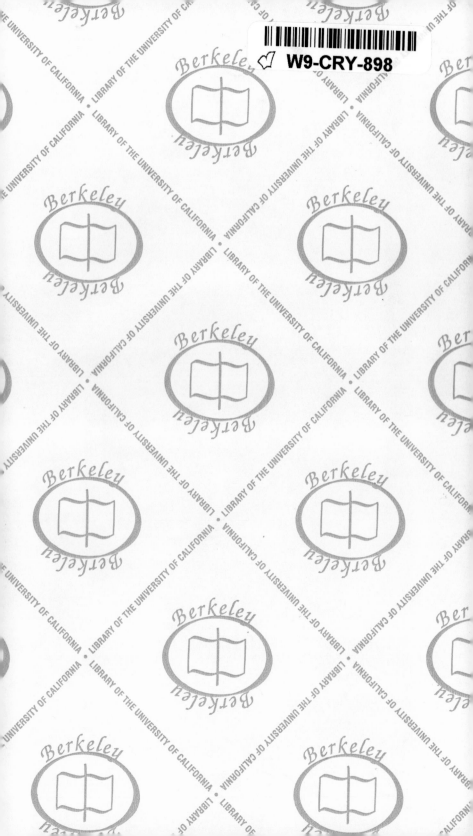

ACS SYMPOSIUM SERIES **343**

Proteins at Interfaces
Physicochemical and Biochemical Studies

John L. Brash, EDITOR
McMaster University

Thomas A. Horbett, EDITOR
University of Washington

Developed from a symposium sponsored by
the Division of Colloid and Surface Chemistry
at the 192nd Meeting
of the American Chemical Society,
Anaheim, California,
September 7–12, 1986

American Chemical Society, Washington, DC 1987

Library of Congress Cataloging-in-Publication Data

Proteins at interfaces.
  (ACS symposium series, ISSN 0097-6156; 343)

"Based on a symposium sponsored by the Division of
Colloid and Surface Chemistry at the 192nd meeting of
the American Chemical Society, Anaheim, California,
September 7-12, 1986."

  Includes bibliographies and indexes.

  1. Proteins—Congresses. 2. Surface chemistry—
Congresses. 3. Biological interfaces—Congresses.

  I. Brash, John L., 1937-   . II. Horbett, Thomas A.,
1943-   . III. American Chemical Society. Division
of Colloid and Surface Chemistry. IV. American
Chemical Society. Meeting (192nd: 1986: Anaheim,
Calif.) V. Series.

QP.P89778   1987   574.19′245   87-14394
ISBN 0-8412-1403-4

Foreword

The ACS SYMPOSIUM SERIES was founded in 1974 to provide a medium for publishing symposia quickly in book form. The format of the Series parallels that of the continuing ADVANCES IN CHEMISTRY SERIES except that, in order to save time, the papers are not typeset but are reproduced as they are submitted by the authors in camera-ready form. Papers are reviewed under the supervision of the Editors with the assistance of the Series Advisory Board and are selected to maintain the integrity of the symposia; however, verbatim reproductions of previously published papers are not accepted. Both reviews and reports of research are acceptable, because symposia may embrace both types of presentation.

Contents

ROLE OF PROTEIN ADSORPTION
IN BLOOD–MATERIAL INTERACTIONS

APPLICATIONS OF PROTEINS AT INTERFACES

Preface

THE CURRENT STATE OF KNOWLEDGE of proteins at interfaces is reflected in this book. Developed from a symposium that was one of a continuing series entitled "Surface Chemistry in Biology, Dentistry, and Medicine," the book is organized around the subtopics of behavior, mechanisms, methods of study, blood-material interactions, and applications of proteins at solid-liquid, air–water, and oil–water interfaces.

We asked authors who had worked in this field for some time to provide minireviews or overviews of their previous work. We also accepted original contributions from them and from those new to the field. Some authors chose to forgo the minireview approach in favor of newer work, but in general many of the contributions provide the broader view we had hoped for.

The content of this book is quite diverse. Many factors contribute to this diversity, including the fact that the contributors' formal training varies widely. Investigators trained as immunologists, biochemists, polymer chemists, chemical engineers, and physical chemists provided chapters. Subjects range from the behavior of prothrombin at oil–water interfaces, to enhancement of albumin binding of certain biomaterials, to studies of protein foam stability. Finally, methods used also vary. Each technique is now recognized as inherently sensitive to certain aspects of proteins at interfaces but insensitive or inapplicable to the measurement of other aspects. For example, in situ ellipsometry is an exquisitely sensitive method but must be used with highly smooth, reflective surfaces and cannot be used easily to detect one protein among other proteins in a mixture.

The book's broad range can make easy understanding of the field difficult for nonspecialists. On the other hand, the book provides a rich source of information for those motivated enough to pursue the topic. We hope that *Proteins at Interfaces* will be of interest and use to both experienced investigators and to newcomers who need to learn more about the field.

JOHN L. BRASH
Department of Chemical Engineering
McMaster University
Hamilton, Ontario L8S 4L7, Canada

THOMAS A. HORBETT
Department of Chemical Engineering, BF-10, and
   Center for Bioengineering
University of Washington
Seattle, WA 98195

December 19, 1986

Chapter 1

Proteins at Interfaces: Current Issues and Future Prospects

Thomas A. Horbett[1] and John L. Brash[2]

[1]Department of Chemical Engineering, BF-10, and Center for Bioengineering,
University of Washington, Seattle, WA 98195
[2]Department of Chemical Engineering, McMaster University, Hamilton,
Ontario L8S 4L7, Canada

The ability of proteins to influence a wide variety of
processes that occur at interfaces is well recognized.
The biocompatibility of clinical implants, mammalian and
bacterial cell adhesion to surfaces, initiation of blood
coagulation, complement activation by surfaces, solid
phase immunoassays, and protein binding to cell surface
receptors all involve proteins at interfaces.
Furthermore, practical problems such as contact lens
fouling, foaming of protein solutions, and fouling of
equipment in the food processing industry, are direct
consequences of the relatively high surface activity of
proteins. In general, any process involving an interface
in which contact with a protein solution occurs is likely
to be influenced by protein adsorption to the interface.
Thus, several reviews of protein adsorption have been
published (1-5).

Since previous reviews provide excellent coverage of
the generally well understood or frequently studied
aspects of the interfacial behavior of proteins, this
chapter will focus on several facets of protein
adsorption that have so far not been examined in much
detail. While this approach is atypical for an overview
chapter, it is in keeping with the intent of this book to
provide information to the reader that reflects more
recent developments in this field. Furthermore, as will
be seen, the topics to be discussed necessitate re-
examination of previous studies and provide some unifying
views of this rather diverse science.

The main topics to be presented include the origins
of the surface activity of proteins, multiple states of
adsorbed proteins, and the competitive adsorption
behavior of proteins. These topics were chosen because it
appears that a better understanding of each is necessary

0097-6156/87/0343-0001$09.25/0

to describe many of the interfacial phenomena involving
proteins, yet the fundamental concepts underlying each
have not been discussed as fully as we hope to do in this
chapter. Finally, we describe some future areas of
research that are likely to yield important advances in
our understanding of protein behavior at interfaces.

On the Origins of Differences in the Surface Activity of Proteins

Molecular Properties Influencing Surface Activity of Proteins.

The molecular properties of proteins that are
thought to be responsible for their tendency to reside at
surfaces are summarized in Table I. The size, charge,
structure, and other chemical properties of proteins that
presumably influence surface activity are all
fundamentally related to their amino acid sequence, which
is fixed for each type of protein but varies greatly
among proteins. Thus, differences in surface activity
among proteins arise from variations in their primary
structure. At this point, further enquiries into the
origin of surface activity differences among proteins
become quite problematical because little detailed
information is available that relates variation in the
primary structure of proteins to changes in the surface
activity of the molecules. However, a discussion of
specific factors will serve to clarify our concepts in
this regard.

Size is presumably an important determinant of
surface activity because proteins and other
macromolecules are thought to form multiple contact
points when adsorbed to a surface. The irreversibility
typically observed for proteins adsorbed to surfaces is
thought to be due to the fact that simultaneous
dissociation of all the contacts with the surface is an
unlikely event. Multiple bonding is also indicated by the
relatively large number of protein carbonyl groups that
contact silica surfaces upon adsorption (6). The bound
fraction of peptide bond carbonyl groups, as calculated
from shifts in infrared frequencies after adsorption, has
been found to be in the range of 0.05-0.20 (6). The bound
fractions correspond to 77 contacts per adsorbed albumin
molecule, and up to 703 contacts per adsorbed fibrinogen
molecule (6). On the other hand, size is clearly not on
overriding factor determining the surface activity
differences among proteins. For example, hemoglobin
appears to be far more surface active than fibrinogen
(7), yet the molecular weight of hemoglobin (65,000) is
approximately 1/5 that of fibrinogen (330,000). While
albumin nearly is the same size as hemoglobin, it is much
less surface active. Finally, slight variations in the
amino acid sequence of hemoglobin make large differences
in surface activity even though these variants have the
same molecular weight (see below).

Table I. Molecular Properties of Proteins
Possibly Influencing Their Surface Activity

1. <u>Size</u>:   larger molecules may have more contact points.

2. <u>Charge</u>: molecules nearer their isoelectric pH may
          adsorb more easily.

3. <u>Structure</u>:

    a. Stability: less stable proteins may be more
                       surface active.

    b. Unfolding rates: more rapid unfolding may favor
                       surface activity.

    c. Cross-linking: -S-S- bonds may reduce surface
                       activity.

    d. Subunits: more subunits may increase surface
                       activity.

4. <u>Other chemical properties</u>:

    a. Amphipathicity: some proteins may have more of the
                       types of side chains favored for
                       bonding.

    b. "Oiliness":   more "hydrophobic" proteins may be
                       more surface active.

    c. Solubility: less soluble proteins may be more
                       surface active.

The charge and charge distribution of proteins are likely to influence surface activity because it is known that most of the charged amino acids reside at the exterior of protein molecules. These charged residues must therefore come into close proximity with the surface in the process of adsorption. Experimentally, proteins have frequently been found to exhibit greater adsorption at or near the isoelectric pH, perhaps because charge-charge repulsion among the adsorbed molecules is minimized under these conditions. However, Norde has concluded that the reduction in adsorption at pH's away from the isoelectric is due to structural rearrangements in the adsorbing molecule, rather than charge repulsion (4). In this context, it is pertinent to note that the isoelectric pH (pI) of hemoglobin is near neutrality (7.2) and that this protein is much more surface active at pH 7.4 than either fibrinogen (pI = 5.5) or albumin (pI = 4.8). It would be of interest to compare the surface activity of these molecules at pH's other than 7.4 to determine whether the ranking of surface activities changed as the isoelectric pH of each protein was approached. The role of protein surface charge is especially important and probably predominant at interfaces with fixed ionic charges, as shown by the ability to adsorb proteins to ionized matrices. Adsorption to this type of surface is strongly affected by the degree of opposite charge on the protein and the degree of competition provided by like charged ions in the buffer. Adsorption to charged matrices is the basis for the widely applied separation of proteins by ion exchange chromatography.

Structural factors important in the surface activity of proteins are not well understood. We may speculate that proteins likely to unfold to a greater degree or that unfold more rapidly would be more surface active because more contacts per molecule could be formed and because the configurational entropy gain favors the adsorption. Thus, disulfide cross linked proteins would be less likely to unfold as rapidly or completely and therefore be less surface active. This prediction is amenable to experimental test since reduction of disulfide bonds can be done specifically and completely with very mild reagents. The only known test of this idea was the observation that disulfide bond reduction by thioglycollic acid increased the number of bonds formed by albumin adsorbed to silica by about 50% (6). On the other hand, additional cross-linking of albumin with diethyl malonimidate did not reduce the number of bonds formed (6), perhaps because native albumin is already heavily cross-linked by 16 disulfide linkages (8). Finally, the existence of non-covalently bonded subunits in a protein may favor surface activity because rearrangements of the inter-subunit contacts to allow more contact of each subunit with the surface can

probably occur more readily than rearrangements within
each subunit. Measurement of the relative surface
activity of the subunits of hemoglobin in comparison to
the tetrameric whole molecule might provide an
interesting test of this idea.

Chemical differences among proteins arising from the
particular balance of amino acid residues in each protein
probably are also important factors influencing the
surface activity of proteins. The amphipathic nature of
proteins, due to the presence of hydrophobic, hydrophilic
and charged amino acid side chains, provides an
opportunity for bonding to sites that vary considerably
in chemical nature. Thus, for a particular surface, some
proteins may have more of the type of residue that favors
bonding to the kind of adsorption sites prevalent on this
surface, and therefore would be more surface active than
other proteins. More generally, the idea that proteins
have a hydrophobic or oily core suggests that proteins
that are more hydrophobic may be preferred on many
surfaces, especially in view of the apparent importance
of hydrophobic interactions in protein interactions with
some surfaces (9). Lastly, since the solubility of a
protein in the bulk phase is a complex function of its
overall chemical composition, and because adsorption to
an interface can be thought of as insolubilization or
phase separation, it could be that differences in
solubility are important indicators of differences in
surface activity. However, the rather high solubility of
hemoglobin (ca 300 mg/ml inside red cells) argues against
this idea because this protein is quite surface active
(7).

Surface Activity of Hemoglobin Genetic Variants. The
best experimental evidence on the molecular origins of
differences in the surface activity of proteins has come
from study of the behavior of hemoglobin genetic variants
at the air/water interface (10-14). The differences in
surface activity of these variants were originally
indicated by the fortuitous observation by Asakura *et al.*
that hemoglobin S solutions tend to form precipitates
when shaken, unlike solutions of the normal hemoglobin A
variant (10). Hemoglobin S is predominant in the red
cells of humans with the sickle cell disease. The rate of
precipitation induced by mechanical shaking is referred
to as "mechanical stability" in this literature.

Since shaking of protein solutions induces bubble
formation, and because agitation without bubble formation
(by slow stirring) causes a much slower rate of
precipitation, the enhanced precipitation rate of
hemoglobin S solutions was attributed to an enhanced rate
of surface denaturation at the air/water-liquid
interface. This idea was confirmed by direct
measurements of the properties of hemoglobin films at the
air/water interface with a surface balance. The surface

balance experiments showed that surface pressure kinetics
(π - t) and isotherms (π - A) for hemoglobin S and other
variants were markedly different from hemoglobin A (14).
The decrease in surface pressure following injection of
hemoglobin solutions into the subphase occurred more
quickly and was greater at steady state for hemoglobin S
than for hemoglobin A. Furthermore, the π - A curves for
the two variants became much more alike when done at
lower temperatures (14), in agreement with the
observation that differences in mechanical stability
among the genetic variants tend to disappear at lower
temperatures (10). The pressure-area isotherms for the
variants, obtained by compression of the protein films,
also showed distinct differences. The sharp increase in
the resistance to further compression (attributed to
monolayer formation) occurred at an area of 8000 $Å^2$/
molecule for hemoglobin S compared to 5000 $Å^2$/molecule
for hemoglobin A. The greater area per molecule suggests
a greater degree of unfolding of the hemoglobin S
molecule compared to hemoglobin A.

        Study of the mechanical stability of other
hemoglobin variants has resulted in the following
ranking: HbA $\approx$ HbC (β6Glu$\to$ Lys) $\approx$ HbF (γ chain
replaces β chain) $\approx$ HbA$_2$ (δ chain replaces β chain) $\approx$ Hb
Deer Lodge (β2 His$\to$Asp)< Hb Korle Bu (β73 Asn $\to$ Asp) <
HbS (β6Glu $\to$ Val) < Hb C$_{Harlem}$ (β6Glu $\to$ Val; β73 Asp $\to$
Asn) (13). The notations in parenthesis indicate the
amino acid substitutions e.g., β6Glu $\to$ Val means the
glutamic acid at position 6 in the β subunit has been
replaced with a valine residue. The majority of these
differences in mechanical stability are attributable to
differences in surface activity i.e., the π-A or π-t
isotherms at the air/water interface have been shown to
vary considerably for these variants. However, some
apparent exceptions to this correlation exist, e.g. no
difference in the surface activity of Hb Korle Bu and HbA
was observed despite their difference in mechanical
stability.

        The large difference in the surface activities of
HbA and HbS apparently arises from a single Glu $\to$ Val
amino acid substitution at position 6 in the β chain.
Similarly, the variant Hb C$_{Harlem}$, which has an
additional Asn $\to$ Asp substitution at β73, is even more
unstable. In contrast, Hb Korle Bu, having only the Asn $\to$
Asp substitution at β73, is much more stable than HbS.
These results clearly indicate that seemingly minor
changes in primary structure can induce large changes in
the surface activity of proteins. On the other hand, the
data also show that the multiple differences resulting

from replacement of the β chain with either the γ chain in HbF or the δ chain in HbA_2 do not introduce significant changes in surface activity. Thus, the amino acid composition of the γ chain varies from the β chain at 38 out of the 146 amino acid positions (15) without apparent effect on the surface activity of HbF compared to the normal variant, HbA.

The differences in surface activity of some of the hemoglobin genetic variants are difficult to rationalize in terms of the effects of size, charge, or chemistry that were discussed previously (see Table I) because these effects presumably operate in an additive or cumulative way. Thus, for example, it is difficult to see how the change of a single residue could change the balance of amphipathicity greatly because all the residues of the protein influence this property. Instead, the large effect of single amino acid substitutions on hemoglobin surface activity points to a very important role for structural stability in the interfacial behavior of proteins. Since structural transitions in proteins occur in a cooperative fashion, e.g., protein "melting" occurs over a narrow range of temperature, structural stability could be strongly influenced by single amino acid substitutions.

The importance of structural transitions on the surface activity of Hb variants is also indicated by the fact that mechanical stability and surface activity differences among the variants tend to disappear at lower temperatures and in the deoxygenated state. The temperature effects have been attributed to stabilization of hydrophobic interactions at lower temperatures (14). The much greater mechanical stability and lower surface activity of the deoxygenated forms of hemoglobin S and other variants has also been interpreted to mean that structural transitions are being influenced by the substitutions, because it is known that the oxy and deoxy forms differ in their three dimensional structures. Furthermore, deuteration of HbS reduces its surface activity, an effect thought to arise from the stronger hydrogen bonding and van der Waals interactions that occur in this solvent (14). The location of the β6 Glu → Val substitution in HbS at the outside of the molecule, where it does not contact other residues, the fact that HbC with a β6 glu → lys substitution retains its stability, and finally the requirement for the oxygenated state to reduce stability, suggest that the surface activity of the hemoglobin molecule (and perhaps other proteins) is very sensitive to minor changes in the conformational states achievable.

Multiple States of Adsorbed Proteins

Background. The idea that adsorbed proteins may exist in
more than one state has been taken into account in more
recent models of protein adsorption (16-18), although
many earlier investigators used a simple Langmuir model
that does not consider such a possibility. However,
review of the literature on protein adsorption reveals
that there is a large amount of experimental evidence
supporting the concept of multiple states. In addition,
it is found that there are quite plausible mechanisms to
explain how and why proteins could reside at interfaces
in more than one way. The motivation for these thoughts
came originally from attempts to understand changes in
the detergent elutability of adsorbed proteins, so these
experiments and an analysis of them are briefly presented
by way of a concrete introduction to the concept of
multiple states.
     The ability to remove fibrinogen or albumin from a
variety of polymeric surfaces with the detergent sodium
dodecylsulfate (SDS) was found to gradually decrease if
the time between adsorption and elution was lengthened
(19). Since the elution conditions were held constant,
the data suggested that the binding strength of the
adsorbed proteins had changed over time, indicating that
they could exist in more than one way, or "state" on the
surface. The rate of loss of SDS elutability was
strongly enhanced at elevated temperatures but was very
low if the samples were stored at 4°C rather than room
temperature. The temperature effects suggest that the
post-adsorption changes in elutability are probably due
to conformational alterations which allow more contacts
per molecule ("molecular spreading"). The studies also
imply that the incomplete elution observed even if
elution was done immediately after adsorption is due to
transitions that occured during adsorption -- that is,
some of the proteins in the adsorbed layer are in one
state (elutable) while other molecules are in the non-
elutable state. This thought led to a more general
analysis of proteins in multiple states of adsorption
because it was clear that more evidence was needed to
support this hypothesis. The possible mechanisms that
could lead to multiple states and the experimental
evidence that supports this idea are presented in the
remainder of this section.

Mechanisms for Multiple States. Possible mechanisms that
could lead to multiple states of adsorbed proteins are
summarized in Table II and illustrated in Figures 1-5.
The adsorption of a protein molecule at a site is likely
to be influenced by the existence of molecules already
adsorbed at nearby sites, by either geometric reduction
of the area available for adsorption as the surface sites
become occupied ("occupancy" effects, Figure 1a) or by

Table II.   Multiple States of Adsorbed Proteins:
Some Possible Mechanisms

1.   "Intrinsic" heterogeneity due to occupancy effects
and adsorbate - adsorbate interactions.

2.   Structurally altered forms of the protein may exist
on the surface due to:

a.   Pre-existing distribution of conformations in
solution phase;
b.   Rapid conformational changes accompanying
adsorption;
c.   Slow conformational changes after adsorption
("residence time" effects).

3.   Orientational states due to "patchiness" of protein
exterior.

4.   Multiple binding modes due to amphipathicity of
protein and mixed site nature of substrate.

5.   A distribution with regard to number of bonds per
molecule may be expected on statistical grounds.

a   Site size greater than molecular size

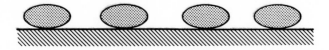

Site size less than molecular size

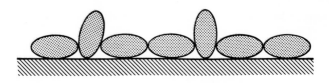

b   Intermolecular distance > effective range of repulsive forces

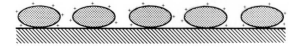

Intermolecular distance < effective range of repulsive forces

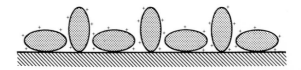

Figure 1. Intrinsic heterogeneity as a cause of multiple states of adsorbed proteins. a. Occupancy effects. b. Adsorbate – adsorbate interactions.

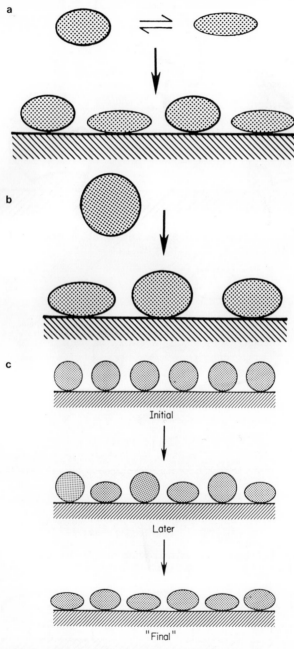

Figure 2. Structurally altered forms as a cause of multiple states of adsorbed proteins. a. Pre-existing forms in solution. b. Rapid formation upon adsorption. c. Slow conformational change after adsorption.

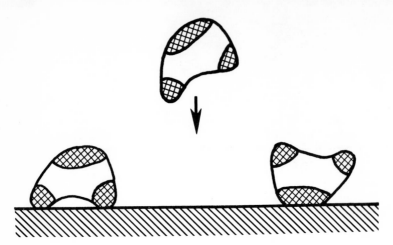

Figure 3. Orientational effects as a cause of multiple states of adsorbed proteins.

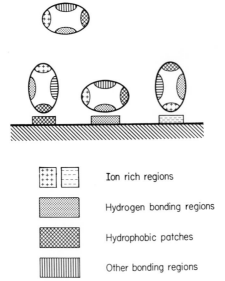

Ion rich regions	
Hydrogen bonding regions	
Hydrophobic patches	
Other bonding regions	

Figure 4. Multiple binding modes as a cause of multiple states of adsorbed proteins.

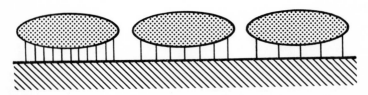

Figure 5. Statistical distribution of bonds per molecule as a cause of multiple states of adsorbed proteins.

repulsive forces that are expected whenever the molecules are close enough ["adsorbate-adsorbate" interactions, Figure 1b]. These factors may cause the adsorbed population to be intrinsically heterogeneous with respect to protein-surface interactions.

The term "intrinsically heterogeneous" is used because limitations on access to the substrate are probably inherent in the process of loading all available surface sites with protein. These limitations in turn probably cause different modes of contact to occur for molecules adsorbed at low concentration than at high concentration, as illustrated schematically in Figure 1. Most monolayer adsorption models applied to proteins at interfaces do not envision this type of heterogeneity because they are based on the Langmuir model for gas adsorption. The Langmuir model was originally developed for adsorption involving very small species that are uncharged and essentially uniform over their surface with regard to their ability to interact with the substrate. Neither assumption applies to proteins.

Several different structurally altered forms of the same protein molecule may co-exist in the adsorbed layer, because of adsorption of these species from an equilibrium distribution pre-existing in the bulk phase (Figure 2a), rapid changes induced by adsorption (2b), or slow re-arrangements leading to forms that are more tightly adsorbed as more contacts are formed (2c). The existence of several conformations of the same protein in the solution phase is suggested by the fact that certain regions of proteins appear to be able to undergo motion, especially in N or C terminal sections of the amino acid sequence. Furthermore, the availability of energetically similar structural states of proteins, as indicated by conformational shifts induced by ligand binding, temperature changes, or pH shifts, makes it likely that a certain fraction of the bulk protein will exist in other than the average or lowest energy state during the adsorption process. Clearly, whether the adsorbed state includes a substantial fraction of altered forms due to this cause will depend on the degree of separation in energy states of the bulk protein population. On the other hand, the ability of proteins to undergo both rapid and slow structural alterations upon adsorption seems to be a plausible source of multiple states because such changes could contribute to increased bonding per molecule. Alteration of the structure could allow closer approach of the external residues to the surface as well as increased exposure of normally internal hydrophobic residues, both of which could increase the number of bonds with the surface. The structural alterations also would contribute to an increase of entropy in the system (see below).

The non-uniform distribution of amino acid residues on the exterior of the protein molecules, the overall

geometric asymmetry of protein molecules, and the
existence of "patches" at the protein exterior enriched
in amino acid residues of a particular type (e.g.,
negatively charged), together with the fact that bonding
to the surface may be favored much more by certain types
of residues, suggests that certain orientations of the
adsorbed molecules are more likely than others. However,
since a variety of orientational states may each bring a
substantial number of the favored residues into close
contact with the surface (see Figure 3), we may
reasonably expect that more than one orientational state
exists in the adsorbed protein layer.

     Recent studies of monoclonal antibody binding to
adsorbed myoglobin have provided some information on the
orientational state of adsorbed proteins. Each of the
antibodies studied bound to specific but different
regions of myoglobin called "epitopes". Some of the
antibodies exhibited the same affinity for myoglobin in
the solution and adsorbed state, but others had very much
lower binding affinity for adsorbed myoglobin than for
myoglobin in solution. Either conformational changes, or
steric blocking of the binding site due to orientation
towards the substrate could explain these differences but
here only the orientational explanation will be
discussed.

     The reduced affinity of some but not all antibodies
with specificity towards various regions of an adsorbed
molecule is exactly what one would expect if the adsorbed
molecules were oriented in such a way as to prevent
access of the antibody to certain regions of the
molecule. However, the data for myoglobin seem to imply a
surprising uniformity in the orientation, since the
affinity would be reduced only if all of the adsorbed
molecules were in a given orientational state. Thus, for
example, if half the adsorbed molecules were oriented
with the epitope fully available to the bulk phase and
the other half with the epitope oriented towards and
partially blocked by the substrate, the antibody binding
should indicate the presence of both high affinity and
low affinity binding, e.g., two slopes in the Scatchard
plot of the binding data. The ability to detect antibody
binding to a fraction of molecules in the adsorbed layer
which retain the same binding constant as the protein in
solution, in the presence of a majority oriented in such
a way as to reduce the binding, depends on the
experimental error encountered. Thus, the data on
monoclonal antibody binding to adsorbed myoglobin may not
mean all molecules are oriented one way, but it does
suggest that most are. On the other hand, the use of
"solid phase" immunoassay techniques in which antibody is
bound first to a solid substrate and then used to measure
ligand uptake has shown that only some of the adsorbed
antibodies (e.g., 1%) bind ligand, clearly indicating the
existence of more than one state (possibly orientational)

of adsorbed antibodies (see chapter by Giaever and
Keese).

Multiple binding modes may also exist due to the
mixed site nature of a real interface (Figure 4). A
perfectly homogeneous, perfectly clean interface is
difficult if not impossible to produce, and in any case
is probably not representative of actual materials used
in adsorption studies. For example, most silicone rubber
samples include silicon dioxide fillers whose surface
properties differ markedly from the polydimethylsiloxane
polymer, the latter of which is itself likely to be
somewhat heterogeneous with respect to molecular weight
and degree of cross-linking. Some polymers (including
polyurethanes) actually phase segregate into markedly
different domains because of differences in their
chemical properties (20). Even chemically simple
materials such as polyethylene are known to have a rather
wide range of structural arrangements (e.g. "crystalline"
and amorphous regions). Thus, surfaces probably have a
variety of sites to which a protein can adsorb, each of
which may favor different types of amino acid residues
for bonding. To occupy all the sites on a surface, it is
thus possible that the proteins may exist in a variety of
adsorbed states, as diagrammed schematically in Figure 4.

Finally, a statistical argument for the existence of
multiple states of adsorbed molecules can be made.
Chemical species are thought to exist in a series of
states governed by a probability function that is related
to the energy of each state --- the Boltzmann
distribution. In the case of an adsorbed protein
molecule, this idea is represented in Figure 5 by
differences in the number of bonds per molecule. If the
total energy released upon the formation of these
variously bonded molecules does not differ greatly, as is
likely in a process involving many bonds per molecule,
then a distribution with regard to the number of bonds
per molecules is expected. The importance of this source
of multiple states is difficult to assess because it
depends on both the strength of individual bonds and the
number involved in adsorption of proteins, neither of
which is presently known.

Evidence for Multiple States of Adsorbed Proteins. The
evidence to be reviewed here supporting the existence of
multiple states of adsorbed proteins derives from many
types of studies (see Table III). In each study,
different techniques were used to study the behavior of
proteins at interfaces but none was specifically designed
nor applied to investigate the possible existence of
multiple states of adsorbed proteins. Furthermore, the
proteins, surfaces, and objectives of each study varied.
Nonetheless, this diverse data base is consistent in
support of the existence of multiple states, and its
diversity may therefore actually provide greater

Table III. Evidence for Multiple States

1. Weakly & tightly bound proteins are indicated by:

    - partial buffer desorption;
    - partial protein exchange;
    - partial protein displacement;
    - partial detergent elution.

2. Freundlich isotherms ($\theta = kC^{1/n}$) imply heterogeneity.

3. Decreases in SDS elutability of proteins as residence time and temperature increase after adsorption.

4. The bound fraction of γ-globulin on silica is lower at high fractional saturation.

5. Calorimetry studies indicate that the molar ΔH for adsorption varies markedly with surface coverage.

6. Fluorescence recovery curves after photobleaching of rhodamine-albumin on quartz indicate presence of three bound states: irreversible, slowly reversible, and rapidly reversible.

7. Adsorption "hysteresis" on derivatized gels: desorption rates for phosphorylase b increase with increasing fractional saturation.

8. IgG desorption from silica and n-pentyl silanized silica occurs at two very different rates. Residence time lowers desorption rate from silica markedly.

9. Protein adsorption may occur in "islands" on some surfaces.

10. Partial losses in enzyme activity occur upon adsorption.

11. BHK cell interactions with adsorbed fibronectin are enhanced by albumin co-adsorption.

12. The sequence of fibrinogen/albumin pre-adsorption influences platelet reactions *in vivo*.

13. Platelet retention in bead columns is well correlated with antifibrinogen uptake but not with the total amount of fibrinogen adsorbed.

confidence in this conclusion than could be derived from a narrower set of experiments.

Evidence supporting multiple states of protein adsorption include observations indicating the presence of weakly and tightly bound proteins. The adsorption of proteins to solid surfaces is typically irreversible in the sense that continued, extensive soaking in the buffer used for adsorption does not remove all the protein. However, some of the protein present after a brief initial rinse is partially or slowly removable in a second, longer buffer rinse. A further fraction (but still not all) is removed when the adsorbed surface is placed in a solution of the protein, a process referred to as exchange (21,22). Similarly, some of the fibinogen adsorbed from plasma is removed when the surface is put in a hemoglobin solution, a process referred to as displacement (23). Finally, still more adsorbed protein (but usually not all) is removable by detergents (24,25). Partial removal by buffer or protein or detergent suggests that some of the adsorbed molecules are more tightly held then others, i.e., that the adsorbed layer is heterogeneous with respect to the reversibility of adsorption. The processes underlying such evident heterogeneity are not clear, however. The studies on elutability losses already discussed seem to shed some light on this question because transitions in the state of the adsorbed proteins appear to occur at different rates, depending on protein or polymer type as well as temperature.

Secondly, protein adsorption isotherms are often not Langmuirian but instead fit a Freundlich isotherm, $\theta = kC^{1/n}$. In this equation, which is a power function, θ is the fractional surface coverage, k is a constant, C is the bulk phase protein concentration, and n is another constant, generally greater than 1.0. The Freundlich isotherm has traditionally been explained in terms of heterogeneous adsorption processes (26). Adsorption from a solution containing only one type of protein is concentration dependent, rising towards a "saturation" amount at higher concentration which approximates the value expected for a close-packed monolayer. However, saturation values are frequently a factor of 2 or more above or below monolayer coverage, and close inspection of the data reveals that saturation is typically never reached but instead the adsorption increases more slowly at higher concentrations than at low concentrations. Nonetheless the Langmuirian monolayer model is frequently applied to protein adsorption isotherms. Deviations from monolayer saturation values are ascribed to lack of knowledge of true surface area or to variations in packing due to orientation of the molecules (e.g., "side-on" versus "end-on" orientations). However, the Langmuir model of hard, non-deformable spheres adsorbing to a homogeneous surface without lateral interaction between the adsorbing molecules, originally devised for gas adsorption, hardly is a realistic model for protein adsorption. If, instead, one considers adsorption to be a molecular version of adhesion, then proteins may exist in a variety of states due to various degrees of achievement of the tightest bonding possible. In this model, the molecules do not achieve maximal and uniform adhesive bonding because of competition for sites by adjacent molecules and because of statistically predictable variations in the number of bonds per molecule.

The fraction of the peptide bond carbonyl groups involved in γ-globulin bonding to colloidal silicia has been measured using infrared difference spectroscopy (6,27). The frequency of the bound carbonyl shifts to a wavelength slightly different from unbonded carbonyl, and so can be measured directly. The fraction of γ-globulin carbonyl groups bonded was higher (ca 0.20) when adsorption occurred from low concentration protein solutions than from high concentration solutions (ca 0.02 in the plateau region of the isotherm). These data were interpreted to mean that the fraction of the γ-globulin molecule that would interact with the surface depended somewhat on the area available to do so. Thus, this example clearly suggests that a molecule may adsorb in more than one way, and implies that at least at intermediate concentrations, one would expect γ-globulin molecules to be present in both the more bonded (low concentration) and less bonded (high concentration) states. The generality of this example is uncertain however, because changes in bound fraction with concentration did not occur for other proteins studied (prothrombin, albumin).

The heat released upon protein adsorption to suspended particulates has been measured by several different groups of investigators (9,28-30; see also the chapter by Norde et al.). In a process in which all the adsorbates interact with the adsorbent in the same way, the heat released per mole should be a constant. The molar heat released upon adsorption of certain organic polymers is independent of fractional surface coverage (9), but molar heats of protein adsorption vary widely in some cases. For example, while the molar heat for fibrinogen adsorption to carbon is essentially invariant as the fractional coverage of the surface with protein increases, the molar heat of adsorption of γ-globulin and fibrinogen to glass were both found to decrease greatly with increasing surface coverage (28,29). Similarly, the molar heat released upon adsorption of albumin to hematite varied with the fractional coverage by protein (9). These data seem to be an unequivocal indication of different types of interactions between protein and the surface at different degrees of coverage, i.e., that molecules adsorbing at low coverage have more interaction with the surface than at high coverage. Presumably, such differences would lead to mixed states of adsorption at any degree of coverage, since it seems unlikely that the transition from conditions of high molar heats to low molar heats would be abrupt.

Some of the clearest evidence for multiple adsorbed states comes from the photo-bleaching experiments of Burghardt and Axelrod (31). Rhodamine labeled albumin adsorbed to quartz can be permanently bleached by a short, high intensity flash of laser light applied to a small area of the adsorbed surface. The reappearance of fluorescence in the bleached spot is a measure of exchange with the (unbleached) bulk phase proteins as well as surface diffusion of adsorbed but unbleached molecules from areas outside the illuminated spot. The fluorescence intensity in the bleached spot did not return to the original value, but was partially restored. The partial restoration of the intensity occurred in two distinct phases, one rapid and one slow. The results therefore indicate an irreversibly bound fraction (accounting for the incomplete restoration of original fluorescence), and both a quickly reversible and a slowly reversible adsorbed state. Thus, at least three states of albumin adsorbed to quartz seem to exist simultaneously.

The term adsorption "hystersis" of phosphorylase b on agarose gels covalently modified with butyl groups described by Jennissen (32) refers to the fact that desorption occurs much more slowly than adsorption. More importantly, the desorption rate was found to increase (by about a factor of 5) as the fractional saturation with adsorbed phasphorylase increased from 0.1 to 0.75. These data indicate that an adsorbed protein may be held very differently, depending on the degree of coverage of the surface with protein. At a given degree of adsorption, it is likely that part of the adsorbed molecules exist in the rapidly removable form while another population of slowly removable molecules also co-exists.

Other, more direct evidence for co-existing populations of adsorbed proteins with very different rates of removal is provided by the behavior of IgG on silica and n-pentyl silanized silica (33). The desorption of this protein from silica, when it had been allowed to reside on the surface less than 100 min, was linear with the square root of time, suggesting a single state existed. However, after longer residence times on silica (1000 min), the desorption curves were distinctly biphasic. Biphasic desorption was observed for IgG adsorbed to n-pentyl silanized silica but in this case it was observed even after short residence times. Thus, under these conditions some of the IgG desorbed much more rapidly than the remainder. It is difficult to attribute

these biphasic curves to anything other than the fact
that the adsorbed IgG molecules exist in at least two
states. It is equally important to note that simply
allowing more residence time after adsorption to silica,
or using a more hydrophobic surface, apparently causes
the adsorbed molecules to convert from a single state to
a mixed population with regard to desorption rates. These
data support the existence of multiple states and also
suggest that they may arise from changes in the
adsorbate/adsorbent interaction after adsorption occurs.
The explanation of the residence time effect on IgG
desorption rates is analogous to the interpretation given
to the loss in detergent elutability, as discussed
previously.

The adsorption of proteins to surfaces in
congregated groups or islands has been suggested by
several different investigations (34-39). These studies
were done with dehydrated protein films and so there is
justifiable skepticism about their ultimate significance.
However, island formation, with empty patches of surface
between, was not observed on all the surfaces studied.
Thus, even if artifacts exist due to dehydration or
staining, important differences in the interaction of
proteins with the various surfaces are being revealed. If
proteins do indeed adsorb in patches or islands on some
surfaces, it may contribute to the existence of multiple
states of adsorbed proteins because the proteins at the
edge of such patches would lack the influences of nearest
neighbors, i.e., adsorbate-adsorbate interactions would
differ. Furthermore, the size and frequency of the
patches appears to be dependent on both the protein
concentration and time of adsorption (see chapter by
Price and Rudee) so that a uniform distribution of
islands is not too likely. Thus, it may be that some
proteins molecules adsorb individually while others are
members of a patch.

Several biological indicators of the behavior of
proteins at interfaces also suggest the existence of
multiple states. Thus, for example, enzymes adsorbed to
many surfaces typically lose some (but not all) of their
activity relative to that in the bulk phase (40-42; see
also chapter by Mizutani). In some studies this has been
found to be due to the loss of the active site on a
certain fraction of the molecules and its retention on
the remainder (43). Grinnell has shown that the ability
of fibronectin to induce BHK cell spreading is strongly
dependent on the substrate and whether co-adsorbed
albumin was present. Thus, BHK cell spreading to
fibronectin coated polystyrene dishes was very low unless
traces of albumin were present in (and presumably co-
adsorbed with) the fibronectin, whereas BHK cell
spreading on "tissue culture" grade polystyrene coated
with fibronectin was high both in the presence and
absence of albumin (25). These data indicate that more

than one functionally important state is available to adsorbed fibronectin. Similarly, platelet interactions with surfaces pre-adsorbed with fibrinogen and albumin are markedly dependent on the sequence of the adsorption, rather than on just the amounts of adsorbed proteins (44). Surfaces having essentially equal amounts of albumin and fibrinogen induced very different platelet deposition when used as shunts placed in contact with dog blood *ex vivo*, depending on which protein was adsorbed first. These results suggest that fibrinogen can be adsorbed in at least two, functionally different states. This idea is also supported by the results of Lindon *et al.* (see their chapter) since platelet retention in polymer coated bead columns was well correlated with the reactivity of adsorbed fibrinogen to antibodies but not with the total amount of fibrinogen adsorbed.

Competitive Adsorption Behavior of Proteins at Interfaces

The nature of competition in multi-protein systems is a question of great interest which is touched on by a considerable number of the papers in this volume. Such interest is understandable in that many of the areas of application involve adsorption from complex media; for example blood, plasma or serum, tear fluid and other body fluids, soil, milk, and food products generally. The information normally sought concerns the concentration profile of the proteins on the surface and how this is related to the concentration "profile" in the bulk phase. In general there is a redistribution of proteins in the surface phase, resulting in an enrichment of some components and an impoverishment of others relative to the bulk phase. The redistribution may also be time dependent and the kinetics as well as the equilibrium aspects are of interest.

Of fundamental importance are the system variables and the properties of proteins and surfaces which determine the surface redistribution. These factors have not been adequately investigated in terms of simply identifying them, and still less in terms of determining the precise way in which the redistribution depends on them. One might also ask how the redistribution can be controlled for a given objective, for example the isolation of a given protein from a mixture, or the preferential adsorption of a "passivating" protein which effectively "turns off" further interfacial activity. The classic example of this type is the preferential adsorption of albumin from blood which, it is believed, would render the blood contacting surface non-thrombogenic. This approach has been investigated for many years and is discussed in this volume in the paper of Eberhart *et al.*

The present state of knowledge in relation to competitive adsorption is certainly meager if not quite

nil. A few scattered studies of blood protein mixtures
and blood plasma have been published and are reviewed in
several of the papers in this book. Of course there is a
wealth of what might be called anecdotal information in
the literature of protein chromatography (e.g., 45-51).
The sum total of these studies is of little help in the
formulation of a generalized predictive model of
competitive adsorption, and in this respect the field is
overdue for a systematic assault.

Some speculation as to the factors which influence
competitive adsorption may be in order. The influence of
the adsorbing surface itself will be at the level of the
possible types of interaction provided by its chemical
properties relative to those of the protein. Charge,
hydrophobicity, and reactive chemical functional groups
are the main properties to be considered. Clearly there
must be reciprocity between surface and protein in this
regard and we need only discuss these factors in relation
to one or the other of the two. To simplify the
discussion we consider the interactions of a multi-
protein system with a given unspecified surface, keeping
in mind that the surface factors must be superimposed to
complete the picture. Some of the more important protein
properties influencing competitive adsorption are
probably as follows: electrical charge, hydrophobicity-
hydrophilicity and available chemical functional groups
at the protein surface, stability/fragility of the
protein conformation, protein-protein interactions in the
adsorbed layer, relative concentration in the bulk phase,
and molecular size. These factors may be classified as
either affinity factors or kinetic factors. Thus charge,
chemical properties, conformational stability and
interlayer interactions may be considered to be affinity
factors since they will influence the binding reactions
themselves. Relative concentration is primarily a kinetic
factor through its effect on the rate of transport and
the rate of adsorption (or binding) of the protein.
Protein molecular size may be considered both a kinetic
factor through its effect on diffusion and an affinity
factor through its effect on the number of binding sites.

As already indicated, protein electrical charge has
a strong effect on protein binding and proteins of
opposite overall charge sign would probably have
different affinities for a surface charged in a single
sense, either positively or negatively. This would be
reflected in a dependence on the isoelectric points
(I.E.P.) of the proteins in the mixture relative to the
pH. Important caveats must always be kept in mind with
respect to charge. First the "point" or "smeared out"
model of charge may not be appropriate. (Norde *et al.*
indicate that this model applies mainly in the
neighborhood of the I.E.P. — see their contribution in
this book). Instead, local areas of both positive and
negative charge which are present on the surface of the

protein may bind as independent entities. Second, divalent cation bridging can occur between negative charges on the protein and the surface.

Other chemical properties of the protein surface such as the presence of hydrophobic and hydrophilic patches and chemical functional groups could obviously give rise to specific forms of binding such as hydrophobic and hydrogen bonding (e.g., $-NH_2$ on protein and $C=O$ on surface). The formation of covalent bonds requiring high activation energy seems unlikely in the temperature range of interest for most applications where protein adsorption occurs.

The conformational fragility of a protein may also affect its ability to compete in adsorption. The occurrence of conformational change upon adsorption will contribute a gain of entropy, thereby increasing the affinity of the protein for the surface (i.e., the decrease in Gibbs free energy accompanying adsorption). Many factors of course contribute to conformational stability, including internal hydrophobic interactions, hydrogen bonding, and disulfide bonds. Clearly also, the entropy effect should be greater for larger proteins, thus contributing to the preferential adsorption of larger proteins as discussed below.

Protein-protein interactions in the surface layer will also affect the redistribution of proteins between bulk and surface. As for the single protein systems the major effect is likely to be charge-charge interactions. However, for single proteins there may be a predominance of repulsive effects, whereas there will also be attractive effects in multi-component systems between proteins which have overall negative and overall positive charge, respectively. These charges will of course depend on the I.E.P.'s of the proteins in relation to the pH of the medium. Again, as indicated above, the point charge model may be inappropriate if local charge effects and small ion bridging are important.

As previously mentioned, the relative concentrations of proteins in the bulk phase will affect the rates of arrival of the proteins at the surface and then the rates of adsorption once there. The rate of arrival at the surface may be expressed as:

$$\frac{dC_{surf}^{diff}}{dt} = C_0 \left(\frac{D}{\pi t}\right)^{1/2}$$

C_{surf}^{diff} is considered to be a 2-dimensional bulk concentration near the surface, t is time, C_0 is the bulk concentration far from the surface, and D is the diffusion coefficient. Arrival rates thus depend on bulk concentration and on diffusivity, the latter of which increases with decreasing size of the protein.

Adsorption rate per se has the general form

$$\frac{d\Gamma}{dt} = f(C, T)$$

where Γ is surface concentration, C is protein concentration near the surface and T is temperature. The dependence on C is likely to be first order. The rate of adsorption will thus depend on the protein concentration near the surface and this will be determined initially by the diffusion rate. The reaction rate constant will also influence adsorption rate, and this should depend on frequency factors and temperature in the usual Arrhenius sense.

The effect of protein size on competitive adsorption has not been investigated to any great extent. By analogy with synthetic polymers, large proteins are expected to adsorb in preference to small proteins. It is well established that in adsorption from a solution of a synthetic polymer having a broad molecular size distribution, the high MW species are preferentially adsorbed (51). The analogy with proteins is, however, not exact because the synthetic polymer species are chemically the same while in a mixture of proteins, chemical as well as size differences must be recognized. It has been shown recently by Zsom (52) that albumin dimers and higher oligomers adsorb in preference to monomeric albumin. Thus it may be predicted that "all other things being equal" there will be a tendency for large proteins to be preferentially adsorbed.

These, then, are the factors which appear to be important in competitive adsorption. The question of how they interact and combine to produce a given concentration distribution in the adsorbed layer from a given concentration distribution in the bulk phase is clearly complex. Some indication of multifactor interactions is given by studies in blood plasma which provides a medium "par excellence" for competitive adsorption. In a number of laboratories it has been found that fibrinogen is adsorbed initially and is later displaced from the surface by other proteins of much lower bulk concentration (see chapters in this book by Brash; Vroman; Horbett). This behavior may be explained by assuming that the displacing proteins have much greater affinity constants than fibrinogen. Initially, however, because of the effect of concentration on diffusion, fibrinogen dominates the adsorbed layer. Later when the bulk concentration of the high affinity trace proteins near the surface is sufficiently great they are adsorbed and fibrinogen is displaced. This explanation also depends on the microscopic reversibility of adsorption which has been invoked to explain protein exchange between surface and solution (<u>32,53</u>). Such a mechanism also provides a basis for sequential adsorption in multi-protein systems as suggested by Vroman (see his chapter).

The above discussion provides some indication of how kinetic and affinity factors may interact in competitive situations. The more complex problem of how to synthesize all of the possible factors into a quantitative predictive model of competitive adsorption remains to be solved, and is certainly worthy of considerable effort. Data for such models are currently unavailable. They could be obtained in two ways. First it may be assumed that affinity constants in single protein systems could be measured and, with a knowledge of diffusion coefficients, could be used to predict adsorption layer compositions in mixtures. This approach may have some validity at low surface concentrations but ignores the effect of interlayer interactions at higher coverages. The only valid approach to the measurement of affinities, for the moment, is to use the mixtures themselves. It may not be too much to hope for the future, however, that "partial" affinities that could be used to estimate affinities in multi-component systems could be determined as sufficient knowledge of such systems develop.

Future Research on Proteins at Interfaces

An overall assessment of the state of research to date on proteins at interfaces is that much more is known about the amount of protein adsorption under various conditions than about the actual nature of the adsorbed layer or about the functional consequences of the adsorption process. Many excellent, well understood methods are available for measuring the quantity adsorbed, but few good techniques exist for determining the qualitative features, e.g., the structure or orientation of an adsorbed protein, or for studying the relationship of a particular aspect of the adsorption process to its influence on other processes at the interface (e.g., cellular interactions). Furthermore, predictive models for any aspect of the adsorption process in terms of specific properties of proteins or the interface are lacking. Perhaps most importantly, readily measurable constants (e.g., the affinity of a protein for the surface) that can be collected and confidently compared between laboratories are notably lacking. Thus, a rather large amount of information about the behavior of proteins at interfaces must be obtained by future research. Some of the topics requiring substantial new efforts are summarized in Table IV and briefly discussed in the remainder of this section.

Comparative studies of the adsorption behavior of closely related proteins and peptides probably

Table IV.   Future Research on Proteins at Interfaces

(1) Molecular understanding of the surface activity of
    proteins.

(2) How heterogeneous is the adsorbed layer with regard
    to the "state" of adsorbed proteins?

(3) Better ways to measure and characterize
    conformational change of adsorbed proteins.

(4) Ways to measure orientation and studies of the
    effect of orientation on protein reactivity at
    surfaces - enzymes, antibodies.

(5) Effects of adsorption generally on protein
    reactivity - enzymes, antibodies, zymogen-enzyme
    conversion.

(6) Protein affinity constants both in single protein
    and multi-component systems, including theories for
    adsorption isotherms.

(7) Predictive models for competitive adsorption
    accounting for displacement (Vroman) effects.

(8) Control of adsorption - selectivity, orientation,
    conformation - through properties of adsorbing
    surface.

(9) Role of adsorbed proteins in cellular interactions
    with surface: composition vs. state.

(10) Modification of adsorbed protein layers by cellular
     interactions, including role in "passivation".

constitutes the best approach to further understanding of the molecular properties influencing surface activity differences among proteins. The very recent comparative studies of hen and human lysozyme (see chapter by Horsley *et al.* in this book) are the first of this type. Studies of the adsorption behavior of hemoglobin genetic and ligand variants to solid surfaces would be a logical and probably productive extension of the studies of these variants at the air/water interface. In addition, however, the availability of families of polypeptides, prepared by either synthetic or digestive methods, provides an opportunity to study the influence of controlled changes in sequence on surface activity. This approach has already been used successfully in analysis of peptide retention behavior on hydrophobic matrices (54). Finally, since the competitive binding of peptide fragments of proteins has proven useful in analysis of proteins binding to cellular receptors (55), a similar approach may prove interesting in analyzing specific contribution of various sections of the sequence to the adsorption process.

The likely existence of multiple states of adsorption of proteins to surfaces also raises several questions requiring new research. For example, the evidence suggesting that two or three states exist with regard to desorption may only reflect the insensitivity of the techniques used to date. The adsorbed layer may actually contain a rather wide range of adsorbed states of protein molecules. In other words, the adsorbed layer may be much more heterogeneous in this regard than evidence to date has shown. Refinement of techniques to elucidate this question are quite possible -- e.g., the use of a graded series of eluting agents varying in surfactant power may show that a continuous series of classes of adsorbed proteins exist, each removable at a given degree of surfactant power. Similarly, extension of antibody binding studies on adsorbed proteins to a wider range of antibody concentrations and the collection of closely spaced experimental data may reveal the existence of more than one binding class.

A major area for new research concerns the structural and functional consequences of adsorption of proteins to surfaces (items 3-5 in Table IV). Measurement of conformational change is still in an early stage of development. Most methods for studying adsorbed protein conformation are restricted to comparison of spectral differences induced by adsorption, without knowledge of the actual type or amount of change these differences reflect. Better methodology, especially on quantitative aspects, is sorely needed in this area. The orientation of adsorbed proteins may prove to be readily explored with the monoclonal antibody method and therefore certainly deserves wider application. Finally, the behavior of enzymes and antibodies at interfaces is not

only important in a range of applications, but it also
provides a rather direct route to understanding proteins
at interfaces because these molecules have natural
properties (active site, ligand binding site) that may be
readily probed to determine the functional consequence of
adsorption. For example, it would be of interest to know
whether all or only some of adsorbed fibrinogen molecules
are available for fibrinopeptide A release by thrombin
cleavage. Quantitation of the degree of enzyme
inactivation, or of the degree of activation of a
zymogen, or of the fraction of adsorbed antibody that
retains its ligand binding ability, all can provide much
information. To date, however, the few studies of this
type have typically not gone beyond reporting some loss
in function without quantitating this more exactly.

     The development of sets of relative or absolute
affinity constants describing the interaction of proteins
with surfaces in a quantitative way is a very important
goal for future research. This goal may possibly be
achieved at least partly through a theoretical
description of adsorption isotherms that allows the
accurate ranking of proteins, even though the process is
operationally irreversible and may not be describable by
the usual equilibrium constant so useful in describing
other binding phenomena. Alternatively, it may prove
necessary to use kinetic studies in which careful
analysis of mass transfer rates allows the calculation of
effectiveness factors, i.e., number of protein molecules
adsorbed per unit time divided by the total number of
protein molecules that contacted the surface per unit
time.

     A concerted effort is presently needed to study the
mechanisms influencing adsorption behavior in protein
mixtures. Does adsorption from mixtures behave as the
sum of independent adsorption events determined by
specific affinity constants charactertistic of each
species? Can such a simple explanation suffice to
explain the peak in adsorption isotherms seen for
fibrinogen from plasma ("the Vroman effect") and also
from binary mixtures (56,57)? If the differences in
adsorption behavior of mixtures compared to single
adsorbates are better understood than at present, a
greater degree of control of the adsorption process to
achieve a desired end (e.g., selection of a desired
protein) may be possible. For example, if the Vroman
effect is a general feature of all protein mixtures, then
clearly there is an optimum concentration for adsorption
to achieve the greatest selectivity.

     Another important area for future research is to
determine more clearly how the properties of the
adsorbing surface influence the adsorbed layer. For
example, is it possible to select for a particular
orientation of an adsorbed protein by changing the
surface from very hydrophobic to more hydrophilic? Is it

possible to change this orientation systematically by using graded series of copolymers varying in the hydrophobic/hydrophilic balance? Alternatively, if it can be shown that certain types of interfacial properties are more likely to induce conformational change than others (e.g., the often mentioned possibility that hydrophobic surfaces may favor unfolding to allow contacts with normally internalized hydrophobic amino acid residues), then one can expect to be able to control the nature of the adsorbed layer in a rather subtle way.

A major area still not clearly understood is the exact role of proteins adsorbed from complex mixtures in cellular interactions with such interfaces. Do cells interact with adsorbed protein layers according to how much of a particular protein is present, or is the conformation or state of the protein more important? This problem has been an important focus of research aimed at understanding the fundamental processes involved in the biocompatibility of implants. There is some recent evidence showing that differences in short term cellular events are dictated by the differences in the state of the adsorbed protein on different surfaces, rather than just on the composition of the adsorbed layer (see Lindon *et al.*'s chapter in this book and reference 58).

The role of protein adsorption in longer term cellular interactions with surfaces is less clear. Thus, for example, a comparison between fibronectin adsorption from serum and the long-term growth of 3T3 cells has suggested that these cells can sometimes overcome a relative lack of adsorbed fibronectin, and grow well, apparently by exuding their own fibronectin (59). Similarly, there are now several diverse indications showing that cellular interactions with surfaces can modify the adsorbed protein layer (see chapter by Feuerstein and reference 60). The frequently noted "passivation" of surfaces placed in the bloodstream, in which an initial period of extensive thrombus deposition is followed by a chronic state with much lower adherent thrombus, may also be an important example of changes in the adsorbed layer induced by cellular interactions (58). In this case also, there is ample reason to wonder at the connection between the initially adsorbed layer (determined by the plasma) and the steady state events. To date, an examination of the proteins at the interface after long blood contact has not been performed.

In summary, even this brief and somewhat incomplete survey of potential areas for future research on proteins at interfaces has revealed a number of important problems about which we are largely uninformed. Furthermore, it is clear that progress in understanding proteins at interfaces is often difficult to achieve and that true breakthroughs that have stimulated other areas of research have been few in this field. On the other hand, many if not most of the papers in this book reflect

original and creative approaches to studying the behavior
of proteins at interfaces, and thus provide an accurate
indication of the diversity of research in this field.
Future research developments will likely continue this
tradition of diversity, in no small part as a response to
the difficulty of the problems remaining. Hopefully, the
required methodological and conceptual breakthroughs will
come about as more and more approaches are developed.

Acknowledgments

The authors received an early and enthusiastic commitment
to include the "Protein at Interfaces" symposium in the
continuing series entitled "Surface Chemistry in Biology,
Dentistry, and Medicine" from the then current chair, Dr.
Robert Baier of the State University of New York at
Buffalo. We thank Dr. Baier for this very helpful early
support which made the organization of the symposium much
easier, especially for the many foreign participants who
were enabled to come because of the early notification of
a definite time and place for the symposium. The
agencies supporting our research in this field are also
gratefully acknowledged: the National Heart, Lung, and
Blood Institute (T.A.H.) and the Medical Research Council
of Canada and the Heart and Stroke Foundation of Ontario
(J.L.B.). A great deal of work in preparing the many
different written documents required to prepare for the
symposium and this book was done by Terry Sochia,
University of Washington, and the staff of the Word
Processing Center, Faculty of Engineering, McMaster
University, and this help is gratefully acknowledged.

Literature Cited

1. Horbett, T.A. In <u>Biomaterials: Interfacial Phenomena
    and Applications, ACS Symposium Series</u>; Cooper,
    S.L.; Peppas, N.A., Eds.; American Chemical Society:
    Washington, D. C., 1982; Vol. 199, p 233.
2. Brash, J.L. In <u>Interaction of the Blood with Natural
    and Artificial Surfaces</u>; Salzman, E.W., Ed.; Marcel
    Dekker: New York, 1981; p 37.
3. Andrade, J.D. In <u>Surface and Interfacial Aspects of
    Biomedical Polymers</u>; Andrade, J.D., Ed.; Plenum
    Press: New York, 1985; Vol. 2: Protein Adsorption, p
    1.
4. Norde, W. <u>Adv. Coll. Interf. Sci.</u> 1986, <u>25</u>, 267.
5. Ivarsson, B.; Lundstrom, I. <u>CRC Crit. Rev.
    Biocompat.</u> 1986, <u>2</u>, 1.
6. Morrissey, B.W.; Stromberg, R.R. <u>J. Coll. Interf.
    Sci.</u> 1974, <u>46</u>, 152.
7. Horbett, T.A.; Weathersby, P.K.; Hoffman, A.S. <u>J.
    Bioeng.</u> 1977, <u>1</u>, 61.

8. Andersson, L.O. In <u>Plasma Proteins</u>; Blomback, B.;
    Hanson, L.A., Eds.; John Wiley and sons: New York,
    1979; p 43.
9. Norde, W. In <u>Surface and Interfacial Aspects of</u>
    <u>Biomedical Polymers</u>; Andrade, J. D., Ed.; Plenum
    Press: New York, 1985; Vol. 2, Protein Adsorption, p
    263.
10. Asakura, T.; Ohnishi, T.; Friedman, S.; Schwartz, E.
    <u>Proc. Nat. Acad. Sci.</u> USA 1974, <u>71</u>, 1594.
11. Adachi, K.; Asakura, T. <u>Biochemistry</u> 1974, 13, 4976.
12. Ohnishi, T.; Asakura, T. <u>Biochim. Biophys. Acta</u>
    1976, <u>453</u>, 93.
13. Roth, E.F., Jr; Elbaum, D.; Bookchin, R.M.; Nagel,
    R.L. <u>Blood</u> 1976, <u>48</u>, 265.
14. Elbaum, D.; Harrington, J.; Roth, E.F., Jr; Nagel,
    R.L. <u>Biochim. Biophys. Acta</u> 1976, <u>427</u>, 57.
15. White, A.; Handler, P.; Smith, E.L. <u>Principles of</u>
    <u>Biochemistry</u>; McGraw-Hill Book Co.: New York, 1964.
    (Third Edition)
16. Beissinger, R.L.; Leonard, E.F. <u>ASAIO J.</u> 1980, <u>3</u>,
    160.
17. Lundstrom, I. <u>Prog. Coll. Polym. Sci.</u> 1985, <u>70</u>, 76.
18. Iordanski, A.L.; Polischuk, A.J.; Zaikov, G.E. <u>J.</u>
    <u>Macromol. Sci. - Rev. Macromol. Chem. Phys.</u> 1983,
    <u>C23</u>, 33.
19. Bohnert, J.L.; Horbett, T.A. <u>J. Coll. Interf. Sci.</u>
    1986, <u>111</u>, 363.
20. Lelah, M.D.; Cooper, S.L. <u>Polyurethanes in Medicine</u>;
    CRC Press: Boca Raton, Florida, 1986.
21. Chan, B.M.C.; Brash, J.L. <u>J. Coll. Interf. Sci.</u>
    1981, <u>84</u>, 263.
22. Weathersby, P.K.; Horbett, T.A.; Hoffman, A.S. <u>J.</u>
    <u>Bioeng.</u> 1977, <u>1</u>, 395.
23. Horbett, T.A.; Mack, K. <u>Trans. Soc. Biomat.</u> 1986,
    <u>IX</u>, 45.
24. Weathersby, P.K.; Horbett, T.A.; Hoffman, A.S.
    <u>Trans. Am. Soc. Artif. Int. Organs</u> 1976, <u>22</u>, 242.
25. Grinnell, F.; Feld, M.K. <u>J. Biomed. Mater. Res.</u>
    1981, <u>15</u>, 363.
26. Adamson, A.W. <u>Physical Chemistry of Surfaces</u>; Wiley-
    Intersciences: NY, 1967.
27. Morrissey, B.W. <u>Ann. N. Y. Acad. Sci.</u> 1977, <u>283</u>, 50.
28. Nyilas, E.; Chiu, T.H.; Herzlinger, G.A. <u>Trans. Am.</u>
    <u>Soc. Artif. Int. Organs</u> 1974, <u>20</u>, 480.
29. Chiu, T.H.; Nyilas, E.; Lederman, D.M. <u>Trans. Am.</u>
    <u>Soc. Artif. Int. Organs</u> 1976, <u>22</u>, 498.
30. Filisko, F.E.; Malladi, S.D.; Barenberg, S.
    <u>Biomaterials</u> 1986, <u>7</u>, 348.
31. Burghardt, T.P.; Axelrod, D. <u>Biophys. J.</u> 1981, <u>33</u>,
    455.
32. Jennissen, H.P. <u>J. Coll. Interf. Sci.</u> 1986, <u>111</u>,
    570.
33. Hlady, V.; Van Wagenen, R.A.; Andrade, J.D. In
    <u>Surface and Interfacial Aspects of Biomedical</u>

Polymers; Andrade, J.D., Ed.; Plenum Press: New
York, 1985; Vol. 2, Protein Adsorption, p 81.
34. Ratner, B.D.; Horbett, T.A.; Shuttleworth, D.;
Thomas, H.R. J. Coll. Interf. Sci. 1981, 83, 630.
35. Paynter, R.W.; Ratner, B.D.; Horbett, T.A.; Thomas,
H.R. J. Coll. Interf. Sci. 1984, 101, 233.
36. Eberhart, R.C.; Prokop, L.D.; Wissenger, J.; Wilkov,
M.A. Trans. Am. Soc. Artif. Int. Organs 1977, 23,
134.
37. Eberhart, R.C.; Lynch, M.E.; Bilge, F.H.; Arts, H.A.
Trans. Am. Soc. Artif. Int. Organs 1980, 26, 185.
38. Eberhart, R.C.; Lynch, M.E.; Bilge, F.H.; Wissinger,
J.F.; Munro, M.S.; Ellsworth, S.R.; Quattrone, A.J.
In Biomaterials: Interfacial Phenomena and
Applications, ACS Advances in Chemistry Series 199;
Cooper, S.L.; Peppas, N.A., Eds.; American Chemical
Society: Wash. D.C., 1982; p 293.
39. Rudee, M.L.; Price, T.M. J. Biomed. Mater. Res.
1985, 19, 57.
40. Mizutani, T. J. Pharm. Sci. 1980, 69, 279.
41. Goldstein, L.; Manecke, G. In Immobilized Enzyme
Principles; Wingard, L.B.; Katchalski-Katzir, E.;
Goldstein, L., Eds.; Academic Press, Inc.: New York,
1976; p 23.
42. Sandwick, R.K.; Schray, K.J. Anal. Biochem. 1985,
147, 20.
43. Venkataraman, S.; Horbett, T.A.; Hoffman, A.S. J.
Biomed. Mater. Res. 1977, 11, 111.
44. Pitt, W.G.; Park, K.; Cooper, S.L. J. Coll. Interf.
Sci. 1986, 111, 343.
45. Goudsward, J.; van der Donk, T.A.; Noordzij, A.
Scand. J. Immunol. 1978, 8, 21.
46. Deutch, D.G.; Mertz, E.T. Science 1970, 170, 1095.
47. Engvall, E.; Ruoslahti, E. Int. J. Cancer 1977, 20,
1.
48. Farooqui, A.A. J. Chromatogr. 1980, 184, 335.
49. Travis, J.; Powell, R. Clin. Chim. Acta 1973, 49,
49.
50. Gianazza, E.; Arnaud, P. Biochem. J. 1982, 201, 129.
51. Cohen Stuart, M.A.; Scheutjens, J.M.M.M.; Fleer,
G.J. J. Polym. Sci., Polym. Phys. Ed. 1980, 18, 559.
52. Zsom, R.L.J. J. Coll. Interf. Sci. 1986, 111, 434.
53. Brash, J.L.; Uniyal, S.; Pusineri, C.; Schmitt, A.
J. Coll. Interf. Sci. 1983, 95, 28.
54. Sasagawa, T.; Ericcsson, L.H.; Teller, D.C.; Titani,
K.; Walsh, K.A. J. Chromatogr. 1984, 307, 29.
55. Pytela, R.; Piersbacher, M.D.; Ginsberg, M.H.; Plow,
E.F.; Ruoslahti, E. Science 1986, 231, 1559.
56. Slack, S.; Horbett, T.A. 1986. (Submitted.)
57. Slack, S.M.; Bohnert, J.L.; Horbett, T.A. Ann. N. Y.
Acad. Sci. 1987. (in press)
58. Pitt, W.G.; Park, K.; Cooper, S.L. J. Coll. Interf.
Sci. 1986, 111, 343.

59. Horbett, T.A.; Schway, M.B.; Ratner, B.D. <u>J. Coll. Interf. Sci.</u> 1985, <u>104</u>, 28.
60. Uniyal, S.; Brash, J.L.; Degterev, I.A. In <u>Biomaterials: Interfacial Phenomena and Applications, Advanced Chemical Series</u>; Cooper, S.L.; Peppas, N.A., Eds.; American Chemical Society: Washington D.C., 1982; Vol. 199, p 277.

RECEIVED April 14, 1987

BEHAVIOR OF PROTEINS AT INTERFACES

Chapter 2

Protein Adsorption at Solid–Liquid Interfaces: A Colloid-Chemical Approach

W. Norde, J. G. E. M. Fraaye, and J. Lyklema

Department of Physical and Colloid Chemistry, Agricultural University, De Dreijen 6, 6703 BC Wageningen, Netherlands

Protein adsorption on solid surfaces is discussed from a colloid chemical and thermodynamic point of view. Information is mainly obtained from adsorption isotherms, (proton)titrations, electrokinetics and calorimetry. The adsorption behavior of human plasma albumin and bovine pancreas ribonuclease at various surfaces is studied. The differences in behavior between the two proteins are related to differences in the structural properties. Furthermore, the essential role of the low molecular weight electrolytes in the overall protein adsorption process is stressed.

Since the beginning of this century the adsorption of proteins at phase boundaries has been investigated for various reasons. A vast amount of literature, including several review articles (e.g., 1,2,3,4), is available now. Most of the published work deals with adsorbed amounts and only during the last few decades have issues such as adsorption mechanisms and structure of the adsorbed molecules been discussed.

Surveying the literature, it appears that the interfacial behavior of proteins is a controversial subject. The main reason is that many studies have been performed under insufficiently defined conditions and/or that conclusions have been drawn on the basis of too scanty experimental evidence. Furthermore, the theoretical description of adsorbed layers of simple, flexible polymers is still in its infancy (5,6). As the structure of proteins is much more complex than that of those simple polymers, theories of polymer adsorption need to be greatly extended to become applicable to proteins. Clearly, our current knowledge of protein adsorption mechanisms and of the structure of the adsorbed layer is far from complete.

In a number of ways the adsorption of polymers, including proteins, differs from that of low molecular weight substances. A polymer molecule attaches to the sorbent surface via several segments. Even if the adsorption free energy per segment is low, say, 1 kT, attachment of tens of segments adds up to a large adsorption free energy of some tens of kT for the whole molecule. As a result, polymer molecules do not readily desorb in the pure solvent but they may be displaced by adding other (macro)molecules that adsorb with a larger (segmental) free energy.

The three-dimensional structure of a protein molecule is the net result of interactions inside the molecule and interactions between the protein and its environment. Adsorption involves the transfer of protein from solution to the sorbent surface and the concomitant displacement of solvent and, possibly, other components from that surface. The resulting environmental change may induce alterations in the protein structure, which, in turn, may affect the biological activity of the protein. Needless to say, structural variations upon adsorption are of great relevance to the various practical applications of immobilized proteins.

It is evident that elucidation of the interfacial behavior of proteins is not a simple matter and requires contributions from several disciplines. In recent years considerable progress has been made in applying spectroscopic techniques to proteins in the adsorbed state (e.g., 7,8,9). In such studies a (small) part of the molecule is analyzed in detail. In our laboratory we study protein adsorption from a more classical, colloid-chemical point of view. Arguments are derived from experimental data referring to whole protein molecules or to layers of them. Information is obtained from adsorption isotherms, proton titrations and both electrokinetic and thermochemical measurements. Recently, topical questions such as reversibility of the adsorption process and changes in the protein structure have been considered. This more holistic approach has produced some insights that could not easily be obtained otherwise.

Adsorption Isotherms

The shape of the equilibrium isotherm (adsorbed amount Γ as a function of the concentration c in solution) yields information about the free energy of adsorption. For most flexible, highly solvated polymers high-affinity isotherms are obtained, i.e. isotherms in which the initial part coincides with the Γ-axis after which a levelling off takes place to a (pseudo-) plateau.

However, isotherms for globular proteins often show a finite initial slope. They develop well-defined plateaus at rather dilute concentrations in solution ($c \lesssim 1$ g dm^{-3}).

In interpreting adsorption isotherms, a distinction should be made between very low coverage (initial part of the isotherm), where the protein molecules interact with the sorbent surface only, and high surface coverage (adsorption plateau), where lateral interactions between the adsorbed molecules play a role as well.

In the literature not much attention has been paid to the initial part of the isotherm. Due to analytical limitations, the trends in this region often are rather uncertain. With human plasma albumin (HPA) adsorbing on either polystyrene latex (10) or single crystals of polyoxymethylene (11) it has been observed that the slope of the initial part of the isotherm becomes less steep with increasing pH. Anticipating the discussion in the following section,

this could be caused by an increased number of carboxyl groups of
the protein oriented towards the sorbent surface.

Furthermore, on polystyrene it was found that the adsorption of
HPA in the low surface coverage region increased with increasing
temperature, except at the isoelectric point (i.e.p.) of the protein
where the adsorption appeared to be independent of the temperature.
According to Clapeyron's law a positive value for $(\delta\Gamma/\delta T)_c$ implies
an endothermic adsorption process under isosteric conditions.
Although with protein adsorption isosteric conditions are difficult
to establish, the qualitative conclusion is that at pH$\neq$ i.e.p. the
adsorption enthalpy is positive. Hence, under those conditions,
adsorption must be entropically driven. We will return to this
subject in section 5.

Usually, the plateau-value, Γ_p, of the isotherm corresponds
roughly to a close-packed monolayer of native molecules in a side-on
or end-on orientation. It indicates that, at least at solution
concentrations that are not excessively high, multilayers are not
formed. In various systems it is observed that $\Gamma_p(pH)$ is at a
maximum in the isoelectric region of the protein molecule (e.g.,
7,12,13,14). For example, $\Gamma_p(pH)$ curves for HPA at several sorbents
are shown in Figure 1. The occurrence of a maximum at the i.e.p.,
independent of the nature of the sorbent, suggests that the charge
of the protein molecule greatly influences Γ_p. In particular, for
HPA on polystyrene surfaces ample evidence has been collected to
conclude that the reduction in Γ_p on either side of the i.e.p. is
due to structural rearrangements in the adsorbing molecules, rather
than to increased lateral repulsion. Since the trends in $\Gamma_p(pH)$ for
the other surfaces are similar, they also may be caused by changes
in the structure of the HPA molecule. It is, furthermore, remarkable
that HPA and many other proteins adsorb spontaneously on hydrophilic
surfaces, even if the surface has the same charge sign as the
protein (e.g. HPA on hematite, pH $>$ 6.8). Under such conditions
dehydration of the sorbent surface and overall electrostatic
interaction oppose the adsorption process.

Therefore another contribution, originating from the protein
molecule, drives the adsorption.

Although the $\Gamma_p(pH)$ pattern, as shown in Figure 1, is quite
common, it is not followed by all proteins. As an example, $\Gamma_p(pH)$
for bovine pancreas ribonuclease (RNase) on different sorbents is
shown in Figure 2. On polystyrene latex, Γ_p is essentially
independent of the pH of adsorption. The value of Γ_p is comparable
with that of a complete monolayer of native RNase molecules. It
suggests that only minor structural changes, if any, occur and that
they are not affected by the protein charge (the i.e.p. of RNase is
pH 9.3). The affinity of RNase for the uncharged, less hydrophobic
polyoxymethylene crystals is so low that no significant adsorption
can be detected. With the hydrophilic hematite, RNase adsorption
occurs only in the pH range where the protein and the sorbent are
oppositely charged. Thus, in contrast to HPA, RNase does not adsorb
on hydrophilic surfaces, except if it is aided electrostatically. It
is concluded that the factor that dominates HPA adsorption is less
strong, or absent, in RNase. It is probable that these differences
in the adsorption behavior between HPA and RNase are related to
differences in structural properties between these two proteins.
This will be discussed further below.

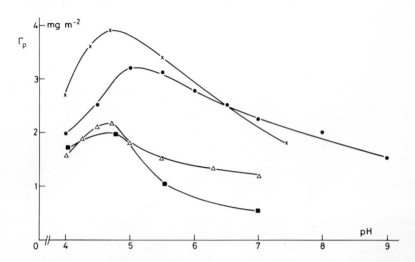

Figure 1. Plateau values for the adsorption of human plasma albumin on polystyrene latex (Δ), silver iodide (×), polyoxymethylene (■) and hematite (●). Electrolyte: 0.01 M KNO$_3$ or 0.05 M KNO$_3$ (for adsorption at polyoxymethylene). T = 22 °C.

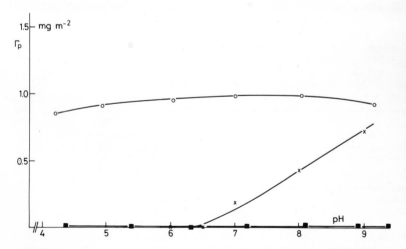

Figure 2. Plateau values for the adsorption of bovine pancreas ribonuclease on polystyrene latex (o), polyoxymethylene (■) and hematite (×). Conditions as in Figure 1.

Adsorption (Ir)reversibility

As indicated, protein adsorption isotherms are often not of the
high-affinity type. However, desorption of the protein into the pure
buffer does not occur significantly within hours or days. This
observation lends support to the suggestion that, after adsorption,
structural changes occur in the protein molecule in order to adapt
its structure to the new environment. Such structural changes lower
the free energy of the system, and, hence, the affinity between
protein and sorbent surface increases after adsorption. The
condition of simultaneous detachment of several segments from the
sorbent surface can be expected to slow the rate of desorption into
the pure buffer, relative to the rate of adsorption. However,
proteins may desorb readily on changing other conditions, e.g. pH or
ionic strength of the medium or by adding a component that has a
larger (segmental) free energy of adsorption (11). Moreover,
relatively fast exchange between adsorbed and dissolved protein
molecules has been observed by various investigators (e.g.,
15,16,17,18).
     For HPA, removed from the sorbents hematite, silica and
polyoxymethylene, the molecular structure has been compared with
that of the native molecule, on the basis of their circular
dichroism spectra (11). It was found that after desorption the helix
content of HPA is some twenty to thirty percent lower. This
reduction is virtually independent of the type of sorbent and the
desorption method. It suggests that the change in the helix content
is related to properties of the protein molecule itself. It is still
not clear to which extent the adsorption and the desorption step
affect the protein structure. It is furthermore interesting that the
helix reduction is larger for the samples with lower Γ_p-values. This
supports the earlier conclusion that a reduced Γ_p value reflects
further structural rearrangements in the protein molecule. It is
noted that the decrease in the helix content of desorbed HPA found
by us is considerably less than that reported by others (19).
     RNase removed from hematite surfaces did not show significant
alteration of its helical content. This is in agreement with our
interpretation of the constant $\Gamma_p(pH)$ for this protein.

Charge Effects

In an aqueous medium, proteins and solid surfaces are usually
charged. In both systems the charge is neutralized by counter- and
co-ions, that are partly physically bound and partly diffusely
distributed. The charge distribution in and around a protein
molecule and at a sorbent surface is schematically represented in
Figure 3.
     When the protein approaches the surface the electrical double
layers overlap, giving rise to a redistribution of charge. This can
have a significant impact on protein adsorption. In some systems,
e.g. RNase with a hydrophilic sorbent surface, overall electrostatic
repulsion between the protein and the sorbent prevents adsorption.
With other proteins, e.g. HPA, interactions between charged groups
do not play a decisive role.
     Interactions between charges on proteins and surfaces are

screened by low molecular weight ions in the system. On increasing the ionic strength, larger values of Γ_p are usually observed (10,20,21). If this is found in the case where the protein and the sorbent are oppositely charged, it indicates that Γ_p is primarily influenced by charge-charge interactions within, or between, protein molecules rather than between protein and sorbent. The influence of ionic strength cannot always easily be interpreted. It has been reported that the effect of ionic strength on HPA adsorption depends on the type of sorbent used (12). Furthermore, adsorption may be sensitive to the type of ion present (22,23). The conclusion is that the role of low molecular weight ions is more complex than just electrostatic screening between interacting charges.

Proton titration data (11,24) point to the uptake of H^+ ions by molecules of HPA and RNase upon adsorption on polystyrene particles. Especially the pK of the carboxyl groups undergoes a considerable shift to higher values. It is inferred that a relatively large fraction of the carboxyl groups faces the negatively (!) charged polystyrene surface. The titration data also reveal the different adsorption behavior between HPA and RNase. For RNase the shift in the titration curve is essentially independent of the pH of adsorption, whereas for HPA the shift increases the further the pH of adsorption is away from the i.e.p. of the protein. These results confirm the conclusions in the foregoing section: the structural perturbation, if any, in RNase is not sensitive to the pH of adsorption and the structure of adsorbing HPA is progressively altered on moving the pH away from the i.e.p.

Electrophoresis experiments (25,26,27) lead to the conclusion that ions, other than H^+, are also transferred between the solution and the adsorbed protein layer. Figure 4 shows the charge alteration due to the transfer of ions (including H^+) for plateau-adsorption of HPA and RNase on various surfaces. The trends are in qualitative agreement with expectations according to electrostatic interaction between the protein and the sorbent surface. The conclusion is that charge-charge interactions are easily neutralized by adsorption of low molecular weight ions.

Based on a model we published some ten years ago (28), the amount of co-adsorbed ions was estimated. Direct determination of ion uptake, using the radionuclides $^{22}Na^+$, $^{133}Ba^{2+}$ and the paramagnetic Mn^{2+} ion has semi-quantitatively confirmed the model predictions (23). According to the model, the ion uptake has an electrostatic reason, i.e. it happens to prevent accumulation of net charge in the low dielectric contact region between the protein and the sorbent surface, which would otherwise result in high values for the electrostatic potential. In addition to transfer of charge, the uptake of ions involves transfer of matter. Hence, the molar Gibbs energy, g, of transfer of an ion i between the solution s and the protein layer p contains both a chemical and an electrical term:

$$g_i^p - g_i^s = (\mu_i^p - \mu_i^s) + z_i F (\psi^p - \psi^s)$$

where μ_i is the chemical potential of i in the phase indicated by the superscript, ψ the electrostatic potential, z_i the valency of i and F the Faraday constant. As water usually is a better solvent for ions than the proteinaceous layer, $\mu_i^p - \mu_i^s > 0$, i.e., the chemical

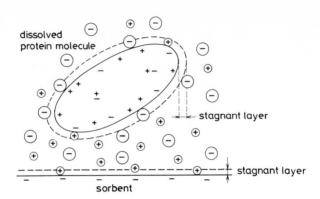

Figure 3. Schematic illustration of the charge distribution in and around a dissolved protein molecule and at a sorbent surface.

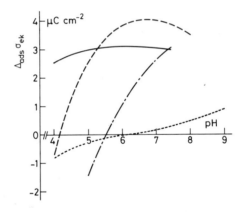

Figure 4. Charge transfer (expressed per unit area of the sorbent surface) between the solution and adsorbed layers of human plasma albumin (HPA) and bovine pancreas ribonuclease (RNase) on various sorbents. Plateau adsorption.
———— HPA on negatively charged polystyrene.
— — — HPA on positively charged polystyrene.
—·— HPA on hematite.
------- RNase on negatively charged polystyrene.
The charge of HPA, RNase and hematite vary with pH, with isoelectric points of 4.8, 9.3 and 6.5, respectively.

contribution to the ion transfer opposes the overall protein adsorption. However, if no ions were incorporated the electrostatic potential in the adsorbed layer would reach very high values. Alternatively, such high potentials would be avoided if the protein would unfold into a very loose structure that was freely permeable to water and small ions and where the dielectric permittivity was not much lower than in the bulk of the buffer. Such loose structures are seldom found with globular proteins. It is, therefore, concluded that the chemical effect of ion transfer is less unfavorable than the exposure of hydrophobic amino acid residues to the aqueous medium, as would occur in a loosely structured, highly solvated, layer.

More detailed information on ion exchange involved in protein adsorption can be derived from titration experiments in systems where the charge of the protein and the sorbent can be varied independently. Currently, we are studying such systems, using bovine plasma albumin (BPA) and cytochrome c as the proteins and silver iodide (AgI) particles as the sorbent. In these systems there are two potential determining ion couples, the H^+/OH^- couple for the protein and the Ag^+/I^- couple for the sorbent. They enable independent control of protein and surface charge. Below, we will briefly discuss some results obtained with the BPA – AgI system. A series of pH static Ag^+/I^- titrations of adsorbed BPA have been performed. From the data it follows that for these rather special systems the extra uptake or release of protons suffices to neutralize the charge in the contact region between the protein and the sorbent surface. Furthermore, under not too extreme conditions (pH $\sim$ 5) the capacitance of the interfacial electrical double layer is fairly high (ca. 0.5 F m^{-2}) and resembles a typical Stern layer capacitance, indicating that a considerable amount of water is still present in the contact region. When the pH is low (pH $\approx$ 4) the capacitance depends on the proton buffering capacity of the adsorbed protein. On increasing the charge contrast between surface and protein it becomes more difficult for a protein molecule to neutralize the surface charge because of an increasing number of already titrated acid-base sites. As a result, the capacitance drops to low values (ca. 0.1 F m^{-2}). This is an interesting phenomenon, because from thermodynamics it follows that a decreasing capacitance may be understood as a decreasing affinity of the protein for the surface, even though a larger charge contrast between the protein and the sorbent exists.

Thermodynamics of Protein Adsorption

It is evident that the protein adsorption process is the result of various interactions mutually occurring between protein molecules, solvent molecules, low molecular weight ions, other solutes and the solid surface. The feasibility of protein adsorption (at constant pressure and temperature) is determined by the overall Gibbs energy of the process, $\Delta_{ads}G$.

Several authors have evaluated $\Delta_{ads}G$ from the adsorption isotherm (e.g., 29,30,31,32). Such an analysis adopts some model assumptions (e.g., Langmuir theory) and assumes that the isotherm represents a reversible process. It has been discussed in section 3 that this assumption is questionable.

        In view of this, a thermodynamic analysis should be based on
the determination of the adsorption enthalpy, $\Delta_{ads}H$, and the
adsorption entropy, $\Delta_{ads}S$. At given temperature $\Delta_{ads}G = \Delta_{ads}H - T\Delta_{ads}S$.
        Because of the almost infinite number of structural variations
that protein and solvent molecules may undergo during adsorption, an
ab initio statistical computation of $\Delta_{ads}S$ is practically
impossible. At constant pressure, which is usually the case for
adsorption from solution, $\Delta_{ads}H$ equals the heat of adsorption.
Therefore, $\Delta_{ads}H$ can be determined directly by (micro)calorimetry.
        We have measured $\Delta_{ads}H$ for plateau adsorption of HPA on
different substrates (11,26,33) and for RNase on polystyrene (33).
With both proteins, and under many conditions, $\Delta_{ads}H > 0$, implying
again that spontaneous adsorption occurs by virtue of an entropy
increase: $\Delta_{ads}S > 0$. In section 2 the same conclusion has been drawn
for the interaction between HPA and polystyrene latex at low surface
coverage.
        Figure 5 shows $\Delta_{ads}H$ for HPA on hematite at low and high
surface coverage. In this way the contribution due to crowding at
the surface can be studied. It appears that this contribution
becomes more positive with increasing distance from the i.e.p. of
the protein, which could be due to increased lateral repulsion
between the adsorbed HPA molecules.
        The calorimetric measurements for HPA and RNase on polystyrene
latexes have been analyzed in great detail (34). Based on additional
data for these systems, it was possible to assign contributions to
$\Delta_{ads}H$ from various interactions. By subtracting the sum of the
estimated contributions from the total enthalpy measured, the most
elusive constituent, $\Delta_{ads}H_{str\ pr}$, resulting from rearrangements in
the protein structure (including hydration), can be deduced. It
appears that the structural rearrangements are endothermic, as would
be expected for a reduction of secondary and/or tertiary structure.
Moreover, for HPA, $\Delta_{ads}H_{str\ pr}$ is larger if the pH of adsorption is
further from the i.e.p. Typically, the values found for $\Delta_{ads}H_{str\ pr}$
are in the range of 0 - 15 J per gram of protein.
        The adsorption of negatively charged HPA (pH 7.4, ionic
strength 0.01 M KNO_3) on hydrophilic, negatively charged surfaces of
hematite and silica deserves additional attention. As discussed in
section 2, under these conditions spontaneous adsorption is due to
some driving 'force' originating from the protein. Since, for
plateau-adsorption, $\Delta_{ads}H$ amounts to +201 kJ mol^{-1} and +136 kJ mol^{-1}
for HPA at hematite and silica surfaces, respectively, this driving
force must be of entropic nature. A lowering of the helix content of
HPA at both surfaces from 55% to ca. 45% (11) (cf. section 3), would
imply a maximum entropy gain of 691 J $K^{-1}mol^{-1}$, which, at 25 °C,
yields a contribution of -205 kJ mol^{-1} to $\Delta_{ads}G$. Hence, it is shown
that even a relatively small change in the structure of the protein
molecule could already account for a lowering of $\Delta_{ads}G$ that could
overcome the unfavorable effects of electrostatic repulsion and
dehydration of a hydrophilic surface. If such structural adaptations
are suppressed, as in the case of RNase, adsorption would not occur
at a hydrophilic surface that is uncharged or like-charged.

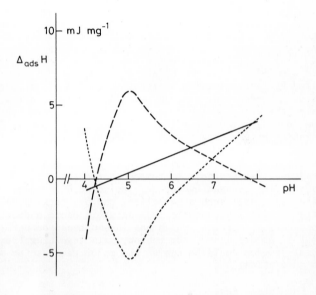

Figure 5. Adsorption enthalpy of human plasma albumin on
hematite surfaces at two different surface coverages, i.e. at Γ
= Γ_p (————) and Γ = 0.1 Γ_p (— — —). The difference curve
(--------) reflects the enthalpy contribution from interactions
between adsorbed albumin molecules. Electrolyte: 0.01 M KNO₃ ;
T = 25 °C (Reproduced with permission from Ref. 4. Copyright
1986 Elsevier Science Publishers B.V.).

Relation between Structural Properties of the Protein and its Adsorption Behavior

Some of the interactions that determine the three-dimensional
structure of a protein molecule support a compact conformation,
whereas others tend to expand the molecule. In aqueous solution
hydrophobic parts of the protein are buried as much as possible in
the interior of the molecule but in the adsorbed state the
hydrophobic residues may be exposed to the sorbent surface, still
shielded from water. Therefore, an expanded structure will be
promoted upon adsorption if the compact structure in solution is
stabilized by intramolecular hydrophobic bonding. More precisely,
whether or not adsorbing protein molecules change their structure
depends on the contribution from intramolecular hydrophobic bonding,
relative to those from other interactions, to the overall
stabilization of the structure in solution. In reference (4) such an
analysis of the structure determining factors has been made for HPA
and RNase. It leads to the conclusion that HPA, more than RNase, is
able to adapt its structure at sorbent surfaces.
    Interaction between hydrophobic amino acid residues stabilizes
secondary structures such as α-helices and β-sheets. A reduction of
the intramolecular hydrophobic bonding would cause a decrease of
such secondary structures, which has indeed been found for, e.g.,
HPA (11,19). As discussed, the entropy gain resulting from increased
rotational freedom could be one of the major reasons for spontaneous
protein adsorption.

Aspects for Future Research

The dawning understanding of some general principles of protein
adsorption helps us to direct further research. One main problem to
be solved concerns the reversibility (exchange and desorption) of
the sorption process. This problem is intimately connected to
structural rearrangements upon adsorption. Such rearrangements, in
turn, are related to the contributions of different structure
determining factors, relative to each other. Therefore, to identify
the adsorption mechanism, it seems fruitful to combine adsorption
data with a characterization of the structure determining factors.
Against this background, studies with a series of proteins of
varying structural properties and with one and the same protein
adsorbing from gradually changing solvents are recommended.

Literature Cited

1. Brash, J.L.; Lyman, D.J. In The Chemistry of Biosurfaces;
    Hair, M.L., Ed.; Marcel Dekker: New York, 1971; p. 177.
2. MacRitchie, F. Advan. Protein Chem. 1978, 32, 283.
3. Ivarsson, B.; Lundström, I. CRC Critical Reviews in
    Biocompatibility 1985, 2, 1.
4. Norde, W. Advan. Colloid and Interface Sci. 1986, 25, 267.
5. Fleer, G.J.; Lyklema, J. In Adsorption from Solution at the
    Solid-Liquid Interface; Parfitt, G.D.; Rochester, C.H., Eds.;
    Academic Press: New York, 1983; p. 153.
6. Van der Schee, H.A.; Lyklema, J. J. Phys. Chem. 1984, 88, 6661.

7. Morrissey, B.W.; Stromberg, R.R. J. Colloid Interface Sci. 1974, 46, 152.
8. Burghardt, T.P.; Axelrod, D. Biochemistry 1983, 22, 1979.
9. Hlady, V.; Reinecke, D.R.; Andrade, J.D. J. Colloid Interface Sci. 1986, 111, 555.
10. Norde, W.; Lyklema, J. J. Colloid Interface Sci. 1978, 66, 257.
11. Norde, W.; MacRitchie, F.; Nowička, G.; Lyklema, J. J. Colloid Interface Sci. 1986, 112, 447.
12. Koutsoukos, P.G.; Mumme-Young, C.A.; Norde, W.; Lyklema, J. Colloids and Surfaces 1982, 5, 93.
13. MacRitchie, F. J. Colloid Interface Sci. 1972, 38, 484.
14. Bagchi, P.; Birnbaum, S.M. J. Colloid Interface Sci. 1981, 83, 460.
15. Lok, B.K.; Cheng, Y.-L.; Robertson, C.R. J. Colloid Interface Sci. 1982, 91, 351.
16. Chan, B.M.C.; Brash, J.L. J. Colloid Interface Sci. 1981, 84, 263.
17. Brash, J.L.; Uniyal, S.; Samak, Q. Trans. Am. Soc. Artif. Int. Organs 1974, 20, 69.
18. Weathersby, P.K.; Horbett, T.A.; Hoffmann, A.S. J. Bioeng. 1977, 1, 395.
19. Soderquist, M.E.; Walton, A.G. J. Colloid Interface Sci. 1980, 75, 386.
20. Shastri, R.; Roe, R.J. Org. Coat. Plast. Chem. 1979, 40, 820.
21. Susawa, T.; Murakami, T. J. Colloid Interface Sci. 1980, 78, 266.
22. Mizutani, T. J. Colloid Interface Sci. 1981, 82, 162.
23. Van Dulm, P.; Norde, W.; Lyklema, J. J. Colloid Interface Sci., 1981, 82, 77.
24. Norde, W.; Lyklema, J. J. Colloid Interface Sci. 1978, 66, 266.
25. Norde, W.; Lyklema, J. J. Colloid Interface Sci. 1978, 66, 277.
26. Koutsoukos, P.G.; Norde, W.; Lyklema, J. J. Colloid Interface Sci. 1983, 95, 385.
27. Norde, W. Colloids and Surfaces 1984, 10, 21.
28. Norde, W.; Lyklema, J. J. Colloid Interface Sci. 1978, 66, 285.
29. Dillman, W.J.; Miller, I.F. J. Colloid Interface Sci. 1973, 44, 221.
30. Lee, R.G.; Kim, S.W. J. Biomed. Mater. Res. 1974, 8, 251.
31. Brash, J.L.; Samak, Q.M. J. Colloid Interface Sci. 1978, 65, 495.
32. Schmitt, A.; Varoqui, R.; Uniyal, S.; Brash, J.L.; Pusineri, C. J. Colloid Interface Sci. 1983, 92, 25.
33. Norde, W.; Lyklema, J. J. Colloid Interface Sci. 1978, 66, 295.
34. Norde, W.; Lyklema, J. J. Colloid Interface Sci. 1979, 71, 350.

RECEIVED February 13, 1987

Chapter 3

Early Stages of Plasma Protein Adsorption

Todd M. Price[1] and M. Lea Rudee[2]

[1]Department of Biology, University of California—San Diego, La Jolla, CA 92093
[2]Departments of Electrical Engineering and Computer Science, University of
California—San Diego, La Jolla, CA 92093

Plasma proteins are adsorbed to the surfaces of carbon, polystyrene, and a series of polyetherurethanes of increasing surface energy from both a static and a flowing milieu. The conformations of the individual protein molecules and the structure of the protein films formed are studied by electron microscopy. The conformation of individual protein molecules and the structure of the adsorbed films are found to be dependent upon the surfaces to which they are adsorbed, and the flow conditions under which the protein solutions contact the materials.

Research in our laboratory has been directed towards understanding the structures of both individual protein molecules and the films these proteins form when they adsorb to the surfaces of blood-contacting, bioengineering materials. The goal is to understand the mechanisms which control the thrombogenic nature of materials that contact the blood.

The first event following the contact of materials with blood is plasma protein adsorption (1). Subsequently, platelets may adhere to this adsorbed protein film. The platelets may then activate, release the contents of their alpha and dense granules, change their shape, increase their rate of adhesion to the surface, and recruit subsequently arriving platelets to the activation reaction. Initiation of the clotting cascade may then ensue with the formation of a thrombus. This thrombus will have three possible fates: (i.) continued growth to occlusion of the device, (ii.) embolization with or without downstream consequences, or (iii.) dissolution via the fibrinolytic pathway. Thrombosis is a major factor limiting the utility of blood contacting devices, especially small bore venous vascular grafts where low flow rates lead to the formation of fibrin rich clots.

The intact, healthy endothelium is the only known nonthrombogenic material. All other materials placed in contact with the blood are thrombogenic. Most researchers agree that the thrombogenic behavior of materials is largely mediated by; (i) the nature of the protein film adsorbed from the blood, (ii) the interactions of this film with both the underlying substrate and the subsequently deposited blood elements, and (iii) the flow characteristics of the device (2).

Due to the fundamental importance of the adsorbed protein film, many methods have been used to characterize its nature. These methods include: ellipsometry (3,4), Fourier transform infrared spectroscopy (FTIR) (5,6), multiple attenuated internal reflection spectroscopy (MAIR) (7,8) immunological labeling techniques (9), radioisotope labeled binding studies (10), calorimetric adsorption studies (11), circular dichroism spectroscopy (CDS) (12), electrophoresis (13), electron spectroscopy for chemical analysis (ESCA) (14), scanning electron microscopy (SEM) (15,16,9), and transmission electron microscopy (TEM) (17-19).

We chose to study the adsorption of plasma proteins to surfaces by using high resolution transmission electron microscopy. This allowed us to assess conformational changes of the protein molecules due to the specific surface to which they are adsorbed, and to examine surfaces following initial protein adsorption. Utilizing this technique, we have been able to observe individual molecules as well as the structure of the protein films adsorbed to these surfaces.

The composition of the adsorbed films can be assessed by both the characteristic shapes of the molecules observed at low surface concentrations, and by the reaction of the films with antibodies conjugated to colloidal gold particles (9). These electron dense antibody-gold-particle conjugates are visible in the electron microscope, and the diameters of the gold particles can be controlled so that two or three different size cohorts can be unambiguously distinguished from one another. This provides a method of compositional analysis of the adherent film which involves neither radiolabeling the protein making up the film under investigation, nor chemically altering the antibody reagent. This approach avoids the possible experimental artifacts due to the labeling chemistry. The adherent films may be simultaneously probed with several different antibody-gold particle conjugates specific to different plasma proteins and therby determine the composition of the film.

EXPERIMENTAL

PROTEINS. Human fibrinogen, supplied by the laboratory of Dr. Russell F. Doolittle of the Department of Chemistry at UCSD, was prepared by a cold ethanol precipitation technique (35). Human serum albumin was obtained from Schwartz-Mann as a lyophile. Plasma was obtained from human volunteers by venipuncture, utilizing an 18 gauge needle and drawing the blood into a Vacutainer tube containing sodium citrate. Following centrifugation, the cellular fraction was discarded and the plasma used immediately.

Colloidal gold particles were prepared by the reduction of chloroauric acid with citric acid or phosphoric acid. Following conjugation at pH 9 with affinity purified antibodies, the colloidal gold-antibody conjugate solutions were stabilized with polyethylene glycol. Following the exposure of the test surfaces in the flow cell, the surfaces were removed and incubated with the antibody-gold conjugate and then washed. The colloidal gold labeled surfaces were then shadowed as described below and examined in the electron microscope.

ELECTRON MICROSCOPY. The samples were prepared for electron microscopy as follows. The test materials with the adsorbed protein were placed in a vacuum evaporator (21). The specimens were dehydrated in vacuo (10^{-6} Torr) for 60 min., shadowed with tungsten, and carbon coated. The plastic substrates were dissolved in an appropriate solvent, either 1,2-dichloroethane (EDC), or N,N-dimethylacetamide (DMAC). The carbon replica, mounted on an microscope grid, was examined in the electron microscope. For observation, we used a Philips 300 electron microscope operated at 80 kV with a 50 micron objective aperture.

MATERIALS. Spectroscopically pure carbon was deposited onto freshly cleaved mica in a Denton DV-502 vacuum evaporator, the carbon films were then stripped off onto water and mounted on microscope grids. Preparation of polystyrene and polycarbonate substrates is described elsewhere (17). A series of polyetherurethanes were obtained from the laboratory of Dr. B.D. Ratner, Department of Chemical Engineering at the University of Washington. They were synthesized from methylene bis(4-phenylisocyanate), three different polypropylene glycols, and ethylene diamine as a chain extender. The polyether segments in these materials were derived from: the monomer, 1,2,-propanediol m.w.76; a polypropylene glycol 7-mer of m.w.425 d., and a polypropylene glycol 34-mer of m.w.2000 d.. These materials are abbreviated peu-ppd, peu-425, and peu-2000 respectively. In order to insure that the surfaces of the polyetherurethanes were sufficiently smooth to allow imaging of the proteins adsorbed to them, these materials were cast from solutions of DMAC against the surface of mercury. The side of the polyetherurethane film exposed to the protein solutions was the one in contact with the mercury.

RESULTS AND DISCUSSION.

Imaging biological macromolecules in the transmission electron microscope requires several manipulations which may lead to difficulty in the interpretation of the results. The molecules imaged in the microscope column are far from the state in which they originally adsorbed to the material surface. The electron microscope operates under a vacuum of approximately 10^{-5} Torr. Material placed in the column is therefore dry.

Electron scattering by a sample is proportional to the atomic number of the scattering material. Biological materials are composed of such low atomic number atoms that they are poor electron scatters, and contrast enhancement techniques must be employed. Typically, heavy metals are employed to render contrast to biological samples. Solutions of heavy metal salt may be used to either negatively or positively stain biological materials, or elemental heavy metals may be vapor deposited at a shallow angle in order to outline molecules adsorbed to smooth surfaces. In the study of protein films adsorbed to surfaces, negative staining (18), and partial gold decoration (19) have been successfully employed to observe the protein film. We have employed unidirectional, vapor deposited tungsten as a contrasting agent. For a review of methods used to image macromolecules, see Slayter (21). All results obtained in the electron microscope must be interpreted in the light of the environment of the high vacuum of the microscope column and its inherent anhydrous environment.

Low angle shadowing requires that the specimen to be imaged is adsorbed to a smooth substrate. Consequently, the polyetherurethane materials were cast against the surface of a mercury puddle. This provided a surface sufficiently smooth to image individual macromolecules. The other plastic substrates were cast against glass microscope slides, and the carbon substrates vapor deposited onto the surface of freshly cleaved mica.

Our earliest studies consisted of exposing amorphous carbon films to citrated whole blood for relatively long periods of time (up to 5 min.) in a static (i.e. no flow) system, washing to remove adherent cellular material, metal shadowing, and observing the apparent increase in roughness as the surface accumulated a layer of adsorbed protein Figure 1. Our interpretations of these images as the backside of a thick carpet of adsorbed protein led us to the use of diluted cell free plasma. The protein solutions were applied to the surfaces by atomization. This insured that a finite amount of protein reached the surface, and that there was no time dependent accretion of protein on the surface that would obscure the details of the molecular structure.

When plasma was diluted one thousand fold, individual protein molecules were apparent, and the characteristic Hall and Slayter (22) trinodular shape of the fibrinogen was discernable in many of the fields, Figure 1C. This led us to look more closely at the structure of the fibrinogen molecule and the location of its postulated domains in collaboration with Dr. Russell F. Doolittle in the Chemistry Department at UCSD.

Specific antisera to the various domains of the fibrinogen molecule were prepared. These affinity purified, polyclonal antibodies were cleaved with papain, and their antigen binding fragments (fab) were isolated. These specific fabs were reacted, in solution, with intact fibrinogen. The solutions were then diluted, sprayed onto ultra thin carbon films, and unidirectionally shadowed with tungsten (27). Figure 2 shows the results. The fabs can be seen to be specifically attached to the domains of the molecule against which they had been raised. Figure 2A shows an unreacted control fibrinogen molecule demonstrating the characteristic Hall and Slayter trinodular morphology. Figure 2B is a micrograph of the reaction of the fibrinogen solutions with fabs directed to a region on the fibrinogen alpha-chain. Figures 2C and 2D are of anti-D domain fabs. Figures 2E and 2F are of anti-E domain fabs.

The carbon support films used in the previous studies were similar, if not identical to LTI carbon (36), a material well tolerated by the circulation and used to coat prosthetic heart valves. These carbon substrates were replaced with other materials in

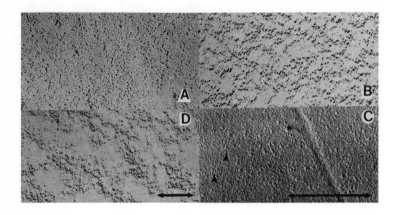

Figure 1. Carbon films are exposed to static, citrated whole blood. Panel A is the control surface with no exposure to blood, Panel B is a 5 sec. exposure, and panel D is a 60 sec. exposure. Panel C is plasma diluted 1:1000 and sprayed onto a carbon surface; the arrows indicate fibrinogen molecules. Scale bars represent 500 nm.

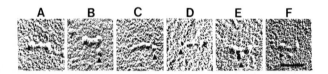

Figure 2. Panel A is purified human fibrinogen sprayed onto the surface of thin carbon films. Panels B-F are of purified human fibrinogen reacted with fabs specific to: a portion of the alpha chain extension (B), the D-domain (terminal nodule) of fibrinogen (C and D), and the E-domain (central nodule)of fibrinogen (E and F). The scale bar represents 50 nm.

order to evaluate the effect of the surface on the structure of the adsorbed protein.

When fibrinogen solutions were sprayed onto polystyrene surfaces, the morphology of the molecule was quite different from that found on the carbon surfaces, Figure 3. The alpha-chain protuberance of the fibrinogen molecule, whose existence had been demonstrated by sequence data (23), was resolved, Figure 3B and C. We believe this observation was due to two factors: (i.) the interaction of the protein with the surface, which apparently extended the alpha-chain protuberance away from the body of the molecule and rendered it available to the shadowing metal, and (ii.) to the improved resolution afforded by the electron microscopic technique employed (24). The alpha-chain protuberance of the fibrinogen molecule has been observed only on polystyrene surfaces and never on the surface of carbon Figure 3D. When the molecule is adsorbed to mica or carbon surfaces (25,26), the alpha-chain protuberance of fibrinogen was thought to lie either coiled at the surface of the D (terminal) domain or associated with the other alpha chain protuberance in the fibrinogen dimer near the E domain.

The alpha-chain protuberance had not been directly observed previously, although we had indirectly shown its existence by imaging anti-alpha-chain fab fragments pre-bound to the fibrinogen molecule when sprayed onto carbon surfaces, Figure 2B (27). The separation of the fab from the fibrinogen D-domain in Figure 2B is a consequence of the antisera having been raised to a proteolytic fragment of the alpha chain, which derived from the central portion of the alpha-chain extension of the molecule. Although the alpha chain extension is not resolved directly on this carbon surface, its existence can be inferred by the localization of the fab near, but not attached directly to, the D-domain. The results we obtain by dispersing the molecules onto polystyrene, and the association of the anti-alpha chain fab near to but not directly with the D domain, lead to the conclusion that the alpha-chain extension exists as a random coil protruding away from the D-domain.

Another phenomenon found to occur when fibrinogen was adsorbed to the polystyrene surface was that the axes of the molecules were not randomly oriented; instead, the axes were aligned and pointed in the direction of the center of the dried droplet (Figure 3A). This observation led to the hypothesis that the physical mechanism responsible for this orientation of the molecules was Blodgett-Langmuir transfer. The adsorption takes place at the air-liquid-solid interface, with the adsorption of the fibrinogen molecule occurring first by the alpha-chain protuberance. As the droplet air-liquid interface recedes, the alpha-chains extensions are drawn out, rendered resolvable by the shadowing process, and the molecules are aligned radially. This alignment of the molecules on polystyrene was not observed when the alpha-chain protuberances were removed enzymatically, Figure 3E, or when fibrinogen, with or without the alpha-chain protuberance, was dispersed onto carbon surfaces, Figure 3D.

Fibronectin was another plasma protein with relevance to the thrombotic reaction that we studied. We found that the structure of the fibronectin molecules dispersed onto carbon was also quite different from those dispersed onto polystyrene surfaces (28). Figure 4 shows that fibronectin, when adsorbed to polystyrene surfaces, appears in an elongated, nodular form, Figures 4 A and B. When it is adsorbed to mica or carbon, however, it exists in coiled pleiomorphic form, Figures 4C and D. This finding has since been confirmed (29-31), and the current thought is that the plasma fibronectin molecule exists in a coiled, inactive form in the circulation, and upon adsorption to a surface it unfolds, exposing its functional domains (fibrin binding, platelet binding, etc.). These results graphically illustrate the influences of the surface directly on the structure of the molecules' tertiary conformation; however, these experiments occurred under static conditions, and the effects of the flowing milieu of the circulatory system was not addressed.

FLOW STUDIES

In order to study the effect of fluid shear on the adsorption of proteins, a flow cell was developed in the laboratory of Dr. Kenneth Keller of the Department of Chemical Engineering at the University of Minnesota. It was designed to produce a constant shear rate over the surface of the test materials. The test materials are mounted in the flow cell as a 90 degree wedge, divided along its axis and oriented so that the division plane is parallel to the direction of flow Figure 5. Due to this design, two different surfaces may be simultaneously exposed to the flowing solution. The test materials may be either manufactured in the shape of a half wedge, or a thin film of the test material may be mounted on the surface of the wedge (32). This latter method was used in these studies. A deceleration cone, mounted immediately upstream of the test surfaces, flattens the velocity profile of the flowing solution. The dependence of the shear rate produced at the surface of the test material upon the fluid flow rate has been calculated and tested for this device, and an approximately linear relationship has been obtained (Kreczmer, M.L., personal communication).

The samples were mounted in the test cell with particular attention that they were not contaminated by contact with any liquids. The flow cell was then connected to a plumbing system that contained a protein solution reservoir, a wash buffer reservoir, and a three-way-valve. The flow cell, with the test materials mounted at the wedge surface, was connected to the system and filled with buffer. The system was purged of air, and degassed under vacuum. After the establishment of a flow of buffer through the cell, the three-way-valve was used to switch the input line to the sample protein reservoir for a selected period of time. Following the exposure of the surfaces to the protein solution, the valve was switched back to the buffer reservoir and the cell flushed with about 300 ml. of buffer.

In the flow cell, the test surfaces were exposed to the flowing solutions under well characterized flow conditions. Carbon, polystyrene, and polycarbonate surfaces were exposed to the flowing (135/sec. shear rate), dilute (100 micrograms/ml.) solutions of intact native human fibrinogen, human fibrinogen with the alpha chain

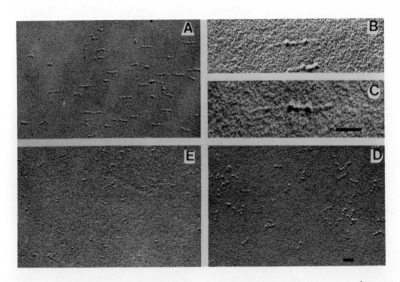

Figure 3. Fibrinogen sprayed onto surfaces of polystyrene (Panels A,B,C,and E), and carbon (Panel D). In panel E the alpha chain extension has been enzymatically removed. The scale bar represents 50nm.

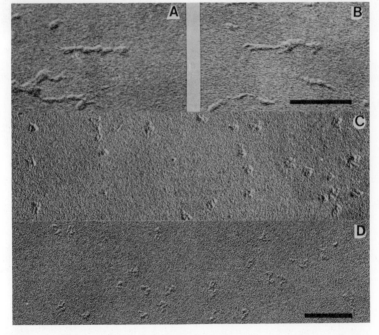

Figure 4.   Human plasma fibronectin sprayed onto the surface of polystyrene (Panels A and B), and the surface of carbon (Panels C and D). The scale bar represents 100 nm.

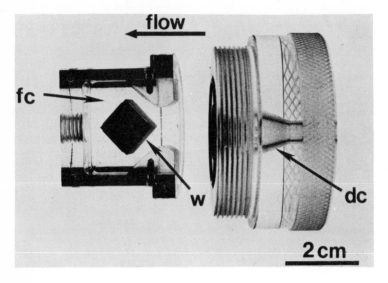

Figure 5. The flow cell, with the cover removed for clarity, showing the test material wedge w, the decelleration cone dc, and the flow chamber fc. (Reproduced with permission from Ref. 17. Copyright 1985 John Wiley and Sons.)

extensions removed enzymatically, and human serum albumin. Several different exposure times were observed.

The results of the adsorption of intact fibrinogen are illustrated in Figure 6. At the shortest times, the typical trinodular structure of the fibrinogen molecule was observed on the surfaces of carbon. However, on polystyrene surfaces, the fibrinogen molecules adsorbed as denatured irregular deposits which were distributed uniformly over the surface and lacked the trinodular structure of the fibrinogen molecule. This is in marked contrast to the behavior of fibronogen sprayed onto polystyrene and adsorbed during the evaporation of the droplet that was described above. In the case of the droplet, the relatively static conditions led to a trinodular structure, and the observation of the alpha-chain, Figures 3B and 3C. The flowing conditions in the cell produced a fundamental difference in the behavior of the single molecule of fibrinogen on polystyrene, but not on carbon. As the exposure times increased, the protein films that formed on carbon surfaces became continuous, while on polystyrene a network developed. With additional exposure, the voids in the network decreased in size and the areas of protein increased in thickness. This process continued until only small isolated islands of uncoated substrate existed. Although the mechanism leading to the network is not understood, the major differences between carbon and polystyrene illustrate the sensitivity of this electron microscope technique in differentiating the behavior of different materials.

When solutions of fibrinogen and human serum albumin were used in the flow cell, Figures 6 and 7 respectively, a network of adsorbed protein formed on the polystyrene substrates. However, the size of the albumin network (i.e., the thickness of the strands of protein and the size of the voids) was much smaller than the network formed by fibrinogen. Colloidal gold conjugated with affinity purified IgG antibodies to serum albumin was employed to identify the composition of the adsorbed protein film. The initial results of these experiments are shown in Figure 8. An interesting result of these experiments is that the gold label is located only on the protein film network and not on the exposed substrate. This demonstrates the specific nature of the binding of the conjugated label to its antigen, as there is no binding to the exposed substrate in the open area of the network.

This observation of the incomplete coverage of the substrate by the adsorbed protein film has been documented by others. Brash and Lyman (18) noted that on the surface of PTFE, gamma globulin was found to cover the surface incompletely. Eberhart (19) observed that even at exposure times (>1 hour) and solution concentrations vastly in excess (30 fold) of those used in our experiments, fibrinogen, albumin and gamma globulin films formed which left substantial portions of the underlying PTFE surface uncovered. Recently Johnston (33) has also observed by ESCA that adsorbed hemoglobin films incompletely cover the surfaces of materials.

POLYETHERURETHANES. Polyetherurethanes are block copolymers (37) with phase domains resulting from the aggregation of the urethane-urea hard segments and the soft polyether segments. As the molecular

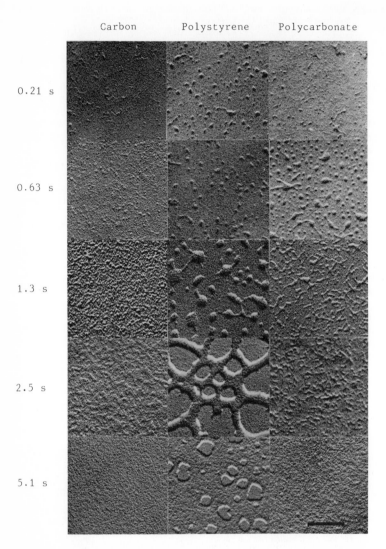

Figure 6. Purified solutions of fibrinogen at 100 micrograms/ml
contacted these surfaces at a shear rate of 135/sec. The scale
bar represents 100 nm. (Reproduced with permission from Ref. 17.
Copyright 1985 John Wiley and Sons.)

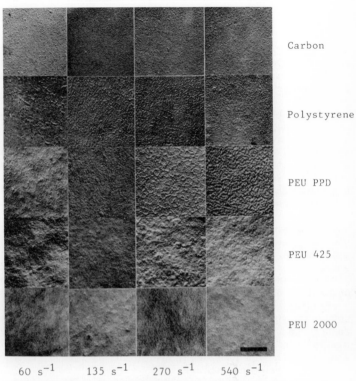

Carbon

Polystyrene

PEU PPD

PEU 425

PEU 2000

60 s^{-1} 135 s^{-1} 270 s^{-1} 540 s^{-1}

Figure 7. Human serum albumin exposed in the flow cell to the above surfaces for 2.5 sec. at increasing shear rates. The scale bar represents 100 nm.

Figure 8. Polystyrene exposed to human serum albumin in the flow cell for 6 sec at 135/sec., then reacted with the anti-HSA antibodies conjugated with 20 nm. colloidal gold particles. The scale bar represents 100 nm.

weight of the polyether segment increases, the surface of these
polymers has been shown, by ESCA (20), and FTIR (38) to become
increasingly different in composition from the bulk of the material.
The low molecular weight polyether tends to concentrate at the
surface, with the urethane (hard segment) buried in the bulk of the
polymer. Merrill (37) has suggested that the relatively good blood
compatibility of the polyetherurethanes is due to the presence of the
polyether segment at the surface. The polyether segment adopts an
amorphous structure at physiological temperatures when linked to the
urethane segments. Further, polymers designed as hard segment
analogs have been shown to adsorb more thrombin than those
polyurethanes which contain the amorphous polyether segment (39). The
peu-ppd, manufactured with the smallest polyether segment (the
monomer), is the hard segment model of these materials and has the
highest surface energy. The other two polyether urethanes are of
intermediate (peu-425) and low surface energy (peu-2000).

These polymers were exposed in the flow cell to albumin
solutions (100 micrograms/ml.) for 2.5 sec. at shear rates 1/2, 1,
and 2 times the rates used to expose the surfaces of carbon,
polystyrene, and polycarbonate in the series of experiments described
above. These shear rate values were 60/sec., 135/sec., 260/sec., and
540/sec. The results are illustrated in Figure 7. Note that the
morphology of the protein films formed varies on a given surface with
increasing wall shear rate except for the soft segment model, peu-
2000. On the soft segment model the existence of a protein film
cannot be visualized at all, and the morphology of the surface is
similar in all of the examples shown. The inability to resolve any
protein film is presumably due to the amorphous character of the
polyether dominated surface.

When the albumin films are adsorbed to the surface of the peu-
425, the formation of a protein film is evident. Some substructure is
seen at the 270/sec and 540/sec shear rates. However, the most
striking observation is that the hard segment model peu-ppd, which
would be expected to adsorb proteins the most readily, adsorbs
albumin in a fashion similar to the polystyrene surfaces. The
albumin adsorbs to the peu-ppd as distinct molecules at the low flow
rates, and at the 270/sec rate the formation of a network is evident.

CONCLUSIONS

We have shown that the morphology of the adsorbed individual molecule
and the protein film is dependent upon the surface to which the
protein is adsorbed and the wall shear rate at which the protein
solution contacts the surface. We have studied the interaction of
proteins with surfaces from two approaches; the first involved
observations of the tertiary structures of individual molecules, and
the second involved the use of a flow cell and the exposure of test
materials to flowing, dilute, purified human plasma protein
solutions. On the hydrophobic surfaces of polystyrene and
polycarbonate, albumin and fibrinogen films form networks. On the
surface of the hard segment model polyurethane, the morphology of the
adsorbed film is very similar to that formed on the hydrophobic
polystyrene and polycarbonate surfaces. On the peu-2000, the lowest

surface energy' material, no images of a protein film could be obtained. On the surface of carbon, fibrinogen and albumin adsorb as individual molecules and eventually form a continuous protein film without network formation.

The formation of a network, or incomplete coverage of the substrata by adsorbed proteins, has been noted by several laboratories utilizing different methodologies. Consequently, these observations must be kept in mind during the design of protocols and in the interpretation of experimental outcomes. In the cases of the materials mentioned above, one may not assume that the substrata are completely coated with a continuous layer of protein. Substantial areas of uncovered surface may continue to remain exposed and unprotected by a passivating layer of protein.

LITERATURE CITED

1. Baier, R. E., and Dutton, R. C., J. Biomed. Mater. Res. 1969, 3: 191.
2. Guidelines for Blood- Material Interactions. U.S. Dept. Health and Human Services, Devices and Technology Branch. NHLBI, NIH Pub. No. 80-2185, Sept. 1980.
3. Vroman, L., Thromb. Diath. Haemorrh. 1964, 10:455.
4. Vroman, L., Adams, A.L., Surface Science 1969, 16: 438.
5. Gendreau, R.M., Leininger, R.I., Winters, S., Jakobsen, R.J., 1982, ACS Advances in Chemistry Series vol. 199, p. 371.
6. Harrick, N.J., Loeb, G.I., Anal. Chem. 1973, 45: 687.
7. Baier, R.E., Loeb, G.I., Wallace, G.T., Proc. Fed. Exp. Biol. Med. 1971, 30: 1523.
8. Stupp, S.I.; Kauffman, J.W.; Carr, S.H., J. Biomed. Mater. Res. 1977, 11: 237.
9. Park, K., Albrecht, R.M., Simmons, S., and Cooper, S.L.,Transactions Eleventh Annual Meeting For Biomaterials 1985, Vol. VII: p. 3.
10. Vroman, L., Adams, A.L., J. Biomed. Mater. Res. 1969, 3: 43.
11. Reichert, W.M., Filisko, F.E., Barenberg, S.A., in Biomaterials: Interfacial Phenomena and Applications. in Advances in Chemistry Series, ed. by Cooper, S. L. et. al., vol. 199, ACS, Washington D.C., 1982, p. 177.
12. Walton, A., Koltisko, B., 1982 ACS Advances in Chemistry Series vol. 199, p. 245.
13. Weathersby, P.K.; Horbett, T.A.; Hoffman, A.S.; Trans. Amer. Soc. Artif. Int. Org. 1976, 22: 242.
14. Ratner, B.D., Horbett, T.A., Shuttleworth, D., Thomas, H.R., Journal of Colloid and Interface Science. 1981 83(2): 630.
15. Eberhardt, R.C., Prokop, J., Wissenger, J., Wilkov, M.A., 1977, Trans. Amer. Soc. Artif. Int. Org. 23: 134.
16. Goodman, S.L., Lelah, M.D., Lambrecht, L.K., Cooper, S.L., Albrecht, R.M., 1984, SEM/1984/1: 279.
17. Rudee, M.L., Price, T.M., 1985, J. Biomed. Mater. Res. 19: 57.
18. Brash, J.L., Lyman, D.J., "Adsorption of Proteins and Lipids to Nonbiological Surfaces " in The Chemistry of Biosurfaces. ed. by Hair, M., 1971, Marcel Dekker Inc., New York, 1: 177.

62   PROTEINS AT INTERFACES

19.   Eberhart, R.C.; Lynch, M.E.; Bilge, F.H.; Wissinger, J.F.; Munro, M.S.; Ellsworth, S.R.; Quattrone, A.J., 1982, ACS Advances in Chemistry Series vol. 199, p. 293.

20.   C.S.P. Sung and C.B. Hu, 1979, J. Biomed. Mater. Res. 13: 161.

21.   Slayter, H.S., 1976, Ultramicroscopy 1: 341.

22.   Hall, C.E., Slayter, H.S., 1959, J. Biophys. Biochem. Cytol. 5: 11.

23.   Strong, D.D., Watt, K.W.K., Cottrell, B.A., Doolittle, R.F.,1979, Biochemistry, 18: 5399.

24.   Rudee, M.L., Price, T.M., 1981, Ultramicroscopy, 7: 193.

25.   Wall, J., Hainfeld, J., Haschemeyer, R.H., Mosesson, M.W., 1983, N.Y. Acad. Sci., 408: 164.

26.   Weisel, J.W., Stauffacher, C.V., Cohen, C., 1985, Thrombosis and Haemostasis 54: 224.

27.   Price, T.M., Strong, D.D., Rudee, M.L., Doolittle, R.F., 1981, Proc. Natl. Acad. Sci. USA, 78 (1): 200.

28.   Price, T.M., Rudee, M.L., Pierschbacher, M., Ruoslahti, E., 1982, Eur. J. Biochem., 129: 359.

29.   Tooney, N.M., Mosesson, M.W., Amrani, D.L., Hainfeld, J.F., Wall, J.S., 1983, J. Cell Biol. 97: 1686.

30.   Kleinman, H. K.; Klebe, R.J.; Martin, G.R., 1981, Journal of Cell Biology. 88: 473.

31.   Klebe, Robert J.; Bentley, Kevin L.; Schoen, Robert C., 1981, Journal of Cellular Physiology. 109: 481.

32.   Lyman, D.J., Kwan-Gett, C., Zwart, H.H.J., Bland, A., Eastwood, N., Kawai, J., Kolf, W.J., 1971, Trans. Artif. Inter. Org. 17: 456.

33.   Johnston, A.B., and Ratner, B.D., in this volume.

34.   Rudee, M.L., Price, T.M., 1985, J. Biomed Mater. Res. 19: 57.

35.   Doolittle, R. F.; Schubert, D.; Schwartz, S.A., 1967, Arch. Biochem. Biophys. 118: 456.

36.   Haubold, A. 1977, Annals of N.Y. Academy of Sci. 283: 383.

37.   Merrill, E.W., Sa Da Costa, V., Salzman, E.W., Brier-Russell, D., Kuchner, L., Waugh, D.F., Trudel III,G., Stoppers, S., Vitale, V., 1982, ACS Advances in Chemistry Series, vol.199, p. 95.

38.   Knutson, K., Lyman, D.J., 1982, ACS Advances in Chemistry Series vol.199, p. 109.

39.   Sa Da Costa, V.; Brier-Russell, D.; Trudel, G.; Waugh, D.F.; Salzman, E.W.; Merrill, E. W.; 1980,J. Colloid. Interface Sci. 76: 594.

RECEIVED January 29, 1987

Chapter 4

Adsorption and Chromatography of Proteins on Porous Glass: Activity Changes of Thrombin and Plasmin Adsorbed on Glass Surfaces

Takaharu Mizutani

Pharmaceutical Sciences, Nagoya City University, Mizuho-ku, Nagoya 467, Japan

The activity changes of thrombin, plasmin, trypsin, phosphatase and peroxidase adsorbed to porous glass have been studied. Some of the factors influencing enzyme inactivation are as follows.
1. Adsorption:the activities of free plasmin, trypsin, and phosphatase were greater than in the bound state, but free thrombin had less activity than bound thrombin.
2. Time:the activity of all bound enzymes decreased with increasing adsorption time.
3. Stability:the bound enzymes thrombin, plasmin and phosphatase were more stable than in the free state, but trypsin and peroxidase were more stable in the free than in the bound state.
4. Temperature:temperature had a strong effect on trypsin and peroxidase, but only a weak effect on phosphatase and thrombin.
Thus inactivation of enzymes on the surface varied from one enzyme to another and did not follow a common pattern.

The adsorption of proteins on the surface of glass is well known in many areas of biochemistry (1-3). For example, blood clotting is promoted on glass surfaces, and siliconization is used to minimize this effect. Clotting starts when factor XII is activated by the anionic charge of the silanol groups on the glass surface (4-5). It is also known that macrophages adhere well on glass surface and this adhesiveness allows separation of macrophages from other cells. The adhesion promoting substance has been isolated from macrophages (6). The adhesion of blood platelets to glass surfaces is known in the field of clinical chemistry, and blood platelets adhere more on surfaces coated with fibrinogen (7). Another general effect of adhesion to glass surfaces is the difference in adhesiveness between cancer cells and normal cells in culture dishes (8). The adsorption behavior of human plasma fibronectin on silica substrates has been studied (9-10).

0097-6156/87/0343-0063$06.00/0

The amounts of proteins adsorbed on the surface of glassware has a significant effect on the quantitative analysis of very small amounts of protein in the case of radioimmunoassay or enzymeimmuno-assay (11).   Columns of glass beads with bound antigens were used for the preparation of lymphocytes to prove the clonal selection theory (12).   Flagellin of Salmonella was found to be adsorbed on glass surfaces (13).   Boone (14) and Kataoka et al. (15) separated cells or proteins based on adhesion to glass beads or synthetic polymers.

Adsorption of proteins on glass surface is not a specific pheno-menon but rather a general phenomenon which is usually neglected because of the low surface area of typical glassware.   However, in the case of porous glass which has a surface area of some hundreds of square meters per g, adsorption cannot be neglected (16).   Porous glass having a large pore size has also been developed for size ex-clusion chromatography and is useful for isolation of cell organelles after treatment with Carbowax 20M to prevent adsorption (17).   In the first part of this paper, the author summarizes his research on the adsorption of proteins to porous glass surfaces, and application to adsorption chromatography.   In the second part, the activity changes of enzymes adsorbed to glass surfaces are discussed.

Studies of Adsorption of Proteins on Porous Glass

Table 1 summarizes the results of our initial studies of protein adsorption on porous glass (18).   The studies contained the yield of albumin from CPG columns in various buffers.   The yields were low in sodium chloride, sodium carbonate, phosphate, borate, acetate, cit-rate and barbital buffer, because of the adsorption of albumin on porous glass.   The yields were higher (30%) in Tris-HCl buffer.   The adsorption of protein and basic amino acids depends strongly on the fact that the former is not inhibited by inorganic salts but the latter is inhibited by sodium chloride.   Elution with solutions of amino acid buffers such as glycine and alanine at pH 8-8.9 resulted in about 70% recovery of the protein.   Although ammonium acetate has carboxyl groups and ammonium groups, it could not be substituted for the glycine of the ampholytes.   This suggested the significance of zwitterions.   As the yield with β-alanine shows, the proximity of amino groups to carboxyl groups is not always necessary.   Such prevention of protein adsorption in amino acid buffers is believed to have its origin in the occupation of the protein-binding site on the glass surface by amino acids.   However the inhibition of adsorption is more effective with basic amino acids such as lysine (19).

Adsorption of proteins on porous glass varied with the protein, the buffer and the pH of the buffer.   Based on these differences, porous glass can be used as an adsorbent for the separation of pro-teins by adsorption chromatography (20, 21).   The amounts of basic amino acids, nucleosides, cations, and proteins adsorbed were about 4-5μmol/100 m^2 (22-23).   These results indicate that these materials adsorb on the surface as monolayers.

The amounts of protein adsorbed on porous glass at various pH were studied (19).   Albumin was adsorbed the most at pH 5, lysozyme at pH 11, chymotrypsin at pH 8; these values correspond to the iso-

Table I. Yields of Bovine Serum Albumin from Columns of
Controlled-pore Glass with Various Buffers

Buffer	pH	Yield (%)
0.05M Phosphate	7.4	0
0.05M Phosphate-0.2M NaCl	7.4	0
0.05M Phosphate-1M NaCl	7.4	0
0.05M Phosphate-3M NaCl	7.4	0
0.05M Phosphate-1M NaCl-2% ethanol	7.4	3
0.05M Phosphate-1M NaCl-5% ethanol	7.4	4
0.05M NaHCO$_3$	8.7	7
0.05M Na$_2$CO$_3$	11.4	8
0.01M NaOH	12	3
0.05M Borate	8.7	21
0.035M Barbitone	8.6	16
0.5M Tris hydrochloride	8.6	33
0.5M Tris acetate	8.6	27
0.1M Ammonium acetate	8.3	5
0.5M Potassium acetate	8.3	17
0.1M Citrate	8.0	15
0.05M Glycine	6.0	17
0.05M Glycine	8.0	68
0.038M Glycine-0.005M Tris-HCl	8.9	76
0.38M Glycine-0.05M Tris-HCl	8.9	76
0.056M Alanine	8.0	68
0.056M Alanine-0.2M Tris-HCl	8.6	13
0.056M β-Alanine	8.0	64
0.056M β-Alanine-0.2M Tris-HCl	8.6	21

Reproduced with permission from Ref. 18. Copyright 1975, Elsevier
Sci. Pub. B.V.

electric points of the proteins and indicate that the proteins are
adsorbed most extensively to a glass surface at their isoelectric
points. Based on the negative charge of the silanol groups on the
glass surface, proteins should be adsorbed at a more acidic pH, where
they would have a positive charge. However interaction between
proteins in the layer would be another important factor. Therefore
it is concluded that the adsorption of proteins onto porous glass was
caused by two factors; one is amine-silanol ionic bonding and the
other is a aggregative property of proteins on silica.

Messing has suggested that the interactions involved in the
adsorption of proteins on glass surfaces are ionic bonding and hydrogen bonding (24). Also, Bresler et al. have suggested the participation of hydrophobic bonding based upon thermodynamic studies (25).
To investigate this aspect, the adsorption of albumin on porous glass
in various detergent solutions was studied (19). Figure 1 shows the
results. Albumin was adsorbed well on porous glass even in 8M urea
and 6M guanidine-HCl both of which affect hydrogen bonding. In urea,
the amount of albumin adsorbed was much the same as in water. The
amount adsorbed in guanidine-HCl was slightly lower than that in
urea; this phenomenon is not due to the inhibition of hydrogen bonding but to the high salt concentration in 6M guanidine-HCl. Since
the helical regions of proteins are denatured in about 2-4M guanidine

(26), hydrogen bonding should not be present in protein molecules in 6M guanidine-HCl.   But under these conditions, proteins adsorb to porous glass; therefore hydrogen bonding does not seem to be impor- tant for protein adsorption on glass.

As for ionic bonding, another main interaction made in adsorp- tion, we showed that for albumin modified with succinate the amount adsorbed was about 3 mg/g of porous glass, about 2% of that in distilled water (Figure 1).   This indicated that ionic amine-silanol bonding between amino groups on the protein and negative silanol on glass surfaces plays an important role.   This phenomenon was con- firmed for glass surfaces through the use of glyceryl porous glass. The amount adsorbed on 1 g(62 m^2) of modified porous glass was 8.7 mg (27), i.e. 9% of that on unmodified porous glass, thus suggesting that the silanol groups on glass surfaces are essential for protein adsorption.   Studies with a hydrophobic probe (28) showed that the probe binds more to proteins having a low affinity for glass sur- faces; this result suggests that hydrophobic interactions are not important for protein adsorption to glass.

Adsorption patterns of albumin on porous glass invarious deter- gents were studied (29).   Albumin was not adsorbed on glass in SDS or in a hard soap (composed of sodium stearate and parmitate) solution, possibly because of repulsion between terminal silanol groups on porous glass and negatively charged dodecyl residues hydro- phobically bound on the protein molecules.   In sodium valerate, this inhibition of adsorption was not observed, suggesting that a sodium salt of a fatty acid having detergent activity was a factor in inhi- biting adsorption.   In contrast, albumin was adsorbed more in alco- holic detergent solutions (Triton X-100 and Brij 35) and a cationic soap solution (benzalkonium chloride); therefore proteins adsorbed on glassware cannot be completely removed with such detergent solu- tions.   Based on our findings that protein in ethylalcohol, octylal- cohol, and alcoholic detergent solutions adsorbed well on porous glass, and that carbohydrates having alcoholic OH groups were not adsorbed on porous glass, it is concluded that alcoholic OH does not affect the adsorption of proteins.

We now discuss our results on adsorption chromatography (21). Figure 2 shows the elution pattern of rabbit serum.   These elution patterns were strongly influenced by pH.   Yields of the fractions eluted at 4°C and 37°C using 20 mg and 100 mg of rabbit serum (one fiftieth to one hundredth of the adsorption capacity of porous glass) at pH 8 are similar to each other.   It is concluded that the repro- ducibility of the elution pattern of proteins is good for protein amounts in the range of one tenth to one hundredth of the adsorption capacity of porous glass.   Yields at 37°C were also similar to those at 4°C, suggesting that the effect of temperature was not strong.

We now consider these data from the standpoint of the separa- tions effected.   Disc gel patterns of the separated fractions of rabbit serum at pH 8 shows that albumin with a mobility of 0.7 rela- tive to BPB (marker dye), was present in fraction I (Figure 2). α-Globulin (transferrin) with a relative mobility of 0.4 was present in fraction II and γ-globulin with a low mobility was present in fraction IV   Most of the γ-globulin in fraction IV was of high mobility and relatively small size.   γ-Globulin of high molecular weight, which did not move on the gel, was absent in fraction IV.

The globulins found in fraction V were eluted with 0.1% SDS.   The recovery with solvent not containing SDS was low.

In a schematic elution pattern of some standard proteins, peroxidase was eluted first with saline, BSA came next with glycine buffer at pH 6.6 and hemoglobin and catalase were eluted at a pH of nearly 8.0.   Aldolase, lysozyme, chymotrypsinogen A, malate dehydrogenase, and cytochrome c were not eluted under these conditions, but were eluted with 0.1% SDS.   The adsorption order does not depend on the isoelectric point, the molecular mass, or the content of basic amino acids.   However, adsorption may depend on the α-helix content, and the secondary structure of those proteins may be important.   We have also reported on protein adsorption and separation on siliconized glass surfaces (30), and on the adsorption and separation of nucleic acids on those same surfaces (31-35).

Recent Results on Activity Changes of Proteins Bound to Glass Surface

From the standpoint of blood clotting on the surfaces of artificial organs, the activity and structural changes of proteins bound on surfaces have been studied (36-41).   The enzyme used in our studies were thrombin, plasmin, trypsin, phosphatase and peroxidase.   Thrombin and plasmin are, respectively, key components of the coagulation and fibrinolytic systems.   Kinetic parameters were obtained by direct measurement of the activities of enzymes bound on porous glass by mixing the porous glass with solutions of the substrate.

Materials and Methods.   The surface used was CPG (Electro-Nucleonics, CPG-10, 240Å, 96% silica, 100μm particles, 1g=2 ml, $97m^2/g$) which was well washed with chromic acid and distilled water and then fully equilibrated with phosphate-buffered saline (0.14M NaCl, 3mM KCl, 8.5mM phosphate, PBS) containing 0.02% NaN_3 as a preservative.   The maximum amount of proteins adsorbed onto CPG was 100-200mg/g.   This value corresponds to monolayer adsorption.   Proteins used were thrombin from porcine plasma (Sigma, 200U/0.3mg/ml), porcine plasmin (Sigma, 5U/1.2mg/ml), trypsin from porcine pancreas (Sigma), alkaline phosphatase from E. coli (PL-Biochemicals), and horse radish peroxidase (Sigma).   Reagents used as substrates were α-N-benzoyl-L-arginyl-ethylester (BAEE), p-nitrophenylphosphate (PNPP) and guajacol.

Assay.   One ml of suspended porous glass, previously equilibrated with PBS and suspended in PBS, was packed in a column made from a pasteur pipet.   Thrombin (200μl), plasmin (0.2ml, 1U), trypsin (20μl, 20μg), or phosphatase (30μg, 1 ml) were applied on the columns and these enzymes completely adsorbed.   Adsorption on the column was confirmed by the measurement of absorbance at 280nm of the eluate from the column.   The porous glass carrying bound enzymes from the column were placed in test tubes and washed with PBS.   The supernatant was then discarded.   The activities of enzymes on CPG were measured by adding one ml of the substrate solution (BAEE for thrombin plasmin and trypsin; PNPP for phosphatase) at various concentrations to the tube and mixing for one, two, or four minutes (42-43).   After standing for 10 seconds, the suspension was centrifuged and the absorbance of the supernatant was measured at 254nm for BAEE or at 405nm for PNPP.   From these data, the mean value of A_{254}/min or

A_{405}/min was estimated and the reaction rate was obtained using the
molar extinction coefficient (ε=808 for BAEE or ε=13000 for PNPP).
These reaction rates were compared with the rate of reaction of the
free enzymes.   The data measured at the bound state or the free
state were used as the apparent reaction rate.   Porous glass with
bound enzymes was stored at 0, 25, or 37°C for 1, 3, 7 and 14 days.
Free enzymes were used as a control.

    In the case of horse radish peroxidase, peroxidase shows the
weakest adsorption on porous glass.   It did not adsorb in saline or
Tris-buffer but only in distilled water.   Therefore, peroxidase was
adsorbed on a porous glass column and stored in distilled water.
After adsorption for an appropriate time, the enzyme was eluted from
the column and the specific activity of peroxidase was estimated (44).
Peroxidase (1 ml, 25μg) in distilled water was applied on the CPG
column (1 ml).   The column was washed with distilled water and stored
at constant temperature.   After adsorption for an appropriate time,
the enzyme on the column was eluted with 0.05M Tris-HCl at pH 8.6.
The activity of the enzyme in the eluate was measured by the guajacol-
H_2O_2 method by measuring absorbance at 470nm.   The specific activity
of the enzyme recovered from the column was compared with that of the
non-bound enzyme which was stored under the same conditions.

Results.   The effect of substrate concentration on the initial rate
of the reaction catalyzed with bound thrombin or free thrombin is
shown in Figure 3.   As can be seen, the initial rate of bound thrombin
increased slightly when compared to the rate of free thrombin.   The
apparent reaction rates of the other bound enzymes were about half
those of the free enzymes described later in this report.   The increased rate for bound thrombin may be due to its activation or stabilization by adsorption on the glass surface, similar to the activation
of the Hageman factor on anionic surfaces.   However, it is likely
that the thrombin binding site for surfaces differs from the active
site.   Adsorption did not change the active structure of thrombin.
    Time-dependent inactivation of bound and free thrombin was
studied at 0 and 37°C as shown in Figure 4.   The reaction was measured at a substrate concentration of 0.031mM at which concentration
the rate indicated the maximum velocity of the enzyme.   In Figure 4,
it is shown that the velocity of bound thrombin after adsorption for
7 days was 70% of that measured immediately after adsorption.   The
velocity of free thrombin after storage for 7 days decreased by 70%.
This tendency toward inactivation was also found after adsorption for
14 and 23 days.   The influence of temperature on the activity of free
enzyme is larger than on bound enzyme.   Bound thrombin is thus more
stable then free thrombin.   These data showing the relative stability
of bound thrombin (Figure 4) are consistent with the data in Figure 3.
The results in Figure 4 also show the trends at 0°C which were similar
to those seen at 25 and 37°C.   The influence of temperature was not
significant for inactivation even though thrombin is slightly more
stable at low temperature; the influence of temperature on the free
enzyme is larger than on the bound enzyme.   The reason for the relative stability of bound thrombin may be that autolysis in the bound
state is more difficult than in the free state.   The apparent Km
value for the substrate BAEE increased in a time-dependent manner as
follows; 25μM immediately after adsorption, 42μM after adsorption for

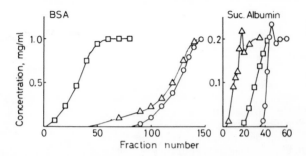

Figure 1: Adsorption pattern of bovine serum albumin (left) and
succinylated albumin (right) in detergent solutions (pH 5.4) on
CPG.   Key: circles, distilled water; triangles, 8M urea; squares,
6M guanidine-HCl.   The fraction volumes were 1.6 ml for bovine
serum albumin and 0.52 ml for succinylated albumin. Reproduced
with permission from Ref. 19. Copyright 1978, American Pharmaceutical Association.

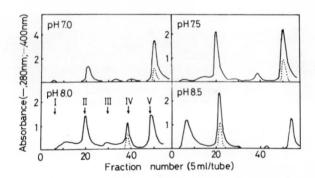

Figure 2: Elution patterns of rabbit serum (100 mg protein) on a
CPG column (20 cm x 1.2 cm) at 4°C and various pH values. Eluants:
I, 0.2M NaCl-0.01M Phosphate; II, 0.01M Tris-HCl; III, 0.2M Tris-
HCl; IV, 0.2M glycine; V, 0.1% SDS in 0.05M phosphate. Reproduced
with permission from Ref. 21. Copyright 1979, Elsevier Sci. Pub.

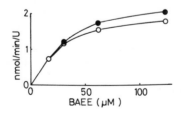

Figure 3: Velocity of free (open circles) and bound (closed)
thrombin at 25°C on CPG.

1 day, 48μM after adsorption for 3 days, and 52μM after adsorption
for 7 days.   However, the change in the apparent Vmax values of
bound thrombin after adsorption for 1-7 days was small.   These Km
and Vmax values indicate that the behavior of bound thrombin shown in
Figure 4 depended on the decrease in the affinity of the thrombin
active sites for the substrate.

The effect of substrate concentration on the initial rate of the
reaction catalyzed with bound plasmin and free plasmin is shown in
Figure 5.   It can be seen that the rate of bound plasmin was about
half that of free plasmin.   This enzyme thus shows behavior opposite
to that of thrombin (Figure 3), whose activity was increased by
adsorption.   The decreased activity of bound plasmin suggests that
the active site and the adsorption site of plasmin on glass may be
relatively close together.   Figure 6 shows the effect of substrate
concentration on the initial rate of the reaction catalyzed with plasmin bound for 1, 3, or 7 days at 0 or 37°C.   The reaction rate
decreased with adsorption time.   The pattern of decreased activity
at 0°C is similar to that at 37°C.   The small effect of temperature
on the activity of bound plasmin is similar to that of thrombin.
The activity of free plasmin at 37°C was lost after standing for 3
days.   Therefore, free plasmin was found to be labile, probably due
to autolysis.

The Km and Vmax values of bound and free trypsin after adsorption for 3 and 7 days at 0, 25, or 37°C are summarized in Tables II
and III.   The Km values of the bound enzyme are similar to those of

Table II.   Km Values of Trypsin Adsorbed onto CPG

Condition	Temperature (°C)	Km value (μM)		
		0 time	3 days	7 days
Free	0	–	–	19
	25	15	–	17
	37	–	–	20
Bound	0	–	24	24
	25	18	22	26
	37	–	22	23

Table III.   Vmax Values of Trypsin Adsorbed onto CPG

Condition	Temperature (°C)	Vmax (μmol/min/mg)		
		0 time	3 days	7 days
Free	0	–	–	64
	25	69	–	62
	37	–	–	57
Bound	0	–	4.7	4.1
	25	4.9	4.4	2.7
	37	–	4.2	2.0

the free enzyme (45) and are almost identical with the values measured
at 0, 25, or 37°C.   There is a tendency to decreased activity with
increasing adsorption time as shown in Table II.   In Table III, it is
seen that the Vmax values of bound trypsin are much lower than those
of free trypsin, with a ratio of about 1 to 10.   Thus, the activity of
the trypsin was decreased by adsorption on the porous glass surface.
The Vmax values of free trypsin were unchanged after adsorption for 7

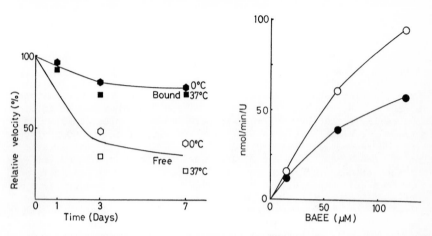

Figure 4 (left). Decrease of activity of free (open) and bound (closed) thrombin on CPG at 0 °C and 37 °C.

Figure 5 (right). Velocity of free (open) and bound (closed) plasmin on CPG at 25 °C.

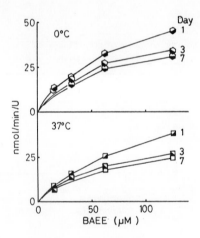

Figure 6. Decrease of activity of bound plasmin on CPG at 0 °C and 37 °C.

days.   The values of bound enzyme were also not changed after adsorption for 3 days at 0°C but after adsorption for 7 days Vmax decreased as a function of temperature during storage.   The Vmax value at
37°C was half that at 0°C.   Thus, the activity decrease of bound
trypsin due to storage depended on both the Km (affinity) and Vmax
(velocity) values.

The effects of substrate PNPP concentration on the initial rate
of the reaction catalyzed with bound phosphatase or free phosphatase
are shown in Figure 7.   The velocity of bound phosphatase at 25°C
was about half that of the free enzyme.   These results are similar
to those for plasmin (Figure 5).   The activity of the bound or free
enzyme decreased with storage time (1, 3 and 7 days).   The activities of bound phosphatase at 0, 25, and 37°C were similar and the
influence of temperature on activity during storage was insignificant.

The specific activities of peroxidase recovered from CPG as a
function of adsorption time are shown in Figure 8.   The free enzyme
was stable but the specific rate of bound peroxidase decreased with
adsorption time.   The bound enzyme was also labile at higher temperature.   This large effect of temperature is different from the
results obtained for alkaline phosphatase, plasmin and thrombin.

In summary, the properties of the five enzymes are shown in
Table IV.   Table IV also shows data sources from the literature (46-
49) for other enzymes similar to those used in the present study.

Table IV. Summary of Some Properties of Enzymes
Adsorbed on Porous Glass

Property	Thrombin	Plasmin Phosphatase	Trypsin	Peroxidase
Activity of bound (B) and free (F)	B=F	B<F	B≪ F	B<F
Effect of temperature	small	small	small	large
Half life of activity of bound enzyme at 25°C	10 days	3-5 days	7 days	3 days
Data sources for similar enzymes (Reference No.)	factor XII (46)	heparin (47) Acyl-CoA hydrolase (49)	glucose isomerase (48)	

The behavior of thrombin resembles that Hageman factor which is activated on anionic surfaces by adsorption.   Generally, proteins are
inactivated or denatured by adsorption (50-53) although the exact
patterns of inactivation are different for the different enzyme
groups.   Thrombin and factor XII which relate to blood clotting appear
to be stable than phosphatase, plasmin, and peroxidase.

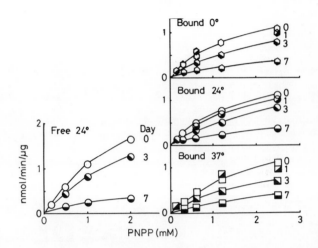

Figure 7: Decrease of activity of free and bound phosphatase.

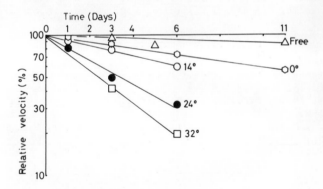

Figure 8: Decrease of activity of bound peroxidase at various temperatures. The lines measured at 0, 14, 25, and 37°C were the results of bound peroxidase. The activity of free enzyme was measured at 25°C.

Acknowledgments

The author thanks John L. Brash for his substantive revision of this
manuscript.

Literature Cited

1. Silman, I.H.; Katchalski, E. Ann. Rev. Biochem. 1966, 35, 873-
    908.
2. Hummel, J.P.; Anderson, B.S. Arch. Biochem. Biophys. 1965, 112,
    443-7.
3. Bull, H.B. Biochim. Biophys. Acta 1956, 19, 464-71.
4. Schoenmakers, J.; Matze, R.; Haanen, C.; Zilliken, F. Biochim.
    Biophys. Acta 1965, 101, 166-76.
5. Ratnoff, O.D.; Saito, H. Proc. Natl. Acad. Sci. USA 1979, 76,
    1461-3.
6. Lehtinen, P.; Pikarainen, J.; Kulonen, E. Biochim. Biophys. Acta
    1976, 439, 393-7.
7. Zucker, M.B.; Broman, N. Proc. Soc. Exp. Biol. Med. 1969, 131,
    318-20.
8. Pouyssegur, J.; Willingham, M.; Pastan, I. Proc. Natl. Acad.
    Sci. USA 1977, 74, 243-7.
9. Jonsson, U.; Ivarsson, B.; Lundstrom, I.; Bergham, L. J. Colloid
    Interface Sci. 1981, 90, 148-63.
10. Grinnell, F.; Feld, M.K. J. Biomed. Mat. Res. 1981, 15, 363-81.
11. Rosselin, G.; Assan, R.; Yalow, R.S.; Berson, S.A. Nature 1966,
    212, 355-8.
12. Wigzell, H.; Anderson, B. J. Exp. Med. 1969, 129, 23-36.
13. Hotani, H. J. Mol. Biol. 1976, 106, 151-66.
14. Boone, C.W. Science 1975, 188, 68-70.
15. Kataoka, K.; Sakurai, Y.; Tsuruta, T. Macromol. Chem. Suppl.
    1985, 9, 53-67.
16. Bettelheim, F.A.; Priel, Z. J. Colloid Interface Sci. 1979, 70,
    395-8.
17. Nagy, A.; Baker, R.R.; Morris, S.J.; Whittaker, V.P. Brain Res.
    1976, 109, 285-309.
18. Mizutani, T.; Mizutani, A. J. Chromatogr. 1975, 111, 214-6.
19. Mizutani, T.; Mizutani, A. J. Pharm. Sci. 1978, 67, 1102-5.
20. Mizutani, T.; Mizutani, A. J. Chromatogr. 1976, 120, 206-10.
21. Mizutani, T.; Mizutani, A. J. Chromatogr. 1979, 168, 143-50.
22. Mizutani, T.; Mizutani, A. Anal. Biochem. 1977, 83, 216-21.
23. Mizutani, T.; Mizutani, A. J. Non-Cryst. Solids 1978, 30, 23-7.
24. Messing, R.A. J. Non-Cryst. Solids 1975, 19, 277-83.
25. Bresler, S.E.; Katushkina, N.V.; Kokikov, V.W.; Potokin, J.L.;
    Vinogradskaya, G.N. J. Chromatogr. 1977, 130, 275-80.
26. Mizutani, T.; Kuo, P.; Mizutani, A. Chem. Pharm. Bull. 1977, 25,
    2821-6.
27. Mizutani, T. J. Chromatogr. 1980, 196, 485-8.
28. Mizutani, T.; Asaoka, A. Chem. Pharm. Bull. 1984, 32, 2395-400.
29. Mizutani, T.; Mizutani, A. J. Non-Cryst. Solids 1978, 27, 437-9.
30. Mizutani, T.; Narihara, T. J. Chromatogr. 1982, 239, 755-60.
31. Mizutani, T. J. Chromatogr. 1983, 262, 441-5.

32. Mizutani, T. J. Biochem. 1983, 94, 163-9.
33. Narihara, T.; Fujita, Y.; Mizutani, T. J. Chromatogr. 1982, 236, 513-8.
34. Mizutani, T.; Tachibana, Y. J. Chromatogr. 1985, 324, 480-3.
35. Mizutani, T.; Tachibana, Y. J. Chromatogr. 1986, 356, 202-5.
36. Kawaguchi, H.; Amagasa, H.; Hagiya, T.; Kimura, N.; Ohtsuka, Y. Colloids and Surfaces 1985, 13, 295-311.
37. Morrissey, B.W.; Stromberg, R.R. J. Colloid Interface Sci. 1974, 46, 152-64.
38. Chuang, H.Y.K.; Mahammad, S.F.; Sharma, N.C.; Mason, R.G. J. Biomed. Mater. Res. 1980, 14, 467-76.
39. Waugh, D.F.; Anthony, L.J.; Ng, H. J. Biomed. Mater. Res. 1975, 9, 511-36.
40. Mizutani, T. J. Pharm. Sci. 1980, 69, 279-82.
41. Mizutani, T. J. Pharm. Sci. 1981, 70, 493-6.
42. Chase, T.; Shaw, E. Methods in Enzym. 1970, 19, 20-7.
43. Castellino, F.J.; Sodetz, J.M. Methods in Enzym. 1976, 45, 273-7.
44. Chance, B.; Maehly, A.C. Methods in Enzym. 1955, 2, 764-75.
45. Widmer, F.; Dixon, J.E.; Kaplan, N.O.. Anal. Biochem. 1973, 55, 282-7.
46. Wiggins, R.C.; Bouma, B.N.; Cochrane, C.G.; Griffin, J.H. Proc. Natl. Acad. Sci. USA 1977, 74, 401-3.
47. Tunbridge, J.J.; Lyoyd, J.V.; Penhall, R.K.; Wise, A.l.; Maloney, T. Am. J. Hosp. Pharm. 1981, 38, 1001-4.
48. Messing, R.A. J. Non-Cryst. Solids 1977, 26, 482-513.
49. Berge, R.K. Biochim. Biophys. Acta 1979, 574, 321-33.
50. Monsan, P.; Durand, G. FEBS Lett. 1971, 16, 39-42.
51. Levin, Y.; Pecht, M.; Goldstein, L.; Katchalski, E. Biochemistry 1964, 3, 1905-13.
52. Hirsh, L.I.; Wood, J.H.; Thomas, R.B. Am. J. Hosp. Pharm. 1981, 38, 995-7.
53. Ogino, J.; Noguchi, K.; Terato, K. Chem. Pharm. Bull. 1979, 27, 3160-3.

RECEIVED January 28, 1987

Chapter 5

Structure and Activity Changes of Proteins Caused by Adsorption on Material Surfaces

Hiroko Sato[1], Takashi Tomiyama[2], Hiroyuki Morimoto[2], and Akio Nakajima[1]

[1]Research Center for Medical Polymers and Biomaterials, Kyoto University, 53 Kawaharacho, Shogoin, Kyoto 606, Japan
[2]Department of Polymer Chemistry, Kyoto University, 53 Kawaharacho, Shogoin, Kyoto 606, Japan

The activity changes of alkaline phosphatase, thrombin, and Factor X_a adsorbed onto glass beads coated with various polymers were studied in the absence and presence of pre-coating with albumin. Activity losses of the adsorbed enzymes were more pronounced than structural changes such as α-helix and β-structure of albumin desorbed from material surfaces. In general, the activity losses of enzymes adsorbed onto polyether urethane nylon (PEUN) and hydrophilic polymer surfaces was smaller than those of enzymes adsorbed to glass and hydrophobic polymer surfaces. Opposite results, however, were obtained by albumin-precoating: thrombin and Factor X_a adsorbed onto the albumin-precoated PEUN surface were more inactivated than those adsorbed onto the albumin-precoated glass surface. That is, albumin adsorbed selectively onto PEUN from plasma could bind more tightly to thrombin and Factor X_a. It was concluded that this is one of the factors responsible for the excellent thromboresistance of PEUN.

It is important to study the adsorption behavior of proteins that are major component of plasma such as albumin, γ-globulin, and fibrinogen in relation to the antithrombogenicity of polymer materials. Fibrinogen is intimately involved in fibrin formation and platelet aggregation (1-3) on artificial surfaces, and albumin molecules in the native state are well-known not to be included in thrombus formation and platelet aggregation. Therefore good thromboresistance of polymer should be achieved by the selective adsorption of albumin onto a polymeric material surface, such as segmented polyether urethanes (4, 5). In our previous paper (6), polyether urethane nylon, abbreviated as PEUN, was reported to have a high adsorption rate of albumin, and was also shown to have good antithrombogenicity.

By adsorption onto an artificial surface, albumin molecules

0097-6156/87/0343-0076$06.00/0

undergo structural changes, which are presumed to be dependent on
the properties of the materials. Such structural changes may be strongly correlated to activity changes of enzymes. Human serum albumin was
found to have an esterase-like activity such that aspirin and *p*-
nitrophenyl esters are hydrolyzed (7). Since the specificity of the
reaction is low, however, a large interface onto which albumin is adsorbed is required to detect the esterase-like activity. Therefore
we investgated the activity changes of coagulation factors II_a
(i. e., thrombin) and X_a , abbreviated as FX_a , and alkaline
phosphatase, abbreviated as AP, adsorbed on various material surfaces (PEUN, glass, a hydrophilic polymer, and a hydrophobic
polymer).

Experimental

Materials. Bovine plasma albumin (Lot. 12F-9365) and bovine plasma
γ-globulin (Lot. 116C-0147), 99 % electrophoretically pure, were
purchased from Sigma Chemical Co., and bovine plasma fibrinogen
(Lot. 26), 95 % clottable, was purchased from Miles Laboratory Inc.
The concentration of these plasma proteins was determined by spectrophotometry, and their absorption coefficients were taken to be
as follows; $E_{1\%}^{280}$ =6.67 for albumin (8); $E_{1\%}^{280}$ =11.2 for globulin (9);
$E_{1\%}^{280}$ =15.06 for fibrinogen (10).
   AP (Lot. 21F-0270), derived from calf intestine, was obtained
from Sigma Chemical Co. The concentration was determined by using
the absorption coefficient $E_{1\%}^{278}$ =7.2 (11). The synthetic substrate
used for AP was *p*-nitrophenylphosphoric acid disodium salt, abbreviated as NPP, obtained from Nakarai Chemical Co., Kyoto. Bovine
thrombin (Lot. T-002) was generously supplied by Mochida Pharmaceutical Co., Tokyo, and had a specific activity of 1400 units/mg
protein. The activated Factor X (Lot. 10F39544) was purchased from
Sigma Chemical Co., and contained 25 mg bovine albumin per unit X_a.
The fluorogenic substrates used for thrombin and X_a were Boc-Val-
Pro-Arg-MCA and Boc-Ile-Glu-Arg-MCA, respectively, where Boc and MCA

PEUN

are tertiary-butoxy carbonyl and 7-amino-4-methyl coumarin amide, respectively. Both substrates were purchased from Protein Research Foundation, Osaka.

Three kinds of polymers were coated on glass beads; PEUN (MW, 50,000) obtained from Toyo Cross Co., polyvinyl alcohol (MW, 83,000), abbreviated as PVA, obtained from Kuraray Chemical Co., and poly-γ-benzyl L-glutamate (MW, 281,000) (12), abbreviated as PBLG. The chemical structure of PEUN is on page 77. Glass beads were purchased from Toshiba Balotini Co., and were ca. 1 mm in diameter and 21.3 cm^2/g in specific surface area. The glass beads, washed in acid and rinsed thoroughly, were coated by solvent evaporation from various polymer solutions as follows: 1 mg/ml PEUN in dimethyl form-amide, 1 mg/ml PVA in distilled water, and 1 mg/ml PBLG in chloro-form. The coated glass beads were dried in vacuo for 15 h. The PVA-coated glass beads were additionally heated at 120°C for 20 min to prevent aqueous dissolution.

Methods. 1) Adsorption of plasma proteins and structure of desorbed plasma proteins: Before adding a protein solution, 90 g beads, packed in a column of 16 mm diameter and 260 mm length (6, 13), were pre-treated by rinsing with a 0.05 M phosphate buffer at pH 7.4, and then aspirated for 5 min. A protein solution was subsequently poured onto the column with a syringe. The isotherm and kinetics of protein adsorption were then investigated using a Hitachi Model EPS-3T or a Hitachi Model 200-20 spectrophotometer by measuring changes in solu-tion concentration of the column effluent. The structural changes of desorbed proteins were measured with a Jasco J-20 CD/ORD spectro-polarimeter after a 2 hour adsorption, removal of protein solution, and incubation with the phosphate buffer.

2) Albumin adsorption on dry beads and activity of AP adsorbed: In the presence of albumin, the amount of water adsorbed on the beads was estimated by using 30 g dry glass beads, either coated or not coated.

For the purpose of the measurement of activity of adsorbed AP, where a 0.05 M TrisHCl buffer containing 0.05 M KCl at pH 8.1 was used, 5 ml of a 0.62 mg/ml AP solution was added to 10 g beads and this was incubated for 2 h at 25°C. The amount adsorbed on the beads was estimated by taking the difference between the AP in solution and the amount of AP adsorbed on the surface of glass containers in the absence of beads.

After the removal of the AP solution with an aspirator, 5.0 ml of 3.5 mM NPP was added to the AP adsorbed on the glass beads. The reaction was stopped after 3 min by adding 7 ml of a 1 M KH_2PO_4 solu-tion adjusted with NaOH to pH 8.1. The amount of hydrolyzed p-nitro-phenol was estimated from the absorbance at 402 nm; $E_{1\ cm}^{1\ M}$ =1.60x10^4.

3) Amount and activity of thrombin adsorbed: The potent fluoro-genic substrate, 7-amino-4-methyl-coumarin, AMC, released from the substrates hydrolyzed by both thrombin and FX_a (14), was detected with a Hitachi 650-10S fluorescence spectrophotometer at 380 nm for excitation and at 450 nm for emission.

One gram of beads in a vial was immersed in 2 ml of a 0.05 M TrisHCl buffer containing 0.05 M KCl at pH 7.4 for 5 min, 20 µl of 0.26 units/ml thrombin was added to the vial, and the thrombin solu-tion was shaken slowly. Five min after the addition, 1 ml of the supernatant was added to 1 ml of a 20 μM substrate solution in a quartz cell to estimate the amount of thrombin adsorbed on the material surfaces by subtracting from the amount of thrombin used. Then, another 1 ml of 20 μM substrate solution was added to the remaining solution and beads, and 1 ml of the supernatant liquid was withdrawn after 2 min. The amount of AMC formed by thrombin in the remaining solution and adsorbed on the beads was estimated after 2 min by back-extrapolation on the recorder. (The 1 ml withdrawn solution was mixed with 1 ml of 10 μM substrate solution, which was used for the adjustment of the base line of fluorescence.) Finally, the activity of the thrombin adsorbed was deduced by subtracting the amount of AMC formed by thrombin in the first-withdrawn supernatant from the amount of AMC formed by thrombin in the remaining solution and on the beads.

   In the case of the experiments for albumin-precoated material surfaces, a 1 mg/ml albumin solution in a 0.05 M TrisHCl buffer and 0.05 M KCl at pH 7.4 was used instead of the same buffer in the absence of albumin.

   4) Relative activity of FX$_a$ adsorbed: As FX$_a$ reactivity with the synthetic substrate was lower than that of thrombin, 40 μl of 0.4 units/ml FX$_a$ solution was added to 2 ml of a 0.05 M phosphate buffer containing 0.05 M KCl at pH 7.4. The FX$_a$ adsorption on glass beads was investigated using almost the same methods as those adopted for thrombin, except that specific activities of FX$_a$ in the remaining solution and on beads were evaluated as values relative to a FX$_a$ solution.

Results and Discussion

Blood plasma is a concentrated protein solution, and the adsorption behavior onto material surfaces may be estimated from the amount and the type of adsorption for major plasma proteins in dilute solution, where the adsorption theory for a monomolecular layer is applied. The adsorption isotherms for albumin, γ-globulin, and fibrinogen at a concentration lower than ca. 1 mg/ml obeyed the Langmuir-type adsorption formula. That is,

$$\frac{1}{Q_p} = \frac{1}{Q_p^\infty} + \frac{1}{KQ_p^\infty} \cdot \frac{1}{c} \tag{1}$$

where Q_p indicates the amount of protein adsorbed at the protein concentration, c, at equilibrium, K is a constant, and Q_p^∞ is the maximum amount of protein adsorbed on the material surface as a monomolecular layer. The Q_p^∞ values of albumin, γ-globulin, and fibrinogen are listed in Table I. The rate constants k_L of the Langmuir-type adsorptions, expressed by Equation 2, are obtained using the Q_p^∞ values listed in Table I and from experiments on the adsorption kinetics of proteins onto the material surfaces. When the amount Q_p of protein is adsorbed on a material surface during the elapsed time, t:

$$\ln \frac{Q_p^\infty}{Q_p^\infty - Q_p} = k_L \cdot t \tag{2}$$

Table I.   Saturated Amounts Q_P^{∞} of Plasma Proteins
Adsorbed on Material Surfaces*

Materials	Q_P^{∞} (mg/m^2)		
	Albumin	γ-Globulin	Fibrinogen
PEUN	10.3	17.9	14.3
Glass	9.5	19.2	13.7
PVA	1.9	6.6	11.6
PBLG	4.9	13.7	11.6

Source: Reproduced with permission from Ref. 6.   Copyright 1984 The
Society of Polymer Science.

The rate constants in Equation 2 include the solution concentration
of the protein which is assumed to be constant in the initial stages
of adsorption. Since the concentrations of plasma proteins albumin:
γ-globulin:fibrinogen are in the ratio 42.0 : 22.4 : 5.6, the rate
constants from Equation 2 were weighted accordingly to give the
values shown in Table II. Thus, the apparent rate constant, $k_{L,pl}$,
for protein adsorption can be estimated in the initial adsorption
process when various material surfaces are put in contact with
plasma. As is obvious from Table II, the rate constant of albumin in
the initial adsorption process in plasma, $k_{L,pl}$, on a PEUN surface
is very large compared with other surfaces. The high rate constant
for albumin may contribute to the excellent thromboresistance of PEUN,
even though the amount of albumin adsorbed onto PEUN was almost the
same as onto glass.

The phenomenon of the rapid adsorption of albumin onto a PEUN
surface may be associated with hydrophobic and hydrophilic interactions of the PEUN surface with some sequences of relatively hydrophobic amino acid residues in the interior of albumin. An albumin molecule is composed of three-subdomains (15). There are two gaps between
the subdomains. One is a hydrophobic pocket with an affinity constant, $K_a=1.1 \times 10^8$ M^{-1} for stearic acid; the other is an intermediate
hydrophobic pocket with $K_a=1.5 \times 10^6$ M^{-1} for bilirubin (16). Perhaps
the structure of adsorbed albumin in contact with a PEUN surface is
composed of hydrophobic and hydrophilic regions corresponding or
complementary to those of the PEUN surface, even though the exterior
of native albumin is rich in hydrophilic amino acid side chains.

When large amounts of coated or uncoated beads were added to a
protein solution of volume V and concentration c_o , the concentration

Table II.   Calculated Rate Constants in the Initial
Adsorption Process in Plasma*

Materials	$k_{L,pl}$ (min^{-1})		
	Albumin	γ-Globulin	Fibrinogen
PEUN	7.27	0.72	0.15
Glass	4.49	1.77	0.45
PVA	2.10	1.08	0.13
PBLG	0.92	1.41	0.13

Source: Reproduced with permission from Ref. 6.   Copyright 1984 The
Society of Polymer Science.

of the protein solution was observed generally to increase owing to
the adsorption of water onto the beads. If the amount of adsorbed
water is represented by Q_w, the following inequality is obtained:

$$Q_w - \frac{V(c - c_0)}{c} = Q_p \geq 0 \qquad (3)$$

If Q_w of Equation 3 is plotted against c, it is found that Q_w decreases hyperbolically with increasing c. On the other hand, Q_p
obeys the Langmuir-type adsorption formula, expressed by Equation 1.
In a monomolecular layer of adsorbed protein, it may be assumed that
the total number, N, of adsorptive sites on a material surface is
constant, as indicated in Equation 4, and that the adsorbed protein
molecules contribute n water-binding sites.

$$N = Q_w + nQ_p = Q_w^\infty = nQ_p^\infty = \text{constant} \qquad (4)$$

where Q_w^∞ is the amount of adsorbed water at surface saturation. Thus,
the relations among the amounts of water and protein adsorbed on
material   surfaces may be represented as follows.

$$\frac{Q_p}{Q_p^\infty} = \frac{ac}{1 + ac} \qquad (1)'$$

$$\frac{Q_w}{Q_w^\infty} = \frac{1}{1 + ac} \qquad (5)$$

where a is a constant. n becomes constant, if no structural changes
occur in the protein.

   In Figure 1 (a) and (b), the amounts of water and albumin adsorbed, respectively, on PEUN and glass are plotted against the concentrations of albumin solutions at equilibrium. Our experimental
values of Q_w^∞ appear to account for the amount of water from capillary
action among the beads as well as the amount of water adsorbed on
glass beads. The hydration of albumin molecules has been estimated to
be 0.0004 g H_2O/mg albumin by isopiestic measurements of water vapor
(17). Although Q_{Alb} values on a PEUN surface are larger than on a
glass surface in the lower concentration region (below 0.4 mg/ml),
Q_{Alb} or Q_w values on both PEUN and glass surfaces are almost the
same in the concenration range from 0.7 to 1.0 mg/ml.

   The Q_{Alb} and Q_w values, and their ratio, Q_w/Q_{Alb}, obtained for
a 1 mg/ml albumin solution, are summarized in Table III. The structural changes of albumin desorbed from material surfaces are also
shown in Table III (6). The Q_w value for a PVA surface seems too
small, perhaps because the small amounts adsorbed are subject to experimental error. The water contents of the adsorbed layer for PEUN
and glass surfaces are almost the same, while the water content of a
hydrophobic surface is small. The degree of disruption of the native
and regular structures such as α-helix and β-structure in albumin
adsorbed on PEUN and glass surfaces are the same within experimental
error, while albumin desorbed from the PVA surface may cause a milder
disruption. Since the structural changes should be closely related
to the enzymatic activity, more pronounced effects may be expected

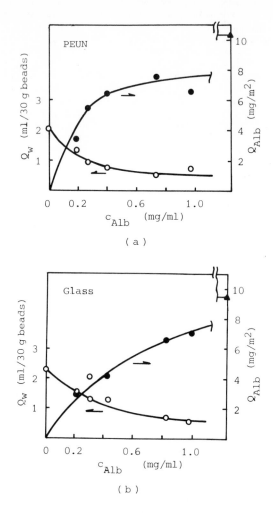

Figure 1. Amounts of albumin Q_{Alb} and calculated amounts of water Q_w adsorbed on PEUN (a) and glass (b) surfaces are plotted against concentrations c_{Alb} of albumin solutions at equilibrium; ▲ corresponds to Q_{Alb}^{∞}.

from the various properties of the material surfaces rather than
from the effects of the water content of the adsorbed layer.
    AP, classified as a cellular enzyme, exists also in plasma (18),
and has a higher reactivity with p-nitrophenyl phosphate than albumin. (The Michaelis-Menten constant, $K_m = 1.13 \times 10^{-5}$ M for AP and NPP
in a 0.05 M phosphate buffer containing 0.05 M KCl at pH 8.1.) Table
IV shows amounts, the specific, and the relative specific activities
of AP adsorbed on material surfaces. Table V shows the amounts and

Table III. Amounts of Albumin (Q_{Alb}) and Water (Q_w) Adsorbed on
            Material Surfaces in a 1 mg/ml Albumin Solution (a),
            and Structural Changes of Desorbed Albumin

Materials	Q_{Alb} (mg/m^2)	Q_w (ml/g beads)	Q_w/Q_{Alb} (ml/mg)	Relative Contents* in Desorbed Albumin for α-Helix;	β-Structure
PEUN	7.4	0.62	1.31	0.73	0.56
Glass	7.1	0.58	1.28	0.67	0.50
PVA	1.8	0.11	0.97	0.85	0.63
PBLG	4.6	0.25	0.85	—	—
native albumin				1 (67%)	1 (16%)

(a) Total area of beads used is 0.0639 m^2.
Source: Reproduced with permission from Ref. 6. Copyright 1984 The
Society of Polymer Science.

Table IV. Amounts and Activity Changes of Alkaline Phosphatase
           Adsorbed on Material Surfaces for 2 h from a 0.62
           mg/ml AP solution

Materials	Q_{AP} (mg/m^2)	Specific Activity (µmoles/mg/min)	Relative Specific Activity
PEUN	2.8	3.72	0.76
Glass	1.6	2.88	0.59
PVA	—	3.76	0.77
PBLG	1.8	3.40	0.70
AP solution	—	4.88	1.00

Table V. Thrombin (in the Absence of Albumin) and FX$_a$ (in
          0.20 mg/ml Albumin) Adsorbed on Material Surfaces

Materials	Thrombin Q_{II_a} (µg/m^2)	Relative Specific Activity	FX$_a$ Relative Specific Activity
PEUN	0.92	0.30	0.963 (a)
Glass	0.92	0.21	0.921 (a)
PVA	0.82	0.29	0.990 (a)
PBLG	1.02 (a)	0.17	0.911 (a)
Solution	—	1.00	1.00

(a) The significant figures are substantially 2 figures.

relative specific activities of thrombin and FX_a adsorbed on material surfaces. Under experimental conditions the same as those in
Table V, $K_m = 7.81 \times 10^{-6}$ M for thrombin and $K_m = 7.69 \times 10^{-5}$ M for FX_a.
AP seems a relatively stable enzyme (19), compared with thrombin,
because of the smaller activity loss of adsorbed AP.

The enzymes used have different reactivities with the same
corresponding substrates. Also, albumin is present in the FX_a
system, which seems to influence the FX_a adsorption on surfaces.
However, it may be concluded that of the three enzymes studied
thrombin loses the most activity on adsorption. Thrombin adsorbed
on cuprophane or PVC (20), and on polyethylene (21) lost activity
completely. The structure of thrombin, therefore, is labile causing
profound inactivation of adsorbed thrombin. The decrease of enzyme
activity due to adsorption on material surfaces depends not only on
the properties of the enzymes, on their affinity and reactivity to
the substrate, on the effects of added albumin, and — especially
for thrombin — on the specific activity but also on the properties
of the material surfaces. It is obvious from Tables IV and V that
the activities of enzymes adsorbed on PEUN and PVA surfaces are
higher than on glass and on PBLG.

Thrombin of a high specific activity loses activity rapidly
in a buffer solution without albumin, but remains active in an
albumin solution (18). For example, the time dependence of thrombin
activity in solutions is plotted in Figure 2, where activity changes
of thrombin are compared in the absence of albumin and in the presence of 1 mg/ml albumin. In the absence of albumin, the effects of
adsorption to a glass container as well as self-hydrolysis of thrombin seem to be detected within the initial 30 min, and then the
effects of self-hydrolysis prevail. Therefore the data on the
relative activities of thrombin in Table V are corrected on the
basis of the data obtained from Figure 2.

In Table VI, the effects of albumin-precoated surfaces on the
enzyme adsorption are compared for PEUN and glass surfaces, because
Q_{Alb} and Q_w values are similar with both surfaces, as is seen in
Table III. Activities of both thrombin and FX_a adsorbed on the PEUN
surface are higher than on the glass surface in the case of the bare
material surfaces, as is shown in Table V. However, the relative
specific activities of both thrombin and FX_a adsorbed on an albumin-
precoated PEUN surface are significantly lower than on an albumin-
precoated glass surface. That is, albumin molecules adsorbed on a
PEUN surface may bind to thrombin and FX_a with higher affinity

Table VI. Thrombin and FX_a Adsorption on Albumin-precoated
PEUN and Glass Beads

Materials	Q_{II_a} ($\mu g/m^2$)	Thrombin Relative Spe- cific Activity	FX_a Relative Spe- cific Activity
PEUN	1.1	0.44	0.73
Glass	1.1	0.54	0.80
Solution	–	1.00	1.00

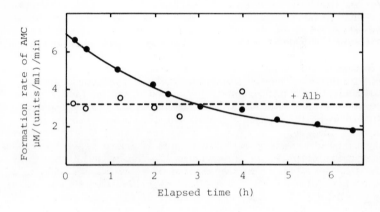

Figure 2. Time dependence of thrombin activity in the presence of 1 mg/ml albumin (O) and in the absence of albumin (●) in a 0.05 M TrisHCl buffer containing 0.05 M KCl at pH 7.4. Concentration of the substrate = 20 μM.

constants than albumin molecules adsorbed on a glass surface. This
fact also may be related to the excellent in vivo thromboresistance
of PEUN (6), for which albumin was adsorbed most rapidly among the
major component plasma proteins, although competing adsorption and
exchange (23) by plasma proteins onto the albumin-covered PEUN
surface under conditions of flow, and attack by blood cells can be
presumed to occur in vivo.

Because albumin can bind thrombin, albumin molecules should
disturb the reaction of thrombin and the synthetic substrate. The
inhibition constant, K_i, of albumin was studied in the system of
thrombin and the fluorogenic substrate in a 0.05 M TrisHCl buffer
containing 0.05 M KCl at pH 7.4; we found that $K_i = 1.79 \times 10^{-4}$ M. From
the K_i value, the inhibitory effect of albumin is negligibly small
in the hydrolysis reaction by thrombin. That is, the effect of albumin is the protection of thrombin from self-hydrolysis rather than
inhibition of thrombin. However, albumin in buffer supposed to have
a higher affinity for thrombin and FX_a than albumin adsorbed and
thus denatured. Finally a schematic model for thrombin adsorbed on
albumin-precoated PEUN and glass surfaces is shown in Figure 3.
Thrombin adsorbed on the albumin-precoated glass surface is more
active toward the substrate than thrombin on the albumin-precoated
PEUN surface.

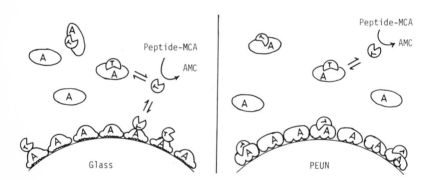

Figure 3. A schematic model for thrombin adsorbed on albumin-
precoated PEUN and glass beads. Thrombin molecules are shown
to be more active on the albumin-precoated glass surface.

Literature Cited

1. Chuang, H. Y. K.; Sharpton, T. R.; Mohammad, S. F. Thromb.
   Haemostas. 1981, 46, 304.
2. London, J.; McManama, G.; Kushner, L.; Merrill, E.; Salzman, E.
   Thromb. Haemostas. 1985, 54, 3.
3. Hawiger, J.; Timmons, S.; Kloczewiak, M.; Strong, D. D.;
   Doolittle, R. F. Proc. Natl. Acad. Sci. USA 1982, 79, 2068-71.
4. Lyman, D. J.; Brash, J. L.; Chaikin, S. W.; Klein, K. G.;
   Carini, M. Trans. Am. Soc. Artif. Intern. Org. 1968, 14, 250-
   55.

5. Lyman, D. J.; Metcalf, L. C.; Albo, D., Jr.; Richards, K. F.; Lamb, J. Trans. Am. Soc. Artif. Intern. Org. 1974, 20, 474-79.
6. Sato, H.; Morimoto, H.; Nakajima, A.; Noishiki, Y. Polymer J. 1984, 16, 1-8.
7. Kurono, Y.; Maki, T.; Yotsuyanagi, T.; Ikeda, K. Chem. Pharm. Bull. 1979, 27, 2781-86.
8. Zurawski, V. R., Jr.; Kohr, W. J.; Foster, J. F. Biochemistry 1975, 14, 5579-86.
9. Lowry, O. H.; Rosebrough, J.; Farr, J.; Randall, R. J. J. Biol. Chem. 1951, 193, 265-75.
10. Mihalyi, E. Biochemistry 1968, 7, 208-23.
11. Malamy, M.; Horecker, B. L. Methods Enzym. 1966, 9, 639-42.
12. Noishiki, Y.; Y. Nakahara, Y.; Sato, H.; Nakajima, A. Artif. Org. 1980, 9, 678-82.
13. Soderquist, M. E.; Walton, A. G. J. Colloid Interface Sci. 1980, 75, 386-97.
14. Morita, T.; Kato, H.; Iwanaga, S.; Takada, K.; Kimura, T.; Sakakibara, S. J. Biochem. 1977, 82, 1495-98.
15. Brown, J. R.; Shockley, P.; Behrens, P. Q. In The Chemistry and Physiology of the Human Plasma Proteins; Bing, D. H., Ed.; Pergamon: New York, 1978; pp 29-37.
16. Anderson, L. O. In Plasma Proteins; Blombäck, B.; Hanson, L. Å., Ed., John Wiley & Sons: New York, 1979; pp 47-53.
17. Bull, H. B.; Breese, K. Arch.Biochem. Biophys. 1968, 128, 488-96.
18. Salthouse, T. N. J. Biomed. Mat. Res. 1976, 10, 197-229.
19. Mizutani, T. J. Pharm. Sci. 1980, 69, 279-82.
20. Chuang, H. Y. K.; Mohammad, S. F.; Sharma, N. C.; Mason, R. G. J. Biomed. Mat. Res. 1980, 14, 467-76.
21. Larsson, R.; Olsson, P.; Lindahl, U. Thromb. Res. 1980, 19, 43-54.
22. Anderson, M. M.; Gaffney, P. J.; Seghatchian, M. J. Thromb. Res. 1980, 20, 109-22.
23. Brash, J. L.; Uniyal, S.; Samak, Q. Trans. Am. Soc. Artif. Intern. Org. 1974, 20, 69-76.

RECEIVED January 28, 1987

Chapter 6

Nonspecific Adhesion of Phospholipid Bilayer Membranes in Solutions of Plasma Proteins
Measurement of Free-Energy Potentials and Theoretical Concepts

E. Evans[1], D. Needham[2], and J. Janzen[2]

[1]Departments of Pathology and Physics, University of British Columbia, Vancouver, British Columbia V6T 1W5, Canada
[2]Department of Pathology, University of British Columbia, Vancouver, British Columbia V6T 1W5, Canada

Recent experimental advances have made quantitation of weak membrane adhesion possible in concentrated solutions of macromolecules. We report direct measurements of the free energy potential for adhesion of phospholipid bilayers in solutions of two plasma proteins (fibrinogen and albumin) over a wide range of volume fraction (0-0.1). The results are consistent with a thermodynamic model for adhesion based on depletion of macromolecules from the contact zone.

Aggregation of cells and other membrane bound capsules in solutions of large macromolecules is generally separated into two catagories: specific and non-specific. Specific adhesion involves identifiable binding reactions between suspended macromolecules and receptor molecules located on the surfaces. Such processes are basic elements of cell agglutination and removal of aberrant organisms and foreign bodies in a living animal. On the other hand, non-specific adhesion cannot be attributed to binding of macromolecules at specific sites on the capsule surfaces. Well known - and always present even in the absence of macromolecules - are the classic colloid forces that act between continuous media, i.e. van der Waals' attraction, electric double layer repulsion, other structural and solvation forces (1-2). In general, these colloidal forces simply superpose on interactions peculiar to the suspended molecules. For biological cells with significant superficial carbohydrate structures, only electric double layer (repulsive) interaction is important; van der Waals' attraction and the shorter range - hydration repulsion can be neglected. However for synthetic membranes with small molecular head groups at the water interfaces, attraction as well as repulsion is present between surfaces. Even with marked differences of surface composition and topography, cells and synthetic membrane capsules often exhibit similar aggregation behavior in solutions of large macromolecules, e.g. red blood cells and phospholipid bilayer vesicles in solutions of dextran polymers (3) or in solutions of plasma proteins (e.g. fibrinogen and macroglobulins, 4). No specific receptors or binding sites for these macromolecules have

0097-6156/87/0343-0088$06.00/0

been demonstrated to exist on red cell or vesicle surfaces; hence, the aggregation is labelled "non-specific". Likewise, no adequate explanation and mechanism(s) have been established to provide an understanding of non-specific adhesion between natural or synthetic membranes due to suspended macromolecules. With recent experimental advances, it is now possible to quantitate adhesion energies, test reversibility, and critically evaluate disparate theories for non-specific adhesion in concentrated solutions of macromolecules. Here, we present direct measurements of the free energy potential for adhesion of phospholipid bilayer membranes in solutions of two plasma proteins (fibrinogen and albumin) over a wide range of volume fraction (0-0.1). The results are consistent with a thermodynamic theory for non-specific adhesion based on depletion of macromolecules from the contact zone.

Experimental Methods and Materials

Solutions of fibrinogen (Imco, Stockholm) or albumin (Calbiochem, San Diego) were formed with 130 mM sodium chloride buffered to pH 7.4 by 20 mM sodium phosphate to give 150 mM PBS. Protein concentrations were determined from optical density at 280 nm (33) and were made-up to give final concentrations in the range of 0-10g% (wt:wt). D-phenylalanyl-L-prolyl-L-arginine chloromethylketone (Calbiochem) was added to the fibrinogen solutions for stabilization. [Note: it was difficult to produce stable solutions with fibrinogen at high concentration (>6% wt:wt) even with the inhibitor present.]

Vesicles were produced by rehydration of an anhydrous lipid (1-stearoyl-2 oleoyl phosphatidylcholine SOPC, Avanti Biochem., Alabama). Although few in number, some vesicles in the final aqueous suspension were of sufficient size (10^{-3}cm or greater in diameter) to be used in micromechanical adhesion tests. The vesicles were formed in non-ionic (sucrose or other small sugars) buffers. Hence when resuspended in iso-osmotic salt solutions, the small difference in index of refraction between the interior and exterior of the vesicle greatly enhanced the optical image as shown in Figure 1. Because of the extremely low solubility of phospholipids in aqueous media and the osmotic strength of the solutes trapped inside the vesicles, vesicles deform as liquid-filled bags with nearly constant surface area and volume (5-6); thus, spherical vesicles are rigid and undeformable. When vesicles are slightly deflated by osmotic increases in the external solution, the bending stiffness is so small that the capsule becomes completely flaccid and deformable (7). Thus, flaccid non-spherical vesicles easily form adhesive contacts with negligible resistance to deformation until the surfaces become pressurized into spherical segments.

We have taken advantage of these deformability properties to establish a sensitive method for measurement of adhesion energy between bilayer surfaces (6, 8-10). Two spherical vesicles are selected and transferred from the initial suspension in a chamber on the microscope stage to an adjacent chamber which contains a slightly more concentrated buffer (0.15M PBS) plus macromolecules. There, the vesicles rapidly deflate to new equilibrium volumes. One vesicle is aspirated by a small

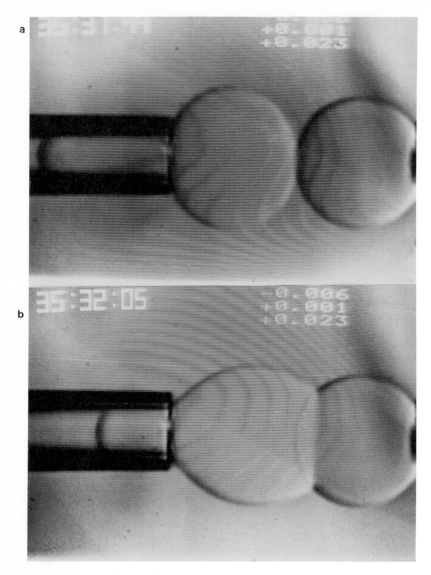

Figure 1. Video micrographs of an adhesion test. (a) Vesicles
in position for contact. (b) Adhesion - equilibrium controlled
by the suction pressure. (Pipet calibre ~1 x 10^{-3}cm; vesicle
diameters ~ 2 x 10^{-3}cm).

micropipet and held with sufficient suction pressure to form a rigid spherical segment outside of the pipet, i.e. the "test" surface for adhesion. The other vesicle is aspirated by a second pipet with a low level of suction pressure controlled to regulate the adhesion process. The second vesicle is then maneuvered into close proximity of the test vesicle surface (Figure 1a) and the adhesion process is allowed to proceed in discrete (equilibrium) steps by reduction of the suction pressure (Figure 1b). This experimental procedure yields the tension in the adherent vesicle bilayer as a function of the extent of coverage of the test vesicle surface, both for the forward process of adhesion and the reverse process of separation. In the tests to be reported here, the adhesion processes were reversible as shown in Figure 2.

Because of the macroscopic dimensions observable in these experiments, it is not possible to determine the interbilayer forces directly; however, cumulation of forces into an integral over distance is measurable. This integral is the negative work or free energy reduction per unit area (i.e. adhesion energy) for assembly of the bilayers from infinite separation to stable contact (where the force between the surfaces is zero).

$$\gamma \equiv -\int_{\infty}^{z_g} \sigma_n \cdot dz \qquad (1)$$

Mechanical equilibrium is established when small reductions in free energy due to formation of adhesive contact just balance small increases in mechanical work of deformation of the vesicle (8,11). This variational statement leads to a direct relation between the free energy potential for adhesion and the suction pressure applied to the adherent vesicle,

$$\frac{\gamma}{P \cdot R_p} = f \text{ (geom)} \qquad (2)$$

When the product of suction pressure P and pipet radius R_p is converted to bilayer tension for the adherent vesicle, this equation takes the form of the Young equation where the geometric function is $(1-\cos\theta_c)$ and θ_c is the included angle between the bilayer surfaces. Based on constraints that the vesicle area and volume remain fixed throughout deformation and the mechanical requirement that the bilayer surface exterior to the pipet is a surface of constant mean curvature, the contact angle can be derived from measurements of either the diameter or polar length of the adhesion zone (cap on the rigid vesicle, 8,11). If adhesion is uniform over the contact zone, then a single curve is predicted for the relationship between pipet suction pressure and the fractional extent of coverage of the test vesicle as shown in Figure 2 (x_c = polar length of the adhesion zone divided by the diameter of the spherical test surface).

To aid in selection of an appropriate model for the adhesion process, we carried out two additional sets of experiments. The first set involved the following sequence of vesicle adhesion-separation: an adherent vesicle pair was assembled in the salt buffer without macromolecules; adhesive contact was

maintained by van der Waals' attraction (6,10). The adherent
vesicle pair was transferred in a few seconds to an adjacent
chamber (with care so as not to disrupt the contact) that
contained salt buffer with high concentration of macromolecules
(5g%, wt:wt). The adhesion energy was then measured by separation
of the contact in the final solution. Similarly, an adherent
vesicle pair was assembled in the solution with macromolecules and
transferred to the pure salt buffer without disruption of the
contact; the adhesion energy was again measured by separation in
the final solution. The rationale behind these tests was the
expectation that macromolecules (via the first procedure) would
likely be prevented from entering the gap or be trapped in the gap
(via the second procedure) because of kinetic restrictions. Thus
if the macromolecules formed cross-bridges, the adhesion energies
would be determined by the composition of the initial solution in
which adhesion was established. However, the results were exactly
opposite, i.e. the adhesion energies at separation were identical
to values determined for reversible assembly and separation in
solutions of composition equivalent to that of the final
solution. Estimates of molecular dimensions (50Å x 450Å for
fibrinogen, 12; 38Å x 150Å for albumin, 13) and values measured
for SOPC bilayer separation in pure salt solutions (26Å from x-ray
diffraction and composition data provided by Dr. P. Rand, Brock
University), indicate that no appreciable concentration of
macromolecules could be present in the gap after vesicles were
first assembled in pure salt buffer and then transferred into the
solution with macromolecules unless macromolecules rapidly
diffused into the thin gap. Likewise, after vesicles were first
assembled in the solution with macromolecules and then transferred
to the pure salt buffer, macromolecules would be trapped in the
gap unless these molecules rapidly diffused out of the gap into
the salt buffer. This kinetic "escape" would be unlikely if
specific cross-bridges existed.

The second set of experiments involved an attempt to
quantitate the number of macromolecules captured in the gap
between the adherent surfaces. Adherent vesicles were assembled
in a solution that contained fluorescently labelled macromolecules
then transferred to an adjacent chamber that contained the same
concentration of the macromolecule but without fluorescent label.
Since the vesicles did not separate, it was expected that the
fluorescent macromolecules trapped in the adhesion zone would be
detectable over a long time period until diminished by exchange
diffusion with the exterior solution. Tests with both
fluorescently labelled fibrinogen and albumin were performed. The
results were negative; we could not detect any fluorescence in the
contact zone except at the exceptional location where there was
obvious invagination formed by liquid trapped during the adhesion
process. Trapped liquid regions were not formed when the vesicles
were assembled carefully in slow-discrete steps; the contact zone
appeared uniform in optical thickness. Based on molecular
dimensions as estimates of the minimum gap thickness, fluorescence
should have easily been detected with our photometric system for
gap concentrations equivalent to 10% of the bulk concentration.
The test clearly showed that there was a significant reduction of
molecules in the gap in comparison to the bulk concentration.

Free Energy Potentials for Lipid Bilayer Adhesion

Adhesion tests were carried out on 5-10 vesicle pairs at specific concentrations of either fibrinogen or albumin in the range of 0-10g% (wt:wt); the results are plotted in Figure 3. Since the vesicle surfaces were uncharged at pH 7.4, there was a threshold level of adhesion energy caused by van der Waals' attraction between the SOPC bilayers (6,10) at zero concentration. Based on protein density, the concentration range represented volume fractions from 0-0.1. Even for these fairly concentrated solutions, the free energy potential for adhesion increased progressively with concentration and showed no tendency to plateau or saturate. This behavior has also been observed for bilayer adhesion in solutions of dextran polymers (9,14) although the levels of adhesion energy were significantly greater in dextran solutions at comparable volume fractions. The effect of surface composition and molecular topography was tested by measuring the free energy potential for adhesion of a red blood cell to a sphered lipid bilayer in concentrated fibrinogen solutions (similar experiments were carried out at low concentrations previously, 15). Because of the steric separation maintained by the superficial carbohydrate structures on the red cell membrane, there was no perceptable level of van der Waals' attraction in pure salt buffer. In 5g% fibrinogen, the level of free energy potential for red cell-vesicle adhesion was equal to the free energy potential in excess of the van der Waals' threshold observed for bilayer-bilayer adhesion in 5g% fibrinogen. Also, the slope of adhesion energy versus fibrinogen concentration derived from Figure 3 is similar to the rate of increase found for red cell-red cell adhesion in fibrinogen solutions (4). These results demonstrate the non-specific character of the adhesion process, i.e. no recognizable dependence on surface composition.

Theoretical Implications and Methods of Analysis

Two diverse views of non-specific adhesion processes form the bases for contemporary theories introduced to rationalize observations of colloidal stability and flocculation in solutions of macromolecules (see 16-18 for general reviews). The first view is based on adsorption and cross-bridging of the macromolecules between surfaces. Theories derived from this concept indicate a strong initial dependence on concentration of macromolecules; there is a rapid rise in surface adsorption for infinitesimal volume fractions (32) followed by a plateau with gradual attenuation of surface-surface attraction because of excluded volume effects in the gap at larger volume fractions (19-20). The interaction of the macromolecule with the surface is assumed to be a short range attraction proportional to area of direct contact. The second - completely disparate - view of non-specific adhesion is based on the concept that there is an exclusion or depletion of macromolecules in the vicinity of the surface, i.e. no adsorption to the surfaces. Here, theory shows that attraction is caused by interaction of the (depleted) concentration profiles associated with each surface which leads to a depreciated macromolecular concentration at the center of the gap. The concentration

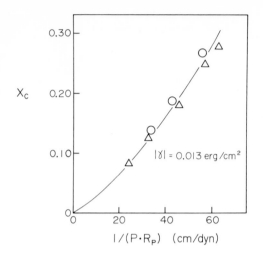

Figure 2. Fractional area x_C of the rigid vesicle covered by the adherent vesicle versus the reciprocal of suction pressure P multiplied by pipet radius R_p. Triangles (Δ) - contact formation; open circles (O) - separation; solid curve - prediction from mechanical analysis for a uniform value of adhesion energy.

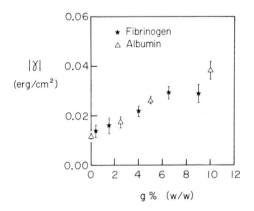

Figure 3. Adhesion energy (erg/cm^2) for SOPC bilayers in 0.15 M salt (PBS) plus either albumin or fibrinogen.

reduction in the gap (relative to the exterior bulk solution) gives rise to an osmotic effect that acts to draw the surfaces together (21-22). When equilibrium exists between the gap and the bulk, stabilization or approach to a plateau level is not anticipated for structureless surfaces (14). The free energy potential for adhesion increases progressively with concentration even at large volume fractions. Thus, adsorption-based and (non-adsorption) depletion-based concepts predict distinctly different adhesion properties: (i) an excess versus a reduction in macromolecular concentration in the contact zone; (ii) a quick rise in free energy potential for adhesion at infinitesimal concentrations which should level-off and eventually attenuate versus an adhesion energy that progressively increases with concentration without stablization. Also, it is expected that adsorption-based phenomena will depend on chemical attributes of the suspended macromolecule and surface whereas (non-adsorption) depletion-based processes will depend only on colligative properties of the macromolecules in aqueous suspension.

Clearly, our results for adhesion of lipid bilayers in fibrinogen and albumin solutions are consistent with the (non-adsorption) depletion type of assembly process. This deduction is based on (i) the null observation that no fluorescently labelled material was detected in the gap between bilayers, (ii) the continuous increase of the free energy potential with concentration even for fairly large volume fractions, and (iii) the transfer of adherent vesicle pairs with subsequent separation which showed that adhesion energy depended only on the composition of the medium exterior to the gap but not the gap composition. Similar results have been obtained for adhesion of lipid bilayers in solutions of high molecular weight dextran polymers (Figure 4, 14). Hence, we have chosen to carefully examine (non-adsorption) depletion-based theories in conjunction with these experiments.

Theoretical development over the past decade or so has focused on analysis of the configurations and distribution of polymer segments in the vicinity of a solid surface or between surfaces to predict the deviation of free energy density from that in the adjacent bulk region (19-25). Even though these studies are very elegant and insightful, little care has been taken to obtain a suitable work potential, the differential of which yields stresses at the surfaces. We will outline a simple thermodynamic approach that provides a formalism for derivation of physical stresses from free energy of mixing and chain configuration (14); then, we will discuss methods for prediction of stationary concentration profiles which result from the proximity of non-adsorbing, impermeable boundaries.

Variations in total free energy associated with adhesion must include both the gap and exterior (bulk) regions,

$$\delta F = \delta F_g + \delta F_B$$

These changes are subject to conservation requirements for the total number of solute molecules and the total volume of gap and exterior regions (which implies conservation of solvent),

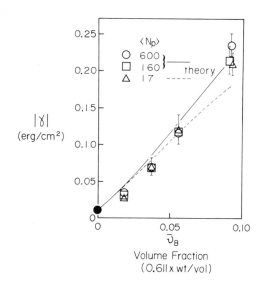

Figure 4. Adhesion energy (erg/cm²) for SOPC bilayers in
0.1 M salt (PBS) plus dextran polymers. Number average polymer
indices - N_p (number of glucose monomers). Solid and dashed
curves - predictions from mean field theory with first and
second virial coefficients from osmotic pressure measurements
(14).

$$\delta\left(\bar{v}_g \cdot V_g + \bar{v}_B \cdot V_B\right) \equiv 0 \qquad , \qquad \delta\left(V_g + V_B\right) \equiv 0$$

V_g, V_B are the volumes of the gap and bulk regions respectively; $\bar{v}_g$, $\bar{v}_B$ are the <u>mean</u> volume fractions of the solute molecules in the gap and bulk regions. The total free energy variation can be expressed in terms of four independent variations: (i) variation with respect to spatial distribution-ϕ and configuration of macromolecules in the gap (where the <u>mean</u> concentration, thickness-z_g, and contact area-A for the gap are held fixed); (ii) variation with respect to <u>mean</u> concentration of macromolecules in the gap (where the spatial distribution, gap thickness and contact area are held constant); (iii) variation with respect to gap thickness (where spatial distribution, <u>mean</u> concentration, and contact area are held constant); and finally (iv) variation with respect to contact area (where spatial distribution, <u>mean</u> concentration, and thickness for the gap are held constant). Symbolically, the total variation is written as,

$$\delta F = \left.\delta F\right|_{(\bar{v}, z_g, A)} + \left.\delta F\right|_{(\phi, z_g, A)} + \left.\delta F\right|_{(\bar{v}, \phi, A)} + \left.\delta F\right|_{(\bar{v}, \phi, z_g)} \tag{3}$$

which must be analyzed in conjunction with the previous conservation requirements.

The first variation yields the optimum concentration profile in the gap (at equilibrium for constant composition and gap dimensions). This variation can be expressed in a form where the requirement that the <u>mean</u> concentration of macromolecules in the gap be held constant <u>is</u> included explicitly, i.e.

$$\left.\delta F_g\right|_{(\bar{v}, z_g, A)} = \left.\delta F_g\right|_{(z_g, A)} - \lambda_c \cdot A \cdot \int_0^{z_g} \delta \cdot v \cdot dz = 0 \tag{4}$$

The Lagrange multiplier λ_c represents a "pressure" which ensures that the constraint,

$$\bar{v} = \frac{1}{z_g} \int_0^{z_g} v \cdot dz \equiv \text{constant}$$

is satisfied appropriate to boundary conditions for exchange of macromolecules between gap and bulk regions.

The second variation represents the free energy associated with exchange of macromolecules between gap and bulk regions and, thus, characterizes the nature of the assembly process. If the process is "true" equilibrium, then this variation is <u>identically</u> zero,

$$\left.\delta F\right|_{(\phi, z_g, A)} = \left\{ \left.\frac{\partial F_g}{\partial \bar{v}_g}\right|_{(\phi, z_g, A)} - \frac{V_g}{V_B} \cdot \left.\frac{\partial F_B}{\partial \bar{v}_B}\right|_{V_B} \right\} \cdot \delta \bar{v}_g \tag{5}$$

and it can be shown that the Lagrange multiplier is also zero (<u>14</u>).

For an assembly (adhesion) process that is "true" equilibrium (both inside the gap and between the gap and bulk regions), the third variation,

$$\delta F \bigg|_{(\bar{v},\phi,A)} = \left\{ \frac{1}{A}\cdot\frac{\partial F_g}{\partial Z_g}\bigg|_{(\bar{v},\phi,A)} - \frac{\partial F_B}{\partial V_B}\bigg|_{\bar{v}_B} + \frac{(\bar{v}_B - \bar{v}_g)}{V_B}\cdot\frac{\partial F_B}{\partial \bar{v}_B}\bigg|_{V_B} \right\} A\cdot\delta Z_g$$

yields the normal stress σ_n that acts on each surface since the first two variations in Equation 3 are identically zero,

$$\delta F \bigg|_A = (\sigma_n\cdot A)\;\delta Z_g$$

where the normal stress is defined by the mechanical work to displace the surfaces at constant area. Further for the "true" equilibrium case, it can be shown that the relation for normal stress reduces to a simple result given by the osmotic pressure change at the mid-point of the gap relative to the bulk region (14).

$$\sigma_n = \Pi_B - \Pi_{Z_g/2} \qquad (6)$$

Equation 6 is equivalent to the classic result obtained for electric double layer repulsion (26). Although this result seems intuitively obvious, it is not consistent with previous derivations (22, 24-25). If exchange of macromolecules between the gap and bulk regions is restricted, then additional equations are necessary to establish the contribution of each of the free energy variations (with respect to gap thickness) to the normal stress which will depend implicitly on the Lagrange multiplier λ_c.

The coefficient in the last variation is the free energy potential γ for creation of contact area at constant gap thickness, composition, and concentration profile,

$$\delta F \bigg|_{(\bar{v},\phi,Z_g)} = \left\{ \frac{1}{Z_g}\cdot\frac{\partial F_g}{\partial A}\bigg|_{(\bar{v},\phi,Z_g)} - \frac{\partial F_B}{\partial V_B}\bigg|_{\bar{v}_B} + \frac{(\bar{v}_B - \bar{v}_g)}{V_B}\cdot\frac{\partial F_B}{\partial \bar{v}_B}\bigg|_{V_B} \right\} Z_g\cdot\delta A$$

For reversible processes, coefficients are related by cross derivatives; hence,

$$\sigma_n = \frac{\partial}{\partial A}(\sigma_n\cdot A)\bigg|_{(\bar{v},\phi,Z_g)} = \frac{\partial\gamma}{\partial Z_g}\bigg|_{(\bar{v},\phi,A)}$$

The stress defined by Equation (6) represents the interaction between surfaces due to the macromolecules in solution. To obtain the total stress, contributions of long range colloidal forces between the bilayers themselves must be added to σ_n. For phospholipid bilayers, the added stress can be expressed as the sum of a strong exponential repulsion due to hydration forces, a weak power law attraction for the van der Waals' force, and exponential repulsion due to electric double layer forces (1-2),

$$\sigma_{ext} = -P_{hyd} \cdot e^{-z_g/\lambda_{hyd}} + \frac{A_H \cdot f(z_\ell/z_g)}{6\pi \cdot z_g^3} - P_{es} \cdot e^{-z_g/\lambda_{es}} - P_f \quad (7)$$

where P_{hyd} is the coefficient for the hydration repulsion and λ_{hyd} is the characteristic decay length; A_H is the Hamaker coefficient for the van der Waals' attraction and $f(z_\ell/z_g)$ is a weak function of the ratio of the bilayer thickness $z_\perp$ to the distance between bilayers z_g; P_{es} is the coefficient for the double layer repulsion and λ_{es} is the characteristic Debye length for the decay. In addition, repulsion may be enhanced by secondary effects due to thermally excited undulations (bending fluctuations) of the bilayers which is represented by the stress component, P_f (27). Equilibrium separation (stable contact) is established where the total stress ($\sigma_{ext} + \sigma_n$) is zero; thus, the free energy potential for assembly of the bilayers is given by the integral of the total stress over the range of infinity to this location.

The next task is to evaluate the concentration profile in the gap in order to calculate the attractive stress given by Equation 6 (the bulk region profile is determined by the gap profile at infinite separation). The simplest approach is to introduce a "constitutive" relation for the free energy as a sum of the free energy of mixing for uniform concentration plus a term that represents the free energy excess due to configurational entropy gradients (24, 28-30). For linear flexible polymers, the relation can be deduced from a statistical equation for pair-wise correlation functions of segment distribution to give a self-consistent, mean-field approximation (21, 23-24, 30). Here, the free energy density is expressed by,

$$F_g = A \cdot \int_0^{z_g} (\tilde{F} + \frac{a_m^2}{6} \cdot \left|\frac{d\psi}{dz}\right|^2) \cdot dz \quad (8)$$

where ψ is like a quantum-mechanical particle distribution function and ψ^2 is the expectation value for local segment density; a_m is the effective length of a rigid segment of the flexible polymer. $\tilde{F}$ is the free energy density evaluated at the local concentration in the absence of gradients and given in terms of chemical potentials as,

$$\tilde{F} = \frac{v \cdot \mu_p}{N_p \cdot \nu_m} + \frac{(1-v) \cdot \mu_s}{\nu_s}$$

or,

$$\tilde{F}' = \frac{v \cdot (\bar{\mu}_p - \bar{\mu}_{p_B})}{N_p \cdot \nu_m} + (\Pi_B - \Pi) \quad (9)$$

where,

$$\frac{\bar{\mu}_p}{N_p \cdot \nu_m} \equiv \frac{\mu_p}{N_p \cdot \nu_m} + \Pi = \frac{\partial \tilde{F}}{\partial v} \qquad \frac{\mu_s}{\nu_s} \equiv -\Pi \quad (10)$$

Chemical potentials for mixing and deformation of polymer

molecules in the gap are described classically by Flory-Huggins theory (31) or by more esoteric scaling theories (23). With the use of the free energy density augmented by segment gradients, Equation 4 leads to an equation that predicts the segment distribution in the gap,

$$\frac{a_m^2}{6} \left|\frac{d\psi}{dz}\right|^2 = \widetilde{F}'_{z_g/2} - \widetilde{F}' + \lambda_c \cdot (\psi_{z_g/2}^2 - \psi^2) \cdot \nu_m \tag{11}$$

and, thereby, the value of the concentration at the mid-point of the gap (which determines the surface stress for equilibrium exchange of macromolecules between gap and bulk regions),

$$z_g/2 = \frac{a_m}{\sqrt{6}} \cdot \int_0^{\psi_{z_g/2}} \frac{d\psi}{\left[\widetilde{F}'_{z_g/2} - \widetilde{F}' + \lambda_c \cdot (\psi_{z_g/2}^2 - \psi^2) \cdot \nu_m\right]^{1/2}} \tag{12}$$

For the "semi-dilute" range of polymer concentrations, this equation takes the form of an elliptic integral.

We have used this, mean-field approach to successfully predict the free energy potential for bilayer-bilayer adhesion in dextran polymer solutions (the theoretical results are shown in Fig. 4 along with the data, 14). The correlation is excellent, based only on the first and second virial coefficients for the polymer (obtained from osmotic pressure measurements) and the segment length (taken simply as the cube root of the monomer volume).

For inflexible (rigid) macromolecules, the free energy density is augmented directly by concentration gradients (29),

$$F_g = A \cdot \int_0^{z_g} (\widetilde{F} + \beta^2 \cdot \left|\frac{dv}{dz}\right|^2) \cdot dz \tag{13}$$

The coefficient β^2 is somewhat phenomenological although theoretical guidelines are established for derivation of this coefficient (29) which show that the coefficient is scaled by molecular dimensions. Again, the free energy density can be used with Equation 4 to establish the concentration profile in the gap,

$$\beta^2 \cdot \left|\frac{dv}{dz}\right|^2 = \widetilde{F}'_{z_g/2} - \widetilde{F}' + \lambda_c \cdot (v_{z_g/2} - v) \tag{14}$$

and the value of the concentration at the mid-point of the gap region,

$$z_g/2 = \beta \cdot \int_0^{v_{z_g/2}} \frac{dv}{\left[\widetilde{F}'_{z_g/2} - \widetilde{F}' + \lambda_c \cdot (v_{z_g/2} - v)\right]^{1/2}} \tag{15}$$

Rod-like shapes will require more careful consideration than globular forms because rotation will be restricted in the vicinity of the surface before translation. The chemical potential will involve the uniform mixing of rod-like molecules and solvent in a gap which depends on the reduction in entropy caused by steric elimination of orientational states. The phenomenological coefficient β^2 will depend on gap width. Hence at large separations, the depletion zones will be scaled by the large dimension of the molecule whereas, at small separations, depletion zones will be reduced to the scale of the shorter molecular dimension. Thus, it is expected that the osmotic pressure between the bulk region and the gap will increase progressively over the range of separations beginning at a value characterized by the long dimension of the molecule down to a value given by the shorter molecular dimension; for separations smaller than the shorter dimension, the osmotic pressure differential will remain constant. In addition, most protein macromolecules contain charged residues that attract counterions; thus, there will be variations in ion concentrations commensurate with gradients in protein concentration. As such, the osmotic effect due to depletion of the protein concentration in the gap could be enhanced.

For fibrinogen and albumin, it is expected that fibrinogen will create larger free energy potentials for adhesion at common molar concentrations because the range of the interaction is much larger for fibrinogen as previously discussed (even though the osmotic pressures of the solutions are inversely proportional to the molecular weights). Obviously, this is consistent with our measurements because the osmotic activity of albumin is about three times greater than that of fibrinogen but adhesion energies are comparable at the same mass concentrations. For large rigid macromolecules, the free energy potential for adhesion would be approximated by the magnitude of the osmotic pressure due to the macromolecules in the exterior solution multiplied by a distance determined by the difference between the size scale of the macromolecule and the strong repulsive barrier. Clearly, the magnitude of the free energy potentials measured in these experiments are consistent with the product of major molecular dimension and the osmotic pressures for fibrinogen and albumin molecules in solution. This preliminary compatibility of numbers for two different protein macromolecules and a wide range of concentrations strongly indicates that careful development of a depletion-based theory will give successful correlations with these theories as for the dextran polymers. It is important to note that specific binding interactions between proteins and the surfaces would greatly modulate this behaviour.

Acknowledgment

This work was supported in part by the Medical Research Council of Canada through grant MT 7477 and the U.S. National Institutes of Health through grant HL 26965.

Literature Cited

1. Verwey, E.J.W.; Overbeek, J. Th. G. Theory of the Stability of Lyophobic Colloids. Elsevier, Amsterdam; 1948, pp. 1-205.
2. Parsegian, V.A.; Fuller, N.; Rand, R.P. Proc. Natl. Acad. Sci. USA 1979, 76, 2750.
3. Evans, E. Colloids and Surfaces 1984, 10, 133.
4. Janzen, J.; Kukan, B.; Brooks, D.E.; Evans, E.A. Biochemistry (submitted, 1986).

5. Kwok, R.; Evans, E. Biophys. J. 1981, 35, 637.
6. Evans, E.; Needham, D. Faraday Disc. Chem. Soc. 1986, 81, in press.
7. Evans, E.; Skalak, R. Mechanics and Thermodynamics of Biomembranes; CRC Press: Boca Raton Fla., 1980; pp 1-254.
8. Evans, E. Biophys. J. 1980, 31, 425.
9. Evans, E.; Metcalfe, M. Biophys. J. 1984, 45, 715.
10. Evans, E.; Metcalfe, M. Biophys. J. 1984, 46, 423.
11. Evans, E.; Parsegian, V.A. In Surface Phenomena in Hemorheology: Theoretical, Experimental, and Clinical Aspects, Copley, A.L., Seaman, G.V.F., Eds.; N.Y. Acad. Sci., 1983; 13.
12. Cohen, C.; Weisel, J.W.; Phillips; G.N., Jr.; Stauffacner, C.V.; Fillers, J.P.; Daub, E. Ann. N.Y. Acad. Sci. 1983, 408, 194.
13. Hughes, W.L. In The Proteins, Neurath, H., Bailey, R., Eds.; Academic Press, New York, 1954, 633.
14. Evans, E.; Needham, D. Macromolecules (submitted, 1986).
15. Evans, E.; Kukan, B. Biophys. J. 1983, 44, 255.
16. Tadros, Th. F. In Polymer Colloids, Buscall, R., Corner, T., Stageman, J.F., Eds.; Elsevier Applied Sci., London, 1985, 105.
17. Vincent, B. In Polymer Adsorption and Dispersion Stability, Goddard, E.D., Vincent, B., Eds.; Am. Chem. Soc., Washington, 1984, 1.
18. Napper, D.H. Polymeric Stabilization of Colloidal Dispersions, Academic Press, London, 1983, pp. 1-428.
19. Ash, S.G.; Findenegg, G.H. Trans. Faraday Soc. 1971, 67, 2122.
20. Scheutjens, J.M.H.M.; Fleer, G.J. Macromolecules 1985, 18, 182.
21. Joanny, J.F.; Leibler, L.; de Gennes, P.G. J. Polym. Sci. Polym. Phys. Ed. 1979, 17, 1073.
22. Fleer, G.J.; Scheutjens, J.H.M.H.; Vincent, B. In Polymer Adsorption and Dispersion Stability, Goddard, E.D., Vincent, B., Eds.; Am. Chem. Soc., Washington, 1984, 245.
23. de Gennes, P.G. Scaling Concepts in Polymer Physics, Cornell University Press, Ithaca, 1979, pp. 1-324.
24. de Gennes, P.G. Macromolecules 1982, 15, 492.
25. Scheutjens, J.M.H.M.; Fleer, G.J. Adv. Colloid Interface Sci. 1982, 16, 361; 1983, 18, 309.
26. Verwey, E.J.W.; Overbeek, J. Th. G. Theory of the Stability of Lyophobic Colloids Elsevier, Amsterdam, 1948, p. 93.
27. Evans, E.; Parsegian, V.A. Proc. Natl. Acad. Sci. U.S.A. 1986, 83, 7132.
28. Cahn, J.W.; Hilliard, J.E. J. Chem. Phys. 1958, 28, 258.
29. Widom, G. Physica 1979, 95A, 1.
30. Edwards, S.F. Proc. Phys. Soc. (London) 1965, 85, 613.
31. Flory, P.J. Principles of Polymer Chemistry; Cornell University Press: Ithaca, NY, 1953, pp. 1-672.
32. Andrade, J.D. In Surface and Interfacial Aspects of Biomedical Polymers: Volume 2 Protein Adsorption; Andrade, J.D., Ed.; Plenum Press, New York, 1985, 1.
33. Kazal, L.A.; Amsel, S.; Miller, O.P.; Tocantins, L.M. Proc. Soc. Exp. Biol. Med. 1963, 113, 989.

RECEIVED January 13, 1987

Chapter 7

Interaction of Prothrombin with Phospholipid Monolayers at Air- and Mercury-Water Interfaces

M. F. Lecompte

Laboratoire d'Electrochimie Interfaciale, Centre National de la Recherche Scientifique, 1 Place A. Briand, 92195 Meudon Principal Cedex, France

The studies on the mode of interaction of prothrombin with phospholipid monolayers, using complementary methods of surface measurement are reviewed. They were investigated at air-water and Hg-water interfaces respectively by radioactivity and electrochemistry. A process more complex than a simple adsorption could be detected. Indeed, the variation of the differential capacity of a mercury electrode in direct contact with phospholipid monolayer, induced by the interaction with prothrombin could be interpreted as a model of its penetration into the layer; this was confirmed by the study of the dynamic properties of the direct adsorption of this protein at the electrode, followed in part by the reduction of S-S bridges at the electrode. It could be also concluded that prothrombin resists complete unfolding at these interfaces.

In order to understand the structure and structural changes of biological components involved in protein-membrane interactions, the surface behavior of proteins must be studied carefully.

In several steps of the blood coagulation cascade, the activity of some of the coagulation factors is enhanced at the surface of the phospholipid membrane. Phospholipids, mainly those which are negatively charged, play a crucial role by accelerating the zymogen-to-enzyme conversions leading to clot formation. Since the importance of thrombin is well known in this process, it was of interest to understand its rather complex formation from the corresponding zymogen, prothrombin. Moreover, conversion of prothrombin into thrombin is a good example of a typical enzymatic activity taking place at a cell/solution interface. The conversion requires a membrane-bound complex of protease, substrate and cofactor.

Vitamin K-dependent proteins, containing γ-carboxyglutamic residues, like prothrombin, are commonly known to bind by Ca^{++} bridges to membranes containing acidic phospholipids (1). Nevertheless, it was important to study whether interactions other than those of an electrostatic nature could also be involved, such as those leading

0097-6156/87/0343-0103$06.00/0
© 1987 American Chemical Society

to penetration of proteins into membranes, since the catalytic process is very specific, during the conversion.

As an example of a membrane model, phospholipid monolayers with negative charge of different density were used. It had already been found (2) and discussed (3) that the physical and biological behavior of phospholipid monolayers at air-water interfaces and of suspensions of liposomes are comparable if the monolayer is in a condensed state. Two complementary methods of surface measurements (using radioactivity and electrochemical measurements), were used to investigate the adsorption and the dynamic properties of the adsorbed prothrombin on the phospholipid monolayers. Two different interfaces, air-water and mercury-water, were examined. In this review, the behavior of prothrombin at these interfaces, in the presence of phospholipid monolayers, is presented as compared with its behavior in the absence of phospholipids. An excess of lipid of different compositions of phosphatidylserine (PS) and phosphatidylcholine (PC) was spread over an aqueous phase so as to form a condensed monolayer, then the proteins were injected underneath the monolayer in the presence or in the absence of Ca^{++}. The adsorption occurs in situ and under static conditions. The excess of lipid ensured a fully compressed monolayer in equilibrium with the collapsed excess lipid layers. The contribution of this excess of lipid to protein adsorption was negligible and there was no effect at all on the electrode measurements.

The Air-water Interface

The adsorption of prothrombin onto the lipid monolayer was followed directly by counting the surface radioactivity of the ^{3}H labelled protein using a gas-flow counter equipped with an ultrathin window as described elsewhere (4). By calibrating the counter as previously described (5), it is possible to determine the surface concentration of the radioactive protein, Γ.

In Figure 1, the kinetics of adsorption of prothrombin at the initial bulk concentration of 5 µg/ml in the presence and in the absence of Ca^{++}, onto a monolayer containing 100 % PS are presented as compared with the adsorption at the pure air-water interface. It is clear that, in the presence of phospholipids, the amount of protein adsorbed is strongly dependent on Ca^{++} concentration, while this is not so at the pure air-water interface. Nevertheless, we see that even in the absence of Ca^{++}, the prothrombin adsorption remains significant. It was shown (6) that even at concentrations as low as 10^{-3} mM, Ca^{++} is coadsorbed with prothrombin. The results showed that the surface concentration of Ca^{++} is proportional to that of the adsorbed prothrombin, about 10 Ca^{++} being coadsorbed with one molecule of prothrombin. The surface concentrations of prothrombin, obtained at equilibrium, were plotted as a function of its initial concentration, while the interaction occurred in the presence of different Ca^{++} concentrations and with phospholipid monolayers of different compositions (5). The Scatchard plots obtained from the adsorption isotherms gave the binding constants, Ka. When the adsorption was onto pure PS monolayer, K_a turned out to be independent of Ca^{++} concentration (around 1.2×10^8 1/mol), while the maximal surface concentration, Γ^{max}, is dependent. At 2 mM Ca^{++}, on the monolayers

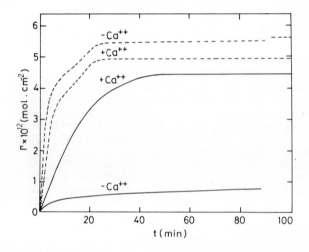

Figure 1.   Time dependence of adsorption of 5 µg/ml of prothrombin, at the air-water interface in the presence of a phosphatidylserine monolayer ——— or in its absence ----, in the presence
or in the absence of 2 mM Ca^{++}.

containing 25% PS - 75% PC, Γ^{max} was 1.75×10^{-12} mol.cm^{-2} as compared with 6.2×10^{-12} mol.cm^{-2} adsorbed on pure PS monolayers; the binding constant was found to be equivalent to the values obtained using pure PS monolayers. These constants were confirmed, in some cases, on multilayers (7) or on monolayers (3) by using ellipsometry, but they are higher than those obtained with phospholipid vesicles (3). These last differences may be ascribed to some causes like the different curvatures used in the two membrane models and the lower concentrations of protein used with the monolayers. Nevertheless, the equilibrium conditions might depend on the technique used and be of importance in the discrepancy in the values of Ka found in the literature. The maximal surface concentration of prothrombin adsorbed on a pure PS monolayer at 2 mM Ca^{++} corresponds to about 27 nm^2/molecule while at 10^{-3} mM Ca^{++}, with a value of 1.5×10^{-12} mol.cm^{-2}, it corresponds to 120 nm^2 per molecule. Since the area occupied by a prothrombin molecule at maximal hexagonal packing oriented with its long axis perpendicular to the surface is 18 nm^2, according to the model given for prothrombin (1), this configuration should be the one approached in the presence of Ca^{++}. At 10^{-3} mM Ca^{++}, a prothrombin molecule lies with the long axis of its ellipsoidal shape parallel to the surface. Indeed, it can cover an area of about 50 nm^2 and thus the total maximal number of adsorbed molecules covers about 50% of the area. Thus, we can distinguish a change in the configuration of the prothrombin molecules relative to the surface of the monolayer, in the absence or in the presence of Ca^{++}.

At the pure air-water interface (Figure 1), the initial rate of adsorption of prothrombin is proportional to the square root of the time, as can be calculated from the curves, indicating that the process, at this interface, is diffusion-controlled, as was observed for native DNA (8). In the presence of phospholipid monolayers the adsorption process is slower; a more complex process must take place at the surface of the membrane, in addition to the conformational change of prothrombin which occurs in the presence of Ca^{++} as was found by Nelsestuen (9). It can be the penetration of prothrombin into the lipid layer, as will be described below.

The surface occupied by a protein molecule, at high Ca^{++} concentration, as obtained from the protein surface concentration, at maximal coverage, is equivalent to the smaller cross-sectional area of a native protein molecule in solution, suggesting that the molecules are bound perpendicularly to the surface. It indicates clearly, in this case, that the protein does not unfold at this interface. The prothrombin molecule contains a high concentration of S-S bonds, and it is known that this is a very important factor in preserving the tertiary structure of proteins (10). At low calcium concentration, this might also be the case, if we take into account the possibility of the protein bound to the surface being able to rotate in the plane.

The Mercury-water Interface

Capacitance, C, provides direct information on the structure of the adsorbed layer (10, 11). The change in the differential capacity of the electrical double layer between a polarized mercury surface and a 0.15 M NaCl solution containing various concentrations of protein

(for example) was used as a measure of its adsorption on a bare mercury surface or on a spread lipid monolayer in contact with the electrode, as will be described below.

As seen in Figure 2, when a compressed spread layer of long
chain lipids is brought into contact with a mercury electrode from
the gaseous phase, the capacitance of the double layer is very low
in comparison with the supporting electrolyte alone and varies between 1.5 and 1.9 μF.cm^{-2} over a wide range of potential. It is characteristic of a hydrocarbon layer, one hydrocarbon chain length
thick, which adheres to the mercury surface, while the polar head
groups orient themselves toward the aqueous solution, and it prevents
access of ions. At some positive or negative potential, a desorption
capacitance peak can be observed. The peak results from the charge
flux following the displacement of the low dielectric layer by the
high dielectric aqueous medium.

If the continuity of the monolayer is perturbed by an interacting molecule of higher polarity, an increase in capacitance proportional to the degree of perturbance or penetration is observed. In
the case where the penetrating molecules contain electroactive groups
undergoing electrode reaction, a pseudocapacitance peak is obtained
which is proportional in size to the ease of access of these groups
(through the lipid layer) to the electrode surface. In the case of
prothrombin and of other proteins, cystine may serve as such an electroactive group. Cystine is strongly adsorbed on the mercury surface
at positive potentials of the redox potential, forming a charge-
transfer complex (12). The surface complex is then reduced at the
redox potential giving rise to the pseudocapacitance peak. The cystine-cysteine transition on the mercury electrode was seen to be a reversible process (12, 13) while the mercury acts as a catalytic
surface following the scheme:

$$RSSR + Hg \rightleftharpoons Hg(RS)_2 \text{ or } Hg_2(RS)_2 + 2H^+ + 2e^- \rightleftharpoons 2 RSH + Hg$$

As seen in Figure 2, the ac polarogram resulting from the interaction
of the monolayer with prothrombin shows that an electroactive group
contributes to the capacitance curve a pseudocapacitance peak, at
around $- 0.7$ V, which is ascribed to the oxy-reduction of the disulfide bridges at pH 7.8. Since the formation of cysteine requires
hydrogen ions, the half-wave potential and thus also the pseudocapacitance peak is shifted with decreasing pH to more positive polarization. Since two contributions, one from the protein and the other
from the phospholipid monolayer, are involved in the capacitance values, it was of importance to study particularly the behavior of the
protein in direct contact with the electrode, in order to be able to
interpret better the data obtained when prothrombin interacts with
phospholipid monolayers.

Direct adsorption of Prothrombin at the Mercury-water Interface.   The
adsorption rate of prothrombin on a hanging mercury drop electrode
(HMDE) was studied by measuring the decrease of the differential capacity with time of contact of the mercury drop with the solution, at
a fixed potential, $- 0.5$ V, in parallel with the increase of the
areas of the voltametric peaks corresponding to the reduction of
some S-S bonds of the adsorbed molecules (14). The representation on

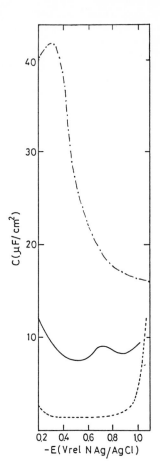

Figure 2.   Effect of prothrombin ———(5 μg/ml) on the capacitance
of a monolayer of 100% PS ---- as compared with the capacitance
of the electrolyte alone -·-·- ; NaCl 0.15M, Tris 1mM, pH 7.8.

the same graphs of the two curves Γ_{S-S} versus t and C versus t, in the presence or in the absence of Ca^{++}, at one concentration of prothrombin for example 4.5 µg/ml, shows clearly in Figure 3 that these functions reach their saturation values simultaneously. Adsorption of prothrombin at the waiting potential of -0.5 V (Figure 3), near the zero charge potential, causes a sharp decrease in capacitance. For each concentration studied (14), about the same lower limit of capacitance was reached, corresponding to a saturation value in the range studied, which was therefore attributed to a completely protein-covered electrode. At about half of the maximum lowering of the capacity, an inflexion point can be distinguished.

From the kinetics of adsorption, the surface coverage could be obtained. In the case of prothrombin, the number of molecules adsorbed on the mercury surface, $\Gamma_{t 1/2}$, could be evaluated from the linear dependence of the capacity on $t^{1/2}$ only at short times (< 50s), when the diffusion layer thickness $(Dt)^{1/2}$ is still smaller than the thickness of the unstirred layer. In this region the concentration is

$$\Gamma_{t 1/2} = (2/\pi^{1/2}) \, c_p \, D^{1/2} \, t^{1/2} \qquad (1)$$

and thus the correlation between the lowering of capacity ΔC, and $\Gamma_{t 1/2}$ is obtained. By extrapolating the plot of capacitance versus $t^{1/2}$ to saturation capacitance, one obtains the limiting saturation surface concentration , $\Gamma^{max}_{t 1/2}$. Around t= 50s, the dependence of capacitance on $t^{1/2}$ starts deviating from linearity and above 100s, a region with linear dependence of capacitance on t is obtained which allowed us, by extrapolation to the saturation capacitance, as described previously (14), to determine the limiting saturation surface concentration, Γ^{max}_{t} .

It was shown in the case of polymers, below surface saturation, that the decrease of capacitance is proportional to their surface concentration (over the whole potential region) (10, 15). It implies that the surface conformation of molecules being adsorbed is established instantaneously and then remains constant. Let us assume that this is also the case for prothrombin. The two maximum surface concentrations $\Gamma^{max}_{t 1/2}$ as Γ^{max}_{t} , equivalent to 1.5×10^{-12} mol.cm^{-2}, represent maximal packing for the initial conformation of the adsorbed molecules and does not take into account a possible change in configuration around the inflexion point on the capacitance versus time curve.

Since the slopes of the linear variation of $\Delta C = f(t^{1/2})$ and $\Delta C = f(t)$ varied with protein concentrations at all concentration as according to the corresponding equations giving Γ (14), the process is diffusion controlled. It also implies immediate adsorption and negligible back reaction.

The saturation surface coverage values can be used for determination of the limiting values of the area occupied per adsorbed protein. The calculated areas were equivalent to 110 nm^2 in absence of Ca^{++} and 95 nm^2 in presence of Ca^{++}, per molecule.

The areas of the peaks obtained by cyclic voltametry allows the calculation of the number of the reduced cystine residues, Γ_{S-S}, or reoxidized cysteine, taking into account that the latter is twice the former. Since the position of the onset of the reduction peak is slightly higher than - 0.5 V, the quantity of S-S groups reduced was

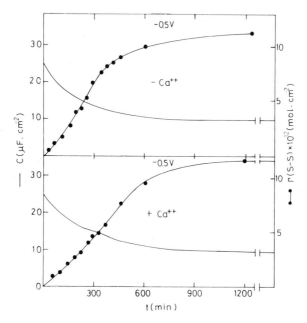

Figure 3.   Comparison of the simultaneous variations with time
of the capacitance C and the surface concentration of electroactive disulfide groups, calculated from the 2nd reduction sweep,
for a concentration of prothrombin of 4.5 μg/ml, in the presence
or in the absence of 2 mM Ca^{++}. Waiting potential: - 0.5 V

determined only at the second negative sweep. In Figure 3, the number of moles of reduced cystine per unit area is plotted versus time. Linear dependence of Γ_{S-S} on time is observed over a relatively large range of coverage and since the slopes obtained at different concentrations were found to vary with each of them according to the correspondent equation (14), we could calculate for each time Γ_q using this equation. The extrapolation of the straight lines drawn from the initial values, obtained from the plot of Γ_{S-S} against time, at the plateau of Γ_{S-S} versus t, allows the determination of the time necessary to complete the surface layer and therefore using the right equation allows the calculation of the actual surface concentration, at equilibrium, Γ_q^{max}, at high coverage, which is about 3×10^{-12} mol.cm^{-2}.

This value, together with the corresponding value of S-S (11×10^{-12} mol.cm^{-2}), gives a mean value of the number of S-S groups reduced per molecule adsorbed when saturation is reached. Only a small fraction (3) of the total cystine residues (12) present in the whole adsorbed prothrombin molecule is available for reduction on the electrode, in spite of the exposure of the molecules to the mercury electrode at positive polarizations, at which cystine tends to be adsorbed on the electrode. This is in agreement with other findings that in the case of proteins only some of the S-S groups are available for electrode reaction (16, 17). The prothrombin molecule, similarly to other proteins, resists complete unfolding, when adsorbed on the mercury electrode, in the range of adsorption potential studied. The degree of partial unfolding depends on the electrode potential during adsorption, on the time of exposure to the surface and on the presence of Ca^{++}.

It was shown that at a more positive polarization, -0.35 V, the area occupied by a molecule is smaller than at -0.5 V and the number of S-S reduced by the molecule higher; this indicates a change in the conformation of the molecules adsorbed at the interface depending on the electrode polarization. The limiting areas in the presence of Ca^{++} are lower than in its absence both in the low and the high surface concentration region, indicating smaller deviation from its globular structure in the bulk. Moreover the number of S-S reduced by a molecule was lower in the presence of Ca^{++} than in its absence. Then it could be concluded that Ca^{++} causes stabilization of the molecular structure of prothrombin at the surface.

We tried to answer the following question: Why is the maximal surface concentration at adsorption equilibrium obtained from the extrapolation of differential capacity against time less than half that obtained from a similar extrapolation of the voltametric peaks?

The plot of Γ_{S-S} as a function of the variation of C from the electrolyte alone, ΔC, at different times of adsorption until saturation is a typical diagram which represents the dynamic picture of the growth of the adsorbed protein layer (Figure 4).

It shows clearly that the contribution of a given amount of adsorbed protein to the changes in Γ_{S-S} and C is different if low or high surface concentrations are considered. This plot shows essentially two linear sections over the whole range of surface concentration. At low surface concentration (region A), the capacitance appears as the more sensitive probe to adsorption, while at high

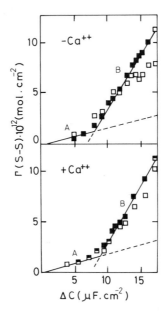

Figure 4.   Surface concentration of electroactive disulfide bonds
from the 2nd reduction ■ , 1st oxidation sweep □ , as a function
of the decrease of the capacitance of the HMDE in contact with a
prothrombin solution at 4.5 µg/ml. Potential of adsorption:-0.5V

surface concentrations the area of the voltametric peak becomes more
sensitive (region B). From this plot an answer could be given to the
above question, by considering the following model for the growth of
the adsorbed layer (14). While extrapolation of Γ_{S-S} versus t gives
informations corresponding to the "A and B" behavior of the system,
extrapolation of C versus t to the equilibrium capacitance value allows an estimation of the hypothetical value of the number of molecules adsorbed at maximal packing in the "A" conformation, identical to
the conformation of each isolated molecule at the surface. The areas
obtained are larger than the cross-sectional area of a native molecule lying in the surface (52 nm^2)(1), indicating different degrees of
unfolding of the molecules which retain a considerable freedom of rotation. In the region of the inflexion in the C(t) curve, the motion
of molecules in the surface is restricted, because of the lateral interaction between the protein molecules, inducing a surface gel formation (18). In region B, adsorption continues to lower the capacitance less efficiently than in part A. In part B, each new molecule
being adsorbed onto the new molecular surface network of the adlayer
occupies its own area, which is lower than in part A, at the Hg surface, causing at the same time a lateral contraction of its neighbour
molecules. The plot of Γ_{S-S} versus ΔC is linear inasmuch as the surface occupied per molecule and the number of S-S groups per adsorbed
molecule are constant. From the change in the slopes in parts A and
B (Figure 4), the adsorption process shows clearly the existence of
two distinct adsorption states for which were defined a molecular a-
area (part A) and a differential one (part B) for the newly adsorbed
molecules which might equilibrate with the already adsorbed ones(14).
The extrapolated values of $(\Gamma_{S-S})_A$ to ΔC^{max} from part A (Figure 4),
divided by the value of Γ^{max} obtained from C versus t at saturation
gives the number of S-S reduced per molecule in part A; it is lower
($\sim$2) than at higher coverage ($\sim$3) where lateral protein interactions
might occur and seems to aid exposure of the sulfide groups. This
diagram was shown to be essentially independent of the bulk protein
concentration and contains characteristic points which are correlated
only with surface concentration.

At low surface concentrations, the redox process is nearly reversible, by taking in account the areas and the potentials of the
peaks. At higher, the differences in the areas between reduction and
oxidation peaks may be attributed to differences in the adsorption of
cystine and desorption of cysteine residues at the positively or negatively charged mercury respectively. At the same time the protein
molecule as a whole remains adsorbed by hydrophobic interactions.

With other biological macromolecules (19, 20), the number of adsorbed molecules was usually calculated from the linear dependence of
the capacitance on $t^{1/2}$ using Equation 1, over the whole range of adsorption. The surface concentration of hormones could also be inferred directly from the calculated number of charges transferred between the electrode and an electroactive group, like S-S, of the adsorbed molecules, each one containing only one S-S (21, 22). This method could not be used for proteins, where only part of the S-S are
available for the electrode reaction,as seen for prothrombin; but in
this case of proteins, the method of exploitation of the data presented above is very useful and quite new.

Only a fraction of the total S-S bonds is reduced in case of

prothrombin, indicating that only the cystine groups in contact with
the mercury are electroactive. It could be concluded from this that
when we get a peak of S-S reduction, while prothrombin interacts with
phospholipids, this protein had to cross the monolayer in order that
the S-S should be reduced; this is in favor of the penetration of
prothrombin into the layer.

Interaction of Prothrombin and some of its fragments with Phospholipid Monolayers. When the adsorption equilibrium of the protein has
been reached, as determined by surface radioactivity, an HMDE was
formed and positioned in order to be in contact with the monolayer
(23). Thus the monolayer transferred onto the mercury electrode-water
interface stayed in equilibrium with the monolayer reservoir on the
air-water interface. Then ac polarograms were recorded after exposure
of the monolayer to the electrode at a given potential, - 0.2 V in
Figure 2 and - 0.5 V in the others which will be presented. Indeed,
at -0.5V, the monolayer is at the capacitance minimum and more stable
and the potential is still remote enough from the cystine-cysteine
redox pseudocapacitance peak potential. The increase in capacitance,
at this potential, was selected to represent the effect of the protein penetration into the lipid layer and is presented in Figures 5
and 6 as a function of the protein concentration in the bulk. We see
clearly distinct behavior between prothrombin and its fragments, depending on the lipid composition and of the presence of Ca^{++}.

The penetration of prothrombin into a monolayer containing 25%
PS- 75% PC, starts only at higher prothrombin concentrations than on
the pure PS monolayer, and a cooperative dependence of the capacitance and of the pseudocapacitance peaks on the prothrombin concentration is observed. However, the limiting capacitances reached at
higher prothrombin concentrations are about the same with both monolayers : it is 7 $\mu F.cm^{-2}$. These high capacitances are obtained instantaneously upon nondamaging contact of the monolayer by the electrode.

Even in the absence of Ca^{++} (Figure 6), there is a significant
increase in capacitance upon addition of prothrombin. This indicates
that other interactions beside the electrostatic ones have to take
place. However, the increase with concentration is less steep in the
low Ca^{++} concentration region.

By taking into account the ratio between the number of molecules
adsorbed on phospholipids and the number of S-S reduced, it was found
(23) that only a small fraction of the total cystine residues of the
adsorbed prothrombin molecules is available for reduction on the
electrode, and is equivalent to the ratio when prothrombin is in direct contact with the electrode. Consequently, there is no gross
conformational change of prothrombin, when interacting with a phospholipid surface. The significant change in this ratio between high
and low Ca^{++} concentrations suggested a conformational change brought
about by the lipid-calcium protein bonds, which is in agreement with
the differences obtained for the areas occupied by molecule at the
air-water interface.

Similarly to prothrombin, Fragment 1 containing the γ-carboxy-
glutamic residues, increased the capacitance of PS-containing monolayers, also giving rise to a pseudocapacitance peak. The effect
increases with Ca^{++} and with Fragment 1 concentration and it is lar-

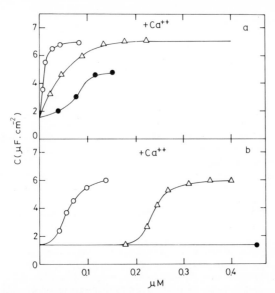

Figure 5. Differential capacity of condensed monolayers containing 100% PS (a) or 25% PS–75% PC (b) at –0.5 V relative to 1N Ag/AgCl electrode as a function of the protein concentrations in the bulk in the presence of Ca^{++}. Prothrombin, ○; fragment 1, △; fragment 2, ●.

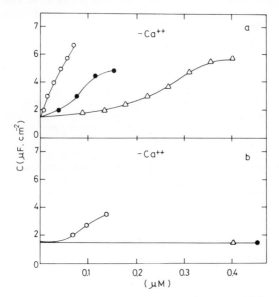

Figure 6. Differential capacity of condensed monolayers containing 100% PS (a) or 25% PS–75% PC (b) at –0.5 V relative to 1N Ag/AgCl electrode as a function of the protein concentrations in the bulk in the absence of Ca^{++}. Prothrombin, ○; fragment 1, △; fragment 2, ●.

ger on a monolayer of pure PS than on a mixed monolayer. The observed
increase in capacitance is also smaller at 10^{-3} mM Ca^{++}. Nevertheless
at high Fragment 1 concentration, an equivalent saturation value for
the capacitance is reached in each case, which is similar to the one
obtained for prothrombin. The sigmoidal curves obtained at 25% PS and
Ca^{++} and at 100% PS with low Ca^{++} concentration indicates a cooperative effect on capacitance, like for prothrombin except that higher
concentrations of Fragment 1 are required in order to penetrate the
lipid layer. While Fragment 1 adsorbs on a monolayer containing 25%
PS till 0.15 μM corresponding to 5 μg/ml, it does not penetrate in
this range of concentration in the bulk. It starts penetrating after
a certain surface coverage has been reached. This shows distinctly
that the capacitance measurements associated with the radioactive
ones allows distinction between adsorbed and penetrated molecules.
In the case of Fragment 2, the dependence of capacitance on concentration remains practically the same whatever the Ca^{++} concentration.
However, there is a dependence on the monolayer composition. Fragment
2 does not affect at all the mixed monolayers till 0.4 μM. As was
shown (23), thrombin penetrates similarly both layers, but as it is
not an intact structural domain of the whole prothrombin (it comes
from the cleavage by factor Xa of prethrombin 2), its effect on capacitance is not represented here. Nevertheless, there is a larger tendency of thrombin to penetrate rather than Fragment 2. This suggested
that the prethrombin domain is probably responsible for penetration
of prothrombin into the lipid layers.

Since S-S bridges are reduced in all cases, the protein molecules have to cross the layer in order to reach the electrode surface
and to be reduced, which implies that the different fragments
penetrate to some extent the monolayer.

Conclusions

The prothrombin molecule does not unfold on the different interfaces.
In the presence of Ca^{++}, some changes of the position of the protein
relative to the lipid layer could be detected. Ca^{++} also induces stabilization of the globular structure, as measured at a bare mercury
electrode. Furthermore, a dynamic picture of the growth of the adsorbed prothrombin layer on mercury electrode was presented and shows 2
distinct adsorption states.
It was confirmed that Ca^{++} increases the adsorption, but it was
found that even in the absence of Ca^{++}, prothrombin interacts with
phospholipids. Consequently, besides the electrostatic interactions,
some Ca^{++}- independent interactions, which might be hydrophobic, are
also involved. The observation that prothrombin and some of its fragments penetrated the layer is in accordance with this idea. This implies that fragments other than Fragment 1 might be involved in the
interaction. It must be noticed, that penetration could not be obtained by the less sensitive technique of surface pressure measurements
(24).
Since it was found that activity occurs either on monolayers or
on vesicles (3), the penetration of prothrombin into monolayers which
was found, and confirmed on vesicles (25), might have a role in the
catalytic transformation of prothrombin into thrombin and its
regulation.

Acknowledgments

The author is particularly very grateful to Professor I.R.MILLER, who provided continual encouragement and helpful advice during the successive stages of this work, part of which was performed in his laboratory in the membrane department , at the Weizmann Institute, Israel
The author thanks his colleagues whose names are clear from the reference list, and financial support from INSERM n° 845016, MRT 85C1094 and A.I.P. 9631/112 grants.

Literature Cited

1.   Lim, T.K.; Bloomfield, V.A.; Nelsestuen, G.L. Biochemistry 1977, 16, 4177-4181.
2.   Blume, A.Biochim.Biophys.Acta 1979, 557, 32-44.
3.   Kop, J.M.M.; Cuypers, P.A.; Lindhout, T.; Hemker, H.C.;Hermens, W.T. J.Biol.Chem. 1984, 259, 13993-13998.
4.   Frommer, M.A.; Miller, I.R. J.Colloid Interface Sci. 1966, 21, 245-252.
5.   Lecompte, M.F.; Miller, I.R.; Elion, J.; Benarous R. Biochemistry 1980, 19, 3434-3439.
6.   Lecompte, M.F.; Miller, I.R. Advances in Chemistry series 1980, 188, 117-127.
7.   Cuypers, P.A.; Corsel, J.W.; Janssen, M.P.; Kop, J.M.M., Hermens W.T.; Hemker, H.C. J.Biol.Chem. 1983, 258, 2426-2431.
8.   Frommer, M.A.; Miller, I.R. J.Phys.Chem. 1968, 72, 2862-2866.
9.   Nelsestuen, G.L. J.Biol.Chem. 1976, 251, 5648-5656.
10. Miller, I.R.; Bach, D. Surface and Colloid Science; Matijevic,E. Ed.; 1973, 6, 185-260.
11. Miller, I.R. Topics in Bioelectrochemistry and Bioenergetics; 1981, 4, 161-224.
12. Kolthoff, I.M.; Barnum, C. J.Am.Chem.Soc. 1941, 63, 520-526.
13. Miller, I.R.; Teva, J. J.Electroanal.Chem. 1972, 36, 157-166.
14. Lecompte, M.F.; Clavilier, J.; Dode, C.; Elion, J.; Miller, I.R. J.Electroanal.Chem. 1984, 163, 345-362.
15. Miller, I.R.; Grahame, D.C. J.Colloid Interface Sci.1961,16, 23-40.
16. Cecil, R.; Weitzmann, P.D.J. Biochem.J. 1964, 93, 1-11.
17. Pavlovic, O.; Miller, I.R. Experientia Suppl. 1971, 18,513-524.
18. Lecompte, M.F.; Rubinstein, I.; Miller, I.R. J.Colloid Interface Sci. 1983, 91, 12-19.
19. Temerk, Y.M.; Valenta, P.; Nurnberg, W. J.Electroanal.Chem. 1982, 131, 265-277.
20. Scheller, F. Bioelectrochem.Bioenerg. 1977, 4, 490-499.
21. Rishpon, J.; Miller, I.R. Bioelectrochem.Bioenerg. 1975, 2, 215- 230.
22. Rishpon, J.; Miller, I.R. J.Electroanal.Chem.1975, 65, 453-467.
23. Lecompte, M.F.; Miller, I.R. Biochemistry 1980, 19, 3439-3446.
24. Mayer, L.D.; Nelsestuen, G.L.; Brockman, H.L. Biochemistry 1983, 22, 316-321.
25. Lecompte, M.F.; Rosenberg, I.; Gitler C. Biochem.Biophys.Res. Commun. 1984, 125, 381-386.

RECEIVED January 30, 1987

Chapter 8

Mixed-Protein Films Adsorbed at the Oil–Water Interface

Julie Castle, Eric Dickinson[1], Brent S. Murray, and George Stainsby

Procter Department of Food Science, University of Leeds, Leeds LS2 9JT, United Kingdom

We report on the use of surface viscosity measurement at the planar oil—water interface to monitor time-dependent structural and compositional changes in films adsorbed from aqueous solutions of individual proteins and their mixtures. Results are presented for the proteins casein, gelatin, α-lactalbumin and lysozyme at the n-hexadecane—water interface (pH 7, 25 °C). We find that, for a bulk protein concentration of 10^{-3} wt %, while the steady-state tension is invariably reached after 5—10 hours, steady-state surface shear viscosity is not reached even after 80—100 hours. Viscosities of films adsorbed from binary protein mixtures are found to be sensitively dependent on the structures of the proteins, their proportions in the bulk aqueous phase, the age of the film, and the order of exposure of the two proteins to the interface.

As part of an experimental investigation into some of the most likely factors affecting the stability of protein-containing emulsions, we are studying the physico-chemical properties of films that have been adsorbed, at the oil—water interface, from aqueous solutions of pure and mixed proteins. After briefly reviewing some of our earlier findings (1-4), we present here some new results—mainly surface viscosities—for binary systems containing two out of the following set of proteins: casein, gelatin, lysozyme and α-lactalbumin.

The now classic experiments of Graham and Phillips showed (5) inter alia that the surface rheology of pure protein films is very dependent on the type and amount of protein adsorbed, as well as on the conditions of adsorption. In this paper, with films formed from mixed protein solutions, we shall show that the surface viscosity is an extremely sensitive probe of the time-dependent structural and compositional changes taking place during competitive adsorption at the oil—water interface. While steady-state tensions can invariably

[1]Correspondence should be addressed to this author.

be reached, steady-state surface rheology is rarely achieved over the normal experimental time-scale. Tension measurements are sensitive to free energy changes occurring just at the fluid interface (within a few solvent molecule diameters of the Gibbs surface), but surface rheological techniques also monitor the associated regions near to the interface, and so are sensitive to macromolecular entanglements and protein—protein electrostatic interactions extending well into the aqueous phase, and also to multilayers if they occur. With disordered proteins like casein or gelatin adsorbing at the oil—water interface, there is a gradual loss of intramolecular hydrophobic interactions as the more surface-active residues arrange themselves at the interface. Globular proteins like lysozyme or α-lactalbumin, on the other hand, are less susceptible to unfolding due to their strong covalent -S—S-cross-links. With both types, the rate of molecular conformational change is dependent on the adsorbed layer density and the prevailing solvent conditions.

There is little information in the literature on competitive protein adsorption at fluid interfaces. Our initial interest in casein + gelatin arose out of Mussellwhite's observation (6) that casein will displace gelatin from an oil—water interface. The high surface activity of casein (especially β-casein) is attributed to its disordered structure and high proportion of hydrophobic residues. Gelatin is also disordered, but is more hydrophilic and much more polydisperse than casein. From the surface rheological viewpoint, casein and gelatin make an interesting pair, with the former giving films which are mechanically much weaker than the latter, when present alone (5,7,8), and gelatin having its own unique capacity for gelation in bulk solution and in thick layers. The casein used here, and earlier by Mussellwhite, is sodium caseinate; this is a heterogeneous mixture of monomeric caseins α_{s1}, β, α_{s2} and $\varkappa$, in the approximate proportions 4 : 4 : 1 : 1 (9). To contrast with these disordered proteins, we are also studying mixtures involving the two globular proteins lysozyme and α-lactalbumin, which are thought to have fairly similar tertiary structures, since their amino-acid sequences are similar and the four disulphide groups located identically (10). But interestingly, at neutral pH, their net molecular charges are respectively +8e and -4e (where e is the electronic charge), which opens up the opportunity for studying the effect of electrostatic interactions between adsorbed protein molecules on the viscosity of mixed protein films. That is, we might expect a qualitative difference in behavior between films containing lysozyme (pI = 10.7) + one of the other proteins (pI = 5—5.5), on the one hand, and films not containing lysozyme (e.g., gelatin + α-lactalbumin), on the other.

Our long-term interest in surface rheology of adsorbed protein is in connection with its implications for the behavior of protein-stabilized oil-in-water emulsions. Several investigators (11-14) have proposed a close correlation between droplet coalescence stability and surface rheology at the oil—water interface, but direct causality has not been properly demonstrated. In fact, it is argued in some quarters (15,16) that film thickness and disjoining pressure are the more important factors. In experiments with emulsions containing casein + gelatin, we have found (3,4,17) that casein dominates the interfacial film around the droplets, as one would expect on the basis of its greater surface activity. However, while interfacial gelatin in a freshly made emulsion is readily replaced by casein added to the continuous phase, the ability to exchange diminishes considerably if the interfacial gelatin is allowed to age for several hours at 25 °C before casein is made available (4). These time-dependent results on mixed protein-stabilized emulsions have prompted us to look further at aging effects in competitive adsorption at a planar oil—water interface, and some new surface viscosity data for casein + gelatin and casein + lysozyme systems are reported in this paper.

The primary aims of this research, therefore, are (i) to study time-dependent changes in surface viscosity and (ii) to compare different protein mixtures. As much as possible, all other relevant experimental variables are kept constant. The oil phase is always n-hexadecane. [One set of measurements with gelatin down to 15 °C (see Figure 3) was taken with n-tetradecane as oil phase.] To limit aggregation in bulk solution whilst ensuring negligible depletion of protein due to adsorption (3), the protein concentration in the aqueous phase is maintained at 10^{-3} wt %. The aqueous phase consists of a phosphate buffered solution of pH 7 and ionic strength 0.005 M or 0.05 M. (Early experiments with casein + gelatin were all in 0.005 M buffer, but a higher ionic strength was later found necessary in the lysozyme-containing mixtures to avoid protein precipitation.) As most of the films are shear-thinning, apparent surface viscosities are mainly quoted under standard conditions of film deformation. Care is taken to ensure that protein solutions are prepared according to standard procedures. Measurements are made at 25 °C except where stated otherwise.

Experimental

Surface viscosities at the n-hexadecane—water interface were measured in a Couette-type torsion-wire surface viscometer with inner and outer radii of 15 and 72.5 mm respectively. Unless stated otherwise, the apparent viscosities were calculated as described previously (1) from the angular deflection of the disk corresponding to a dish rotation speed of 0.73 revolutions per hour (1.3×10^{-3} rad s^{-1}). Based on earlier work (5) it seems reasonable to assume a maximum surface concentration of ca. 5 mg m^{-2}, from which it can be estimated that there is negligible lowering of the bulk protein concentration (10^{-3} wt %) during the course of an experiment. The contribution of the bulk phases to the observed viscous drag on the disk was negligible, except at the very earliest stages of film formation, when the surface viscosity was less than 0.1 mN m^{-1} s (18). Further details about the surface viscometer can be found elsewhere (1,3).

Experiments in which aged protein film A was exposed to fresh protein B were performed as follows. With a glass tube (ext. diam. 8 mm) positioned in the Couette gap near the dish surface, a film of protein A was formed at the oil—water interface. (It was separately demonstrated that the presence of the tube had no measurable effect on the surface viscosity.) After aging the film for 24 h, a 2.5 ml sample of a concentrated solution of protein B was added to the aqueous phase (volume 375 ml) via the guide tube. Satisfactory mixing was achieved in 2—3 min by gently sweeping the lower region of the aqueous phase, now containing proteins A + B, with an L-shaped glass rod that had been present throughout the experiment. Separate tests showed that there was no detectable change in viscous drag on the disk associated with the small increase in level of the interface

following addition of protein B, but that stirring in the absence of
addition did tend slightly to increase (by about 10 %) the measured
viscosity.

Tensions at the n-hexadecane—water interface were measured at
short times ($\leqslant 30$ min) by drop-volume and pendant-drop techniques (2),
and at longer times using a Wilhelmy-plate torsion balance (3). Under
conditions for which protein concentration, aqueous phase volume and
surface area were similar to those existing in the surface viscometer,
all the pure proteins gave a steady-state tension within 5—10 h.

Measured surface viscosities (>1 mN m^{-1} s) were reproducible
generally to within 5—10 %. Tensions have an estimated precision of
± 0.3 mN m^{-1}. In the case of surface viscosity, these errors refer to
solutions made from the same batch of protein. With lysozyme, we have
found differences of as much as 50 % in apparent viscosity between
different commercial batches, and substantial changes (say 20—30 %) on
storage, but with little associated change in the interfacial tension.

Gelatin was food grade (pI 5.7) with a weight-average molecular
weight of 2.4×10^5 daltons (19). Sodium caseinate of low calcium
content (0.1 g kg^{-1}) was obtained from the Scottish Milk Marketing
Board; α_{s1}-casein (2.36×10^4 daltons, pI 5.0), β-casein (2.40×10^4
daltons, pI 5.2) and $\varkappa$-casein (1.90×10^4 daltons, pI 5.5) were
separated from whole casein and purified by standard procedures (20,
21). Highly purified α-lactalbumin (1.42×10^4 daltons, pI 5.0) and
β-lactoglobulin (1.83×10^4 daltons, pI 5.3) were gifts from AFRC Food
Research Institute, Reading (formerly N. I. R. D.). Lysozyme (1.46×10^4 daltons, pI 10.7) and n-hexadecane (>99 wt %) were obtained from
Sigma Chemicals. Buffer solutions were made with AnalaR grade
reagents and doubly distilled water.

Results and Discussion

Let us first consider the surface viscosity behavior of the single-
protein films at 10^{-3} wt % bulk protein concentration. Figure 1 shows
apparent viscosity plotted logarithmically against time of adsorption
at the n-hexadecane—water interface (pH 7, 25 °C). The films of
gelatin (ionic strength 0.005 M) and α-lactalbumin (0.05 M) are about
ten times as viscous as the caseinate film (0.005 M); the lysozyme
film (0.05 M) is about a hundred times as viscous as the caseinate
film. There is a gradual build up in viscosity with time, and no
limiting value is reached over the experimental time-scale (ca. 80 h).
Only with β-casein, which gives films more than ten times less viscous
than caseinate, do we find an apparent steady-state viscosity. (With
β-casein, the experimental scatter in Figure 1 is high since it
roughly represents the lower limit capability of the apparatus.)

Surface viscosities of films of the globular proteins, lysozyme
and α-lactalbumin, are more sensitive to ionic strength than those of
the disordered proteins, casein and gelatin. Data for caseinate and
gelatin in Figure 1 refer to an ionic strength of 0.005 M, but values
for 0.05 M buffer are essentially the same (within 10 %). But, with
α-lactalbumin, when ionic strength is reduced to 0.005 M, the apparent
viscosity becomes two or three times larger than that for gelatin (3);
after 30 h, for instance, the values are 160 and 300 mN m^{-1} s at 0.05
and 0.005 M, respectively. It appears that electrostatic interactions
influence the surface rheology to a greater extent with globular
proteins than with disordered proteins.

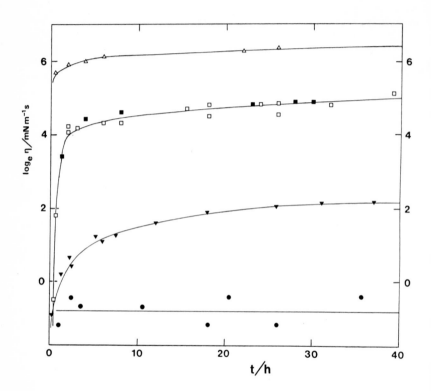

Figure 1:   Time-dependent behavior of various proteins adsorbed
at the n-hexadecane—water interface (10^{-3} wt % protein, pH 7,
25 °C).   The logarithm of the apparent surface viscosity η is
plotted against the time t following exposure of protein solution
to fresh interface: •, β-casein (ionic strength 0.005 M); ▼,
sodium caseinate (0.005 M); □, gelatin (0.005 M); ■, α-lactalbumin
(0.05 M); △, lysozyme (0.05 M).   Lines are drawn to guide the eye.

The individual caseins have very different surface viscosities (4). The values after 30 h under standard conditions are 0.5 mN m^{-1} s for β-casein, 5 mN m^{-1} s for α_{s1}-casein, and 200 mN m^{-1} s for κ-casein (as compared with 7.5 mN m^{-1} s for sodium caseinate). So, although κ-casein makes up only ca. 10 % of milk casein, its contribution to caseinate surface rheology is probably quite substantial. The only other important milk protein not mentioned so far is β-lactoglobulin: this gives films that are more viscous than those of lysozyme, and also more strongly time dependent. It was particularly difficult to obtain reproducible results with β-lactoglobulin, but this is not too surprising in view of its tendency to dimerize and to undergo pH-dependent conformational changes (22). We have speculated (3) that it is the free —SH group on each subunit of β-lactoglobulin that mainly distinguishes its behavior from that of α-lactalbumin: through -S—S-linking of subunits, the film becomes more coherent in structure, and therefore grows more viscous with time. Polymerization through -S—S-linkages may be the reason why κ-casein forms films which are so much more viscous than either α_{s1}- or β-casein, although other factors could also be involved (e.g., surface packing density, charge group frequency, ability to form intermolecular hydrogen bonds, etc.).

The data in Figure 1 were obtained under conditions of a constant dish rotation rate of 1.3 X 10^{-3} rad s^{-1}. Figure 2 shows that the caseinate film is close to Newtonian, whereas the lysozyme film is extremely shear-thinning (d ln w/d ln $\tau \approx$ 9). Film properties are not, however, affected by the application of the shear field itself, at least at these shear-rates; that is, with an aging film, the same results are obtained when the viscosity is measured intermittently as when the dish rotates continuously. Protein surface rheology is rather sensitive to temperature, particularly so for gelatin, as shown in Figure 3. Temperature-dependent changes in the viscosity of gelatin films are reversible over a time-scale of ca. 1 h, and data for films of various ages (and viscosities) do seem roughly to scale on a universal curve (Figure 3) when normalized with respect to the viscosity at 25 °C [i.e. η(298)] for that age of film. Apparent surface viscosity is also sensitive to bulk protein concentration. We find that increasing the bulk protein concentration leads to an increase in measured viscosity for the same adsorption time—as one would expect, of course, on thermodynamic grounds. With lysozyme at 25 °C in 0.1 M phosphate buffer (pH 7), apparent viscosities after 30 h are 0.6, 0.85 and 1.2 N m^{-1} s for 1, 3 and 10 X 10^{-3} wt % protein.

Table I lists values of the steady-state tensions for the various individual proteins at the n-hexadecane—water interface. There is no obvious relationship between interfacial tension and apparent surface viscosity. Gelatin and α-lactalbumin have similar viscosities but very different tensions; β-casein and κ-casein have similar tensions but very different viscosities. It is interesting to note that, although at short times (≤ 15 min) the surface activity of sodium caseinate lies intermediate between that for α_{s1}-casein and that for β-casein (2), the limiting value for caseinate in Table I is the same as that for β-casein. It seems that the two major caseins, α_{s1} and β, adsorb together in the early diffusion-controlled stage, but that β-casein predominates at the interface in the steady (equilibrium?) state after several hours. This is consistent with the recent observation (23) that β-casein can displace α_{s1}-casein from the surface of emulsion droplets over the same sort of time-scale.

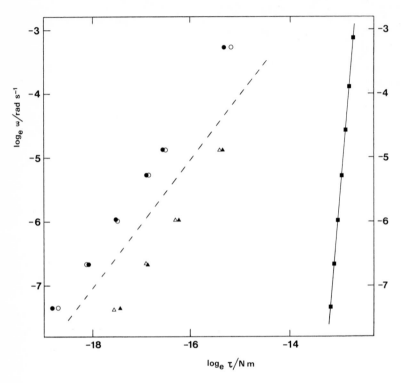

Figure 2: Comparison of the surface rheology of casein and
lysozyme at the n-hexadecane—water interface (10^{-3} wt % protein,
pH 7, 0.005 M, 25 °C). The logarithm of the angular rotation rate
ω (of the dish) is plotted against the logarithm of the torque τ
(on the disk): ●, ○, casein (duplicate runs, 8 h old, η = 3.55 ±
0.05 mN m^{-1} s); ▲, △, casein (duplicate runs, 50 h old, η = 12.6 ±
0.1 mN m^{-1} s); ■, lysozyme (η = 0.55 N m^{-1} s). Dashed line
represents Newtonian behavior (slope = 1); solid line represents
highly non-Newtonian behavior (slope ≈ 9).

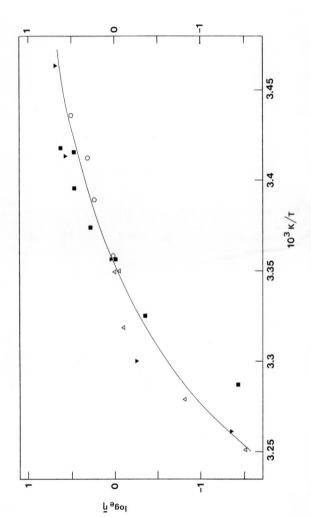

Figure 3. Temperature dependence of gelatin films of various ages at the n-hexadecane-water interface (10^{-3} wt % protein, pH 7, 0.005 M). The logarithm of the reduced surface viscosity $\tilde{\eta} = \eta(T)/\eta(298)$ is plotted against the reciprocal of the absolute temperature T: ■, $\eta(298) = 88$ mN m^{-1} s; △, $\eta(298) = 155$ mN m^{-1} s; ○, $\eta(298) = 253$ mN m^{-1} s; ▼, $\eta(298)$ (n-tetradecane) = 55 mN m^{-1} s.

Table I. Values of Steady-State Tension γ for Individual Proteins at n-Hexadecane—Water Interface (10^{-3} wt %, pH 7, 0.005 M, 25 °C) as Determined by the Wilhelmy Plate Method (Numbers in brackets refer to an ionic strength of 0.05 M)

Protein	γ/mN m^{-1}
(none)	53.3
gelatin	35.2
lysozyme	(30.3)
α-lactalbumin	28.5 (24.0)
α_{s1}-casein	24.0
κ-casein	23.0
β-casein	22.4
sodium caseinate	22.4 (19.5)

Let us now consider adsorption from mixed solutions of casein + gelatin. As far as the tension measurements are concerned, there are apparently three time régimes—short, medium and long. At short times (up to a few minutes at these protein concentrations), the change in tension is determined primarily by the protein's concentration and not its composition (2). At long times, the limiting tension approaches that of casein(ate), irrespective of the composition of the bulk protein solution (3). In the intermediate régime, which may last from a few minutes to several hours, there is a complex time-dependent change in tension reflecting the fact that, as time passes, more and more casein is available near the surface to compete with gelatin which has already adsorbed during the earlier stages. The less casein there is in the bulk, the longer it takes to reach the limiting tension—assuming, that is, that there is enough in the whole bulk phase to saturate the interface. At a total protein concentration of 10^{-3} wt %, the limiting tension is reached after 7—8 h when 25 % of bulk protein is casein, but it takes some 20 h when just 5 % of bulk protein is casein (3).

As casein(ate) is so much more surface-active than gelatin (see Table I), it is perhaps not surprising that it predominates in mixtures of the two. What, however, about mixtures of more well-matched partners like casein + lysozyme ($\Delta\gamma \approx 10$ mN m^{-1}) or casein + α-lactalbumin ($\Delta\gamma \approx 5$ mN m^{-1}) ? As $\Delta\gamma$ becomes reduced, we would expect the competitive edge of casein(ate) to become less marked, and indeed recent measurements in this laboratory (24) do confirm the expected trend. Under identical adsorption conditions (10^{-3} wt % total protein, pH 7, 0.05 M, 25 °C), the steady-state tension for a mixture of 25 % casein + 75 % lysozyme is the same as that for 100 % casein, whereas the steady-state tension for a mixture of 25 % casein + 75 % α-lactalbumin is closer to that of 100 % α-lactalbumin. So, while casein dominates the interface in casein + lysozyme (as in casein + gelatin), it does so to a much lesser extent in casein + α-lactalbumin, a mixture in which the surface activities of the two proteins are considerably closer together.

Relative surface activities of proteins go some way towards explaining the composition dependence of surface viscosity for films adsorbed from mixed solutions, but it is clear that other factors are also involved. Previously, we reported (1) that, while gelatin forms much more viscous films than casein (see Figure 1), when casein in the bulk is replaced <u>pro</u> <u>rata</u> by gelatin up to a level of <u>ca</u>. 90%, the surface viscosity at the oil—water interface is actually reduced. Figure 4 shows the composition dependence of apparent viscosity after 30 h for casein + gelatin, together with similar data for the other five binary mixtures formed from the set of casein, gelatin, lysozyme and α-lactalbumin. The minimum in the surface viscosity <u>versus</u> bulk composition graph for casein + gelatin could be due to a lowering in the protein surface concentration as gelatin replaces casein in the bulk. Possibly also there is a weakening of intermolecular forces in the adsorbed layer when interfacial casein is partially replaced by gelatin, even at very low levels. Eventually, of course, when most of the casein has been replaced, one expects gelatin—gelatin interactions to predominate and the viscosity to approach that of a pure gelatin film. Because casein is so much more surface-active than gelatin, however, this apparently does not occur until the bulk gelatin weight fraction is relatively high (say, 90—95%).

For clarity of presentation, the data in Figure 4 are separated into two groups: type I mixtures (Figure 4a), in which there is a minimum in surface viscosity <u>versus</u> bulk composition, and type II mixtures, where there is not. The first thing to note is that, although the tension measurements suggest that casein dominates the interface after 30 h in mixtures with both gelatin and lysozyme, only in the former case is there a minimum in the viscosity—composition plot. As with the proteins alone, there appears to be no simple correlation between surface activity and surface rheology. It may be significant, however, that type I behavior occurs in the non-lysozyme-containing mixtures (where each of the two proteins carries a net negative charge), whereas type II behavior is found with the lysozyme mixtures (where the two proteins are oppositely charged). It is tempting to infer that viscosity in a mixed protein film is increased by intermolecular electrostatic interactions between adsorbed protein molecules of opposite net charge. This view is prompted by the fact that casein + lysozyme and α-lactalbumin + lysozyme are incompletely soluble in 0.005 M phosphate buffer solution, and by the recent report (25) that addition of lysozyme to solutions of albumin or other acidic proteins greatly improves stability of foam lamellae at pH values between the isoelectric points. That having been said, our surface rheology results do not provide unequivocal evidence for strong complex formation at the oil—water interface, as only for gelatin + lysozyme is there a maximum in the viscosity—composition plot, and even then only a modest one at that. In addition, preliminary experiments at other ionic strengths in the range 0.01—0.05 M suggest that the type II plots are not very sensitive to the ionic strength, which is perhaps not what one might have anticipated if electrostatic interactions were indeed the overwhelming factor affecting the viscosity of the mixed films.

With regard to the surface rheology of films adsorbed from mixed protein solutions, the situation is, in fact, even more complex than has been indicated so far. With 10^{-3} wt% casein + gelatin solutions, while there is certainly a minimum in the viscosity—composition plot over many hours (well beyond the time at which the limiting tension is reached), it disappears at very long times. Take the case of a

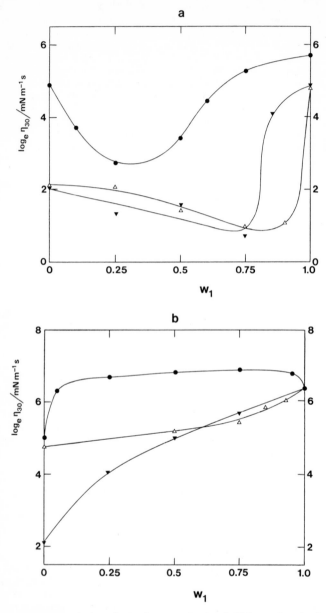

Figure 4: Composition dependence of mixed films adsorbed at the n-hexadecane—water interface (10^{-3} wt % protein, pH 7, 25 °C). The logarithm of the apparent viscosity η_{30} after 30 h is plotted against weight fraction w_1 of component 1. (a) Type I mixtures: ●, α-lactalbumin (1) + gelatin (2) (0.005 M); Δ, gelatin (1) + casein (2) (0.005 M); ▼, α-lactalbumin (1) + casein (2) (0.05 M). (b) Type II mixtures: ●, lysozyme (1) + gelatin (2) (0.05 M); Δ, lysozyme (1) + α-lactalbumin (2) (0.05 M); ▼, lysozyme (1) + casein (2) (0.05 M).

mixture of 50% casein + 50% gelatin: the film viscosity is lower than that for 100% casein for more than 50 h, but eventually it does cross the pure casein curve (see Figure 5), and after 80 h it is three times as large as the pure casein value. Much the same happens in a mixed system containing 25% casein, but it takes a little longer. One must infer, then, that type I behavior is a kinetic phenomenon; it does not necessarily reflect the equilibrium balance of protein—protein interactions in the mixed layer. We note, therefore, that, while the approach to steady-state surface rheology is slow in pure protein systems, it will tend to be even slower, as a general rule, in mixed protein systems.

In the last part of this paper, we consider what happens when an aged film of protein A is exposed to new protein B in the bulk aqueous phase. Figure 5 shows results for the case of gelatin injected (to give 10^{-3} wt%) below a 24 h old casein film which had previously been adsorbed from a 10^{-3} wt% casein solution. We observe no significant change in viscosity for about 50 h following addition of gelatin, but thereafter the viscosity increases strongly. Most interestingly, the data obtained after ca. 40 h are superimposable upon those found with 'normal' competitive adsorption from a mixture of 50% casein + 50% gelatin (10^{-3} wt% total protein); in each case zero time (t = 0) in Figure 5 is when the interface is first exposed to gelatin. These results strongly suggest that, in mixed casein + gelatin systems, the gelatin accumulates slowly at the primary casein film—under the influence of weak casein—gelatin interactions—to form a thickening gel-like secondary layer with rheology resembling that of a time-dependent bulk gelatin gel.

When gelatin is injected (to give 10^{-3} wt%) below a 24 h old casein film that has been adsorbed from a more dilute casein solution (2.5×10^{-4} wt%), we see from Figure 5 that there is a much steeper rise in measured viscosity after ca. 50 h than was found with the 10^{-3} wt% aged casein film. The implication is that, the more casein there is present in solution (and presumably therefore at the fluid interface), the harder it is for gelatin to accumulate in multilayers at the interface. Aging effects in pure gelatin films have been known for many years (7,26). What we have shown here is that they occur also in protein mixtures containing gelatin, even if the other component (casein) is much more surface-active. Whether this type of aging is unique to gelatin mixtures is still an open question.

In turning to the converse case of casein addition to an aged gelatin film, we move, rheologically speaking, from the sluggish to the catastrophic. Figure 6 shows results for the exposure of a 24 h old gelatin film to a solution containing 10^{-3} wt% casein. Just after the addition, the viscosity drops to very low values; but then it recovers fairly rapidly to produce a film more viscous than with 10^{-3} wt% casein alone. Sixty hours after the addition, the mixed film has a higher viscosity than the original gelatin film had had immediately prior to the event, but a lower viscosity than the gelatin film would have had, had the event not taken place (i.e., no casein added). When the same experiment is done with an aged lysozyme film replacing the gelatin one, there is again a sudden drop in viscosity, but this time only by ca. 40–50%, as shown in Figure 7. It is noticeable, for the case of the lysozyme film exposed to casein, that long-term changes in viscosity are relatively minor, especially in 0.005 M

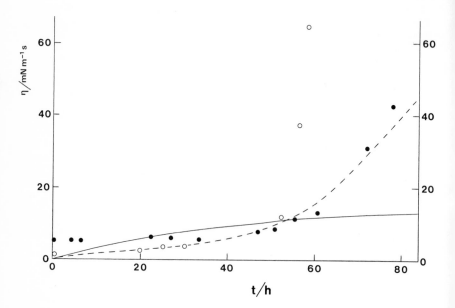

Figure 5: Gelatin addition to 24 h old casein films at the n-hexadecane—water interface (pH 7, 0.005 M, 25 °C). The apparent viscosity η is plotted against time t following the addition: $\bullet$, 10^{-3} wt % gelatin to 10^{-3} wt % casein film; o, 10^{-3} wt % gelatin to 2.5×10^{-4} wt % casein film. Solid curve represents 10^{-3} wt % pure casein film, and dashed curve represents 5×10^{-4} wt % casein + 5×10^{-4} wt % gelatin film, both as a function of adsorption time.

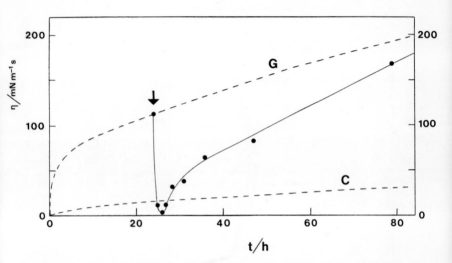

Figure 6:   Casein addition to a 24 h old gelatin film at the n-hexadecane—water interface (pH 7, 0.005 M, 25 °C).   The apparent viscosity η is plotted against time t following exposure of 10^{-3} wt % gelatin solution to fresh interface.   The arrow indicates the point at which 10^{-3} wt % casein is added.   The two dashed lines represent the behavior of pure casein (C) and pure gelatin (G).

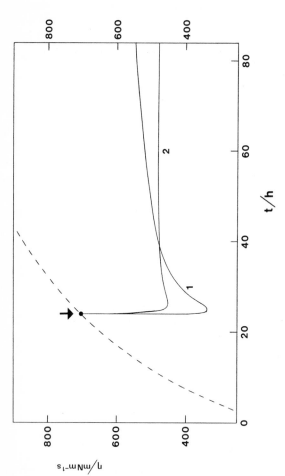

Figure 7. Casein addition to a 24 h old lysozyme film at the n-hexa-decane-water interface (pH 7, 25 °C). The apparent viscosity η is plotted against time t following exposure of 10^{-3} wt % lysozyme solu-tion to fresh interface: (1) 0.05 M, (2) 0.005 M. The arrow indicates the point at which 10^{-3} wt % casein is added. The dashed line represents behavior of pure lysozyme.

buffer, where protein aggregation in the bulk phase was indicated by a visibly turbid solution. This would seem to imply that lysozyme does not in fact form a gel-like secondary layer analogous to gelatin.

A detailed molecular interpretation of the protein displacement experiments must await further information on the packing density and structure of the molecules in the adsorbed layer. All we can say at present is that casein is able to displace gelatin completely from the primary interfacial layer, as one would expect on the basis of the interfacial tension results (4), and that gelatin from the film aged prior to the addition of casein is able to reform at the surface via casein—gelatin interactions to produce a composite film of higher viscosity than would have occurred without prior aging before addition of casein. Lysozyme films, on the other hand, are less easily disrupted by casein, perhaps due in part to some electrostatic complex being formed at the interface. Finally, it is interesting to note that, whereas in pure protein films the surface viscosity is a strongly increasing function of the surface pressure (27), in mixed films containing casein the surface viscosity may tend to become increased by reducing the surface pressure (of casein) (3).

Acknowledgments

We acknowledge financial support from the Chief Scientist's Group at the Ministry of Agriculture, Fisheries and Food (U.K.). The results are the property of the Ministry and are Crown Copyright.

Literature Cited

1. Dickinson, E.; Murray, B. S.; Stainsby, G. J. Colloid Interface Sci. 1985, 106, 259-262.
2. Dickinson, E.; Pogson, D. J.; Robson, E. W.; Stainsby, G. Colloids Surf. 1985, 14, 135-141.
3. Castle, J.; Dickinson, E.; Murray, A.; Murray, B. S.; Stainsby, G. In Gums and Stabilisers for the Food Industry; Phillips, G. O.; Wedlock, D. J.; Williams, P. A., Eds.; Elsevier Applied Science: London, 1986; Vol. 3, pp. 409-417.
4. Dickinson, E.; Murray, A.; Murray, B. S.; Stainsby, G. In Food Emulsions and Foams; Dickinson, E., Ed.; Royal Society of Chemistry: London, 1987; pp. 86-99.
5. Graham, D. E.; Phillips, M. C. J. Colloid Interface Sci. 1980, 76, 227-239, 240-250.
6. Mussellwhite, P. R. J. Colloid Interface Sci. 1966, 21, 99-102.
7. Kislalioglu, S.; Shotten, E.; Davis, S. S.; Warburton, B. Rheol. Acta 1971, 10, 158-162.
8. Izmailova, V. N. Progr. Surf. Membr. Sci. 1979, 13, 141-209.
9. Swaisgood, H. E. In Developments in Dairy Chemistry; Fox, P. F., Ed.; Applied Science: London, 1982; Vol. 1, pp. 1-59.
10. Brew, K.; Vanaman, T. C.; Hill, R. L. J. Biol. Chem. 1967, 242, 3747-3749.
11. Cumper, C. W. N.; Alexander, A. E. Trans. Faraday Soc. 1950, 46, 235-253.
12. Biswas, B.; Haydon, D. A. Proc. Roy. Soc. (London) 1963, A271, 296-316, 317-323.
13. Boyd, J. V.; Parkinson, C.; Sherman, P. J. Colloid Interface Sci. 1972, 41, 359-370.

14. Rivas, H. J.; Sherman, P. Colloids Surf. 1984, 11, 155-171.
15. Graham, D. E.; Phillips, M. C. In Theory and Practice of
    Emulsion Technology; Smith, A. L., Ed.; Academic Press: London,
    1976; pp. 75-98.
16. Phillips, M. C. Food Technol. 1981, 35, 50-57.
17. Chesworth, S. M.; Dickinson, E.; Searle, A.; Stainsby, G.
    Lebensm. Wiss. Technol. 1985, 18, 230-232.
18. Goodrich, F. C. Progr. Surf. Membr. Sci. 1973, 7, 151-181.
19. Dickinson, E.; Lam, W. L.-K.; Stainsby, G. Colloid Polym. Sci.
    1984, 262, 51-55.
20. Annan, W. D.; Manson, W. J. Dairy Res. 1969, 36, 259-268.
21. Zittle, C. A.; Custer, J. H. J. Dairy Sci. 1963, 46, 1183-1188.
22. Walstra, P.; Jenness, R. Dairy Chemistry and Physics; Wiley:
    New York, 1984; p. 117.
23. Dickinson, E.; Whyman, R. H.; Dalgleish, D. G. In Food Emulsions
    and Foams; Dickinson, E., Ed.; Royal Society of Chemistry:
    London, 1987; pp. 40-51.
24. Pickering, J. P. B. Sc. Final Year Project Thesis, University of
    Leeds, 1986.
25. Poole, S.; West, S. I.; Walters, C. L. J. Sci. Food Agric.
    1984, 35, 701-711.
26. Pouradier, J. J. Chim. Phys. 1949, 46, 627-634.
27. MacRitchie, F. J. Macromol. Sci., Chem. 1970, A4, 1169-1176.

RECEIVED January 28, 1987

Chapter 9

Human Erythrocyte Intrinsic Membrane Proteins and Glycoproteins in Monolayer and Bilayer Systems

M. N. Jones and R. J. Davies[1]

Department of Biochemistry, University of Manchester, Manchester M13 9PL, United Kingdom

The interfacial properties of human erythrocyte membrane components including glycophorin, the anion transporter (Band 3) and the glucose transporter in monolayer and bilayer systems is reviewed. Glycophorin in monolayers at the air-water interface undergoes reversible aggregation and in glycophorin-phospholipid monolayers can eliminate the liquid expanded (L_1) to intermediate state (I) lipid phase transition. Methods of incorporation of the anion and glucose transporters into planar bilayer lipid membranes (BLMs) are considered. Membrane proteins are shown to facilitate the transfer of lipid from proteolipid vesicles into monolayers at the air-water and oil-water interfaces.

Within the last 25 years there have been major advances in the characterisation of membrane proteins and glycoproteins particularly with regard to the structure and topography of proteins and glycoproteins of the human erythrocyte membrane (1-3). Applications of polyacrylamide gel electrophoresis in sodium n-dodecylsulphate (SDS-PAGE) (4) to the human erythrocyte plasma membrane reveals at least a dozen bands heavily stainable by Coomassie blue (protein bands) and approximately half as many bands stainable with periodic acid - Schiff's (PAS) reagent (glycoprotein bands). Although there are numerous faintly stained bands, relative to many plasma membranes the gel pattern of the erythrocyte is well characterised and there is a reasonable degree of concordance between quantitative estimates of particular membrane components (5).

Amongst the major components of the erythrocyte membrane are spectrin (designated by Fairbanks et al. (6) as Bands 1 and 2 on SDS-PAGE), the anion transporter (Band 3) and glycophorin, the major sialoglycoprotein of the membrane (7) first isolated by Marchesi and Andrews (8). Spectrin and Band 3 account for ~

[1]Current address: Department of Pediatrics, Addenbrooke's Hospital, Hills Road, Cambridge CB2 2QQ, United Kingdom

0097-6156/87/0343-0135$06.00/0

30% and ~ 25% of membrane protein by weight corresponding to ~ 4 x 10^5 and ~ 10^6 copies per cell respectively (9). There are 2.5 x 10^5 - 5 x 10^5 copies of glycophorin per erythrocyte membrane (1,6,10). Of the minor components the glucose transporter has received considerable attention (9,11). Although it has been designated a component of Band 3 having the same molecular weight (~90,000) and comigrating with the anion transporter (12), it is more generally regarded as having a molecular weight of ~ 55,000 and to be a component of region 4.5 of the gel profile (11).

The identification, isolation and characterisation of erythrocyte membrane proteins have made possible the study of reconstituted systems in which a particular membrane protein can be studied in a monolayer or bilayer environment free of the complications which arise in multicomponent membrane systems. Not only do reconstituted systems enable the interfacial behaviour of particular membrane proteins to be investigated but the properties of reconstituted systems can be used as a means of identification of membrane protein function and hence of identifying specific proteins.

A pre-requisite of reconstitution is the isolation of the protein (or glycoprotein) in relatively pure form; a process which for intrinsic membrane components requires the use of mild detergents (13) which will separate and solubilize the protein without initiating denaturation. In contrast to solubilization, reconstitution requires the protein to penetrate an aqueous monolayer or bilayer interface and hence some degree of detergent removal is required to facilitiate the reconstitution. It is the conflicting requirements of solubilization followed by protein pentration which makes reconstitution studies difficult. *In vivo* these problems are overcome by the synthesis and subsequent cleavage of hydrophobic signal sequences (14) however *in vitro* the process must be carried out by manipulation of the hydrophobic-hydrophilic balance in the protein-detergent complex.

The work discussed here is concerned with three intrinsic membrane components, glycophorin, the anion transporter (Band 3) and the glucose transporter, in monolayers, vesicles or liposomes and planar bilayer membranes (BLMs).

Glycophorin in Monolayers.

Glycophorin can be isolated from the erythrocyte membrane by partitioning extracts produced with either lithium diiodosalicylate (8) or deoxycholate (15) between phenol and water during which the glycophorin passes into the aqueous-rich phase. After removal of the lithium diiodosalicylate or deoxycholate by dialysis and further purification by precipitation with ethanol the glycophorin, which is soluble although partially aggregated (16,17) in aqueous media, can be isolated in a pure form. The material so produced is designated as glycophorin A although it also contains the structurally similar genetic variants B and C (18,19). Glycophorin A has a molecular weight of 31,000 and an amino acid sequence of 131 residues. The N-terminal domain (residues 1-74) is glycosylated and carries 16 oligosaccharide chains terminated by sialic acid residues and is largely responsible for the surface charge on the red blood cell[20]. The trans-membrane domain (residues 73-95) consists of 22 mainly hydrophobic amino acid residues and contains α-helical structure ([21-23]). The cytoplasmic C-terminal domain (residues 96-131) contains a large number of imino and acidic amino acids.

<u>Glycophorin at the Air-Water Interface</u>.   Glycophorin was prepared by the deoxycholate method ([15]). The tripartite structure of glycophorin gives no specific indication as to the surface properties of the molecule. By using ^{14}C-glycophorin we found that glycophorin adsorbs at the air-water interface and that the amount of adsorption increases with ionic strength of the substrate in the series H_2O < 0.1M NaCl < 1.0M NaCl < 1.6M $(NH_4)_2SO_4$ ([24]). Figure 1 shows the time course for adsorption at several aqueous interfaces at pH 7.4. Analysis of this data in terms of a model for diffusion-controlled adsorption and the equation ([25]).

$$ \Gamma \; = \; 2Cp \; \left(\frac{Dt}{\pi} \right)^{\frac{1}{2}} \qquad\qquad (1) $$

where Γ is the surface concentration, Cp the glycophorin concentration and t the adsorption time gives a diffusion coefficient $\sim$ 2 x 10^{-10} m^2 s^{-1}. The diffusion coefficient of monomeric glycophorin (molecular weigh 31,000) would be $\sim$ 1.1 x 10^{-10} m^2 s^{-2} while an aggregate of 10^6 molecular weight ([16,17]) would have a diffusion coefficient $\sim$ 0.4 x 10^{-10} m^2 s^{-1}. Thus like globular proteins ([25]) glycophorin diffuses to the air-water interface somewhat faster than expected for its size which may in part be a consequence of convective stirring. However, these observations give no indication that there is any potential barrier to adsorption at the interface.

Glycophorin can be spread at the air-water interface and desorption occurs on low ionic strength substrates and on high ionic strength substrates at high surface concentrations in excess of 2-3 mg m^{-2}, however at low surface concentrations in the range 0.04-0.5mg m^{-2} desorption is negligible and it is possible to obtain reproducible surface pressure-area isotherms as shown in Figure 2. The isotherms are unusual in exhibiting reproducible and reversible non-quantitative spreading; they become more expanded as the initial spreading concentration decreases in the range 0.5 to 0.05 mg m^{-2}. The results can be interpreted in terms of a surface aggregation model. If the limiting areas are extrapolated to zero initial surface concentration a surface area (A_0) of $\sim$ 80 nm^2 molecule^{-1} is obtained which can be used in the Boltzmann equation

$$ A_0 \; - \; A_i \; = \; A_0 \; e^{-\Delta G_{ass}/RT} \qquad\qquad (2) $$

where A_i is the limiting area at a given surface concentration, to obtain the Gibbs energy of association (ΔG_{ass}) in the monolayer. The Gibbs energies of association lie in the range

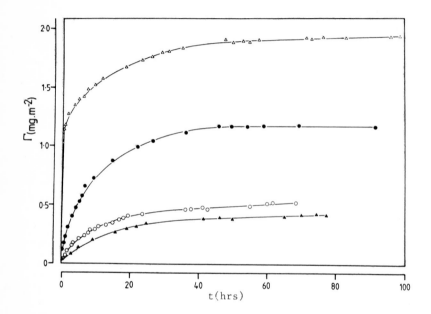

Figure 1.   Adsorption of ^{14}C-glycophorin as a function of time at
the aqueous-air interface at 22 °C. ▲, distilled water; O, 0.1M
NaCl, 20mM sodium phosphate pH 7.4; ●, 1.0M NaCl, 20mM sodium
phosphate pH 7.4; △, 1.6M ammonium sulphate pH 7.4. The glycophorin
concentration in the subphase was initially 0.303μg cm⁻³. (Reproduced
with permission from Ref. 24. Copyright 1983 Elsevier Science.)

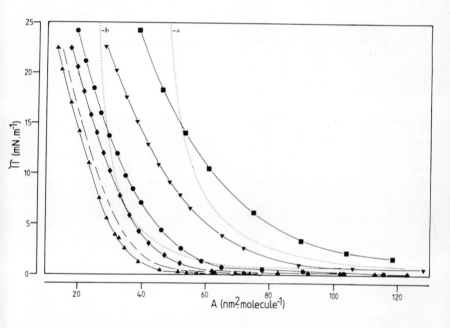

Figure 2.   Surface pressure-area isotherms for glycophorin on 1.6M ammonium sulphate, pH 1.2 at 20 °C as a function of initial surface concentration. ■, 0.055mg m^{-2}, ▼, 0.083 mg m^{-2}; ●, 0.221 mg m^{-2}, ◆, 0.227 mg m^{-2}, O, 0.332 mg m^{-2}, ▲, 0.498 mg m^{-2}.   Curve a is the Singer isotherm for glycophorin including contributions from the oligosaccharides.   Curve b is the Singer isotherm for the poly-peptide backbone of glycophorin. (Reproduced with permission from Ref. 24.   Copyright 1983 Elsevier Science.)

1-5 kJ mol^{-1} in the surface concentration range 0.5-0.05 mg m^{-2} corresponding to association constants of 0.66 to 0.13.

The structure of the glycophorin aggregates at the interface is not known, it might be envisaged that the association occurs between the trans-membrane hydrophobic domains. Surface pressure-area isotherms based on the Singer equation (26) give limiting areas of 46 nm^2 molecule^{-1} for the whole molecule assuming the 131 amino acid residues are α-helical and the oligosaccharide side chains are adsorbed at the interface (curve (a) in Figure 2) and 24 nm^2 molecule^{-1} if it is assumed that the oligosaccharide side chains are orientated into the substrate (curve (b) in Figure 2). Although these calculations are only crude what is clear is that the molecules must be held at the interface by more than the 22 amino acid trans-membrane domain which assuming it to be α-helical would have a limiting area of only 4nm^2 molecule^{-1}.

Glycophorin-Lipid Monolayers at the Air-Water Interface. Further to the study of pure glycophorin monolayers we investigated the interaction between the glycophorin and dipalmitoylphosphatidylcholine (DPPC) in mixed monolayers at the air-water interface (27). Pure DPPC undergoes the characteristic liquid expanded (L$_1$) to intermediate state (I) transition in monolayers at temperatures below the chain-melting temperature ($\sim$ 42°C) of DPPC. This phase transition is no longer observable in mixed monolayers at glycophorin to DPPC molar ratios in the range 1:355 to 1:44.

Figure 3 shows two isotherms for mixed monolayers on 1.6M ammonium sulphate at pH 7.0. From the surface pressure-area isotherms of glycophorin and DPPC alone under the same conditions, isotherms were calculated assuming additivity according to the equation.

$$A(\text{area per lipid}) = (N_{GP}A_{GP} + N_LA_L)/N_L \tag{3}$$

where N_{GP} and N_L are the numbers of glycophorin and lipid molecules having areas per molecule A_{GP} and A_L respectively at a given surface pressure π. The dotted lines in Figure 3 show the resulting calculated isotherms which display a DPPC phase transiton. Clearly the interaction between glycophorin and DPPC is sufficiently strong to eliminate the phase transition and give rise to marked deviations from simple additive behaviour. Furthermore, at low areas per lipid molecule the deviations from addivity are negative and the area occupied by the DPPC molecules is much lower than that calculated from equation (3).

At very high molar ratios of DPPC to glycophorin in the range 1100-1700:1 the phase transition is still observable which suggested that there is a critical ratio of lipid to glycophorin above which excess lipid is capable of undergoing the phase transition. To determine the critical ratio the contraction in the area per lipid at a given surface pressure can be plotted as a function of glycophorin to lipid molar ratio. Such plots are linear and when extrapolated to zero contraction gave critical ratios in the range 1300 ($\pi \simeq$ 30 mN m^{-1}) to 210 (π = 15

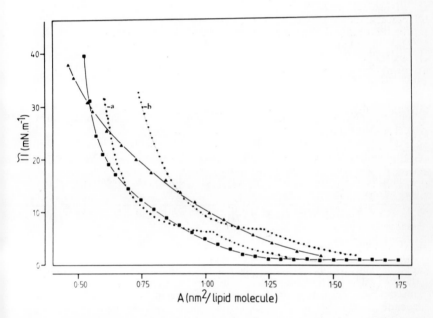

Figure 3. Surface pressure (π) as a function of area per lipid molecule (A) for mixed monolayers of glycophorin and L-α-dipalmitoyl-phosphidylcholine at molar ratios of 1:355.(■) and 1:178 (▲) on 1.6M ammonium sulphate, pH 7.0 at 20 °C. The initial surface concentration of glycophorin was 0.083 mg m^{-2}. The dotted curves a and b were calculated from the π-A isotherms of the pure components, assuming additivity. (Reproduced with permission from Ref. 27. Copyright 1984 Elsevier Science.)

mNm^{-1}). These figures correspond to the number of DPPC molecules immobilized by a glycophorin molecule and are of the same order of magnitude found from calorimetric studies on DPPC-glycophorin multilamellar phases (see below).

Glycophorin in Bilayer Systems

Thermotropic Properties. Glycophorin can be incorporated into multilamellar liposomes by hydration of a glycophorin-lipid film prepared by rotary evaporation of a mixture of glycophorin and lipid in an azeotropic solvent (chloroform/methanol/water, 81:15:14% v/v bpt 52.6°C). We found that the azeotropic solvent mixture facilitated uniform dispersion of the glycophorin in the lipid (28). The thermotropic properties of multilamellar liposomes of dimyristoylphosphatidylcholine (DMPC), dipalmitoylphosphatidyl choline (DPPC) and distearoylphosphatidylcholine (DSPC) incorporating glycophorin have been studied by differential scanning calorimetry (DSC) in the glycophorin to lipid molar ratio (GP/L) range upto approximately 5 x 10^{-3} (28,29).

The incorporation of glycophorin into these lipid multilamellar systems results in a decrease in the enthalpy (ΔH_0) of the gel to liquid crystalline phase transition at constant temperature. Interpretation of the data in terms of the relationship

$$\Delta H = \Delta H_0 \ (1\text{-}N \ (GP/L)) \tag{4}$$

gives values for the number of lipid molecules (N) withdrawn from participation in the phase transition. The values found (29) were as follows, 42 ± 22 (DMPC), 197 ± 28 (DPPC) and 240 ± 64 (DSPC). The extent of disruption in the gel phase of the lipid caused by the incorporation of glycophorin molecules is thus dependent on the lipid acyl chain length. Although there are more sophisticated treatments of the thermotropic behaviour of protein lipid systems (30-32) this approach was found to be helpful in interpretation of the agglutination of glycophorin containing liposomes discussed below. It should be stressed that the use of equation (4) does not imply that the thermotropic behaviour necessarily supports a boundary lipid model (30).

Lectin Mediated Agglutination of Liposomes Containing Glycophorin. Using a similar procedure to that described above glycophorin can be incorporated into sonicated phospholipid liposomes. The orientation of the glycophorin in the bilayer is not known with certainty although van Zoellen *et al.* (33) found that 75-80% of the glycophorin molecules are orientated with their N-terminal oligosaccharide chains on the outside of small unilamellar liposomes. Photon correlation spectroscopy showed that the liposome size increases with glycophorin incorporation (29).

Sonicated liposomes incorporating glycophorin can be aggregated by wheat germ agglutinin (WGA) (29). WGA, a lectin with molecular weight 36,000, has multiple binding sites for N-acetylneuraminic which is the terminal sugar residue in the oligosaccharide chains of glycophorin. The agglutination process is easily followed spectroscopically, the rate is second order in lipid concentration and can be reversed on swamping the lectin binding sites with N-acetylglucosamine. A particularly interesting feature of the agglutination process is that the rate of agglutination decreases linearly with temperature below the gel to liquid-crystalline phase transition of the lipid (Figure 4) (28,34), but becomes approximately independent of temperature above the phase transition. Below the chain-melting temperature (T_c) the rates of agglutination decrease in the series DSPC > DPPC > DMPC, whereas above T_c the rates are all similar. These effects would seem to be predominantly related to glycophorin-acyl chain interactions rather than to glycophorin-lipid head group interactions.

It is well established that proteins in bilayers are clustered below T_c and hence it follows that the clustering of glycophorin (29) increases the agglutination rate. However, the data require a model in which there is an optimum packing of glycophorin molecules to give the fastest rate. From the DSC data the extent of gel phase disruption follows the same sequence as the rate of agglutination. The amount of disrupted lipid in the glycophorin-rich phase will influence the packing of the glycophorin and "tight" packing as in the case of DMPC would restrict binding of WGA and hence reduce the rate of agglutination by a bridging mechanism between liposomes.

The geometric constraints on these bilayer systems are shown in Figure 5. It follows that of the lipids considered DMPC has acyl chain dimensions which are closest to the hydrophobic trans-membrane domain of glycophorin. To bring the acyl chains into register with the trans-membrane sequence requires the introduction of trans to gauche kinks in the chain, two such conformation changes introduces a "2gl kink". As shown in Figure 5 as the acyl chain length increases in the sequence DMPC > DPPC > DSPC more 2gl kinks would have to be introduced with a concomitant increase in disruption of the surrounding lipid and a greater separation between the glycophorin molecules, favouring WGA binding and increasing the rate of agglutination.

Penetration of Erythrocyte Membrane Proteins into Planar Bilayers

Band 3 and the Glucose Transporter. The advent of methods of forming bilayer lipid membranes (BLMs) across an orifice separating two aqueous phases pioneered by Mueller *et al.* (35) and Tien (36) made possible the reconstitution of membrane transport systems *in vitro*. In principle the idea of preparing erythrocyte plasma membranes, solubilizing and fractionating membrane proteins and assaying, for example, transport function in BLMs has great attraction. In practice the experiments were far from easy. An inherent disadvantage of BLMs is their incipient instability particularly when exposed to surface active materials. Solubilization of membrane proteins requires detergents so that there were considerable difficulties to be overcome.

Our first attempts to incorporate erythrocyte membrane proteins into BLMs were made using extraction procedures based on organic solvents specifically acetic acid-pyridine, n-butanol and

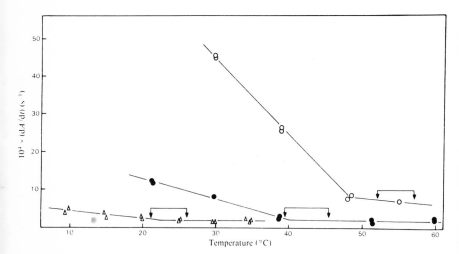

Figure 4. Temperature dependence of the initial rate of liposome agglutination by wheat-germ agglutinin (WGA). The liposomes contained glycophorin at a glycoprotein:lipid molar ratio of 2×10^{-3}: 1.○ , Distearoylphosphatidylcholine; ●, Dipalmitoylphosphatidylcholine △, Dimyristoylphosphatidylcholine. Arrows denote the temperature range over which chain-melting is observed by differential scanning calorimetry. (Reproduced with permission from Ref. 34. Copyright 1982 The Biochemical Society, London.)

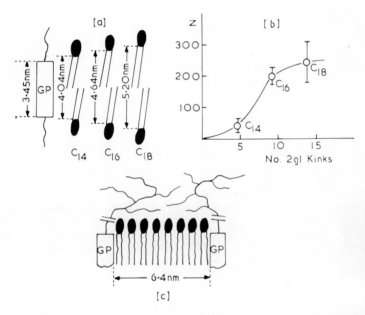

Figure 5. (a) Relative dimensions of the transmembrane section of glycophorin and phospholipid bilayers (to scale). (b) The number of disordered lipid molecules (N) determined by DSC as a function of the number of 2gl kinks required to bring the acyl chains into register with the dimensions of the transmembrane section of glyco-phorin. (c) Relative spacing of glycophorin molecules in clusters in DMPC bilayers. (The N-terminal oligosaccharide-bearing segment of glycophorin is not to scale.) (Reproduced with permission from Ref. 29. Copyright 1982 Elsevier Science.)

pentanol (37). The objective was to reconstitute stereospecific D-glucose transport in a BLM, but none of the organic extracts expressed any significant D-glucose transport activity. However, development of detergent extraction methods based on Triton X-100 particularly by Yu *et al.* (38) enabled us to achieve stereospecific D-glucose transport across BLMs (39-44).

Triton X-100 extracts of erythrocyte membrane proteins can be incorporated into BLMs provided excess Triton X-100 is removed by gel filtration (Sephadex G50) to bring the free detergent level to approximately $1\mu g$ cm^{-3}. By monitoring the electrical characteristics of the BLM on exposure to Triton X-100 solubilized protein extracts the penetration process can be followed. Figure 6 shows the capacitance and resistance of a BLM (equimolar egg lecithin and cholesterol) on exposure of one interface to a protein extract obtained by incubation of the erythrocyte membrane with 8mM Tris-HCl buffer containing 0.5% (w/v) Triton X-100 (40). The extract contains an appreciable fraction of Band 3 the anion transporter and on penetration into the bilayer there is a permanent decrease in the resistance. The permanence of the resistance change is proof that the change is due to transporter protein and not detergent (40).

Initial experiments were done with relatively crude extracts containing numerous membrane proteins. As well as demonstrating ion transport, D-glucose transport could also be demonstrated with the same membrane extraction procedure and BLMs having a D-glucose permeability of upto approximately 50 times that of the passive permeability (P_0 = 4.51 ± 0.95 x 10^{-8} cm s^{-1}) were obtained. More refined membrane extracts containing only the sugar transporter can be obtained and the penetration of this membrane protein (Band 4.5) into a BLM can be followed by monitoring the increase in permeability (P) to D-glucose based on the equation

$$C_0V_0/C_i \; A = Pt \qquad\qquad (5)$$

where C_0 and C_i are the sugar concentrations on the trans and cis sides of the bilayer of area A at time t. Figure 7 shows a permeability plot based on equation (5). The slope of the plot of $(C_0V_0/C_i \; A)$ vs. t gives the permeability of the BLM to radiolabelled D-glucose. The change from passive permeability (P_0) to carrier mediated permeability (P) on penetration of the bilayer by the transporter protein is clearly seen. No such change occurs for L-glucose confirming that the process is stereospecific.

The dependence of BLM permeability on the concentration of protein extract (μg cm^{-3}) in the solution bathing the BLM obeys the linear relationship (43).

$$P_{relative} = P/P_0 = 5.83 \; (\pm 0.65)[Protein] + 1.14 \; (\pm 0.63) \quad (6)$$

suggesting that the solubilized extract partitions into the BLM. It should be noted that in these experiments the protein penetration was from only one side of the bilayer. The bilayers become unstable when attempts were made to add the protein symmetrically to both interfaces. A number of questions remain

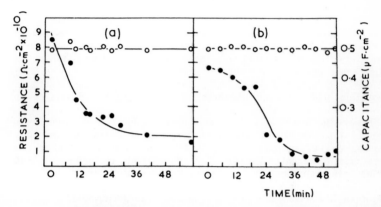

Figure 6. Time dependence of resistance (●) and capacitance (○) of bilayer lipid membranes on asymmetric addition of Triton X-100 membrane protein extracts at 27 °C. (a) Bathing solution 0.1M Kcl, pH 6.4. Protein concentration 18 μg cm^{-3}, Triton X-100 concentration 12.5 μg cm^{-3}. (b) Bathing solution 0.1M NaCl, pH 6.4. Protein concentration 17.2 μg cm^{-3}, Triton X-100 concentration 10.9 μg cm^{-3}. (Reproduced with permission from Ref. 40. Copyright 1978 Elsevier Science.)

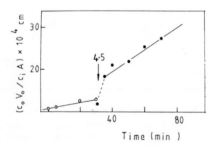

Figure 7. Sugar permeability plot for bilayer lipid membrane (egg lecithin-cholesterol in n-decane) at 25 °C. The slope of the plot before and after addition of extract is equal to the permeability coefficient. Passive diffusion of D-[^{14}C]glucose (○) and facilitated diffusion (●) on addition of band 4.5 (sugar transporter) at a concentration of 0.99 μg cm^{-3} to the trans side of the bilayer. (Reproduced with permission from Ref. 44. Copyright 1982 Elsevier Science.)

to be answered relating to this type of experiment; specifically the form of Triton X-100 solubilized protein extracts, the mechanism of the penetration process, the concentration of the protein in the bilayer interface and finally the conformation within the bilayer. Taking cognizance of the inherent instability of BLMs, it was considered that at least some of the factors relating to the penetration of the solubilized extracts might be studied more easily at a monolayer interface.

Penetration of Solubilized Membrane Extracts into Monolayers

Studies on the penetration of erythrocyte membrane proteins were carried out using Triton X-100 extracts fractionated by ion-exchange chromatography followed by gel filtration to remove excess free detergent. The protein, cholesterol, phospoholipid and Triton X-100 content of the extracts were determined and the protein composition was established by SDS-PAGE (45). The major protein in these extracts was the anion transporter, Band 3, but they also contained glycophorin which copurified with Band 3. A typical extract had compositon: protein ($200\mu g$ cm^{-3}), cholesterol ($35\mu g$ cm^{-3}), phospholipid ($40\mu g$ cm^{-3}) and Triton X-100 ($\leqslant 5\mu g$ cm^{-3}). The lipid to protein molar ratio in the extracts was ~ 50-60:1 It was not possible to reduce the lipid in the extracts to lower levels with prolonged detergent washing implying that the residual lipid is tightly bound to the protein. Electron microscopy of the extracts revealed approximately spherical particles with diameters ~ 200nm which suggested that the extract could be regarded as proteolipid vesicles.

Penetration into Monolayers at the Air-Water and Oil-Water Interface.
Membrane extracts as described above penetrate monolayers at the air-water interface reaching equilibrium surface pressures in 10-30 minutes depending on the protein concentration in the substrate. Figure 8 shows the equilibrium spreading pressures as a function of protein concentration for penetration into monolayers with an initial surface pressure of 10mN m^{-1}. The curves are similar for DPPC, cholesterol and cholesterol:DPPC (2:1 molar ratio) monolayers and reach a limiting spreading pressure of 42-43 mN m^{-1}. The limiting spreading pressures approach the collapse pressures of the pure lipid monolayers; DPPC (50mN m^{-1}), cholesterol (44mN m^{-1}) and cholesterol:DPPC (46mN m^{-1}).

It was important to establish that the observed increases in pressure could not be accounted for by penetration of free lipid or Triton X-100 present in the substrate at comparable concentrations as would arise on introducing the protein extracts. With this objective control experiments demonstrated that free Triton X-100 would lead to surface pressure increases of 0.2-0.5mN m^{-1}. It follows that the large increases in surface pressure as demonstrated in Figure 8 could not arise in the absence of the intrinsic membrane protein.

Measurements of surface pressure and surface potential isotherms of monolayers penetrated by membrane extracts showed that the presence of the protein increased the compressibility of

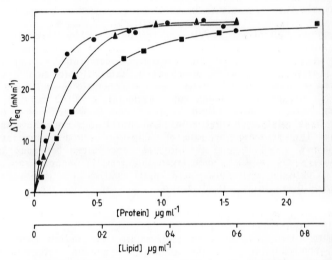

Figure 8. The equilibrium surface pressure increases reached after interaction of protein extract with films of cholesterol (●), DPPC (■), and cholesterol:DPPC (2:1 molar ratio,▲) at an air-water interface. The initial surface pressure of the monolayers was 10mN m^{-1}; the aqueous substrate was 20mM sodium phosphate/0.1M NaCl (pH 7.4). (Reproduced with permission from Ref. 45. Copyright 1986 Elsevier Science.)

the monolayer by a factor of 8 and the surface potential by approximately 50mV.

The monolayer at the oil-water interface has been argued to be a better model of one half of a lipid bilayer (46,47). Penetration of membrane extracts at the n-decane-water interface occurred in a similar manner to penetration at the air-water interface. The limiting equilibrium spreading pressure was however lower, 33 mN m^{-1} compared with 43 mN m^{-1}. Triton X-100 adsorbs more readily at the n-decane-water interface however, it would account for no more than an increase of 2mN m^{-1} in surface pressure. Liposomal lipid adsorbs less readily than at the air-water interface and would account for less than 0.2mN m^{-1} surface pressure. Thus as at the air-water interface protein is responsible for the increase in equilibrium spreading pressure.

Mechanism of Proteolipid Vesicle Penetration into Monolayers. The principle conclusion from the penetration studies at the air-water and oil-water interfaces is that intrinsic membrane protein in vesicles greatly facilitates the transfer of material into monolayers. In marked contrast lipid vesicles do not penetrate monolayers to any appreciable extent although some exchange of lipid between a monolayer and the outer lipid layer of a liposome can occur (48,49). It is established that both glycophorin (50) and the anion transporter (51) increase the rate of "flip-flop" when incorporated into bilayers. Thus in the initial encounter between the proteolipid vesicles and the monolayer the protein-enhanced rate of "flip-flop" between the inner and outer halves of the vesicle bilayer would facilitate lipid transfer to the monolayer. The process of redistribution of lipid between vesicle and monolayer would bring the protein into intimate contact with the monolayer leading to penetration.

Future studies in this area will require the radiolabelling of intrinsic membrane so that the amount of protein entering the monolayer can be accurately measured. More information should also be obtained on the dependence of the initial monolayer surface pressure on protein mediated fusion of vesicles with monolayers. The surface pressure at which the properties of a monolayer most closely mimic the properties of a bilayer is known to be relatively high. It is clear from our studies that protein extracts penetrate monolayers upto equilibrium surface pressures approaching the monolayer collapse pressure which suggests that data can be obtained from monolayer studies at surface pressures which are directly applicable to bilayers.

Literature Cited

1. Steck, T.L. J. Cell Biol. 1974, 62, 1-19.
2. Marchesi, V.T.; Furthmayr, H.; Tomita, M. Ann. Rev. Biochem. 1976, 45, 667-698.
3. Quagliariello, E.; Palmieri, F. Editors. "Structure and Function of Membrane Proteins". Developments in Bioenergetics and Biomembranes, 1983, 6.
4. Laemmli, U.K. Nature, 1970, 227, 680-685.
5. Jones, M.N.; Nickson, J.K. FEBS Letters. 1980, 115, 1-8.

6. Fairbanks, G.; Steck, T.L.; Wallach, D.F.H. Biochemistry, 1971, 10, 2606-2616.

7. Lux, S.E. Nature, 1979, 281, 427-429.

8. Marchesi, V.T.; Andrews, E.P. Science, 1971, 174, 1247-1248.

9. Jones, M.N.; Nickson, J.K. Biochem. Biophys. Acta. 1981, 650, 1-20.

10. Marchesi, V.T. Semin. Hematol. 1979, 16, 3-20.

11. Carruthers, A. Prog. Biophys. Molec. Biol. 1984, 43, 33-69.

12. Shelton, R.L.; Langdon, R.G. Biochemistry, 1985, 24, 2397-2400.

13. Helenius, A.; Simons, K. Biochim. Biophys. Acta. 1975, 415, 29-79.

14. Houslay, M.D.; Stanley, K.K. in "Dynamics of Biological Membranes" John wiley & Sons, Chichester, 1982, pp. 235.

15. Segrest, J.P.; Wilkinson, T.M.; Sheng, L. Biochim. Biophys. Acta. 1979, 554, 533-537

16. Springer, G.P.; Nagi, Y.; Tetmeyer, H. Biochemistry, 1966, 5, 3254-3272.

17. Morawiecki, A. Biochim. Biophys. Acta, 1964, 83, 339-347.

18. Furthmayr, H. J. Supramol. Struct. 1978, 9, 79-95.

19. Fruthmayr, H. Nature, 1978, 271, 519-524.

20. Eylar, E.H.; Madoff, M.A.; Brody, O.V.; Oncley, J.L. J. Biol. Chem., 1962, 237, 1992-2000.

21. Segrest, J.P.; Kohn, L.P. in H. Peeters (Ed.) "Protides of the Biological Fluids", Pergamon Press, New York, 1973, p. 183.

22. Segrest, J.P. in A. Jamieson and D. Robinson (Eds) "Surface Membranes of Specific Cell Types" Butterworths, New York, 1977 3, 1.

23. Welsh, E.J.; Thom, D.; Morris, E.R.; Rees, D.A. Biopolymers, 1985 24, 2301-2332.

24. Davies, R.J.; Goodwin, G.C.; Lyle, I.G.; Jones, M.N. Colloids and Surfaces, 1983, 8, 29-43.

25. Graham, D.E.; Phillips, M.C. J. Colloid Interface Sci., 1979, 70, 403-414.

26. Singer, S.J. J. Chem. Phys. 1948, 16, 872-876.

27. Davies, R.J.; Goodwin, G.C.; Lyle, I.G.; Jones, M.N. Colloids and Surfaces, 1984 8, 261-270.

28. Goodwin, G.C.; Jones, M.N. Biochem. Soc. Trans. 1980, 8, 323-324.

29. Goodwin, G.C.; Hammond, K.; Lyle, I.G.; Jones, M.N. Biochim. Biophys. Acta. 1982, 689, 80-88.

30. Jones, M.N.; Skinner, H.A. Ann. Rep. Royal Soc. Chem. 1982, 3-39.

31. Kapitza, H.G.; Ruppel, D.A.; Galla, H.J.; Sackmann, E. Biophys. J. 1984, 45, 577-587.

32. Mouritsen, O.G.; Bloom, M. Biophys. J. 1984, 46, 141-153.

33. Van Zoellen, E.J.J.; Verkleij, A.J.; Zwaal, R.F.A.; van Deenen, L.L.M. Eur. J. Biochem. 1978, 86, 539-546.

34. Goodwin, G.C.; Hammond, K.; Lyle, I.G.; Jones, M.N. Biochem. Soc. Trans. 1982, 10, 47-48.

35. Mueller, P.; Rudin, D.O.; Tien, H.T.; Wescott, W.C. Recent Progress in Surface Science, 1964, 1, 379-393.

36. Tien, H.T. in "The Chemistry of Biosurfaces. Ed. M.L. Hair, Marcel Dekker Inc., New York, 1971, vol. 1, p. 282.

37. Lidgard, G.P.; Jones, M.N. J. Membrane Biol. 1975, 21, 1-10.
38. Yu, J.; Fischman, D.A.; Steck, T.L. J. Supramol. Struct. 1973, 1, 233-247.
39. Nickson, J.K.; Jones, M.N. Biochem. Soc. Trans. 1977, 5, 147-149.
40. Jones, M.N.; Nickson, J.K. Biochim. Biophys. Acta. 1978, 509, 260-271.
41. Jones, M.N.; Nickson, J.K.; Phutrakul, S. Proc. of the 25th Colloquium "Protides of the Biological Fluids". ed. H. Peeters. Pergamon Press Oxford, 1978, 43-46.
42. Phutrakul, S.; Jones, M.N. Biochem. Biophys. Acta. 1979, 550, 188-200.
43. Jones, M.N.; Nickson, J.K. Biochem. Soc. Trans. 1982, 10, 5-6.
44. Nickson, J.K.; Jones, M.N. Biochim. Biophys. Acta. 1982, 690, 231-40.
45. Davies, R.J.; Jones, M.N. Biochim. Biophys. Acta. 1986, 858, 135-144.
46. Ohki, S.; Ohki, C.B. J. Theor. Biol. 1976, 62, 389-407.
47. Gruen, D.W.; Wolfe, J. Biochim. Biophys. Acta. 1982, 688, 572-580.
48. Schindler, H. Biochim. Biophys. Acta. 1980, 55, 316-336.
49. Jahnig, F. Biophys. J. 1984, 46, 687-694.
50. Gerritson, W.J.; Henricks, P.A.J. Biochim. Biophys. Acta. 1980, 600, 607-619.
51. DeKruijff, B.; van Zoelen, E.J.J.; van Deenen, L.L.M. Biochim. Biophys. Acta. 1978, 509, 537-542.

RECEIVED January 13, 1987

MECHANISMS OF PROTEIN-INTERFACE INTERACTIONS

Chapter 10

Why Plasma Proteins Interact at Interfaces

L. Vroman and A. L. Adams

Interface Laboratory, Veterans Administration Medical Center, 800 Poly Place, Brooklyn, NY 11209

We found certain proteins able to deposit multimolecular layers between 2 surfaces separated at a dis-
ance inversely proportional to protein concentration.
"Excess" deposited in wider gaps soon redissolves,
perhaps by "self-removal". Blood plasma injected between a convex lens and a clot-promoting (glass or
oxidized metal) surface leaves concentric rings of
proteins. These apparently displace each other in the
following sequence: albumin, immunoglobulin G, fibrinogen and fibronectin, high molecular weight kininogen
and factor XII - i.e. from most concentrated to least
concentrated protein. Computer programs based on "self-
removal" and physiological protein concentrations only,
do yield realistic models; experiments with binary
protein solutions indicate that more specific properties
of each protein are involved in this interaction.

Each of our proteins evolved along with the structures that are
reproducing it and along with the mobile cell membranes and sub-
cellular objects that it must, from its birth to its death, exist
with.

In sharp contrast to things in vivo, most laboratory surfaces
are created to be a) uniform, so that they will allow very few
choices of scenery for a protein molecule to attach itself any-
where on them, and b) unchanging, remaining so rigid and so constant with time that they will not adapt to the submolecular de-
sires of the protein they adsorb: the protein molecule will have
to adapt to the artificial substrate instead. Such large, uniform
and uncompromising material surfaces force the protein solution to
deposit a measurably large number of uniformly adapting molecules.
Their synchronized behavior will thus be amplified enough to be
measurable but may not represent physiological events.

Cell membranes have evolved to be most unlike any manmade
surface not only by becoming variable and adaptable on a molecular

and submolecular scale, but also by their becoming dotted with receptors for specific proteins. Thus, living cells may adsorb and
retain only specific proteins out of their physiological liquid environment and only under very specific conditions.

Caught within the relatively dense structure of a cell, or in
the varying and often narrow spaces among cells, a protein molecule
must be expected to encounter a number of situations, each of which
deserves to be imitated with simplifying models in the laboratory.
Limiting ourselves to 2 proteins (P1 and P2), and 2 kinds of surfaces (Sa and Sb), we may list the following choice of possible
interactions (commas separating non-preadsorbed proteins).

1) P1, Sa and P1, P1Sa
2) P1, P2Sa (where P2 is adsorbed onto Sa)
3) P1, P2, Sa (competition for 1 surface)
4) Sa, P1, Sa (1 protein in a narrow space between identical
surfaces)
5) Sa, P1, P2, Sa (2 proteins competing in a narrow space between
identical surfaces)
6) Sa, P1, Sb (2 different surfaces competing for 1 protein)
7) Sa, P1, P2, Sb
8) SaP1, SaP1 (interaction among 2 mobile identical particles coated
with the same protein)
9) SaP1, SaP2
10) SaP1, SbP1 (interaction among 2 surfaces carrying the same
protein)
11) SaP1, SbP2
12) SaP1, P2, SbP1

Somewhat more physiological conditions can be created by replacing at least 1 of the 2 proteins by blood plasma. The ability
of a surface to adsorb a certain protein, e.g. a clotting factor,
out of normal plasma, appears to be demonstrated by the following
deceptively simple manipulations. Incubate normal citrated "intact"
plasma (plasma that had never been in contact with a negatively
charged, high free energy surface such as glass) in a glass test
tube. Rinse the tube out and place citrated plasma in it from a
patient who lacks clotting factor XII. After adding calcium
chloride, we will find the clotting time of this deficient plasma
to be corrected by the adsorabte that the normal plasma had left
on the glass (1). We cannot help but conclude that normal plasma
deposits factor XII on glass.

Finding one protein deposited by plasma should not lead us to
believe - as I unfortunately did for many years - that no other
protein will be present at other locations on the surface at the
same time, or at the same location at other times. Having detected
factor XII at coagulation-activating surfaces, it took us 5 years
plus Dr. S. Witte's suggestion before we started looking for fibrinogen being deposited by plasma. We did find it, and another 12
years later we found evidence of an entire shifting population of
proteins replacing one another at the solid/plasma interface.

In the following description of early studies as well as of our
recent findings, I will use the classification of protein/surface
interactions given above. Among the proteins to be discussed
fibrinogen will be prominent, not only because of our own work, but
also because this work fits rather well within the context of others'
studies showing protein, plasma and cellular (platelet) interactions.

ONE PROTEIN ON ONE SURFACE

a) The air interface
     Most protein solutions tend to form a monomolecular layer at
their air/liquid interface at a rate depending on their concentration. Apolar amino acid residues will become exposed on the surface
of the film that faces air. Therefore, one can imagine that a drop
of protein solution suspended in air will rapidly form a skin in
which it will be trapped. If placed onto a hydrophobic solid, such
a drop will be held by its hydrophobic skin surface, and where it is
forced to roll along, the drop will "break out of its bag" of skin,
leaving pieces of skin stuck in its trail while forming a new bag.
We detected the phenomenon e.g. by observing the interference color
pattern left by protein drops landing on and rolling over ferric
stearate polished anodized tantalum slides.
     This Blodgett-Langmuir transfer is almost instantaneous (30).
To observe adsorption from bulk solution, we must insert the hydrophobic substrate into the aqueous medium before adding protein.
Under such conditions, the recording ellipsometer shows adsorption
rates out of bulk solutions onto hydrophobic substrates that are not
unlike those on hydrophilic ones (2). Where the hydrophobic solid
slide is then dipped deeper into the container of protein, data and
recordings show that the slide has gained thickness instantaneously
and must therefore have dragged a surface film down into the light-
path. This film may appear not much thinner than the one that had
been more slowly adsorbed out of the bulk solution (2).
     Proteins adsorbed onto a hydrophobic solid tend to render it
more hydrophilic, while the same proteins adsorbed ont a hydrophilic
solid tend to render it more hydrophobic. On the dried film of adsorbate, the contact angle of a water droplet, or the scattering of
light by condensing water vapor, appears to demonstrate this property (3), but the shape of these slowly spreading droplets is affected by protein being "scooped up" at their advancing air/water/
solid interfaces (4).
     Among plasma proteins we studied at solution/air interfaces,
fibrinogen behaved most remarkably. Compressed in a surface film
balance (5), fibrinogen tended to form a cohesive film that could be
lifted off the interface with a hydrophobic slide (6). The finding
suggests that fibrinogen may form polymer-like complexes at certain
interfaces even in absence of thrombin (7), the enzyme normally
needed to polymerize fibrinogen into fibrin.

b) One protein at the liquid/solid interface (Pl, Sa and Pl, PlSa)
     Though we realized early (2) that adsorption of proteins may be
less reversible than we had expected, other techniques than ours
were required to show that at least fibrinogen adsorbed onto certain
surfaces (8) will be exchanged with fibrinogen molecules in solution.
Since elutability appears to decrease with increased time of residence at the interface (9), some gradual change in conformation may
occur as the adsorbed molecule perhaps increases its area of contact
with the immobile substrate. We have some evidence that fibrinogen,
adsorbed onto certain substrates, will be able to adsorb a mono-

clonal antibody to part of its E domain, while the same antibody does not bind fibrinogen in solution. This indicates that adsorption can expose a group of amino acids that is normally not exposed in solution (10).

ONE PROTEIN ON TWO IDENTICAL SURFACES (SaPl, Pl, Sa etc.)

The fact that a film of protein adsorbed onto 1 surface can still adhere to another surface is proved simply by exposing the film to a suspension of solid particles, e.g. of a metal oxide suspension. Many protein films are intensely "stained" (coated) by chromium oxide or iron oxide suspensions (2, 29).

We studied the effect of a much wider space between two identical surfaces upon adsorption of a protein as follows. We placed a planar-convex lens (radius of curvature between about 150 and 2000 mm) on a glass slide, and injected a solution of the protein between them. After incubation at room temperature for about 10 min, we tilted the slide and rinsed it with buffer while the lens was allowed to slip off. The density distribution pattern of adsorbed protein could be (but often need not be) enhanced by exposure to a matching antiserum; after rinsing again the slide was usually stained with concentrated Coomassie Blue. Under these conditions albumin, certain preparations of purified fibrinogen and all preparations of purified immunoglobulin G (IgG) we tested, deposited not only a presumably monomolecular layer, but also a ring of much greater thickness at a distance from the lens center corresponding to an about 10 to 20 microns thickness of liquid for solutions containing about 8 mg/ml of the protein (11). More diluted solutions deposited rings of greater diameter.

Direct observation of IgG injected as described, and serial experiments interrupted from 1 to about 4 sec after injection, showed that a thick layer is deposited out of an approximately 8 mg/ml buffer solution (pH 7.4) within about 2 to 3 sec except near the center, where the liquid layer is so thin that the amount of IgG per surface area is barely sufficient to deposit a monolayer and is then depleted. This creates the "hole in the doughnut". At its immediate periphery, the liquid layer is thick enough to supply the amount of protein needed for "excess" deposition of the thick ring, at which point the liquid above this area is also depleted and the thick ring is stabilized. More peripherally, additional protein molecules are available to collide with the thick deposit and desorb the excess. It should be noted that the conditions required to yield a rather stable multimolecular layer are strictly limited to relatively high concentrations of protein in spaces over a narrow range (e.g. more that 10 and less than 20 microns for solutions of about 5 mg/ml for at least several proteins).

These findings would support those of others that proteins such as fibrinogen (12) and phosphorylase b (13) can interchange adsorbed molecules with those in solution even though, facing a protein-free solution, the adsorbed molecules do not desorb spontaneously.

TWO PROTEINS ON ONE SURFACE

a) One protein preadsorbed (SaPl, P2)

Turnit demonstrated that ellipsometry is suitable for showing the interactions of a preadsorbed enzyme with a solution of its protein substrate (13). We found that preadsorbed thrombin will adsorb what appears to be a monomolecular layer of fibrinogen out of fibrinogen solution, but will continue to create fibrin monomer so that the solution will clot (2), mainly or most rapidly at the site of adsorbed thrombin.

Antibodies are deposited out of solution onto preadsorbed antigens as part of countless routines, including ellipsometry, in methods intended to identify the preadsorbed antigen or to quantitate antibody potency or antigen concentration. Where antigen is more densely packed than the detecting immunoglobulin can be, the proportionality between surface concentration of antigen and its ability to adsorb antibody is disturbed. In general, preadsorbed antigens retain their ability to adsorb polyclonal antibodies, and we rely on this ability routinely (4, 11, 14).

In our experience, unless the preadsorbed protein has a specific (enzyme/substrate, antigen/antibody) relationship with the dissolved one, the preadsorbed protein will adsorb little or no other protein. Where the ellipsometer shows increased optical film thickness upon exposure of, e.g., a preadsorbed layer of albumin to fibrinogen, such interactions occur at the expense of the ability of the preadsorbed albumin to adsorb antibody to albumin (14). This indicates that exchange - in this case by the larger species - took place. A notable exception may be the ability of plasma to deposit a thus far unidentified protein onto preadsorbed immunoglobulins (IgG) (15), but here too, some specific interactions may take place.

Out of the hundreds of proteins that our plasma contains, we must expect that for a random choice of most pairs, the dissolved species of protein facing a surface completely occupied with the preadsorbed species of protein, can only be deposited by displacing the preadsorbed species. Many experiments have been published to show how one protein can "inhibit" surface activation of a clotting factor such as high molecular weight kininogen or factor XII, while in fact the surface is merely blocked nonspecifically by preadsorption of a protein that is not easily displaced.

b) <u>Both proteins dissolved initially (Sa, P1, P2).</u>

In several series of binary mixtures, relative concentration of each protein in solution determined which was adsorbed most (15), but relationships were not simply predictable. For example, judged by optical thickness of deposited antibody, adsorption of fibrinogen in presence of IgG rapidly rose to its maximum as the relative concentration of fibrinogen was increased, while in presence of albumin fibrinogen adsorption increased almost linearly as the proportion of fibrinogen to albumin in solution was increased (15).

Those of us venturing over the edge of pure protein studies into the field of life by studying more than one protein at a time, will soon face criticism from workers in both fields: that our systems are not simple enough to apply known laws of adsorption, and that they are not complex enough to mimic the behavior of the same proteins in presence of blood or blood plasma. Most work published recently by others, some of which is included in this

issue, uses more sophisticated tools and analyses than ours,
allowing a view of increasingly submolecular detail and revealing
more and more individual differences in rate and extent of conformation changes upon adsorption and interaction.

ONE OR TWO PROTEINS ON TWO SURFACES (SaP1, SbP1 and SaP1, SbP2)

We found that one surface (Sa) such as Fe_2O_3 powder will adhere to
fibrinogen that is adsorbed out of a solution onto a second surface
(Sb) such as a glass slide. Fibrinogen adsorbed onto a flat surface
even coats itself to some degree when the coating powder also had
been coated with either fibrinogen or another protein. The powders
we tested (iron oxides, chromium oxide and ultramarine) adhered most
to surfaces carrying fibrinogen, less to those carrying IgG and
least to those carrying albumin, and this ability to coat surfaces
appeared to be reduced by coating the particles with a protein.
        Though there is a possibility that the protein adsorbed on one
surface "sees through" the protein coating of the other surface, it
seems more likely that it sees through holes in the imperfect coating. Holes may affect studies of SaP1, SaP1 interactions (16), and
cause aggregation of particles while being coated (17).
        Very simple experiments are able to show the rather complex
events occurring when a protein is exposed to 2 surfaces at once.
For example, when fibrinogen solution is introduced to a glass test
tube contining an iron oxide suspension, the suspension will tend
to aggregate and both single and aggregated particles will adhere to
the wall. Using our notations (Sa representing the particles, Sb
the test tube wall, and P1 the fibrinogen), we may assume the following interactions take place.
Sa, Sb, P1 -> SaP1, SbP1, P1 ->
SaP1Sb (Sa adheres to P1Sb and SaP1 adheres to Sb),
SaP1Sa (particles clump) ->
SaP1SaP1Sb (clumps adhere to protein-coated test tube wall, P1Sb).

PLASMA AT INTERFACES

As we expand our observations from the behavior of a few proteins
to that of plasma, we may begin to feel justified in asking why the
system performs rather than how it does. The question "Why do
plasma proteins interact at interfaces?" can then be interpreted to
mean: "what aspects of the behavior and interactions among purified
plasma proteins can be seen as well in their natural habitat - the
plasma -, and can these aspects be 'explained' as being beneficial
to our survival?" What thus far had appeared as senselessly complex behavior of purified proteins at interfaces may become more
reasonable in the context of many plasma proteins interacting at
the mottled surfaces of cells in a way that will allow the survival
of their host. Meanwhile, we will discover that purified proteins
behave unlike their sibblings in vivo and that in the eyes of our
plasma a purified protein adsorbed out of an artificial solution
will not look like a film of the same protein deposited by the
plasma itself.

a) Plasma at one interface.
Our own studies of human citrated intact plasma at activating
surfaces (glass, anodized tantalum, oxidized silicon) resulted in
the following findings.

In the ellipsometer, plasma even when diluted about 150-fold,
deposits immunologically identifiable fibrinogen but the film loses
its ability to adsorb antibody to fibrinogen within minutes under
these conditions. The loss is probably not caused by the plasma's
proteolytic activity (18) and occurs more slowly at 10C than at
37C (19). At room temperature, undiluted plasma deposits fibrinogen within about 2 sec and the loss occurs within 30 sec.

More recently, we found that the fibrinogen is replaced by
high molecular weight kininogen (HMK) (20, 21). Studies with radiolabelled fibrinogen rather than with antibodies to fibrinogen recently confirmed that plasma deposits fibrinogen and then removes
it (22, 23). We found the event does not occur on various hydrophobic surfaces: others found it did (24).

These experiments imply that fibrinogen, a protein present at
a concentration of about 2 to 3 mg per ml plasma, is deposited and
then replaced by a protein (HMK) present at a concentration about
2 orders of magnitude less. The possible significance of trace
proteins and their inability to cover or displace coatings of relatively abundant proteins in narrow spaces is discussed below.

b) Plasma between two surfaces.
   In spaces between activating surfaces that are less than about
20 microns but more than about 3 microns apart, normal intact plasma
will still contain enough fibrinogen to create a confluent carpet,
but will lack the concentration of HMK per unit of surface area to
displace this fibrinogen film. As a result, the plasma when injected between a flat activating surface and a convex lens resting
belly-down upon the surface will leave a disk of fibrinogen where
the plasma occupies a space of about 3 to 20 uM wide (25). Dilution of plasma causes it to leave a correspondingly larger disc of
fibrinogen (requiring a thicker layer of plasma to contain enough
HMK per surface area) with a central hole that contains other proteins (26).

More detailed studies (using lenses of about 147mm radius of
curvature) indicated that a set of concentric rings of proteins is
left by plasma under these conditions. Diluted to about .02 to
.01% it will leave albumin centrally, surrounded by immunoglobulins
(IgG): at about 1%, it will leave a central disk of IgG surrounded
by fibronectin and fibrinogen, while at about 20%, it will leave a
solid disc of fibringoen with a central hole containing IgG surrounded by a fine line of fibronectin (11).

c) Plasma at a protein-coated interface
    An overview of our data on the effect of plasma on pre-adsorbed
fibrinogen (e.g. 14, 15, 28) shows intact rather than activated
plasma is able to displace preadsorbed fibrinogen, but under certain
conditions (degree of packing?) pre-adsorbed fibrinogen is removed
more slowly than the fibrinogen that had been deposited by the
plasma itself.

Most remarkable were the interactions we found occurring when
plasma came in contact with IgG preadsorbed onto a wettable (oxi-

dized silicon crystal) surface in the ellipsometer. The plasma deposited a layer about as optically thick as was the original IgG
film; then, if the plasma was intact, it would remove more than the
amount it had deposited (15). The matter deposited was not removed
if the plasma had been activated, and could not be distinguished
immunologically from the underlying IgG. In retrospect, we believe
the event may be related to the deposition of IgG on top of IgG that
we observed in our lens-on-slide experiments described above: in
presence of sufficient IgG, it removed itself. This would not explain why activated plasma was unable to remove its own deposit from
the IgG substrate.

d. More complex systems involving plasma
     When contact of plasma is interrupted after it has   depos-
ited fibrinogen but before it replaced the fibrinogen with HMK, and
the surface is then coated with a metal oxide (one of the iron
oxides, or chromium oxide) powder, subsequent re-exposure to intact
plasma will cause the oxide to be lifted off wherever it had been
deposited on the fibrinogen film. This lift-off can serve to demonstrate displacement of one protein by another as follows. Intact
plasma is injected between lens and slide, incubated and rinsed off
as described earlier. At this point, the entire slide will have
been exposed to plasma under various conditions. Where the plasma
had resided in the peripheral (wide) area under the lens, there will
be HMK. Centrally (in the narrow area), as well as where contact
was brief (during the rinse, beyond the area of residence and in the
refillable "scratch" left by the lens while sliding off), fibrinogen
will be present. Such a slide will coat itself entirely when exposed to a suspension of the metal oxide. Subsequent exposure of
the entire coated slide to plasma as described above will cause removal of oxide where it had resided on fibrinogen but not where it
had resided on HMK. As a result, only a ring of oxide remains,
corresponding to the location of HMK that had been deposited by the
original (injected) plasma. Plasma lacking kininogens was unable
to remove the oxide (29).

e. Plasma in flow between surfaces
     When a 1% solution of normal intact plasma in Veronal buffer
was injected between an anodized tantalum-sputtered slide and a
convex lens of about 2000mm radius of curvature, the plasma left
a teardrop-shaped ring of albumin on the slide, its "tail" pointing
away from the site of injection (26). In a stagnation point flow
chamber plasma created the following pattern. In the narrow areas
around the metal tips that supported the cover slip, and along lines
("tails") pointing to the periphery of the chamber thus indicating
the direction of flow, undiluted intact plasma took longer than it
did elsewhere to deposit fibrinogen and then took longer (more than
5 min) to remove it (27). Under identical conditions, heparinized
blood deposited platelets where plasma would leave fibrinogen.

WONDERING HOW AND WHY PROTEINS IN PLASMA AT SURFACES INTERACT

If our observations are correct, plasma deposits a sequence of proteins, the more abundant being adsorbed first and then being re-

placed by less abundant ones - at least on surfaces that can activate clotting. At the highest dilutions tested (.05 to .1%), normal
plasma appears to deposit albumin first, and slowly replaces it with
immunoglobulins (IgG). We suggest that in undiluted plasma too,
albumin would be deposited first, to be replaced by IgG, which is
replaced by fibrinogen and fibronectin, all within a few seconds:
and finally, the latter are replaced by HMK and factor XII in a
measurable amount of time. Thus, the behavior of a very complex
mixture such as plasma differs markedly from that described for more
simple mixtures (31).

On this rather narrow basis of the 4 to 6 proteins studied so
far, we can build the hypothesis that there is a correlation between
the concentration of each protein in the plasma and its rates of
both adsorption and desorption. Two possible causes of this correlation can then be considered.
a) There is an inverse correlation between the normally occurring
concentrations of plasma proteins and their desorption constants.
This would mean our proteins evolved in such a way that the rare
ones would stick latest but longest: it would imply that there is
some purpose of this dictated sequence to the survival of the host.
b) Instead, there may be a single mechanism that cannot help but
link each protein's concentration with its rates of adsorption and
desorption, e.g. by means of two rules:
1) Each adsorbed protein molecule can only be removed by one or
more molecules of the same species present in the bulk solution,
perhaps briefly forming a dimer or larger aggregate with the adsorbed mate after which the pair or complex is desorbed. This
model would be suggested by our experiments on presumably pure proteins forming thick rings between lens and slide, as described
(under "one protein on two identical surfaces").
2) The longer a protein molecule resides on a surface, the stronger
it is bound and the more energy (e.g. collisions or time spent with
a mate) is required for desorption.

We have written a simple home computer program incorporating
these 2 rules, so that at more or less normal concentrations of
albumin, IgG, fibrinogen, HMK and factor XII these proteins will
displace each other in the sequence found experimentally. The program displays the proteins as if gradually collapsing with increased
time of residence, and recovering when hit and joined with a mate
until the pair is desorbed. This model represents the simplest of
possible situations: plasma on a uniform, flat surface with minimal
flow. If plasma dilution, narrow spaces and less "activating"
(more apolar?) surfaces all delay the sequence of protein interactions at the interface, then these events will be less synchronized
on physically rough and chemically non-uniform devices even when
introduced rapidly to blood, and even if the 2 simple rules given
above were the only ones governing these interactions.

It appears most likely that the relationship we presumed to
exist between the physiological concentrations of proteins in our
plasma and their surface properties does not hold true for most of
the proteins we have not yet studied. The fact that plasma injected
between a lens and a slide leaves concentric rings of all proteins
we have looked for - even including traces of complement factors,
plasminogen and prealbumin (unpublished findings) - suggests that

many proteins we have not yet looked for are also deposited, in
their own time, conditions of flow and micrometers of space between
interfaces. The physiological significance of their presence at
the interface will lie in their time of arrival: it may coincide
with the arrival of a cell carrying receptors for that protein. The
adsorbed protein molecule, like a word of text, will then be
stressed by the cell and the context of the text will change. An
increasedly complex scenery will grow out of more or less synchronized events that occur upon introduction of a uniform but rough
surface into the blood. This growth can be depicted (fig. 1) as
if progressing from near to far distance, with its distant complexity of cell behavior and fibrin formation mercifully blurred by
perspective and environmental haze.

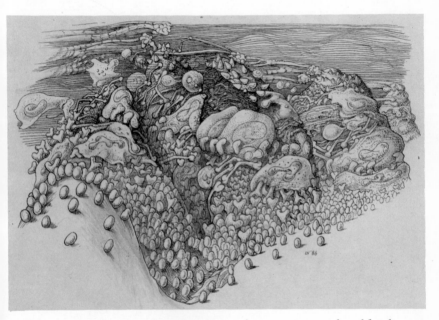

Figure 1. A drawing showing events that may occur when blood
comes in contact with a rough, activating surface. Time is
shown as if progressing from nearest to distant scenery.
Albumin (drawn egg-shaped) begins to be deposited in the most
open areas, then also in the narrow valleys. It is followed by
immunoglobulins (IgG) (drawn Y-shaped). On open areas where IgG
has replaced albumin, granulocytes adhere. Farther away, where
IgG is displaced by fibronectin (drawn as a narrow band of fat
hooks), monocytes adhere, while more distantly fibrinogen has
displaced IgG, platelets become attached and begin to liberate
their products. On the far mountain peaks, high molecular weight
kininogen has displaced fibrinogen and interacts with other
factors and with platelet products to form strands of plasma-
blown fibrin.

Unfortunately, it is this blissful haze that will keep us from seeing the details we need for understanding the beauty of this scenery.

Literature Cited

1. Vroman, L. Ph.D. Thesis, Rijksuniversiteit Utrecht, Netherlands, 1958.
2. Vroman, L. Thrombos. Diathes. Haemorrh. 1964, 10, 455-493.
3. Vroman, L. Nature 1962, 196,476-477.
4. Adams, A.L.; Fischer, G.C.; Vroman, L. J. Colloid Interface Sci. 1978, 65, 468-478.
5. Vroman, L; Kanor, S.; Adams, A.L. Rev. Sci. Instrum. 1968, 39, 278-279.
6. Vroman, L.; Adams, A.L.; Klings, M.; Fischer, G.C. DCIEM Conf. Proc. #73-CP-960, Downsview, Ont., 1973, 49-70.
7. Copley, A.L. Ann. New York Acad. Sci. 1983, 416, 377-395.
8. Brash, J.; Davidson, V.J. Thromb. Res. 1981, 9, 249-259.
9. Bohnert, J.L.; Horbett, T. J. Colloid Interface Sci. 1986, 111, 363-377.
10. Kudryk, B.J.; Vroman, L. To be published.
11. Vroman, L; Adams, A.L. J. Colloid Interface Sci. 1986, 111, 391-402.
12. Brash, J.L; Uniyal, S. Plastics Med. Surg. 1979, 3, 29-1-29-7.
13. Jennissen, H.P. J. Colloid Interface Sci. 1986, 111, 570-577.
14. Vroman, L.; Adams, A.L; Klings, M. Feder. Proc. 1971, 30, 1494-1502.
15. Vroman, L.; Adams, A.L.; Klings, M.; Fischer, G.C. Adv. Chem. Series 1975, 145, 255-289.
16. Klein, J. J. Colloid Interface Sci. 1986, 111, 305-313.
17. Silberberg, A. J. Colloid Interface Sci. 1986, 111, 486-495.
18. Vroman, L.; Adams, A.L. Thromb. Diathes. Haemorrh. 1967, 18, 510-524.
19. Vroman, L.; Adams, A.L. J. Biomed. Mater. Res. 1969, 3, 43-67.
20. Vroman, L.; Adams, A.L.; Fischer, G.C.; Munoz, P.C. Blood 1980, 55, 156-159.
21. Schmaier, A.H.; Silver, L.; Adams, A.L.; Fischer, G.C.; Munoz, P.C.; Vroman, L. Thromb. Res. 1983, 33, 51-67.
22. Brash, J.L.; ten Hove, P. Thromb. Haemostas. 1984, 51, 326-330.
23. Horbett, T.A. Thromb. Haemostas. 1984, 51, 174-181.
24. Horbett, T.A.; Cheng, C.M.; Ratner, B.D.; Hoffman, A.S.; Hanson, S.R. J. Biomed. Mater. Res. 1986, 20, 739-772.
25. Adams, A.L; Fischer, G.C.; Munoz, P.C.; Vroman, L. J. Biomed. Mater. Res. 1984, 18, 643-654.
26. Vroman, L.; Adams, A.L.; Fischer, G.C.; Munoz, P.C; Stanford, M. Adv. Chem. Series 1982, 199, 265-276.
27. Stanford, M.F.; Munoz, P.C.; Vroman, L. Ann. N.Y. Acad. Sci. 1983, 416, 504-512.
28. Vroman, L. Bull. N.Y. Acad. Med. 1976, 48, 302-312.
29. Vroman, L.; Adams, A.L; Brakman, M. Haemostasis 1985, 15, 300-303.
30. Blodgett, K.B.; Langmuir, I. Phys. Rev. 1937, 51, 964.
31. Horbett, T.A. Personal communication, 1986.

RECEIVED January 3, 1987

Chapter 11

Consequences of Protein Adsorption at Fluid Interfaces

F. MacRitchie

Wheat Research Unit, Commonwealth Scientific and Industrial Research Organization, P.O. Box 7, North Ryde 2113, Australia

The consequences of protein adsorption are discussed in terms of (a) the changes in structure, properties and reactivity of adsorbed protein and (b) the role played by adsorbed protein in phenomena where interfaces exert an influence. Discussion is mainly restricted to the air/aqueous interface. The changes in configuration of proteins on adsorption and the topics of interfacial coagulation, desorption from monolayers and the applicability of the Gibbs Adsorption Equation are considered with reference to the fundamental question of reversibility of adsorption. Some of the evidence which has been used in support of irreversibility can be rationalized on the basis of a reversible process. Studies of protein monolayers reflect a flexible chain configuration where the behavior is governed by segments of the molecule, usually of 6-10 amino acid residues in size, rather than whole rigid molecules. Some of the phenomena in which protein adsorption is implicated include the fluidity of interfaces, the precipitation of proteins from solution by shaking, the formation and stability of foams and emulsions, bacterial adhesion and the reactivity of enzymes.

The consequences of protein adsorption may be conveniently discussed in terms of
(a) the effects of adsorption on the structure and properties of the protein itself, and
(b) the influence on various phenomena in which interfacial behavior plays an important role.
This review will focus on studies at the air/aqueous interface, although conclusions will in many cases be applicable to other types of interface. The air/aqueous interface has certain advantages experimentally. In contrast to interfaces with solids, it is possible to manipulate molecules by use of a film balance and measure film properties and the kinetics of various processes as a

0097-6156/87/0343-0165$06.00/0
© 1987 American Chemical Society

function of precisely determined two-dimensional concentrations. In
particular, the surface pressure, which is a measure of changes in
surface free energy that result from adsorption, is a fundamental
parameter that can be measured simply and accurately.

Effects of Adsorption on Protein Structure and Function

One of the striking features of protein adsorption is that many
proteins, although highly soluble in aqueous solution, form films
which are extremely stable and difficult to desorb. This property
was highlighted by Langmuir and Schaeffer ([1]) who applied the Gibbs
Adsorption Equation in its simple form

$$d\Pi/d \ln c_b = c_s kT \tag{1}$$

to calculate the increase in solubility that should accompany
compression of a surface film of protein. Here, Π is the surface
pressure, c_s the number of molecules per unit area of the surface
and c_b the concentration in solution in equilibrium with the surface
film. For a protein of molecular weight 35,000 (ovalbumin), it was
calculated ([1]) that an increase in Π of 15 mNm^{-1} should increase
the solubility of the film by a factor of 10^{95}. Because little
tendency for proteins to dissolve is normally observed on
compression of their surface films to surface pressures of the order
of 20 mNm^{-1}, this was interpreted to mean that protein adsorption
was not a reversible process and therefore the Gibbs Equation was
not applicable.

Since the question of whether protein adsorption can be
considered to be a reversible process is basic to an understanding
of protein behavior at interfaces, it is proposed to discuss the
problem in some detail. Apart from the loss of solubility on
adsorption, other observations that have been interpreted as
evidence for irreversibility are:

1. Proteins, on adsorption at fluid interfaces, undergo a change
    from their globular configuration in solution to an extended
    chain structure. This has often been referred to as surface
    denaturation.

2. An insoluble coagulum is frequently formed when protein mono-
    layers are compressed to high surface pressures, when proteins
    adsorb at quiescent interfaces or when protein solutions are
    shaken.

3. In certain cases, losses of biological activity (enzymatic,
    immunological) have been reported as a result of adsorption.

On the other hand, well-defined adsorption isotherms of
proteins have been reported. Figure 1 shows one example, that for
chymotrypsin in pure water at 20°C. The attainment of steady
surface pressure values, which increase with increasing protein
concentration in solution indicating true equalization of bulk and
surface chemical potentials, argues in favor of a reversible
adsorption process. In addition, desorption from protein monolayers
has been measured. How to rationalize these apparently conflicting
results therefore presents an intriguing challenge.

Structure of Protein Films.   Spread monolayers of proteins have
pressure-area (Π -A) curves which are generally similar,
extrapolating to an area close to 1.0 m^{2}mg^{-1} at Π =0.   It can be
calculated that the thickness of the monolayer at this area
corresponds to about 10Å.   The exact configuration has not been
resolved.   On energetic grounds, it is expected that the
polypeptide backbone lies in the plane of the surface with polar and
non polar side chains directed towards and away from the aqueous
phase respectively.   The surface in this way acts as a good
solvent , in this case a two-dimensional one.   When a protein
molecule adsorbs, an area of high surface free energy is replaced by
interfaces of low free energy; i.e. polar side chains/water and non
polar side chains/air. The achievement of this free energy lowering
accounts for the unfolding of the molecule at the surface.   A
number of experimental techniques have been introduced to
investigate the conformation of protein monolayers after removal
from the surface, either as collapsed or uncollapsed films.
Malcolm (2) has presented evidence to show that surface films of a
range of polypeptides were in the α-helical conformation, using
infrared spectroscopy, electron diffraction and deuterium exchange.
Other studies of removed protein films have also supported the
presence of the α-helix, using infrared spectroscopy (3), optical
rotary dispersion and circular dichroism (4,5).   It must be
remembered, however, that whereas polypeptide chains may exist
exclusively in the α-helix form, proteins in their solution state
generally have their chains only partially in this form with
relatively larger or smaller proportions in the beta form or random
structure.   Furthermore, because the adsorption step represents a
transition from a relatively poor solvent (water) to a relatively
good solvent (air/water interface), there does not appear to be a
strong driving force for retention of helical structure.   A number
of experimental results point indirectly toward a predominantly
random chain structure;   in particular, the surface flow and the
configurational changes on film compression.

Surface Viscosity of Protein Monolayers.   Figure 2 shows the
surface viscosity (η_s) of a number of proteins and one polyamino
acid as a function of Π (6).   The extremely high surface viscoelasticity of protein monolayers appears to be more characteristic
of an interacting random chain system than an array of rigid
helices.   The theory of surface viscosity of Moore and Eyring (7),
based on the Theory of Absolute Reaction Rates, postulates that the
flow of a monolayer consists of movements of flow units, normally
molecules, from one equilibrium position to another, over an intermediate activation energy barrier.   The equation derived for the
coefficient of surface viscosity (η_s) is

$$\eta_s = \frac{h}{A} \exp\left(\frac{\Delta G + \Pi \Delta A}{kT} \right) \qquad (2)$$

where h is Planck's constant and ΔG is the activation free energy
for flow at Π = 0. ΔG is made up of two terms: (a) the work
required to form a hole in the surface sufficiently large for the
molecule to enter, and (b) the work required to move the molecule
into the hole;   this term includes the work necessary to break all

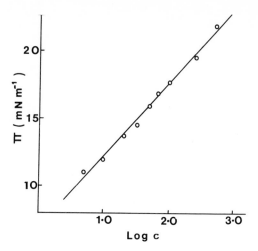

Figure 1. Adsorption isotherm plotted as surface pressure (Π) against the logarithm of the bulk concentration (c) for chymotrypsin in pure water at 20°C.

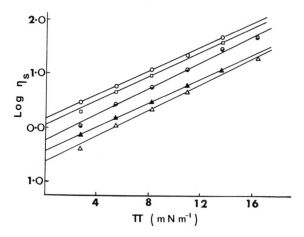

Figure 2. Logarithm of surface viscosity as a function of surface pressure for several proteins and one polyamino acid at pH 5.5.   o, poly-DL-alanine;   □, human γ-globulin;   ⊖, pepsin; ▲, bovine serum albumin;   △, lysozyme. (Reproduced with permission from Ref. 6.   Copyright 1970 Marcel Dekker).

bonds formed with neighbouring molecules. $\Pi\Delta A$ is then the additional work that has to be done against the surface pressure Π to create the hole of area ΔA.   Equation 2 predicts that a plot of log η_s vs Π should be linear, permitting the calculation of ΔG from the intercept at $\Pi=0$ and ΔA from the slope.   Table I summarizes data for the results of Figure 2.

It is apparent from the data that ΔA, the area occupied by the flow unit is similar for all proteins (100-120Å^2) and corresponds to segments of 6-8 amino acid residues.   Joly (8), using an independent approach, also calculated a similar value for the area of the elementary flow units of protein monolayers.   These calculations suggest that molecules in the monolayer are sufficiently flexible that segments of this size, on the average, move as units.   This resembles the manner in which long chain hydrocarbons appear to diffuse in solution (9).   As a result, ΔG and η_s are practically independent of molecular weight.

Table I.   Calculated Values of ΔG and ΔA for Proteins and one Polyamino acid

Protein	Molecular weight (x10^{-3})	ΔG (kJmole^{-1})	ΔA (Å^2)
Polyalanine	1.5	69	105
γ - Globulin	160	69	110
Pepsin	34	67	120
Bovine Albumin	70	66	100
Lysozyme	15	65	115

The influence of adsorbed proteins on interfacial viscosity is relevant to the fluidity of biological membranes.   An unusual effect is observed when lipid molecules are incorporated into protein monolayers, first reported by Schulman and Rideal (10).   As the mixed film is compressed, η_s increases normally but then goes through a maximum, thereafter decreasing sharply with further increase of Π(6).   Evidently, above a certain surface density, lipid molecules disrupt interactions between protein chains.   The behavior is completely reversible, η_s increasing as the surface area is expanded.   Alteration of the composition of the mixed film and the surface density thus provides a very sensitive means for varying the surface fluidity, a method that may well be utilized in biological systems.   Formation of brittle monolayers that greatly increase the surface viscoelasticity also may occur when protein films are spread on subsolutions containing low concentrations of divalent metal ions (6) or silicic acid (11), again suggesting mechanisms for the deleterious effects of these species on membrane function.

<u>Configurational</u> <u>Changes</u> <u>in</u> <u>Compressed</u> <u>Films</u>.   When protein monolayers are compressed, relaxation processes occur which are manifested by decreases in surface area, A (if Π is kept constant) or decreases in Π (A constant).   Providing the effects of

desorption or surface coagulation (discussed below) are separated,
the decreases are totally recoverable on expansion of the monolayer.
At high surface pressures ($\Pi > 20$ mNm^{-1}) these reversible changes
are large and can only be rationalized if it is assumed that
portions of the molecules leave and re-enter the surface. It has
been well established that adsorbed linear polymer molecules may
exist as trains (segments attached to the surface), loops (segments
displaced into the adjacent phase) and tails (segments at the ends
of molecules which also tend to be displaced from the surface).
This model appears to be appropriate for protein monolayers. By
using a film balance to study protein monolayers at the air/water
interface, it is possible to quantitatively evaluate the equilibrium
distribution between attached and displaced segments at a given
surface pressure. Because of the high surface viscoelasticity in
these monolayers, the relaxation processes are sufficiently slow to
enable them to be followed experimentally.

If, at a given surface pressure Π, A_o is the initial area of
the protein monolayer before any relaxation has occurred and A is
the area after equilibrium has been established, then the ratio of
attached to displaced segments (r) is given by

$$ r \quad = \quad A/A_o - A \qquad\qquad (3) $$

This assumes that displaced segments make no contribution to the
surface area occupied by the molecules. The variation of A and A_o
as Π is increased is illustrated in Figure 3 for a monolayer of
bovine serum albumin (BSA). The variation of r with Π should be
given by an equation of the form (12)

$$ r \quad = \quad \exp \left(\frac{\Delta G_s - \Pi \Delta A_s}{kT} \right) \qquad\qquad (4) $$

where ΔG_s is the difference in free energy between attached and
displaced segments at $\Pi = 0$ and $\Pi \Delta A_s$ is the additional free energy
required by a displaced segment to enter the surface at a finite
value of Π, ΔA_s being the area occupied by the segment. Plots of
log r vs. Π have been found to be linear for a number of proteins
(13), permitting ΔG_s and ΔA_s to be evaluated from the intercept at
$\Pi = 0$ and the slope respectively. Values of ΔG_s varied from 4.4
to 8.8 kT and ΔA_s ranged from 90Å^2 to 160Å^2. The size of the
segment is similar to that of the flow unit estimated from surface
viscosity data and corresponds to 6-10 amino acid residues.

Surface Coagulation. During the preparation of solutions of
certain proteins, it is observed that an insoluble precipitate may
form. This occurs as a result of interfacial action and is
accentuated by excessive shaking. The mechanism appears to be
roughly as follows. As fresh interface is formed, protein
molecules adsorb and unfold to an extended chain conformation. At
a given surface density, the solubility limit of the protein at the
surface is reached and precipitation occurs. This is a special
case of precipitation since it involves a transition from a two-
dimensional (monolayer) to a three-dimensional state (coagulated
protein). The solubility limit is usually characterized by a

coagulation surface pressure, Π_c, corresponding to the equilibrium spreading pressure of monomeric compounds.   At a quiescent interface, the coagulum reduces the available interface by its presence and also prevents diffusion of molecules to it;   thus, the coagulated film tends to a limiting thickness.   In the case of a solution that is being agitated, fresh interface is continually being created so that surface coagulation proceeds at a rate dependent on the degree of agitation. 'In this way, much of the protein can be reduced to an insoluble form.

An equation similar to that used to describe duplex films can be used to predict the coagulation pressure, Π_c, at a given interface (14).

$$\text{i.e.}\quad \Pi_c = \gamma_a - \left(\gamma_b + \gamma_{ab} \right) \tag{5}$$

where γ_a is the initial surface free energy, γ_b the interfacial free energy between non polar groups of the monolayer and the non aqueous phase and γ_{ab} is the interfacial free energy between polar groups of the monolayer and the aqueous phase.   It can be seen from this equation that, other conditions being equal, Π_c is lowered as γ_a, the initial interfacial free energy decreases.   This is consistent with the result that coagulated films form more easily at oil/aqueous than at air/aqueous interfaces.   A wide variation in the ease of surface coagulation is observed from one protein to another and, for a given protein, from one set of conditions (pH, ionic strength) to another.   In a comparative study (15), ovalbumin was most susceptible followed by β-lactoglobulin, γ-globulin hemoglobin, myoglobin and lysozyme.   Least susceptible were cytochrome C, α-casein and BSA.   Those proteins that are least susceptible tend to have higher average hydrophobicity as defined by Bigelow (16), although β-lactoglobulin appears to be an exception.   It may be that the failure of the monolayer to present a sufficiently non polar surface to the non aqueous phase (signifying a high value for γ_b) is an important factor causing susceptibility to surface coagulation.   The sensitivity of surface coagulation to protein structure is shown by the report that fully oxygenated sickle cell hemoglobin (HbS) is much more vulnerable to precipitation by shaking than normal hemoglobin (HbA) despite the otherwise great similarity in structure and properties (17,18).

Desorption and the Question of Reversibility

A number of important questions revolve around the central one of reversibility of protein adsorption.   Are proteins, once adsorbed, able to desorb?   Why are proteins so difficult to remove from a surface?   What happens in the situation where there is competitive adsorption between different proteins or between proteins and other species?   Do the different points on an adsorption isotherm correspond to dynamic equilibria? If a protein molecule can desorb, does it revert to its original solution configuration and recover its original biological activity?   Can the Gibbs Adsorption Equation be applied to protein adsorption?   Some of these questions may be effectively tackled by studies with the film balance.

Langmuir and Waugh (19) were the first to study the effects of compression on the stability of protein monolayers. They distinguished between pressure displacement and pressure solubility. Pressure displacement refers to the reversible expulsion of chain segments, discussed above, while pressure solubility is the desorption of complete molecules. Pressure solubility of the pure proteins, insulin and ovalbumin, was very small. However, after enzyme proteolysis, pressure solubility increased with increasing digestion time and increasing surface pressure. Based on a theory of pressure solubility (20), it was estimated that the degradation products of insulin had molecular weights in the range 1,000-2,000. As a result of the reversible relaxation processes in protein monolayers, desorption cannot be measured as for monomeric compounds by the loss of surface area, maintaining Π constant. However, because losses of area of protein monolayers are totally recoverable at low surface pressures (10 mNm^{-1} and below), Gonzalez and MacRitchie (21) introduced a method for measuring permanent area losses based on monitoring the area at a reference pressure of 5 mNm^{-1}. This method has subsequently been used (22,23) to obtain data on protein desorption. The results of these studies may be summarized as follows.

Confirmation of Desorption. Permanent area losses are observed for protein monolayers when kept at high Π (> 15 mNm^{-1}). A number of independent checks have been used to confirm that these losses are caused by desorption. Rates of area loss are enhanced by stirring and diminished by the presence of protein in the sub phase (21). Using radiolabelled BSA, the compressibility and the specific radio-activity of the film remained unchanged after extensive losses of area and the radioactivity subsequently measured in the sub phase accounted well for the amount of material lost from the surface (22).

Analysis of Desorption Kinetics. Several features are evident from the kinetics of desorption. Plots of the logarithm of the area are a linear function of the square root of time in agreement with a diffusion controlled process governed by the equation:

$$n = 2C_0 \left(\frac{Dt}{\pi} \right)^{\frac{1}{2}} \tag{6}$$

where n is the number of molecules that desorb in a time t, C_0 is the concentration (in molecules per unit volume) of a thin subsurface layer assumed to be in equilibrium with the monolayer, D the diffusion coefficient and $\pi = 3.14$. This is illustrated in Figure 4 for desorption of β-lactoglobulin at different surface pressures. From Figure 4, an induction period of 1-2 min is evident at the commencement. During this period, the monolayer is rapidly approaching its equilibrium configuration with respect to displacement of molecular segments. The rate of desorption is given by differentiation of Equation 6:

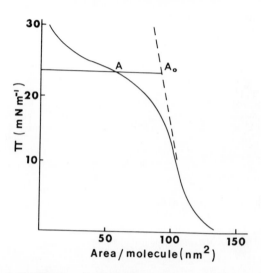

Figure 3. Equilibrium Π-A curve of BSA showing areas occupied by attached segments (A) and displaced segments (A_o-A). Dashed line is the instantaneous Π-A curve; i.e. that which would be obtained if no relaxation of monolayer. (Reproduced with permission from Ref. 25. Copyright 1977 Academic Press).

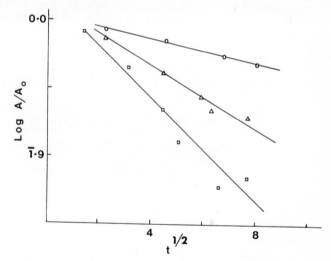

Figure 4. Plots of the logarithm of the ratio of the area at time t to the initial area as a function of square root of time for β-lactoglobulin monolayers; o, 20 mNm^{-1}; Δ, 25 mNm^{-1}; □, 30 mNm^{-1}. (Reproduced with permission from Ref. 23. Copyright 1985 Academic Press).

$$dn/dt = C_o \left(\frac{D}{\pi} \right)^{\frac{1}{2}} t^{-\frac{1}{2}} \tag{7}$$

The other characteristic that can be deduced from experimental data is that plots of the rate as a function of $t^{-\frac{1}{2}}$ (Equation 7) do not extrapolate to zero rate at infinite time as required by a purely diffusion controlled process (21). This indicates the presence of a barrier to the desorption step at the interface. This is confirmed by the very much smaller values for the equilibrium sub surface concentrations calculated from desorption kinetics than those expected from the adsorption isotherm at the same pressures.

If we assume that the rate of desorption (dn/dt) is controlled by two barriers, the diffusional resistance R_1 (equal to δ/D, where δ is the thickness of the diffusion layer near the surface and D is the diffusion coefficient) and the interfacial resistance R_2, then by substituting the equilibrium concentration C_o, obtained from the adsorption isotherm, we may calculate the magnitude of R_2 from the equation

$$dn/dt = \frac{C_o}{R_1 + R_2} \tag{8}$$

A value of 2.4×10^8 sec cm^{-1} was calculated for R_2 (21) from data for a BSA monolayer desorbing under steady state conditions at a pressure of 25.6 mNm^{-1}. We may conclude that, although the kinetics of desorption show the influence of the diffusional resistance (1.3×10^4 sec cm^{-1}), the absolute rate is governed by the much larger interfacial resistance.

Effects of Molecular Weight. Molecular weight appears to be a governing factor in the ease of desorption of proteins. Results for several proteins spanning a range of molecular weights are summarized in Table II.

Table II. Rates of Desorption of Proteins at Different Surface Pressures

Protein	Mol. wt ($\times 10^{-3}$)	15	20	25	30	35	40	45
				(Π, mNm^{-1})				
Insulin	6	56	530					
β –Lactoglobulin	17.5		20	50	90			
Myoglobin	17			34	67	144		
γ –Globulin	160				9	20	40	
Catalase	230					30	70	110

Rates of Desorption (min^{-1} x 10^4)

In mixed monolayers of insulin with γ-globulin or catalase, insulin may be completely desorbed from the monolayer by suitable choice of the surface pressure (23). Based on these results, some predictions may be made about situations where there is competition for the interface by a mixture of proteins. After creation of a new interface, the composition of the adsorbed film would tend to favor the smaller molecules because of their higher diffusion coefficients. However, the final composition would favor the high molecular weight proteins.

An interesting result is that the absolute rate of desorption for the desorbing protein is the same in the mixed film as in the pure monolayer providing the area occupied is over 30% of the total surface area (23). Under these conditions, the concentration of protein in the sub-surface layer is evidently determined by the surface pressure and not the surface density. This result has implications for systems where proteins are transported across interfaces or membranes.

Theory of Desorption. Based on measurements of kinetics of adsorption (24), it has been estimated that protein molecules require reduction of their areas in the surface to a critical value of about $100 Å^2$ for them to become unstable and desorb. This may occur either (a) at constant pressure by fluctuations of molecules about their mean configuration (the probability of such a fluctuation is negligibly low when practically all segments are in the form of trains at low Π), or (b) by compression of the film to a sufficiently high pressure Π^*, corresponding to the critical area. Assuming that the path is the same for each case, we can use the equilibrium Π-A curve of the protein to evaluate the free energy of activation for desorption (ΔG_{des}). At a given pressure, Π, this is given by:

$$\Delta G_{des} = \int_{\Pi}^{\Pi^*} A d\Pi \qquad (9)$$

The integral may be simply evaluated from the equilibrium Π-A curve of the protein (see Figure 3). Calculations based on Figure 3 for BSA showed that ΔG_{des} decreased steeply from a value of 106 kT at $\Pi = 20$ mNm^{-1} to 9 kT at $\Pi = 28.8$ mNm^{-1}(25). This coincided with the appearance of measurable desorption of BSA over this range of surface pressure. The theory also explains the effect of molecular weight on desorption since for a given ratio of displaced to attached segments, the area occupied by the attached segments will be proportional to molecular weight and it is this parameter that determines ΔG_{des}.

Application of Gibbs Adsorption Equation. By substituting $C_s = 1/A$ (where A is the area per molecule in the adsorbed monolayer), in the simple form of the Gibbs Adsorption Equation (Equation 1), we obtain

$$d\Pi/d\ln C_b = kT/A \qquad (10)$$

A number of independent studies have shown that well-defined adsorption isotherms can be obtained for proteins (21, 26-28) and plots of Π against the logarithm of C_b are linear over wide ranges. The most notable feature is that the slopes are relatively similar for most proteins and values of A calculated from Equation 10 are very much smaller than those of whole protein molecules estimated from their Π-A curves.   For example, from the Π-log C_b plot for chymotrypsin (Figure 1), A is calculated to be 170A^2 .   It appears that the adsorption behavior reflects the flexible linear chain structure of proteins in the surface so that we are not observing the contributions of rigid whole molecules but of partially independent molecular segments.   Joos (27) has proposed a theory of protein adsorption based on the Frisch-Simha model for flexible polymers (29).

The Role of Proteins in Interfacial Phenomena

Some consequences of the interaction of proteins with interfaces have already been mentioned; for example, the high viscoelasticity imparted to interfaces and the precipitation of soluble protein caused by shaking solutions.   As a result of their ubiquitous presence at interfaces, a wide range of apparently unrelated phenomena are affected by protein adsorption.   The role that proteins play at interfaces in biomedical systems is extensively discussed in other contributions to this book.   Here it is intended to focus more on areas related to food and agriculture.   Some of the relevant topics are:

1.   The formation and stability of dispersed systems (foams, emulsions etc.).
2.   Bacterial adhesion.
3.   Reactivity of enzymes.

Because of the great complexity of these systems, it is proposed only to briefly indicate some of the more basic concepts that might be usefully applied in these areas.

Formation and Stability of Emulsions and Foams.

The contributions of protein adsorption to the properties of dispersed systems encompass problems of great diversity ranging from the manufacture of food emulsions (mayonnaise, butter) and foams (meringues, whipped cream, bread) through the preparation of pharmaceutical suspensions to bloat in cattle.   It is useful to separate the ease of formation of foams and emulsions from the problem of their stability.   Ease of formation requires rapid adsorption and therefore requires relatively high concentrations of proteins, a low net charge so that the electrical potential barrier to adsorption is minimal and the absence of destabilizing agents.   Once a foam or emulsion is formed, the different processes of creaming and flocculation (emulsions) and coalescence begin to break down the dispersed system.   Proteins alone are very effective stabilizers for foams and oil/water emulsions.   Contrary to what might be predicted from some colloid theories, the stability of protein foams and emulsions is generally a maximum at or near the isoelectric point of the protein.   This is probably a result of the greater adsorption under these conditions.   It also indicates that factors other than electrical repulsion are important for conferring stability.   Some factors that may be relevant to stability are suggested by the previous discussion of fundamental aspects of protein interfacial behavior.

The coalescence of two emulsion drops proceeds with a reduction of interfacial area. This signifies that, during coalescence, there is a compression of the adsorbed stabilizing films surrounding the droplets if the film material cannot desorb sufficiently quickly. The resistance to compression by the films leads to an elastic restoring mechanism, thus creating an energy barrier which tends to oppose coalescence (30,31). This energy barrier, of the form $\int Ad \, \Pi$ probably operates only in the early stage of coalescence. Once a sufficiently large puncture is made in the colliding droplets, coalescence would then proceed spontaneously. The tensiolamininometer is an instrument that has been used to measure the energy opposing reduction of liquid lamellae (32) and this quantity has been correlated with foam stabilizing properties of different surfactants, including proteins. A high energy barrier is achieved by a low compressibility and a low tendency to desorb. For proteins, short circuiting of the compressional energy barrier by desorption does not occur to a significant extent. The compressibility of protein films depends on the speed of compression because of the time-dependent relaxation processes. For colliding droplets, the speed of compression is expected to be high and thus the film compressibility correspondingly low. Proteins conform to the general rule (Bancroft rule) that the phase in which the stabilizer is soluble becomes the continuous phase; i.e. they invariably form oil/water emulsions. A possible mechanism for determining emulsion type may be as follows. When two droplets approach, the continuous medium near the point of contact is displaced so that if the stabilizer is soluble in the continuous phase, it is prevented from desorbing (or molecular segments are prevented from being expelled in the case of proteins), thus contributing to a sharp increase in Π, producing a high compressional energy barrier. On the other hand, if molecules (or segments) of the stabilizer are soluble in the dispersed phase, they are free to be displaced from the interface, thereby short-circuiting the energy barrier.

The pushing out of segments into the aqueous phase (formation of loops and tails) could conceivably have two main effects on stability. The interaction between protein segments, as droplets approach, produces a steric barrier (33), a result of the free energy increase accompanying a rise in concentration of protein in the interstitial liquid. It is also possible that loops and tails can form bridges between droplets (34), thus promoting flocculation.

The tendency for proteins to form thick membranes, as a result of interfacial coagulation, needs to be taken into account, especially in emulsions since it occurs more easily at oil/water than at air/water interfaces. These thick films can present a steric barrier simply by preventing the dispersed phase in the droplets from coming into contact. On the other hand, because of their gelatinous and cohesive nature, they are likely to produce flocculation.

Bacterial Adhesion. Bacteria often adhere to interfaces and the resulting biofilms can cause a nuisance on ship hulls, to water reticulation and hydro-electric pipelines and in heat exchangers

(35).   Bacteria also colonize surfaces in soils, streams, oceans and sediments as well as plant roots and other plant surfaces and surfaces of higher organisms, including the skin.   Mixtures of proteins and glycoproteins with other compounds exist in solution at low concentrations in aqueous habitats.   Adhesion of bacteria to a surface is usually preceded by adsorption to form a "conditioning film" (36).   Adsorbed proteins are found to affect the electrical potential and hydrophobicity of the surface and vary in their effects on adhesion of bacteria.   With Pseudomonas NCMB2021, attachment on tissue culture dishes was reduced by both BSA and bovine glycoprotein whereas attachment to petri dishes was only reduced by BSA (36).   Bubble contact angles on the petri dishes were much lower for BSA (< 15°) than for the glycoprotein (64°). This experiment indicates the different effects that adsorbed proteins have on the hydrophobicity of surfaces.

Reactivity of Enzymes.   Enzyme systems occur naturally in plants and animals and their effects can become important in food processing and nutrition.   Enzyme catalysis at the interfaces of dispersed systems is of great interest.   Fundamental surface studies have been mainly done on lipase activity in relation to the digestion of triglycerides.   The results of Verger and co-workers (37-39) are particularly interesting and give new insights into the behavior of proteins at interfaces.   They have used a zero-order trough to study the kinetics of hydrolysis of lipids in monolayers. The trough consists of two compartments connected by a small canal, enzyme being present in only one compartment.   When lipid monolayers are hydrolyzed by lipases present in the sub-phase, the reaction products become soluble and diffuse away, leading to a decrease of area at constant surface pressure.   From the linear slope of the area-time graph, the velocity of the enzyme reaction is directly evaluated.   It is found that proteins in solution at relatively low concentrations may inhibit the hydrolysis of di- and triglycerides by lipases (37).   Marked differences are found in the susceptibility of different lipases to inhibition and also between the inhibiting effects of different proteins.   Experiments using mixed lipid-protein film transfer showed that the inhibition of pancreatic lipase is due to the protein associated with the lipid at the surface and not caused by direct protein-enzyme interaction in the aqueous phase (38).   Using radiolabelled enzymes and proteins, it was found that the inactivation of the pancreatic lipase was correlated with a lack of lipase binding to the mixed lipid-protein film (39).   Since a large fraction of the lipid film remained potentially accessible to the enzyme in the presence of the inhibiting protein, it was believed that the role of the protein was to modify the properties of the interface in such a way as to either cause desorption of the lipase or prevent it from attaching at the interface.   The challenge of how this is achieved is bound to stimulate work which will greatly increase our understanding of interfacial reactions.

Literature Cited

1.   Langmuir, I.; Schaeffer, V.J.   Chem. Rev. 1939, 24, 181-202.
2.   Malcolm, B.R.   Prog. Surface & Membrane Sci. 1973, 7, 183-229.

3. Loeb, G.J. J. Colloid Interface Sci. 1969, 31, 572-574.
4. Cornell, D.G. J. Colloid Interface Sci., 1979, 70, 167-180.
5. Cornell, D.G. J. Colloid Interface Sci., 1984, 98, 283-285.
6. MacRitchie, F. J. Macromol. Sci.-Chem. 1970, A4, 1169-1176.
7. Moore, W.J.; Eyring, H. J. Chem. Phys. 1938, 6, 391-394.
8. Joly, M. Biochem. Biophys. Acta 1948, 2, 624-632.
9. Van Geet, A.L.; Adamson, A.W. J. Phys. Chem. 1964, 68, 238-246.
10. Schulman, J.H.; Rideal, E.K. Proc. Roy. Soc. 1937, B122, 29-45.
11. Minones, J.; García Fernández, S.; Iribarnegaray, S.; Sanz
    Pedrero, P. J. Colloid Interface Sci. 1973, 42, 503-515.
12. MacRitchie, F. Adv. Protein Chem. 1978, 32, 283-326.
13. MacRitchie, F. J. Colloid Interface Sci. 1981, 79, 461-464.
14. MacRitchie, F.; Owens, N.F. J. Colloid Interface Sci. 1969,
    29, 66-71.
15. Henson, A.F.; Mitchell, J.R.; Mussellwhite, P.R. J. Colloid
    Interface Sci. 1970, 32, 162-165.
16. Bigelow, C.C. J. Theoret. Biol. 1967, 16, 187-211.
17. Asakura, T.; Ohnishi, T.; Friedman, S.; Schwartz, E. Proc.
    Nat. Acad. Sci. USA 1974, 71, 1594-1598.
18. Asakura, T.; Minakata, K.; Adachi, K.; Russell, M.O.; Schwartz,
    E. J. Clin. Invest. 1977, 59, 633-640.
19. Langmuir, I.; Waugh, D.F. J. Amer. Chem. Soc. 1940, 62,
    2771-2793.
20. Langmuir, I. J. Amer. Chem. Soc. 1917, 39, 1848-1906.
21. Gonzalez, G.; MacRitchie, F. J. Colloid Interface Sci. 1970,
    32, 55-61.
22. MacRitchie, F.; Ter-Minassian-Saraga, L. Colloids and
    Surfaces 1984, 10, 53-64.
23. MacRitchie, F. J. Colloid Interface Sci. 1985, 105, 119-123.
24. MacRitchie, F.; Alexander, A.E. J. Colloid Sci. 1963, 18, 458-
    463.
25. MacRitchie, F. J. Colloid Interface Sci. 1977, 61, 223-226.
26. Benhamou, N.; Guastalla, J. J. Chim. Phys. 1960, 57, 745-751.
27. Joos, P. Proc. Int. Congr. Surf. Act., 5th, 1968, 2, 513-519.
28. Phillips, M.C.; Evans, M.T.A.; Graham, D.E.; Oldani, D.
    Colloid Polym. Sci. 1975, 253, 424-427.
29. Frisch, H.; Simha, R. J. Phys. Chem. 1957, 27, 702-706.
30. MacRitchie, F. Nature 1967, 215, 1159-1160.
31. MacRitchie, F. J. Colloid Interface Sci. 1976, 56, 53-56.
32. Eydt, A.J.; Rosano, H.L. J. Amer. Oil Chem. Soc. 1968, 45,
    607-610.
33. Vincent, B. Adv. Colloid Interface Sci. 1974, 4, 193-277.
34. Smellie, R.H.; La Mer, V.K. J. Colloid Sci. 1958, 13, 589-599.
35. Characklis, W.G. In Fouling of Heat Transfer Equipment;
    Somerscales, E.F.C.; Knudsen, J.G. Eds.; Hemisphere:
    Washington, D.C.; p 251.
36. Fletcher, M.; Marshall, K.C.; Appl. Environ. Microbiol. 1982,
    44, 184-192.
37. Gargouri, Y.; Julien, R.; Sugihara, A.; Verger, R.; Sarda, L.
    Biochim. Biophys. Acta 1984, 795, 326-331.
38. Gargouri, Y.; Piéroni, G.; Rivière, C.; Sugihara, A.; Sarda,
    L.; Verger, R. J. Biol. Chem. 1985, 260, 2268-2273.
39. Gargouri, Y.; Piéroni, G.; Rivière, C.; Sarda, L.; Verger, R.
    Biochem. 1986, 25, 1733-1738.

RECEIVED January 28, 1987

Chapter 12

Plasma Proteins at Natural and Synthetic Interfaces: A Fluorescence Study

E. Dulos[1], J. Dachary[1,2], J. F. Faucon[1], and J. Dufourcq[1]

[1]Centre de Recherche Paul Pascal, Centre National de la Recherche Scientifique, Domaine Universitaire, 33405 Talence Cedex, France
[2]Laboratoire d'Hématologie, Université de Bordeaux II, 146, rue Léo Saignat, 33076 Bordeaux Cedex, France

Blood clotting proteins bind to charged surfaces mainly by ionic forces often reinforced by hydrophobic contribution. Cardiotoxin binding to heparin, phospholipids and polymers illustrates both possibilities.

The adsorption of the thrombin-antithrombin complex (T-AT) on various anionic anticoagulant polystyrene derivatives results in a quenching of the T-AT fluorescence. The stability of the resulting complexes depends on the polymer, as shown by desorption by polybrene. From competition experiments between heparin and polymers, it is proposed that the T-AT desorption parallels the anticoagulant activity of the polymers.

The binding site of blood clotting proteins on phospholipids involves both charged and zwitterionic lipids. By fluorescence energy transfer, no selectivity is demonstrated for factor Va while its light chain is highly selective for the negatively charged lipids. In contrast, the complex Va - Xa has some selectivity for the zwitterionic lipids. The binding of the vitamin K-dependent factors does not induce phase separation in lipids. A calcium-independent binding is demonstrated and allows to propose a new model for the interaction.

Proteins generally contain aromatic fluorescent residues which are very sensitive to local events. Their interactions with surfaces can therefore be followed by fluorescence techniques.

This study deals with the formation of complexes between blood clotting proteins and natural and artificial surfaces. As these surfaces are generally charged, the behavior of a basic protein, cardiotoxin (CTX), the interaction of which is strictly charge-dependent, is also reported for comparison. Two types of interface have been investigated.

First, phospholipid bilayers which mimic cellular membrane and platelet factor 3, on which several blood clotting factors bind in order to generate the more efficient cascade of enzymatic reactions (1). They are the vitamin K-dependent proteins II, X and IX which are

0097-6156/87/0343-0180$06.00/0

supposed to bind to membranes via Ca^{++} bridges (2) between the Gla residues and the negative charges of the interface and factor V which is not vitamin K-dependent but interacts with membranes by both electrostatic and hydrophobic forces (3).

Second, synthetic polymers designed to be used as parts of intra- or extracorporeal blood circulation. These materials must be able to remain in contact with blood without promoting coagulation. They are polystyrene derivatives (4,5) bearing functional acidic groups of heparin, the natural inhibitor of coagulation which catalyses the inactivation of Thrombin (T) by Antithrombin (AT). Three polymers with different anticoagulant activities (6), the so-called "PSSO$_2$Glu", "PAOM" and "PSSO$_3$", are compared with respect to their adsorption properties towards these plasma proteins.

In the first part of this study, we focus on the binding of CTX to the two types of interface. Secondly, we discuss the binding and desorption of the 1:1 T-AT complex from the polymeric interfaces. Finally, we investigate the charge distribution within the plane of the membrane when blood clotting factors II and V are bound.

Materials and Methods

Materials. Human blood clotting factors II, X and IX were purified from prothrombin concentrates, according to the method of Di Scipio (7). Human factors V and Va light chain (VaLC) were the generous gift of Dr Lindhout, Maastricht University. CTX, (MW 6840) was a gift from Pr Rochat and Dr Bougis, Marseille University. Human Thrombin, T, (MW 38,000) and human Antithrombin (AT), (MW 65,000), were obtained from the Centres de Transfusion Sanguine, respectively of Paris and Lille. Heparin (mean MW $\cong$ 15,000) was from Choay (France). Polybrene was from Serva (F.R.G.).

The polymers used herein were the generous gift of Pr M. Jozefowicz and Dr C. Fougnot, Paris-Nord University (4,6). The polymers consist of a polystyrene backbone bearing, statistically, acidic groups of heparin. In the "PSSO$_3$" derivative, only sulfonate groups are present. Both sulfonate and sulfamide aminoacid groups are present in the other polymers. Specifically, glutamic acid and arginyl-methyl-ester respectively are present in the "PSSO$_2$Glu" and "PAOM" derivatives. PSSO$_2$Glu, PAOM, PSSO$_3$, are respectively very active, active and only slightly active as anticoagulant materials. They are used as fine particles with mean diameter lower than 0.1 µm, giving stable suspensions. Similar light scattering of all polymers at the same concentration indicates that they have almost similar particle size.

Natural beef brain phosphatidylserine (PS) was obtained from Lipid Products (U.K.). Egg lecithin (PC) was prepared in the laboratory. Synthetic lipids, namely dimyristoyl- and dipalmitoylglycerophosphocholine (respectively DMPC and DPPC) and dipalmitoylglycerophosphoserine (DPPS), were obtained from Medmark (F.R.G.). The fluorescent probes, namely 1-acyl-2-(6-pyrenylbutanoyl)-sn-glycero-3-phosphocholine (PBPC) and 1-acyl-2-(6-pyrenylbutanoyl)-sn-glycero-3-phosphate (PBPA) were synthesized in the laboratory. Other analogous probes, with pyrenyldecanoyl chains (respectively PDPC and PDPA), came from KSV (Finland). Their concentrations were determined using $\varepsilon_{347} = 50,000$ $M^{-1}.cm^{-1}$. The Forster distance R_o for the Trp-pyrene pair is 4 nm (9).

Methods. Intrinsic fluorescence measurements were carried out on an
SLM 8000 spectrofluorometer. "Titration" was performed by adding
either phospholipids, heparin or polymers into the reference cuvette
and into the measure cuvette which contains protein solutions at
fixed concentrations. All the corrected spectra were obtained under
the following conditions : excitation wavelength 280 nm, slits 8 mm,
temperature 25°C,background subtraction at every stage of the titration experiments. Moreover, in the presence of polymers, the fluorescence intensities were corrected taking into account the absorbance of the samples at the excitation and at the emission wavelengths,
according to the method of Parker and Barnes (10).

For energy transfer experiments, fluorescence spectra were obtained by subtracting the fluorescence spectra of the phospholipids,
in the absence or in the presence of the energy acceptor, to correct
for light scattering and for the fluorescence of the probe. Energy
transfer efficiencies (Et) were calculated as previously described
(11). The absorbance of samples in the presence of either heparin or
labelled lipids never exceeded 0.1, therefore, the inner filter effect was negligible.

Fluorescence polarization measurements were performed with an
apparatus built in the laboratory and connected to a minicomputer
(Digital LSI II). This was, in turn, connected to a Vax 11/780, thus
allowing complete automatization of the measurements. Diphenylhexa-
triene (DPH) in tetrahydrofuran (6 mM) was added to phospholipids in
amounts never exceeding 1 %.

Results

Interactions of Cardiotoxin with Polyanionic Surfaces

Cardiotoxin-lipid complexes. Due to its effect on membranes, the binding of CTX to lipid interfaces has been extensively studied in the
recent past (8,12). We first demonstrated that it can be followed by
fluorescence changes of Trp11, the emission being blue-shifted by
15 nm from 350 to 335 nm, and the relative intensity increased by
about 200 % when CTX is bound to PS (13) (Fig. 1). This interaction
occurs only with negatively charged species, even in lipid mixtures
consisting of charged and zwitterionic lipids. Such behavior implies
a strict selectivity towards charged groups (13).

Seven charged groups have been shown to be involved at the interface (13). They have been tentatively attributed to the Lys residues in the N-terminal region and at the end of the second loop of
the molecule, mainly based on the structure of analogous protein established by X-ray studies (14). By Raman spectroscopy, it has been
demonstrated that the toxin has very similar secondary and tertiary
structures when in solution or bound to lipids. This is due to the
cross-linking by disulfide bridges. Indeed, when these bridges are
reduced and methylated, the resultant toxin still binds, but becomes
flexible and adopts a different β sheet structure at the interface
(15). Comparison of different iso-toxins leads to the conclusion that
Arg5 residue of CTX IV has a strong stabilizing effect on the protein
at the interface (12).

Because of the strong stability of the protein, the effects of
extreme pH can be investigated. The CTX-lipid complexes previously
formed at pH 7.5 (13) dissociate only at pH values higher than 10,

when the net charge of the protein is severely decreased or, conversely, at pH values lower than 4, when the lipids are protonated. The binding of constitutive peptides of CTX with blocked COO-terminal, namely the peptides 6-12 and 5-12, the sequence of which is Arg5-Leu-Ile-Pro-Pro-Phe-Trp-Lys12, demonstrates that Arg5 plays an important role in maintaining the peptide at the interface up to pH 11 (8). Since charged groups are involved at the interface, it is possible to dissociate the complex by increasing the ionic strength. The release of the protein, inferred from the recovery of the fluorescence of protein free in solution and centrifugation experiments (13), occurs at concentrations of NaCl higher than 1 M (12) (Table I). In contrast, the small constitutive peptides investigated have a lower stability at interface since lower amounts of salt, or calcium in the mM range, are able to dissociate the lipid-peptide complexes (8).

Although the major effect is clearly neutralization of charges involved at the interface, one has to keep in mind that the burying of Trp11, detected by fluorescence, can be interpreted as a passage of this group from a highly polar water environment to a less polar one where solvent relaxation cannot occur. Therefore, hydrophobic forces also are involved and we proposed that the first loop (residues 6-11) may serve as hydrophobic anchor (12). The protein must penetrate through the polar group region of the bilayer and compress the lipid molecules. This has been inferred more directly from a monolayer study by Bougis et al. who demonstrated that at high lateral pressures, compatible with those in a bilayer, the protein occupies at the interface an area of about 500 Å^2 (16).

Cardiotoxin-heparin complexes. The strong binding of CTX to anionic lipids, as indicated above, suggests that the protein would interact with other strongly acidic surfaces or polyanions. Therefore, the fluorescence of CTX in the presence of heparin, one of the more relevant polyanionic compounds involved in coagulation, was studied. As shown in Fig. 1, one can follow the increase of fluorescence intensity of the protein which parallels the addition of heparin. The total increase is relatively small (about + 45 %), and only a slight blue shift of the emission wavelength occurs (about 2 nm). Similarly, it has been found that the antithrombin binding to heparin results in a 25 % increase of the protein fluorescence (17-19). These features immediately indicate that a complex is formed between CTX and heparin , giving rise to an increase of the quantum yield without serious burying of Trp. The stoichiometry of these complexes can be estimated as about 3 to 5 CTX bound per heparin molecule and the dissociation constant as about 0.5 x 10^{-6} M^{-1}. Addition of Ca^{++} in the mM range which allows a quick recovery of the fluorescence of the toxin free in solution (Fig. 2), is interpreted as a dissociation of the complexes and thereby demonstrates the ionic pairing effect.

Cardiotoxin-polymer complexes. Finally, the interaction of CTX was also studied with a polymeric surface, $PSSO_2Glu$, the composition and properties of which resemble those of heparin. As shown in Fig. 1, a result quite different from those for heparin or lipid binding is observed. When the surface is maximally covered, at a polymer to protein ratio of 12 mg of polymer per µmole of CTX, the quantum yield of Trp11 is decreased by about 50 %, while the emission wavelength remains almost unchanged at about 345 nm. This can be interpreted

Table I.   Binding characteristics of cardiotoxin
to different polyanionic interfaces

	PS	PI	Heparin	PSSO$_2$Glu
K_D	$<10^{-7}$	$<10^{-7}$	$0.5\ 10^{-6}$	
n	7	7		
Half dissoc. [NaCl]	1M	0.7M		
[Ca^{++}]	12mM	35mM	2mM	140mM

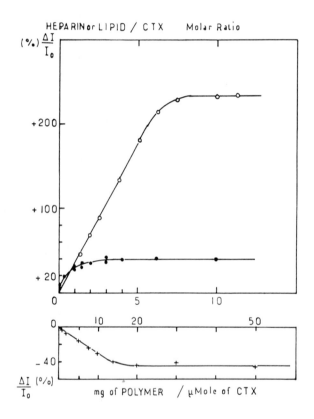

Figure 1.   Changes in the relative fluorescence intensity of CTX
upon binding to : o lipid vesicles (PC–PS 1:1), ● heparin,
+ polymers (PSSO$_2$Glu).

either as a local change in the fluorophore environment or, more probably, in terms of close proximity and contact with quencher groups
at the interface, the polystyrene network being a good candidate for
such an effect. Similarly to what already found with other polyanionic compounds, calcium addition to the CTX-polymer complexes allows
to recover the fluorescence of the free protein as shown on Fig. 2.

In Table I, are compared the stabilities of the complexes between CTX and the three types of interface, as reflected by the concentration of cations required to recover 50 % of the fluorescence
intensity of the protein free in solution. Calcium dissociates the
CTX-heparin complexes at concentrations of a few mM, while concentrations greater than 100 mM are required in order to significantly
displace the same protein from the polymer interface. One has to conclude that very different forces are involved. Heparin can develop
polar forces implying ion pairing between Lys groups of the protein
and sulfate groups of heparin. Calcium is then able to compete effectively through heparin-Ca^{++} interaction. Since the polymer $PSSO_2Glu$
bears acidic groups similar to those of heparin, the better stability
of the CTX-polymer complexes implies other interactions, most likely
hydrophobic interactions with the polystyrene network . However, the
fluorescence emission wavelength of the bound protein does not indicate an hydrophobic environment of Trp, in contrast to what occurs
with lipid interfaces (8,12,13).

Then, as a first step analysis, it can be proposed that CTX can
adsorb onto the polymer surface where it occupies negatively charged
sites. Assuming that the area occupied at such an interface is the
same as at the lipid surface, i.e. 500 $\overset{o}{A}^2$, the area available for negative protein can be roughly estimated at about 2500 cm^2 per mg of
polymer. This area is compatible with the total covering of homogeneous spherical beads of about 0.02 µm mean diameter.

The Thrombin-Antithrombin Complex at the Polymer Interface

Adsorption on polymers. Thrombin and Antithrombin contain, respectively, 7 and 4 tryptophan and several tyrosine residues. The fluorescence spectrum of the T-AT complex, already known (20), is centered around 332 nm, due to Trp residues in a non polar environment.

Addition of polymer to a solution of T-AT complex induces a fluorescence quenching, as seen in Fig. 3, without significant shift of
the emission wavelength. Very similar features are observed when the
excitation wavelength is shifted from 280 to 295 nm, thereby indicating that the fluorescence changes essentially involve Trp residues.
This quenching may originate in the vicinity of quencher groups in
the polystyrene backbone, as already proposed for CTX.

The curves of Fig. 3 are independent of the protein concentration in the range investigated (2 x 10^{-6} to 4 x 10^{-8} M). These binding
curves are very similar for the three polymers. The quenching effect
is saturable and the plateau values are at about - 33 ± 3 % for the adsorption of T-AT on $PSSO_2Glu$, PAOM or $PSSO_3$. The plateau intercepts
the tangent at the origin at a point, the abscissa of which allows to
evaluate as about 0.12 µmole of T-AT per mg of polymer the binding
capacity of the polymers. Similarly, for the adsorption of thrombin
alone, this value has been estimated as about 0.17 µmole of T per mg
of polymer (data not shown). These results have to be compared to

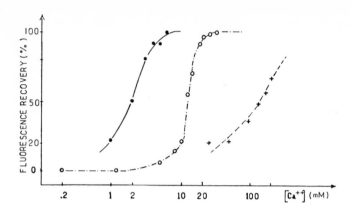

Figure 2. Effects of the calcium concentration on the fluorescence of CTX bound to various interfaces : ● heparin, ○ lipids, + polymers (PSSO$_2$Glu).

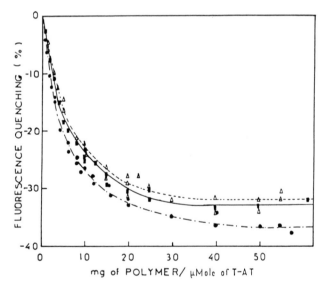

Figure 3. Quenching of the relative fluorescence intensity of T-AT versus the quantity of polymer added. Adsorption of T-AT onto : ■ PSSO$_2$Glu, △ PAOM, ● PSSO$_3$.

those of biochemical studies where the enzymatic activity of thrombin was recorded during its reversible adsorption onto the same polymers (21,22). Taking into account the very different mean diameters of the polymer beads used in both studies (0.02 μm and 25 μm respectively in fluorescence and in biochemical studies) and making the crude hypothesis of spherical beads, rather similar amounts of proteins adsorbed are calculated for both independent techniques.

At this stage of knowledge of these systems, one has to conclude that binding experiments do not allow to discriminate between the different polymers since they have similar binding capacities both for plasma proteins and for CTX used as reference.

Desorption from polymers. The stability of the protein-polymer complexes was investigated in the presence of polycations which have already been shown to dissociate the complexes (21). Polybrene addition to T-AT previously adsorbed on $PSSO_2Glu$, PAOM or $PSSO_3$ induces a partial recovery of the fluorescence of the free protein, as shown in Fig. 4. Since polybrene has no direct effect on the protein fluorescence, these data provide a clear indication that a partial protein release from the polymer surface is occurring. However, this dissociation is not total. Assuming a simple two state equilibrium, one can estimate that about 70 % of the complex is released from $PSSO_2Glu$ and from $PSSO_3$ and only 47 % from PAOM. Maximum desorption from $PSSO_2Glu$ and PAOM is obtained with 45 and 50 moles of polybrene per mole of T-AT, while 80 moles of polybrene are necessary for maximum desorption from $PSSO_3$. Thus, desorption induced by polybrene allows to show differences in the stabilities of T-AT bound to the various polymer interfaces.

Competition between heparin and polymers. In order to better document differences in the abilities of the various polymers to adsorb T-AT, competition experiments between natural and synthetic anticoagulants were performed in two different ways.

First, the effect of heparin on preformed (T-AT)-polymer complexes was investigated. Heparin addition results in a partial recovery of the protein fluorescence of about 17 to 20 % for all polymers. Since the binding of heparin to the 1:1 T-AT complex does not modify the T-AT fluorescence (17), the data were interpreted as reflecting a partial protein release from the polymer surface. However, no quantitative difference between polymers can be shown in these experiments.

Conversely, the effects of polymers on heparin-bound protein were investigated. Polymer addition results in a strong quenching of the initial fluorescence of the heparin-(T-AT) complexes. The formation of ternary complexes between heparin, T-AT and polymer was ruled out by centrifugation experiments on equivalent mixtures. It was shown that heparin is released while T-AT becomes adsorbed onto the added polymer, resulting in the fluorescence quenching observed. Adsorption curves obtained in this case, in the presence of heparin, are almost identical to those of Fig. 3, without heparin. Plateau values are almost at the same level as in Fig. 3 but a slight difference in the maximum effects of the various polymers can be detected. Plateau values systematically decrease in the sequence $PSSO_3$ > PAOM > $PSSO_2Glu$ and the same sequence was found for polymer effects on heparin-thrombin and on heparin-antithrombin complexes (data not shown). This sequence can be related to that of the anticoagulant effects of polymers, namely, $PSSO_3$ < PAOM < $PSSO_2Glu$.

The experimental results of T-AT adsorption on $PSSO_3$ and $PSSO_2Glu$ in the presence and in the absence of heparin were further examined by looking at the normalized curves shown in Fig. 5. Making again the hypothesis of simple equilibria, the amounts of protein bound can be compared using these curves. For the same quantity of polymer added (dashed vertical line in Fig. 5), it can be seen that : i) In the absence of heparin, the fraction of protein adsorbed on $PSSO_2Glu$ (point a) is higher than on $PSSO_3$ (point b). ii) The fraction of protein adsorbed on $PSSO_3$ is the same (point b) in the absence and in the presence of heparin. iii) In the presence of heparin, the fraction of protein adsorbed on $PSSO_3$ is higher than on $PSSO_2Glu$. In other words, in the absence of heparin, the T-AT adsorption occurs more readily on $PSSO_2Glu$ than on $PSSO_3$. Heparin reverses this order so that, in the presence of heparin, the T-AT adsorption occurs more readily on $PSSO_3$ than on $PSSO_2Glu$. This result may be expressed as follows : in the presence of heparin, the T-AT complex can be released more easily from the $PSSO_2Glu$ than from the $PSSO_3$ surface. This observation could be of some significance for the functioning of polymers as anticoagulant materials, if the quantitative difference between their anticoagulant activities is taken into account.

Search for Topological Changes in the Distribution of Charged Groups at the Phospholipid Interface, Induced by Blood Clotting Factors

Due to their requirement for negatively charged groups at the phospholipid interface, one can expect that, upon binding, blood clotting proteins change the lateral distribution of lipids in the membrane at their binding site. In order to document such an effect, the composition of the binding sites of blood clotting factors was estimated in two ways : i) Through changes in the thermotropic properties of lipids. ii) Using fluorescence energy transfer data under isothermal conditions.

Thermotropic behavior of phospholipids in the presence of blood clotting factors. The changes in the transition temperature (Tm) of phospholipid mixtures induced by the presence of the proteins were monitored by fluorescence polarization of the hydrophobic probe, DPH, inserted in the bilayers. These changes are generally interpreted as indicative of phase separation, as already demonstrated for the effect of calcium, at concentrations higher than 10 mM in PS-rich bilayers (23), and in the case of some proteins (24).
    For CTX, it has already been shown that, on pure negatively charged lipids, phase separation is induced between areas of pure lipid and aggregated CTX-lipid complexes whose thermal transitions disappear (24). Moreover, in binary PC-PS mixtures, CTX always strongly shifts the melting temperature Tm, towards that of the pure PC component (25).
    The presence of the clotting factor II (prothrombin) at a lipid to protein molar ratio of 20, induces a shift in Tm of pure lipids such as PS, DMPC, DPPC, from 8 to 12°C for PS, from 21 to 23.5°C for DMPC and from 38 to 43°C for DPPC, as seen on Fig. 6a. These results are indicative of the binding of factor II with the pure lipids. With DMPC and DPPC, they further suggest that, at the very least, an aggregation of the vesicles does occur, the transition profile being

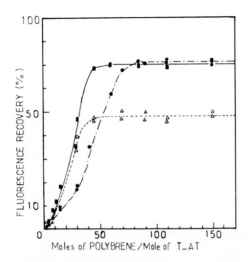

Figure 4. Effects of polybrene on the fluorescence of T-AT initially adsorbed onto polymers : ■ $PSSO_2Glu$, Δ PAOM, ● $PSSO_3$. ($[\text{T-AT}] = 10^{-6}$ M, 40 mg of polymer per μ mole of T-AT.)

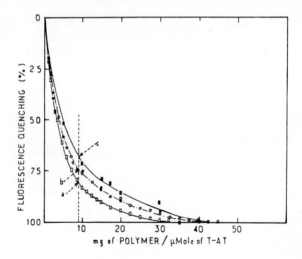

Figure 5. Quenching of the relative fluorescence intensity of T-AT versus the quantity of polymer added. In the absence of heparin, adsorption of T-AT onto :□ $PSSO_2Glu$, o $PSSO_3$. In the presence of heparin, adsorption of T-AT onto :■ $PSSO_2Glu$, ● $PSSO_3$. ($[\text{T-AT}] = 0.5 \times 10^{-6}$ M.)

sharpened and resembling that of large vesicles. This was recently confirmed by Lentz et al.(26)who showed by freeze fracture electron microscopy that fragment I which is the N-terminal part of prothrombin induces a fusion of DMPC vesicles, while negatively charged vesicles remain intact.

With binary mixtures such as DPPC-PS (50-50), the shift of Tm was from 32 to 35°C in the presence of 5 mM calcium. Blood clotting factors II, X and IX induced a further small shift. With the DMPC-DPPS binary mixture, calcium shifts the Tm from 37.5°C down to 33.5°C, this temperature being again decreased by 1 to 2°C in the presence of blood clotting factors (Fig. 6b).

These effects are quite weak when compared to those induced by CTX on the same lipid mixtures : an increase of 8°C for DPPC-PS and a decrease of 9°C for DMPC-DPPS (25). For this protein, which only binds to PS, the Tm observed is that of a PC-enriched phase.

There appears to be no reason to conclude, therefore, that phase separation occurs as a consequence of blood clotting factor binding, in contrast to the conclusion of Mayer and Nelsestuen (27,28) from similar experiments. The lack of phase separation in similar systems was recently confirmed by Lentz (26) describing the whole phase diagram of a PC-PG mixture.

The binding site for blood clotting factors can be enriched in negative lipids but, if domains of different composition compared to the bulk are formed, they must be very small in size and/or their composition must be very close to that of the initial mixture.

Energy transfer experiments with vitamin K-dependent factors. Equimolar mixtures of PC-PS were labelled with either PBPC or PBPA, at the same concentration, namely 3.8 %. The transfer efficiency values (Et) are shown in Table II. Since the two probes have identical spectroscopic features (9), the comparison of Et values for the membranes having similar amounts of label, allows a direct comparison of the lipid environment of the factors. The higher Et value indicates a lower statistical distance and/or a greater number of labelled phospholipids in the neighborhood of the protein.

Results show that the energy transfer process can occur with both types of lipid probe for factors II and IX, indicating that both types of lipid are involved in the binding sites of these factors. Energy transfer efficiency values were lower for PBPC- than for PBPA-labelled membranes in the case of factor II. This indicates a PS selectivity for factor II. On the other hand, for factor IX, the efficiencies are very similar (0.35 and 0.45 for PBPC and PBPA respectively), so that both types of lipid seem to be present in similar proportions in the factor IX binding site.

Experiments were also done with PC-PS (80-20) mixtures. Comparison of the Et values shows that the binding sites for both factors have a higher PS content in the PC-PS (50-50) membrane compared to that in the PC-PS (80-20) membrane. This indicates that the factors do not have a single, well-defined binding site, independent of the initial membrane composition. Another interesting result from our work is that the transfer process was also observed, with both probes, in the absence of calcium. This again leads to the conclusion that the two types of lipid are present in the Ca^{++}-independent binding site of the factors.

Since these experiments are direct studies of binding, the asso-

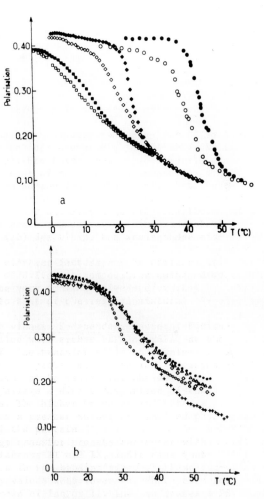

Figure 6.   a) Phase transitions of PS, DMPC, DPPC in the presence of factor II :□ PS alone,■ PS, factor II, R_i = 50 ;◇ DMPC alone, ◆DMPC, factor II, R_i = 20 ; o DPPC alone, ● DPPC, factor II,R_i=50.
          b) Phase transition of DMPC-DPPS (1:1) : + alone ; or with : o CTX ;◇ Ca^{++} 5mM; Δ Ca^{++} 5mM, factor IX,R_i=29; ● Ca^{++} 5mM, factor X, R_i = 20.

ciation parameters can be obtained from the transfer experiments. In
the presence of calcium, the values of Kd are about 10^{-8} M^{-1} for each
factor. These values agree with those of some authors (29,30), but
are lower by one or two orders of magnitude than those determined by
others (31,32). In the absence of calcium, the Kd values we obtained,
which are the first reported in the literature, are of the same order
of magnitude, suggesting that this ion has no great influence on the
binding process.

Energy transfer experiments with factor V, factors Va and VaLC. Similar experiments have been performed with another factor involved in
prothrombin activation but which is not vitamin K-dependent, namely
factor V. The experiments were done with PC-PS (50-50) mixture labelled with 5 % PDPC or PDPA, at a single phospholipid-to-protein molar
ratio, R_i = 530. Fluorescence quenching values are reported in Table
III. They are of the same magnitude with both probes, for factor V,
factor Va and for factor VaLC in the absence of calcium. We conclude,
therefore, that no significant selectivity for a particular class of
lipid occurs.
     On the contrary, in the presence of Ca^{++}, some specificity for
the negatively charged PS occurs with VaLC, since the fluorescence
quenching is greater with PBPA- than with PBPC-labelled membranes.
When the complexes V-Xa and Va-Xa were the energy donors, the fluorescence quenching was greater with PBPC than with PBPA. This indicates some selectivity for PC, in contrast to each individual component
where no selectivity was detected.
     Finally, when the influence of calcium was studied, the results
indicated that the fluorescence intensity corresponding to the fraction of bound protein decreases as calcium concentration increases
(Fig. 7a), reflecting the decrease of the affinity of VaLC for the
membrane (3). However, transfer efficiencies are constant up to 8 mM
calcium, suggesting that calcium does not change the composition of
the binding site (Fig. 7b). At 15 mM, the Et value is greater for
PBPA and decreases for PBPC. This suggests that the binding site is
then enriched in PS molecules. This result runs parallel to the higher degree of phase separation induced by high concentration of Ca^{++}.
     In conclusion, even though electrostatic forces between factors
Va and VaLC with membranes are now well documented, the present results demonstrate that hydrophobic contacts between these proteins
and the hydrocarbon region of the phospholipids are probably also important.

Conclusion

The overall features developed in this study all imply that the
strong binding of the proteins studied requires ion pairing with negatively charged groups located at interface. Such effects are clearly dominant in the case of cardiotoxin and are probably sufficient to
stabilize CTX-heparin complexes documented for the first time in this
report. The binding of CTX and VaLC to lipid interface seems also to
be relevant to such a type of interaction . However, in this case,
the occurrence of hydrophobic "burying" is clearly documented. In the
case of CTX, this is well proved and hydrophobic effect provides a
better stability of the toxin at the lipid interface. This leads to
the suggestion that, although the use of heparin as an agonist against

Table II. Final fluorescence intensities in the absence (I_{FD}) and in the presence of acceptor (I_{FDA}) extrapolated for infinite phospholipid concentration, and energy transfer efficiency values (Et).

For PC-PS (80:20) membrane : $[\text{factor II}] = 0.165\ \mu M$, $[\text{factor IX}] = 0.205\ \mu M$.
For PC-PS (50:50) membrane : $[\text{factor II}] = 0.080\ \mu M$, $[\text{factor IX}] = 0.088\ \mu M$.

Lipids	Label PBPC %	Label PBPA %	Fluorescence of factor II — $Ca^{++}=5$ mM I_{FDA} or I_{FD}	Et	Fluorescence of factor II — EDTA I_{FDA} or I_{FD}	Et	Fluorescence of factor IX — $Ca^{++}=5$ mM I_{FDA} or I_{FD}	Et	Fluorescence of factor IX — EDTA I_{FDA} or I_{FD}	Et
PC-PS 80-20	–	–	62		77		86		87	
	–	3.84	49	0.21	59	0.23	59	0.32	65.5	0.25
PC-PS 50:50	–	–	58		65		62		72	
	3.84	–	46	0.20	56	0.14	40	0.35	67	0.07
	–	3.84	21	0.64	47	0.28	34	0.45	47	0.33
	–	3.04					38	0.38		

Table III. Fluorescence intensities of the proteins in the presence of unlabelled PC-PS (50:50) membrane (I_D), in the presence of labelled PC-PS (50:50) membrane (I_{DA}), and of labelled membrane (I_{DA}). Δ is the percentage of fluorescence decrease in relation to I_D. Concentration of the proteins : $22.5 \cdot 10^{-9}$ M.

With Ca^{++} 3 mM	PC-PS I_D	PC – PS + 5 % PDPA I_{DA}	Δ %	PC – PS + 5 % PDPC I_{DA}	Δ %
V	95.5	86	– 10	85.5	– 10
Va	87.5	82.5	– 5.7	83.25	– 4.8
VaLC	90	56	– 37.7	62	– 31
V–Xa	100	90	– 10	80	– 20
Va–Xa	100	90	– 10	86	– 14
Without Ca^{++}					
Va	100	83.5	– 16.5	83.5	– 16.5

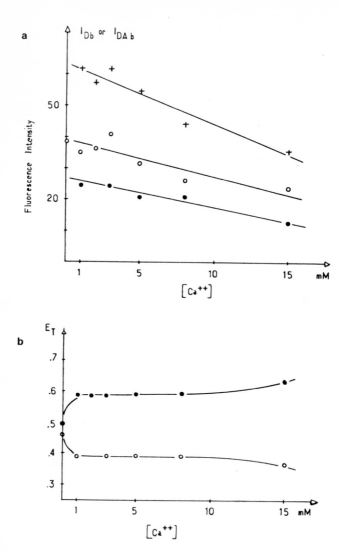

Figure 7.   a) Fluorescence intensities corresponding to the fractions of VaLC bound to the membrane, with (I DAb) and without
(I Db) acceptor, as a function of the Ca^{++} concentration.
($[VaLC] = 22.5 \times 10^{-9}$ M.)
          b) Energy transfer efficiencies as a function of Ca^{++}
concentration.

the toxic action of CTX has been proposed (33), it probably would not be very efficient when dealing with membrane effects. On the other hand, we propose that the polymeric "heparin-like" materials here investigated, would be much more efficient, due to the higher stability of CTX bound to these interfaces.

The binding of CTX to lipid mixtures strongly modifies the thermotropic behavior of the lipids and induces phase separation in the gel phase, as a result of ion pairing between the Lys residues of the peptide and the negative charges of the lipids (25,34). Such an ion pairing may be involved in the binding of the vitamin K-dependent blood clotting factors, particularly, in the absence of Ca^{++}. However, the perturbation of the thermal transition of the lipids is weak. One can then propose that the main effect of Ca^{++} would not be to form bridges between the proteins and the polar head groups of the lipids. More probably, as recently documented from lanthanide luminescence (35), Ca^{++} binding induces a conformational change which shields the Gla negative charges and allows a better approach to the interface by Lys and hydrophobic residues. The lack of strict selectivity of the vitamin K-dependent factors for charged lipids is strongly supported by FET results presented here above and leads to the conclusion that neutral phospholipids are involved in the binding site. The composition of the binding site depends on the initial composition of the mixture, suggesting that the clotting factors are not able to modify the distribution of the lipids in the plane of the membrane, at constant temperature, in order to provide a more favorable binding site.

Comparing now these features to that observed on anticoagulant interfaces, it appears that the binding of proteins is probably stronger and less reversible. This is again well demonstrated by the behavior of CTX. The stronger binding to polymers is not only due to the different nature of ion pairing with sulfate or sulfonate groups, but also to an important hydrophobic contribution, as suggested by desorption experiments. When comparing the various polymers which only differ in their charged groups, anticoagulant activity seems to be correlated to the easier release of T-AT from the interface. This conclusion makes more attractive the idea that these polymers act as catalysts in the inactivation at thrombin by antithrombin, as already proposed (36). Thus, active polymers would be endowed with heparin-like properties. However, it must be pointed out that the experimental conditions used herein are far from physiologic where polymers are in contact with a complex mixture of proteins and cells carried by the blood stream.

Acknowledgments

We are grateful to Pr H. Rochat and Dr P. Bougis, Marseille-Nord University, for providing cardiotoxin, to Pr M. Jozefowicz and Dr C. Fougnot, Paris-Nord University, for providing polymeric heparin-like compounds, to Pr H.C. Hemker and Dr Th. Lindhout, Maastricht University, for providing factor V and to J. Lalanne for the synthesis of the pyrene-labelled phospholipids.

This research was supported by a C.N.R.S. grant, GRECO 130048 "Polymères Hémocompatibles". One of us, J. Dachary, is recipient of an I.N.S.E.R.M. grant n° 835003.

Literature Cited

1. Zwaal, R.F.A. Biochim. Biophys. Acta 1978, 515, 163-205.
2. Nelsestuen,G.L. Fed. Pro.,Fed. Am. Soc. Exp. Biol. 1978,37,2621-25.
3. Van De Waart, P. Ph.D. Thesis, Maastricht University, Holland, 1984.
4. Fougnot, C. ; Jozefonvicz, J. ; Samama, M. ; Bara, L. Annals of Biochem. Eng. 1979, 7, 429-39.
5. Boisson, C. ; Gulino, D. ; Jozefonvicz, J. ; Fischer, A.M. ; Tapon-Brétaudière, J. Thrombosis Research 1984, 34, 269-76.
6. Fougnot,C.;Dupillier, M.P.; Jozefowicz,M. Biomaterials 1983,4,101-4.
7. Di Scipio, G. ; Hermodson, A. ; Yates, G. ; Davie, E.W. Biochemistry 1977, 16, 698-706.
8. Bougis, P. ; Tessier, M. ; Van Rietschotten, J. ; Rochat, H. ; Faucon, J.F. ; Dufourcq, J. Mol. Cell. Biochem.1983, 55, 49-64.
9. Freire, E. ; Markello, T. ; Rigell, C. ; Holloway, P.W. Biochemistry 1983, 22, 1675-80.
10. Parker, C.A.; Barnes, W.J.,in Fluorescence Spectroscopy;Pesce,A.J.; Rosén,C.G.;Pasby,T.L., Ed.; M. Dekker : New York, 1971, pp. 166-68.
11. Prigent-Dachary, J. ; Faucon, J.F. ; Boisseau, M.R. ; Dufourcq, J. Eur. J. Biochem. 1986, 155, 133-40.
12. Dufourcq,J.;Faucon,J.F.;Bernard,E.;Pézolet,M.;Tessier,M.;Bougis, P.;Delori,P.;Van Rietschotten,J.;Rochat,H. Toxicon 1982,20,165-74.
13. Dufourcq, J. ; Faucon, J.F. Biochemistry 1978, 17, 1170-76.
14. Tsernoglou,D.;Petzko,G.A. Proc. Nat. Acad. Sci. USA 1977,74,971.
15. Pézolet, M. ; Duchesneau, L. ; Bougis, P. ; Faucon, J.F. ; Dufourcq, J. Biochim. Biophys. Acta 1982, 704, 515-23.
16. Bougis, P. ; Rochat, H. ; Pieroni, G. ; Verger, R. Biochemistry 1980, 20, 4915-20.
17. Dachary, J. ; Dulos, E. ; Faucon, J.F. ; Boisseau, M.R. ; Dufourcq, J. Colloid and Surfaces 1984, 10, 91-99.
18. Einarsson,R.;Anderson,L.O. Biochim. Biophys. Acta 1977,490, 104-11.
19. Olson,S.T. ; Shore, J.D. J. Biol. Chem. 1981, 256, 11065-72.
20. Wong,R.F.;Windwer,S.R.;Feinman,R.D. Biochemistry 1983,22,3994-99.
21. Fougnot,C.;Jozefowicz,M.;Rosenberg,R.D.Biomaterials 1983,4,294-98.
22. Boisson,C.;Gulino,D.;Jozefonvicz,J. Thrombosis Research in press.
23. Hui,S.W.;Boni,L.T.;Stewart,T.P.;Isac,T. Biochemistry 1983,22,3511-16.
24. Dufourcq, J. ; Faucon, J.F. ; Maget-Dana, F. ; Piléni,M.P.;Hélène, C. Biochim. Biophys. Acta 1981, 649, 67-75.
25. Faucon, J.F. ; Bernard, E. ; Dufourcq, J. ; Pézolet, M. ; Bougis, P. Biochimie 1981, 63, 857-61.
26. Lentz, B.R. ; Alford, D.R. ; Jones, M.E. ; Dombrose, F.A. Biochemistry 1985, 24, 6997-7005.
27. Mayer, L.D. ; Nelsestuen, G.L. Biochemistry 1981, 20, 2457-63.
28. Mayer, L.D. ; Nelsestuen, G.L. Biochim. Biophys. Acta 1983,734,48-53.
29. Lecompte, M.F. ; Miller, I.R. ; Elion, J. ; Bennarous, R. Biochemistry 1980, 19, 3434-39.
30. Mersten,K.;Cupers,R.;VanWinjgaarden,A. Biochem J.1984,223,599-605.
31. Nelsestuen,G.L.;Kisiel,W.;Di Scipio,G. Biochemistry 1978,17,2134-8.
32. Dombrose, F.A. ; Gitel, S.N. ; Zouvalich K. ; Jackson, C.M. J. Biol. Chem. 1979, 254, 5027-40.
33. Lin,S.Y.S.;Hsia,S.;Lee,C.Y. J. Chinese Biochem. Soc. 1973,2,38-44.
34. Faucon, J.F.; Dufourcq, J. ; Bernard, E. ; Duchesneau, L. ; Pézolet, M. Biochemistry 1983, 22, 2179-85.
35. Rhee,M.J.;Horrocks,W.D.;Kosow,D.P. Biochemistry 1982,21,4524-28.
36. Fougnot,C.;Jozefowicz,M.;Rosenberg,R.D. Biomaterials 1984,5, 94-99.

RECEIVED February 18, 1987

Chapter 13

Interaction Mechanism of Thrombin with Functional Polystyrene Surfaces: A Study Using High-Performance Affinity Chromatography

X.-J. Yu[1], D. Muller[1], A. M. Fischer[2], and J. Jozefonvicz[1]

[1]Laboratoire de Recherches sur les Macromolécules, Greco 130048, Centre Scientifique Polytechnique, Université Paris—Nord, Avenue J. B. Clement, 93430 Villetaneuse, France

[2]Département d'Hématologie, Greco 130048, C.H.U. Necker-Enfants Malades, 156, Rue de Vaugirard, 75015 Paris, France

Insoluble sulfonated polystyrenes ($PSSO_3$) and their L—arginyl methyl ester derivatives (PAOM) have been investigated by high—performance affinity chromatogra—phy. Highly active thrombin(Th) was found to be strongly adsorbed on both surfaces and was eluted by increasing ionic strength of the eluent. Compared to the $PSSO_3$ surface, desorption from the PAOM resin occurred at a lower salt concentration and with a more specific enzyme—resin interaction. Masking of the enzyme's active site by its natural inhibitor antithrombin III (AT III) caused a total disappearence of its affinity for both types of surface. It was demonstrated that the active seryl residue of thrombin was involved in enzyme—PAOM interactions. In contrast, this same site of the enzyme was always available after binding to $PSSO_3$ resin. This allowed a biospecific desorption procedure using AT III, heparin (Hep) and Hep—AT III complex. The present work provides further confirmation of the "AT III—like" and "heparin—like" properties, respectively, of PAOM and $PSSO_3$ materials.

In recent years, a great deal of effort has been devoted to the study of antithrombogenic polymers (1-3).It has been shown in the present authors' laboratory (4-7) that modified insoluble polysty—renes substituted either with sulfonate or amino acid sulfamide groups, exhibit anticoagulant activity, when suspended in plasma. This property can be attributed to the adsorption of thrombin and antithrombin III, at the plasma—polymer interfaces (7-9).

These anticoagulant polymers may be divided into two major categories, namely "AT III—like" or "heparin—like" materials, according to their different interaction mechanisms with thrombin

($\underline{1},\underline{2},\underline{7}$). Sulfonate- and L-arginyl methyl ester-substituted polystyrenes have a specific affinity for thrombin, similar to that of its natural inhibitor AT III ($\underline{10}$). In contrast, sulfonated polystyrenes act as a catalyst in Th-AT III complex formation ($\underline{1},\underline{2}$), which is generally considered to be an important property of heparin ($\underline{11}$). Thus, they have been designated heparin-like materials. It has therefore been postulated that the thrombin site involved in the thrombin-PAOM interaction may be different from that involved in the thrombin-PSSO$_3$ interaction.

Based on their specific and reversible interactions with thrombin, PAOM resins have been used as stationary phases for affinity chromatography of the protease ($\underline{12}$). Thus, in a simple one-step chromatographic procedure, human thrombin was isolated from activated prothrombin complex concentrate in high purity and yield ($\underline{13}$). Because of their excellent mechanical properties the resins can also be used as supports for high-performance liquid affinity chromatography (HPLAC). Thrombin adsorption and desorption processes have been investigated using this method. Since a smaller amount of sample can be chromatographed in a shorter time ($\underline{14}$), this HPLAC method would be useful for further studies of the thrombin-polymer interaction mechanisms.

In the present paper, we report high-performance affinity chromatography of thrombin in presence of AT III and Hep, using two types of resins as stationary phases : either heparin-like PSSO$_3$ or AT III-like PAOM. In order to differentiate their mechanisms of interaction with thrombin, we examined the chromatographic behavior of thrombin in the presence, or in the absence of AT III and/or heparin. Finally, thrombin was injected on the columns at low ionic strength. The desorption of bound thrombin from the two solid surfaces was then carried out using AT III, heparin and the AT III-Hep complex, to elucidate the specificity of the interactions involved.

Experimental

<u>Reagents</u> Human α-thrombin (3000 NIH U/mg) and antithrombin III (3 IU/mg) were purchased from the Centre National de Transfusion Sanguine (Paris, France) and from the Centre Régional de Transfusion Sanguine (Lille, France). Hog intestinal heparin(H 108),(173 IU/mg, MW 10700 daltons) was supplied by Institut CHOAY Paris, France). The column loadings generally used were 90 units of thrombin in 45 µl of elution buffer (0.05 M phosphate, 0.1 M NaCl, pH= 7.4) or 0.25 unit of antithrombin III in 10 µl of buffer.

<u>Preparation of Complexes</u>. 25 µl of thrombin solution (50 NIH U) were mixed with AT III (1 IU) or heparin (25 µg) in a 4:1 molar ratio of inhibitor or polysaccharide to enzyme ($\underline{15}$). The mixture was incubated at 37 °C for 15 min($\underline{11},\underline{16}$). Heparin-AT III complex was prepared in a 1:1 molar ratio under similar conditions.

<u>Preparation of Chromatographic Supports</u>. Styrene-divinylbenzene copolymers (Bio-Beads SX2, 200-400 mesh) from BioRad, France, were first chlorosulfonated ($\underline{17}$). Further reaction of the chloro-

sulfonate groups with L-arginyl methyl ester via formation of
sulfamide bonds or total hydrolysis of this chlorosulfonated
polymer in 2 M NaOH solution leads to PAOM or $PSSO_3$ resins, respectively (4,12). The percentage of the unsubstituted, sulfonated and
L-arginyl methyl ester-substituted monomer units are represented
respectively by X,Y,Z in Figure.1.

Chromatographic Procedure   Before packing, the resins were washed
in 2 M aqueous sodium chloride solution and then in Michaelis
buffer 0.026 M sodium barbital, 0.026 M sodium acetate, 0.1 M
sodium chloride,(pH = 7.3). The fine particules were eliminated by
flotation. The chromatographic columns (25 x 0.4 cm ID) were packed
using the slurry method, with a suspension of about 3g of resin in
a high ionic strength solution phosphate buffer 0.05 M, NaCl 2 M,
(pH=7.4). The column was equilibrated for several hours by washing
with the chromatographic eluent buffer at a flow-rate of 1ml/min.

The HPLC apparatus consisted of a three-head (120°) chromatographic pump (Merck LC 21B), connected to a Rheodyne 7126 injection valve (Sample loop 100 µl). A variable-wavelength UV-Visible
detector (Merck-LC 313) and the gradient system were connected to
an Epson QX-10 computer. The chromatographic signal was monitored,
integrated and stored by the computer. These various components
were obtained from Merck-Clevenot (Nogent-sur-Marne, France).

In a first series of chromatographic experiments, thrombin, Th-
AT III or Th-Hep samples in 0.1 M NaCl phosphate buffer were
injected on the columns and then eluted by increasing the salt
concentration in a linear gradient 0.1 M to 2 M NaCl in 0.05
phosphate buffer (pH=7.4). In a second series of chromatographic
experiments, thrombin was first injected at low ionic strength.
Desorption of the bound thrombin from the solid surfaces was
performed by injecting an excess of AT III, Hep or AT III-Hep
complex. The flow-rate was decreased to 0.2 ml/min. The elution was
stopped for 5 minutes just after injection of the desorbing
substance. This procedure was designed to minimize possible kinetic
effects in this process.

Results and Discussion

Elution of   α-thrombin. Highly active enzyme was chromatographed
on the two resins under gradient elution conditions (Figures 2A and
2B). The chromatograms show that thrombin in 0.1 M sodium chloride
in phosphate buffer was strongly adsorbed on both supports.
Desorption of the bound enzyme from PAOM resin occurred at around
1.7 M salt concentration (Figure 2A). As previously reported (14),
the ionic strength required for desorption varied with the chemical
and physical properties of the resins and with the elution conditions, but never exceeded 2 M NaCl. However, thrombin adsorbed on
$PSSO_3$ surface could not be eluted at ionic strength lower than 2 M
$NaCl$ (Figure 2B). This result suggests that thrombin-$PSSO_3$ interactions are stronger than those of thrombin -PAOM and is consistent
with the relative magnitudes of the affinity constants of thrombin
for these polymers. These were previously determined from Langmuir
isotherms, to be 10^7 1 M^{-1} and 10^6 1 M^{-1} for $PSSO_3$ and PAOM
resins, respectively (1,2,7,18).

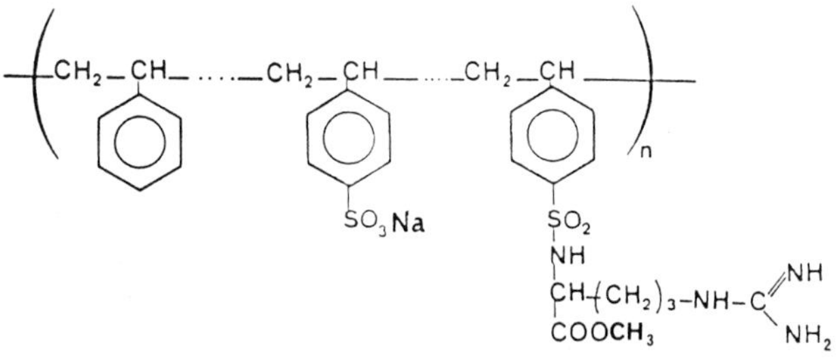

	X	Y	Z
PAOM	11.7%	83.7%	4.6%
$PSSO_3$	11.7%	88.3%	0%

Figure 1 : Structure and percentage of different monomer units of sulfonated polystyrene ($PSSO_3$) and its L-arginyl methyl ester derivative (PAOM).

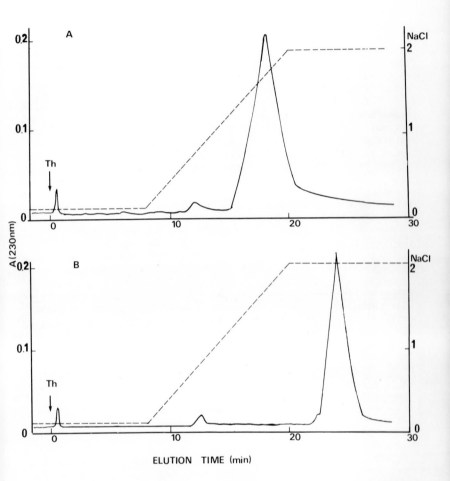

Figure 2 : Elution of human α-thrombin at 25 °C on PAOM (A) and
$PSSO_3$ (B) resins with a linear gradient from 0.1 to 2 M sodium
chloride in 0.05 M phosphate buffer, pH = 7.4 ; flow-rate,
1 ml/min.

<u>Stepwise Elution of Antithrombin III or Heparin</u> Both types of
material have affinities for several other plasma proteins (6,10).
Antithrombin III, for example, has an affinity constant of about
10^5 l M^{-1} for either PAOM or PSSO$_3$ solid surface (8,18). Consequently, retention on these resins was observed in low-pressure
affinity chromatography, and only at a low ionic strength (10). It
should be noted that the adsorption of thrombin was far greater
(100-fold) than that of its inhibitor under the same experimental
conditions.

Under HPLAC conditions, similar elution curves for AT III and
Heparin were observed on PAOM and PSSO$_3$ stationary phases. Antithrombin III was immediately eluted by 0.1 M sodium chloride
solution and no detectable protein desorption was observed at 2 M
ionic strength (Figure 3A). Thus the inhibitor was not retained by
either PAOM or PSSO$_3$. Compared with the findings mentioned above,
this can be explained in terms of kinetic effects. By decreasing
the elution flow-rate and simultaneously overloading the columns,
retention of a small amount of AT III (1%) was observed at the same
ionic strength, i.e. 0.1 M NaCl.
Figure 3B illustrates that heparin was not retained on either of
the two supports in 0.1 M NaCl. This effect may be attributed to
electrostatic repulsion at the interface between the anionic
mucopolysaccharide and the strongly anionic sulfonated resins.

<u>Stepwise Elution of Th-AT III Complex.</u> It has been widely reported
that thrombin inhibition by antithrombin III requires a number of
critical amino acid residues in each of the two proteins (11,20).
These binding sites are essentially an active-center serine of the
enzyme and an arginyl residue of the inhibitor. In addition, the
inherently slow formation of thrombin-antithrombin III complex is
accelerated by heparin (11). Although the mechanism of this
catalysis is still under investigation (11,21,22), it has been
shown that the binding sites involved in the thrombin-antithrombin
III interaction differ from those of heparin for the two proteins
(23,24,25).

The PAOM resins were designed to mimic antithrombin III
affinity for thrombin by partially substituting, on the backbone of
the synthetic polymer, one of the major binding sites of the
inhibitor, namely L-arginyl methyl ester. This strategy would be
justified if a real involvement of the active seryl residue of
thrombin were found in the enzyme-PAOM interactions and not in the
enzyme-PSSO$_3$ interactions.

Injection of the Th-AT III complex on PAOM or PSSO$_3$ supports
was performed at 0.1 M salt concentration. The chromatograms on
both supports show only one peak which appears at 0.1 M NaCl
solution. No peak is detected at high ionic strength (2 M NaCl),
indicating that the Th-AT III complex and the excess of antithrombin III are eluted together at 0.1 NaCl buffer(Figure 4A).
Consequently, it is concluded that thrombin with a masked seryl
site loses its affinity for both resins.

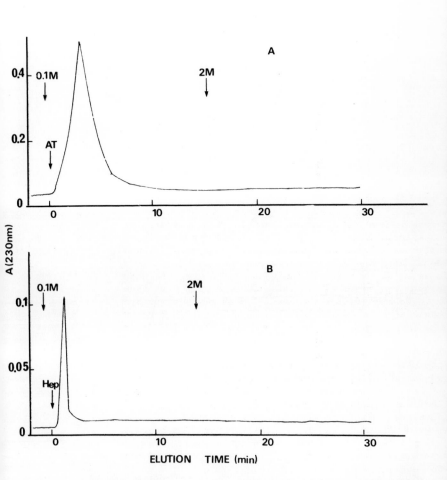

Figure 3 : Stepwise elution of antithrombin III (A) and heparin (B) on PAOM or PSSO$_3$ resins at 25 °C. Both resins have same curve. Eluent : 0.05 M phosphat buffer (pH = 7,4); NaCl 0.1 M and 2 M; flow rate, 1 ml/min.

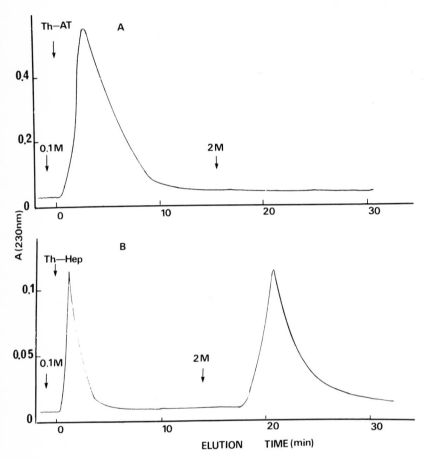

Figure 4 : Stepwise elution of Th–AT (A) and Th–Hep (B) complexes on PAOM or PSSO$_3$ resins at 25 °C. Conditions as in Figure 3. Both resins have same curve.

Assuming that the active seryl residue of thrombin is responsible for thrombin–resin interactions, a lack of affinity of the
inactivated enzyme for these resins would be expected. However,
the thrombin binding site(s) involved in the enzyme–resin interactions may be different from the active seryl center, but may be
located nearby. A simple steric restriction could thus prevent
adsorption.

The thrombin–heparin complex was also injected onto both the
PAOM and $PSSO_3$ supports, to investigate Th/Hep/resin interactions.
The chromatographic profiles on each of the solid phases were
almost identical as shown in Figure 4B. To identify the components
in each peak, the enzyme and heparin were chromatographed separately but under the same experimental conditions (Figures 1 and 3).
The first peak was thus attributed to free–heparin molecules and
the second, eluted at a higher ionic strength, to the enzyme
alone. In contrast to the Th–AT III complex, the Th–Hep complex
was dissociated by both solid phases.

Desorption of Bound Thrombin by Antithrombin III, Heparin or Hep–AT
III Complex. Thrombin was injected onto PAOM or $PSSO_3$ chromatographic supports in 0.1 M NaCl solution. To minimize possible kinetic
effects, the flow-rate was reduced to 0.2 ml/min. After washing
with 0.1 M NaCl buffer, an excess of either AT III, Hep, or AT III–
Hep complex was injected. The flow was then stopped for five
minutes in order to obtain a better exchange of macromolecules
between the mobile and the stationary phases. The objective was to
determine whether AT III, Hep or AT III–Hep complex could act as
eluents for the "biospecific desorption" of the bound enzyme at
0.1 M NaCl.
     The elution curve obtained with PAOM resins (Figure 5A)
demonstrates that none of the three species was able to desorb the
bound enzyme, which was only eluted at 2 M ionic strength. In
contrast, thrombin was totally eluted from $PSSO_3$ resins by AT III
and AT III–Hep complex (Figures 5B and 5C).Only partial desorption
was observed when heparin was used alone (Figure 5D).
     The comparison between the chromatogram of Th–AT complex
(Figure 4A) and the desorption of bound thrombin by antithrombin
III from the PAOM resins (Figure 5A) indicates that the active
seryl residue of thrombin and possibly other binding sites are
blocked by the PAOM resins. However, the same amino acid residues
of bound thrombin on the $PSSO_3$ support are always available to AT
III, in the presence or in the absence of heparin. These comparisons of the "biospecific desorption" behavior of PAOM and $PSSO_3$
resins demonstrate the different mechanisms according to which AT
III–like or heparin–like properties can be attributed to PAOM or
$PSSO_3$ materials, respectively.
     Finally, when the same experiment was performed under hydro-
dynamic elution conditions without any flow stoppage, the bound
enzyme was not subsequently desorbed either by AT III or heparin.
In constrast, AT III–Hep complex was able to effect complete
desorption of thrombin. It is concluded that the well–known
catalytic mechanism of heparin in Th–AT III complex formation also
occurs in the Th/AT III/resin system under our experimental conditions.

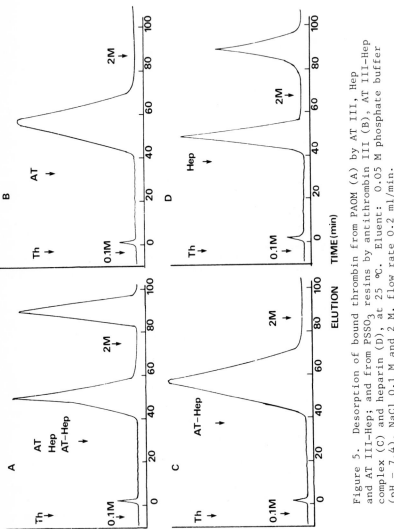

Figure 5. Desorption of bound thrombin from PAOM (A) by AT III, Hep and AT III-Hep; and from PSSO₃ resins by antithrombin III (B), AT III-Hep complex (C) and heparin (D), at 25 °C. Eluent: 0.05 M phosphate buffer (pH = 7.4), NaCl 0.1 M and 2 M, flow rate 0.2 ml/min.

Literature Cited
1.   Jozefowicz, M.; Jozefonvicz, J.; Fougnot, C. and Labarre.
      D., in Chemistry and Biology of Heparin, Lundblad, R.;
      Brown, W.V.; Mamm, K. GH. and Roberts, H.R. Editors),
      Elsevier, Amsterdam 1981, 475.
2.   Fougnot, C.; Labarre, D.; Jozefonvicz, J.; and Jozefowicz,
      M. in Macromolecular Biomaterials ; Hastings, G.W. and
      Ducheyne, P. Editors) CRC Press, Inc., 1984, 215.
3.   Jozefowicz, M.; and Jozefonvicz, J., Pure & Appl. Chem.,
      1984, 56, 1335-1344
4.   Fougnot, C.; Jozefonvicz, J.; Bara, L. and Samama, M., Ann.
      Biomed. Eng., 1979, 7, 429.
5.   Fougnot, C.; Jozefowicz, M.; Bara, L. and Samama, M., Anna.
      Biomed. Eng., 1979, 7, 441.
6.   Fougnot, C.; Jozefowicz, M.; Bara, L., and Samama, M.,
      Thromb. Res., 1982, 28, 37.
7.   Boisson, C.; Gulino, D.; Jozefonvicz, J.; Fischer, A.M.,
      and Tapon-Bretaudiere, J., Thromb. Res., 1984, 34, 269.
8.   Boisson, C.; Gulino, D.; Jozefonvicz, J. in press.
9.   Fougnot, C.; Jozefowicz, M. and Rosenberg, R.D.,
      Biomaterials 1984, 5, 85.
10.   Fischer, A.M.; Tapon-Bretaudiere, J.; Bros, A.; Boisson,
      C.; Gulino, D. and Jozefonvicz, J. Thromb. Haemost., 1985,
      54, 22.
11.   Rosenberg, R.D. and Damus, P.S., J. Biol. Chem., 1973, 248.
12.   Yu, X.J.; Fischer, A.M.; Muller, D.; Bros, A.; Tapon-
      Bretaudiere, J. and Jozefonvicz, J., J. Chromatog., 1986,
      376, 429.
13.   Fischer, A.M.; Yu, X.J.; Tapon-bretaudiere, J.; Muller, D.;
      Bros, A.; Jozefonvicz, J., J. Chromatog. 1986, 363, 95.
14.   Muller, D.; Yu, X.J. Fischer, A.M.; Bros, A. and
      Jozefonvicz, J.; J. Chromatog. 1986, 359, 351-357.
15.   Danielsson, A. and Bjork, I. Febs letters, 1980, 110, 241-
      244.
16.   Machovich, R.; Blasks, G. and Palos, L., Biochim. Biophys.
      Acta, 1975, 359, 153-200.
17.   Petit, M.A. and Jozefonvicz, J., J. Appl. polym. Sci.,
      1977, 2589.
18.   Fougnot, C.; Jozefowicz, M. and Rosenberg, R.D.
      Biomaterials, 1983 4, 294.
19.   Boisson, C.; Brash, J.L. unpublished.
20.   Bjork, I; Danielsson, A.; Fenton II, J.W. and Jornavvall,
      H., Febs Letter, 1981, 126, 257.
21.   Machovich, R., Biochim. Biophys. Acta 1975, 412, 13-17.
22.   Holmer, E.; Soderstrom, G. and Andersson, L-O., Eur. J.
      Biochem 1979, 33, 1-5.
23.   Rosenberg, R.D. and Rosenberg J.S., J. Clin. Invest., 1984,
      74, 1-6.
24.   Li, E.H.M.; Orton, C. and Feimman, R.D., Biochem., 1974,
      13, 5012.
25.   Hatton, M.W.C. and Regoeezi, E.Thromb. Res., 1977, 10, 645

RECEIVED January 30, 1987

Chapter 14

Kinetics of Protein Sorption on Phospholipid Membranes Measured by Ellipsometry

Peter A. Cuypers[1], George M. Willems, Jos M. M. Kop, Jan W. Corsel, Marie P. Janssen, and Wim Th. Hermens

Department of Biophysics, Biomedical Centre, University of Limburg, P.O. Box 616, 6200 MD Maastricht, Netherlands

The sorption kinetics of prothrombin, fibrinogen and albumin on phospholipid bilayers were studied by ellipsometry. Using an unstirred layer model, it is possible to detect the presence of a transport limitation in sorption kinetics and to estimate the thickness of the unstirred layer.

Prothrombin sorption is reversible and calcium-dependent. The prothrombin association constant K_a is dependent on the surface concentration of protein and on the composition of the phospholipid bilayers, indicating interacting binding sites. The initial rate of prothrombin adsorption is transport limited in all conditions studied. Values of the sorption rate constants k_{on} and k_{off} are dependent on the surface concentration. The rate of adsorption decreases for higher surface concentration and the intrinsic values of k_{on} and k_{off} can be estimated as soon as the adsorption rate drops below the diffusional limit. Similar effects are seen for the adsorption of albumin and fibrinogen.

Prothrombin adsorption remains reversible on pure phosphatidylserine (PS) bilayers and on a mixture of 80% PS and 20% phosphatidylcholine (PC). For PS/PC mixtures with less than 80% PS the initial reversible prothrombin adsorption is followed by a slow second surface reaction which causes irreversible adsorption. A similar slow surface reaction is seen for fibrinogen on 100% PS. Elimination of calcium after adsorption of fibrinogen gives a fast desorption of part of the adsorbed layer, possibly due to increased negative charge of the fibrinogen molecules.

The present research program in Maastricht on protein adsorption

[1]Current address: Dutch State Mines Research BV, P.O. Box 18, 6160 MD Geleen, Netherlands

0097-6156/87/0343-0208$06.00/0
© 1987 American Chemical Society

started in 1976 with the modification of a manual Rudolph ellipsometer such that the adsorption of proteins at solid-liquid interfaces
could be followed automatically under well defined conditions (1).
Initial studies with this instrument showed that the refractive index
and thickness of the adsorbed proteins differed considerably depending on the underlying surfaces. It was also shown that the structure
of a fibrinogen layer after adsorption on the surface changed with
time (2,3). In the following years several techniques were developed
for the deposition of phospholipid mono- and doublelayers on solid
substrates and the stability and temperature-dependent behavior of
these membranes in buffer solutions was studied (4). It was also
shown that the thickness and refractive index of the adsorbed protein
layers can give structural information like the water content, swell-
ing and shrinking of the layer and penetration of the proteins into
the underlaying phospholipid layers (5-10).

   In 1983 we were able to validate two exact formulae, based on
the Lorenz-Lorentz equations, allowing the calculation of the mass of
the adsorbed layer from the refractive index and thickness. This
experimental validation was performed by measuring stacked multilayers of known mass of phosphatidylserine and by the adsorption of
radiolabeled albumin and prothrombin on these multilayers (9).

   Using these equations the dissociation constants K_d and the
number of binding sites of prothrombin on 14:0/14:0 PS and 18:1/18:1
PS (DOPS) monolayers could be measured (9,10). Working in the field
of blood coagulation, this study was extended to the adsorption on
double layers of different mixtures of phospholipids. Binding sites
were found with at least two different dissociation constants for
prothrombin on DOPS double layers. Values of the dissociation constants were strongly influenced by adding phosphatidylcholine (DOPC)
to the bilayers (7,10). Until then, conclusions were based on equilibrium measurements. The prothrombin adsorption on DOPS layers is
reversible and offers an exceptional model for investigation of adsorption as well as desorption kinetics. Using an unstirred layer
model a simple graphical representation of the results allowed detection of transport limitation in sorption kinetics (12). The initial rate of prothrombin adsorption is transport-limited under all
conditions studied. Adsorption of albumin and fibrinogen is much
slower and is determined by the intrinsic rate of protein binding
(12).

   The unstirred layer adsorption model can be generalized by the
introduction of surface concentration dependent sorption rate constants k_{on} and k_{off}. This subject is currently being studied as well
as the existence of a second, irreversible, surface reaction following reversible initial adsorption for fibrinogen and prothrombin on a
60% DOPS/40% DOPC mixture.

   A recent modification of the ellipsometer with a rotating reflecting surface, instead of stirring the buffer, allows much better
control of the hydrodynamics in this system (in preparation).

Materials and Methods

The ellipsometer. Ellipsometry is an optical technique for the measurement of changes in the polarization of light caused by reflection. These changes are strongly influenced by the presence of very
thin (0.1 - 100 nm) films of phospholipids and proteins deposited on

the reflecting surface. The instrument is a modified Rudolph & Sons
ellipsometer, type 4303-200E shown in Figure 1. It is equipped with a
He-Ne laser and computer controlled stepping motors on adjustable
optical components, the polarizer (P) and the analyzer (A). Measurement of the changes in polarization during protein adsorption gives
the values of the refractive index and thickness of the adsorbed
protein layer. This measurement is repeated every 4 seconds. A complete description of the instrument is given in refs. (1,3).

Figure 2 shows the three reflecting systems analyzed in each
experiment. First the reflection of a chromium slide is measured in
buffer solution (Figure 2 upper panel) giving the real and complex
part of the refractive index of the chromium surface. Next the slide
is covered with a phospholipid bilayer (Figure 2 middle panel) and
the change in polarization is measured again, giving the refractive
index and thickness of the phospholipid layer. Protein adsorption on
the phospholipids is analyzed according to the system presented in
the lower panel of Figure 2.

The surface concentration Γ, expressed as the mass of the substance adsorbed on the slide per unit surface area, can be calculated
from the refractive index n and the thickness d of the adsorbed
layer:

$$\Gamma = 3d \ (n^2 - n_b^2)/[(n^2+2)(r(n_b^2+2)-v(n_b^2-1))] \tag{1}$$

where r and v are the specific refractivity and the partial specific
volume of the substance deposited on the slide and n_b is the refractive index of the buffer solution (9).

Stacking of the monolayers or multilayers.

Stacking was done with a
preparative Langmuir-trough (Lauda, Type FW-1) according to the method of Blodgett and Langmuir (11,12). The reflecting surface was dipped into the trough with a specially developed dipping machine allowing regular and exactly adjustable dipping- and withdrawing speeds of
about 2 mm/min.

The surface pressure of the phospholipid monolayer on the trough
is critical and depends on the phospholipid. Transport from the
trough to the ellipsometer cuvette was done in a special sample hol-
der such that exposure of the bilayer to air was prevented.

Materials.

The following phospholipids were used:
1,2 dimyristoyl-sn-glycero-3-phosphoserine (DMPS),
1,2 dioleoyl-sn-glycero-3-phosphoserine (DOPS) and
1,2 dioleoyl-sn-glycero-3-phosphocholine (DOPC).

Bovine prothrombin, human fibrinogen and bovine albumin were
used. The proteins were either obtained commercially or prepared according to established procedures (12). Unless mentioned otherwise a
0.05 M Tris-HCl buffer of pH=7.5 was used, containing 0.1 M NaCl and
1.5 mM CaCl$_2$.

Sorption experiments.

Experiments were performed in a cuvette filled
with buffer. The buffer was continuously stirred by a rapidly rotating magnetic stirrer and the temperature was controlled by a Peltier
element. After adding the protein to the cuvette, the adsorption was
followed by measuring the new positions of P and A every 3-5 seconds.
Equilibrium was usually attained after 10-60 min, depending on the

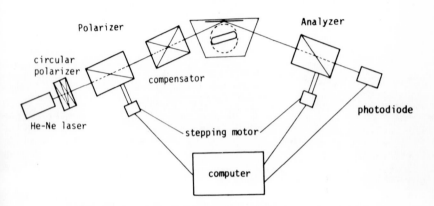

Figure 1. Schematic representation of the automated ellipsometer.

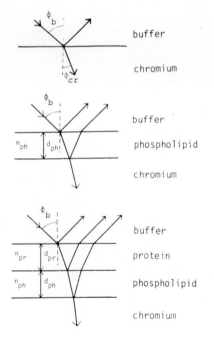

Figure 2. Sequential steps in the determination and calculation of protein adsorption.

protein concentration. Desorption was measured after rapidly flushing
the cuvette with fresh buffer (t=0) and subsequent constant slow
flushing in order to maintain zero bulk concentration of protein.
Figure 3 presents an example of the experimental data obtained during
adsorption and desorption of prothrombin on a DOPS doublelayer. The
positions of the polarizer P and the analyzer A as a function of
time, and the parameters calculated from these data, the refractive
index n, the thickness d and the adsorbed mass Γ are shown.

Analysis of sorption kinetics. During adsorption a concentration
profile C(x,t) of protein is established in an unstirred layer separating the adsorbing surface, situated at x=0, from the buffer
solution. It is assumed that initially no protein is present in the
system and that at time t=0 the bulk concentration of protein in the
buffer is changed to a fixed value C_b. It is also assumed that the
adsorption rate is proportional to the number of free binding sites
and to the protein concentration at the surface. The rate of desorption is assumed to be proportional to the surface concentration. For
this binding model one has:

$$\frac{d}{dt} \Gamma(t) = k(\Gamma)_{on}^{int} (\Gamma_{max}-\Gamma(t)) C(o,t) - k(\Gamma)_{off}^{int} \Gamma(t) \qquad (2)$$

where $\Gamma(t)$ is the adsorbed quantity of protein per unit surface area
at time t. The intrinsic adsorption and desorption rate functions
$k(\Gamma)_{on}^{int}$ and $k(\Gamma)_{off}^{int}$, are dependent on surface coverage, including
effects as: the interaction forces of the protein with the surface,
interaction between the adsorbed protein molecules, orientation effects of the proteins on the surface etc. For this case the diffusion
equation

$$\frac{\partial}{\partial t} C(x,t) = D \frac{\partial^2}{\partial x^2} C(x,t)$$

was solved numerically with the boundary conditions

$$D \frac{\partial}{\partial x} C(o,t) = k(\Gamma)_{on}^{int} (\Gamma_{max}-\Gamma(t)) C(o,t) - k(\Gamma)_{off}^{int}\Gamma(t) = \frac{d}{dt}\Gamma(t)$$

$$C(\delta,t)=C_b$$

where D is the diffusion constant of the protein and δ is the thickness of the unstirred layer.

As a linear concentration gradient of protein builds up in the
unstirred layer within a second after changing the buffer concentration to C_b (12) the boundary conditions can be written

$$\frac{d\Gamma}{dt} = D \frac{\partial}{\partial x} C(o,t) = D(C_b-C(o,t))/\delta \quad \text{or} \quad C(o,t) = C_b - \frac{\delta}{D} \frac{d\Gamma}{dt} .$$

Inserting this expression for C(o,t) in equation 2 one obtains:

$$\frac{d}{dt} \Gamma(t) = k(\Gamma)_{on}^{app} (\Gamma_{max}-\Gamma(t))C_b - k(\Gamma)_{off}^{app} \Gamma(t), \text{ with} \qquad (3)$$

$$k(\Gamma)_{on}^{app} = k(\Gamma)_{on}^{int} D/(D+\delta k(\Gamma)_{on}^{int}(\Gamma_{max}-\Gamma)) \text{ and} \qquad (3a)$$

$$k(\Gamma)_{off}^{app} = k(\Gamma)_{off}^{int} D/(D+\delta k(\Gamma)_{on}^{int}(\Gamma_{max}-\Gamma)). \qquad (3b)$$

It follows from this equation that the presence of an unstirred layer

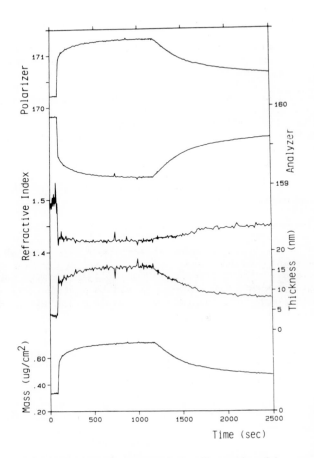

Figure 3. Adsorption and desorption of prothrombin on a DOPS doublelayer. The adsorbed mass is calculated from the changes in the refractive index and thickness, which in their turn are calculated from the changes in polarizer and analyzer readings (see text).

leaves the binding equation 2 apparently unaltered with apparent rate functions depending on the transport parameters as well as on the intrinsic rate functions. During desorption ($C_b = 0$) one has $d\Gamma/dt = -k(\Gamma)_{off}^{app}\Gamma$ and $k(\Gamma)_{off}^{app}$ can thus be directly estimated from the desorption curve. The value of $k(\Gamma)_{on}^{app}(\Gamma_{max}-\Gamma)$ can then be calculated from the adsorption curve as $(\frac{d\Gamma}{dt} + k(\Gamma)_{off}^{app}\Gamma)/C_b$. Plotting this expression as a function of Γ it follows from equation 3a that if $\delta k(\Gamma)_{on}^{int}\Gamma_{max}$ is much larger than D, the plot will approach a constant value of D/δ for small values of Γ. This allows direct determination of δ from the initial part of the plot if the value of D is known.

Results

The equilibrium binding constant as a function of surface concentration. We measured the equilibrium values of the adsorbed amount of prothrombin on a double layer of DOPS in Tris-HCl buffer as a function of the prothrombin concentration (10,12). A plot of the reciprocal concentration of bound protein as a function of the reciprocal protein concentration in buffer allows determination of the association constant K_a defined by $K_a = \Gamma_{eq}/(\Gamma_{max}-\Gamma_{eq})C_b$. Figure 4 shows K_a as a function of the surface concentration. The lowest surface concentration for which this relation could be measured was about 0.08 $\mu g/cm^2$ giving a K_a of about 10^9 M^{-1}. Although the ellipsometric detection limit is much lower, such high K_a values imply almost irreversible adsorption and the accuracy of the measurements becomes insufficient. The highest surface concentration of 0.2 $\mu g/cm^2$ is close to Γ_{max} in this system. The value of K_a here is 5.6×10^7 M^{-1}. Figure 4 suggests an exponential dependence of K_a on the surface concentration.

The sorption rate constants as a function of surface concentration. In a previous paper (12) it was shown that the adsorption of prothrombin on a double layer of DOPS in Tris-HCL buffer is transport-limited for low surface concentrations. In Figure 5 this adsorption is analyzed for higher surface concentrations. At a surface concentration of 0.1 $\mu g/cm^2$ the adsorption rate drops below the diffusion limit and the value of k_{on}^{int} can be calculated. First, the value of $k_{on}^{app}(\Gamma_{max}-\Gamma)$ is calculated from the data as explained in the Methods.

using the values of $D = 4.8\times10^{-7}$ cm^2 s^{-1} and $\delta = 4.4\times10^{-4}$ cm (12) the value of k_{on}^{int} can then be calculated from equation 3a. Figure 6 shows the same calculations for fibrinogen-adsorption. In this case the desorption rate can be neglected. For the adsorption at pH=6.0 the adsorption of fibrinogen is transport limited up to a surface concentration of 0.16 $\mu g/cm^2$. For increasing surface concentration k_{on}^{int} thereafter is lowered by about three orders of magnitude. At pH=7.5 the total adsorption of fibrinogen is much less and $\Gamma_{max}=0.14$ $\mu g/cm^2$.

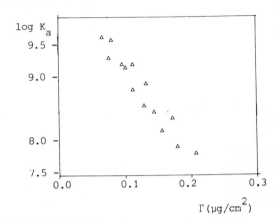

Figure 4. The equilibrium association constant K_a, as a function of surface concentration, for prothrombin adorption on a DOPS doublelayer.

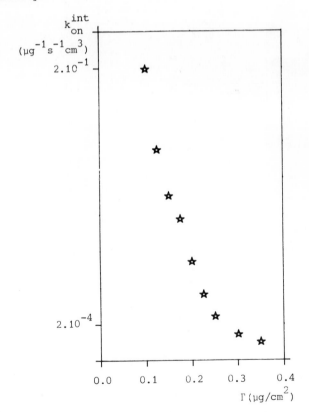

Figure 5. The intrinsic adsorption rate function, as a function of surface concentration, for prothrombin adsorption on a DOPS doublelayer.

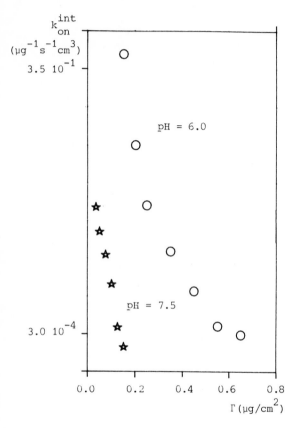

Figure 6. The intrinsic adsorption rate function, as a function of
surface concentration, for fibrinogen adsorption on a DOPS doublelayer.

The initial value of $k_{on}^{int}=5.5\times10^{-3}$ μg^{-1} s^{-1} cm^3 for $=0.05$ $\mu g/cm^2$ is much lower in this case and drops to $k_{on}^{int}=8\times10^{-5}$ μg^{-1} s^{-1} cm^3 for $\Gamma=$

0.14 $\mu g/cm^2$. These results show that the drop in k_{on}^{int} is much steeper for pH=7.5 than it is for pH=6.0 and starts at very low surface concentrations.

Figure 7 shows the intrinsic adsorption rate constants of albumin for two conditions: near the isoelectric point and at a more physiological pH. In all conditions tested, albumin adsorption on DOPS double layers was never transport limited and so the intrinsic constants equal the apparent constants. Near the iso-electric point (pH=6.0) of albumin Γ_{max} = 0.15 $\mu g/cm^2$ and the drop in k_{on}^{int} is only a factor 10 and less steep than it is at pH=7.5 where Γ_{max} is only 0.03 $\mu g/cm^2$. From the adsorption at pH=7.5 it is apparent however that even at a surface concentration below Γ=0.005 $\mu g/cm^2$ values of k_{on}^{int} are already decreasing.

<u>Time-dependent reversibility of prothrombin and fibrinogen adsorption.</u> The equilibrium binding constant of prothrombin on DOPS bilayers can be calculated both from equilibrium measurements and dynamic measurements ($K_a=k_{on}/k_{off}$), indicating that this system is completely reversible (12). If prothrombin is adsorbed on a mixture of 60% DOPS and 40% DOPC however the system is no longer reversible. Figure 8 shows the adsorption from a buffer with 20 $\mu g/cm^3$ prothrombin on a 60% DOPS/40% DOPC doublelayer. In situation A desorption was started 100 seconds after the addition of the protein to the buffer; in situation B the desorption started after 1000 seconds. In the first case there is a rapid complete desorption whereas in the second case only 25% desorbs rapidly and subsequent desorption is much slower.

This behavior was also seen for fibrinogen on a 100% DOPS double layer at pH=7.5 and NaCl 0.1 M. as shown in Figure 9. Here desorption started at t=60 seconds results in considerable desorption. If the desorption is started after 600 seconds, however, hardly any desorption occurs.

<u>Calcium-dependent reversibility of fibrinogen adsorption.</u> Figure 10 shows the adsorption and desorption of fibrinogen on a double layer of DOPS in Tris-HCl buffer 0.05 M pH=7.5 with 20 $\mu g/cm^3$ fibrinogen. Without Ca^{2+} the value of Γ_{max} = 0.27 $\mu g/cm^2$. After the maximum value was reached, 10 mM calcium was added and the surface concentration increased to Γ=0.33 $\mu g/cm^2$ (not shown in the figure). Adsorption of 20 $\mu g/cm^3$ fibrinogen in the presence of 10 mM $CaCl_2$ gives a surface concentration of Γ = 0.50 $\mu g/cm^2$. When equilibrium was reached, EDTA was added to remove the Ca^{2+} ions and a rapid desorption was seen to at a value of Γ = 0.27 $\mu g/cm^2$.

Discussion

For the interpretation of the results we used the following simple model: The proteins at the surface have an attractive interaction A with the surface and a repulsive interaction B between adsorbed molecules. The ratio of A and B will determine the total amount of the protein adsorbed at the surface. The intrinsic adsorption rate func-

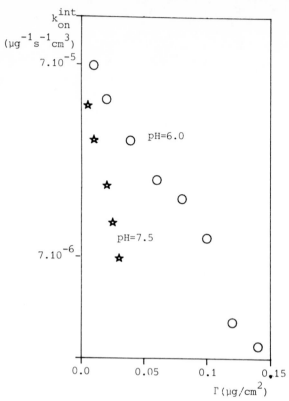

Figure 7. The intrinsic adsorption rate function, as a function of surface concentration, for albumin adsorption on a DOPS double-layer.

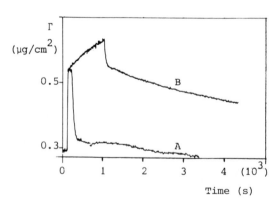

Figure 8. Time-dependency of the desorption of prothrombin on a 60% DOPS/40% DOPC mixture (see text).

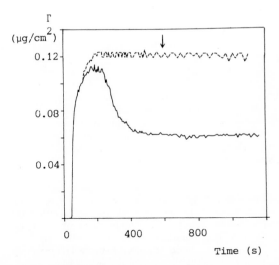

Figure 9. Time-dependency of the desorption of fibrinogen on a DOPS doublelayer (see text).

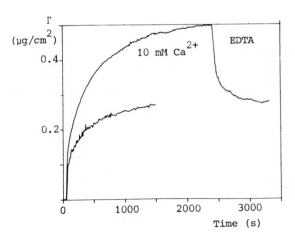

Figure 10. Effect of calcium, and addition of EDTA, on the adsorption of fibrinogen on a DOPS doublelayer (see text).

tion of prothrombin in Figure 5 drops three orders of magnitude if the surface concentration increases from $\Gamma = 0.1$ µg/cm^2 to $\Gamma = 0.3$ µg/cm^2. This is consistent with the model if we assume that the interaction A does not change, whereas the distance between the protein molecules becomes smaller and so the repulsive force B between the negatively charged molecules will increase, resulting in a lower k^{int}_{on}. Measurements on fibrinogen give the same decrease of k^{int}_{on} values. Near its isoelectric point, at pH=6.0, the negative charge of the molecules is much smaller than at pH=7.5 resulting in a much smaller repulsion and so the decrease in values of k^{int}_{on} is less steep than at pH=7.5. As a result Γ_{max} at pH=6.0 will be much higher than at pH=7.5. As the adsorption of fibrinogen at pH=7.5 is not diffusion limited, k^{int}_{on} can be measured even at a surface concentration as low as $\Gamma = 0.005$ µg/cm^2. At this low Γ the decrease in k^{int}_{on} already starts. This behavior is even better demonstrated by the adsorption of albumine where the lowest measurable surface concentration at pH=7.5 is 0.003 µg/cm^2 which is only about 3% of the total coverage. The change in the equilibirum association constant K_a as a function of surface coverage (Figure 4) also fits the model of interaction between adsorbed molecules. All these data are suggestive of a repulsion model. However, more quantitative results will be needed to test which repulsion model will give surface concentration-independent constants (13).

Figures 8 and 9 suggest that in the case of irreversible adsorption, a reversible initial adsorption is followed by a second irreversible reaction. This behavior was clearly seen at 60% PS/40% PC with prothrombin. For fibrinogen adsorption (Figure 9) preliminary experiments show that the conditions of adsorption can be also modified in such a way that an initial reversible adsorption is followed by a second irreversible reaction. This initial reversible reaction was seen on a doublelayer of DOPS at pH=7.5 and a sodium chloride concentration of 0.1 M. Making the conditions more favorable for adsorption, like lowering the pH, lowering the ionic strength or using different surfaces like chromium or silicon surfaces, the reversible phase disappears, probably because of a much faster second reaction. The effect of calcium on the fibrinogen adsorption also fits the model as calcium will decrease the repulsive force B caused by the negative charges of the molecules, thereby facilitating adsorption. This experiment also shows that under these conditions fibrinogen adsorption can be partly reversible. The addition of calcium at the end of the adsorption gives Γ_{max}=0.33 µg/cm^2 whereas the adsorption in the presence of calcium gives Γ_{max}=0.50 µg/cm^2. This might indicate that the structure of the layer and the orientation of the molecules in the layer are dependent on rate of adsorption. In the presence of 0.1 M NaCl much less adsorption was observed (Γ=0.18 µg/cm^2) and the calcium effect was not seen. The observation that adsorption is maximal at the iso-electric point and that the pH dependence of Γ_{max} is symmetrical around the i.e.p. (unpublished results) also supports the repulsion model. This pH dependency was also observed by others (14).

Another complication is the effect of multiple stepwise additions of protein in adsorption experiments. If one adsorbs prothrombin on a 80% PS/20% PC double layer by repeated addition, for instance of 0.1 µg/cm^3 after previous adsorption of 0.1 µg/cm^3, the resulting surface concentration is much lower than after a single addition of 0.2 µg/cm^3. Similar effects were observed by other authors (15). These observations also may indicate that the structure of the adsorbed layer is dependent on the adsorption rate, as we already saw with fibrinogen, and that the second surface reaction is also surface concentration dependent. Other techniques, like calorimetry (14), using completely different surface/volume ratios with low adsorption rates and low surface concentrations may yield results in which the presence of an initial reversible phase is not observed.

References

1. Cuypers, P.A. Ph.D. Thesis University of Limburg, Maastricht, The Netherlands 1976.
2. Cuypers, P.A., Hermens, W.Th., Hemker H.C. Annals N.Y. Acad. Sci. 1977, 283, 77-85.
3. Cuypers, P.A., Hermens, W.Th., Hemker, H.C. Anal. Biochemistry 1978, 84, 56-67.
4. Cuypers, P.A., Janssen M.P., Kop, J.M.M., Hermens, W.Th. Surface Science 1980, 96, 555-563.
5. Cuypers, P.A., Corsel, J.W., Janssen, M.P., Kop, J.M.M., Hemker, H.C., Hermens, W.Th. In Surfactants in Solution; Mittal, K.L., Lindman, B. Ed.; Plenum Press, 1984; Vol. 2, 1301-1312.
6. Cuypers, P.A., Corsel, J.W., Kop, J.M.M., Janssen, M.P., Hermens, W.Th. Journal de Physique, Colloque C10 1983, 44, C10-491-494.
7. Kop, J.M.M., Corsel, J.W., Janssen, M.P., Cuypers, P.A., Hermens, W.Th. Journal de Physique, Colloque C10 1983, 44, C10-495-498.
8. Cuypers, P.A., Hermens, W.Th. In Sensors & Actuators; Lodder J. Ed.; Kluwer Technical Books, Deventer, 1986, 143-154.
9. Cuypers P.A., Corsel, J.W., Janssen, M.P., Kop, J.M.M., Hermens, W.Th., Hemker, H.C. J. Biol. Chem. 1983, 258, 2426-2431.
10. Kop, M.J.M.M., Cuypers, P.A., Lindhout, T., Hemker, H.C., Hermens, W.Th. J. Biol. Chem. 1984, 259, 13993-13998.
11. Blodgett, K.B., J. Am. Chem. Soc., 1935, 57, 1007-1022.
12. Corsel, J.W., Willems, G.M., Kop, J.M.M., Cuypers, P.A., Hermens, W.Th. J. Colloid Interface Sci. 1986, 111, 544-554.
13. Karolczak M. J. of Colloid Interface Sci. 1984, 97, 284-290.
14. Norde, W., Lyklema, J. J. of Colloid Interface Sci. 1978, 66, 257-295.
15. Jönsson, U., Ivarsson, B., Lundström, I., Berghem, L. J. of Colloid Interface Sci. 1982, 90, 148-163.
16. Ivarsson, B., Lundström, I. Critical reviews in Biocompatibility 1986; 2, 1-95.
17. Cuypers, P.A., Hemker, H.C., Hermens, W.Th. In Blood compatible materials and their testing; Davids, S., Bantjes, A. Ed.; Publishers Martinus Nijhoff, Dordrecht, 1986, 45-55.

RECEIVED January 13, 1987

Chapter 15

Adsorption and Desorption of Synthetic and Biological Macromolecules at Solid–Liquid Interfaces: Equilibrium and Kinetic Properties

J. D. Aptel[1], A. Carroy, P. Dejardin, E. Pefferkorn, P. Schaaf, A. Schmitt, R. Varoqui, and J. C. Voegel[1]

Institut Charles Sadron (Centre de Recherches sur les Macromolécules—Ecole d'Applications Hauts Polymères), Centre National de la Recherche Scientifique, Université Louis Pasteur, Strasbourg 6, rue Boussingault, 67083 Strasbourg Cedex, France

Adsorption, desorption and exchange processes occurring at the interface between solid adsorbents and solutions of synthetic (polyacrylamide) or biological (albumin, fibrinogen) macromolecules are described and discussed. Experimental data on thermodynamic, structural and kinetic interfacial properties were obtained using radiolabeling, hydrodynamic and optical (reflectometry) techniques. An attempt is made to present an unified description of the coupling between surface chemical kinetics and bulk transport. As a general rule, flexible synthetic macromolecules display interfacial equilibrium and dynamic properties wich are significantly different from those of more rigid protein molecules.

Considerable theoretical and experimental results have been published during the two last decades wich describe the adsorption/desorption process of synthetic or biological macromolecules at solid/liquid interfaces (1–4). In this context, the work developed in our laboratory has been, and is, aimed at: (i) setting up experimental techniques adapted to the description of the main aspects of macromolecular adsorption, namely thermodynamics, structure and kinetics, (ii) analyzing, from a theoretical standpoint, the significance of various structural and kinetic parameters, and also exploring the possibilities and limits of the corresponding experimental techniques, (iii) comparing, where possible, the adsorption/desorption behavior of synthetic flexible polymers with that of proteins.
   The present paper focuses on results obtained recently on equilibrium and kinetic properties of macromolecular adsorption, at various "model" interfaces. We have worked mainly on polyacrylamide as an example of a flexible, water-soluble polymer, and on albumin and fibrinogen as typical plasma proteins. For a number of

---

[1]Current address: CTR Odontologies, Institut National de la Santé et de la Recherche Médicale U.157, 1, Place de l'Hôpital, 67000 Strasbourg, France

0097-6156/87/0343-0222$06.00/0
© 1987 American Chemical Society

studies, these polymers were radiolabeled ($\beta/^3H$ labeling for poly-acrylamide, and $\gamma/^{125}I$ labeling for the proteins). Details may be found in the original publications (5,6).

Equilibrium Properties

After having established contact between a solid surface and a polymer solution, equilibrium is usually attained within hours (sometimes days, in the case of a dispersed solid phase and/or a poor solvent). A comprehensive description of this equilibrium requires both thermodynamic and structural investigations. Therefore, we consider first adsorption isotherms, then we describe the methods and discuss the results of measurements of the adsorbed layer thickness. Finally, we discuss occurrence of interfacial conformational transitions, which have been observed in a number of cases.

Adsorption isotherms. There are many experimental methods adapted to determine adsorption isotherms, but three observations must be kept in mind: (i) macromolecular isotherms are usually of the high-affinity-type; consequently, the sensitivity of the radiolabeling technique is well adapted to the investigation of the initial part of the isotherms; (ii) polydispersity effects have to be taken into account for systems with broad molecular weight distribution, since they generate "round-shaped" isotherms and adsorbance values which depend on the surface/volume ratio (3). In the case of proteins, however, these effects may be neglected at the low concentrations usually investigated (dimers or multimers may appear at higher concentrations); (iii) with an adsorbent in the form of colloidal particles, floculation due to polymer bridging may lead to non-homogeneous adsorption at low bulk concentrations, prior to the establishment of steric stabilization. In the following discussion, we will focus our attention on three main topics: (i) adsorption at low surface coverage (evaluation of the Gibbs free energy of adsorption), (ii) analysis of the overall shape of the adsorption isotherm and (iii) adsorption at high surface coverage, in the plateau domain. It should be mentioned here that a similar, but more detailed analysis of these problems may be found in the work published by Norde et al (7-9).

To describe the interaction of a bulk macromolecule with a surface, a single Langmuir equation may be written:

$$B + S \rightleftharpoons A \qquad (1)$$

B and A symbolize the dissolved and adsorbed polymer states, respectively, while S represents a free site on the surface, i.e. an interfacial microdomain (including solvent and other solutes) having the lateral dimensions of an adsorbed and isolated macromolecule. If we postulate that the only interactions arising between adsorbed macromolecules are surface excluded interactions, due to short-range (hard core) forces, the thermodynamic equilibrium condition leads to the Langmuir equation,

$$K\Phi = \frac{\theta}{1-\theta} \qquad (2)$$

where Φ is the bulk volume fraction of the solute, θ the dimension-less fraction of surface occupied by adsorbed molecules and K the adsorption equilibrium constant.

Let us first analyze the initial part of the isotherm, when lateral interactions among adsorbed molecules remain negligible. Equation (2) simplifies to Henry's law:

$$K = \exp\left(\frac{-\Delta G^\circ}{RT} \right) \cong \frac{\theta}{\Phi} \qquad (3)$$

ΔG° is the standard molar Gibbs free energy of adsorption, and RT the molar quantum of thermal energy. We studied, under physiological conditions of pH and ionic strength, the adsorption of fibrinogen and albumin on adsorbents displaying two different geometries: (i) glass tubes coated internally with complexes of polycations and polyanions derived from polyacrylonitrile (6) and (ii) small columns filled with polystyrene beads chemically modified to obtain so-called "heparin like" polymers (10). The standard Gibbs free energy ΔG° was estimated from the data at low surface coverage. It was found that $-\Delta G^\circ$, which represents the sum of several monomeric binding energies, is not very high, of the order of a few RT. Also, for a given family of polymers, the value of $-\Delta G^\circ$ do not vary much with chemical side chain structure. A tentative explanation is that interactions with the polymer backbone dominate, and those are probably of hydrophobic origin. Finally, preferential adsorption of fibrinogen over albumin was anticipated and indeed observed, but the differences in the binding energies are not great (10) Pefferkorn and coworkers studied the adsorption of a neutral polyacrylamide ($M_w \leq 1.2 \times 10^6$ g.mole^{-1}) on Na-kaolinite (a natural aluminosilicate) and on glass beads which had been surface-modified, to replace a fraction of the silanol (= Si-OH) groups by aluminol (= Al-OH) groups (11). The polymer does not adsorb on the unmodified glass beads, but does adsorb on the modified glass through site-binding, via hydrogen bonds, to the aluminol groups. The equilibrium contants K measured in the low-concentration limit are two orders of magni-tude higher than for the proteins. There are two main explanations for this difference: the polymer, which is both flexible and of high-molecular weight, is able to form a greater number of bonds per molecule with the adsorbing sites scattered on the surface; in addition, hydrogen bonds are more energetic than Van der Waals interactions or entropically driven interactions.

Considerable progress has recently been made in the struc-tural description of adsorbed flexible polymer layers, under θ or good solvent conditions (12,13). However, no analytic formulation exists to relate, at equilibrium, surface and bulk concentrations over the entire bulk concentration range. For protein molecules, displaying different flexibilities and quaternary strutures, the loss of generality is a priori much greater. Hence, to unravel what appears to be a complex process, considerable experimental data must be obtained. Langmuir adsorption isotherms have been reported by several authors (14,15), and we have observed Langmuir isotherms for albumin adsorbed on five different polymer surfaces, all of which display anticoagulant properties (10). The Langmuir model implies, as already explained, that the adsorption energy of a molecule does not depend on surface coverage, the superficial entropy of mixing being proportional to $\ln(\theta/(1-\theta))$. As a consequence, saturation corresponds to a monolayer, when $\theta \cong 1$. Other experimental evidence however, shows that in some instances, a simple Langmuir approach is inadequate or even invalid. The question of reversibility, often raised, will be discussed later. Moreover, multilayer adsorption, or isotherms showing "kinks", are often observed (16-18). With a protein molecule like fibrinogen, which has a complex trinodular structure, we observed at least two adsorption domains. We hypothesized that side-on adsorption occurs at low bulk and surface concentrations. Once the surface becomes saturated with side-on molecules, additional adsorption involves a change in the structure of the adsorbed layer and subsequent binding is less energetic. Beyond the cross-over domain, end-on adsorption or the building up of a second layer, may be postulated. A log-log representation of the data clearly illustrates the existence of two adsorption regimes. Oreskes and Singer (19), for a different system, used a Langmuir representation and arrived at similar conclusions. In the next section, devoted to layer thickness measurements, additional arguments are proposed. With the polyacrylamide/kaolinite system (11), it appears that a Langmuir isotherm provides only a first approximation to fit the data.

If saturation in the adsorption isotherm corresponds to a close-packed layer of specifically-oriented macromolecules, then, for rigid protein molecules, the amount adsorbed should not depend on the nature of the surface. What are the experimental facts in this regard? First it should be recognized that the model of a flat adsorbing surface may not always be appropriate on the scale of macromolecular dimensions. This problem was particularly critical in our own work with chemically modified polystyrene beads, since the accessible adsorption area could only be approximated. Once the plateau amount is estimated, it is possible to calculate the mean area occupied by one macromolecule, and this can be compared to its dimensions in solution. With fibrinogen adsorbed on complexes of polyelectrolytes (6), we found that both dimensions were in reasonable agreement, except with the positively charged surface, where the existence of a bilayer had to be postulated. It should also be kept in mind that the driving force exerted by the surface is able to create a concentrated interfacial gel which offers the possibility of new interactions among the adsorbed molecules, interactions which are much less probable in solution. This is possibly one source of "irreversibility" (see "Kinetic Properties"). The situation is both similar, but different, with a flexible polymer. To illustrate how flexible polymers may interact laterally in the plateau domain, we calculated, for different samples, the ratio between the mean square radius of gyration in solution and the mean area Σ occupied by one molecule at the surface. We see, in Table I, that this ratio remains close to 10, indicating a significant lateral interaction and overlap between the adsorbed chains.

Table I

Comparisons between interfacial and bulk dimensions, for different polyacrylamide samples dissolved in aqueous solutions at ph 4.5. Molecular weights M_w and radii of gyration R_G were determined by elastic light scattering

$M_w \times 10^{-6}$ g.mole^{-1}	ℓ_H Å	R_G Å	R_G/ℓ_H	$R_G{}^2/\Sigma$	$R_G{}^3/\Sigma.\ell_H$
0.61	795	443	0.56	8.4	18.6
0.85	920	542	0.59	9.8	23.0
1.08	1030	625	0.61	10.5	25.4
1.14	1090	646	0.59	10.7	25.3
1.34	1160	712	0.61	11.3	27.6

Layer thickness. To gain deeper insight into the structure of the interfacial layers, we developed two methods to measure their thicknesses: hydrodynamic techniques and the reflectometry of visible light. In the hydrodynamic method, one studies the resistance to fluid flow by an adsorbed, almost impenetrable layer. The process may involve the flow of fluid through a porous and regular structure, the viscous flow of a bead suspension or simply Brownian diffusion of the beads. The related transport parameter (hydraulic permeability, reduced viscosity or diffusion coefficient) is measured before and after adsorption. Its variation is related to the characteristic dimension of the system by a power law (fourth power of the pore radius, third power of the bead volume or inverse of the bead diameter). The higher the absolute value of the exponent, the higher the experimental precision. Therefore, solution or solvent flow through a porous medium appears to be the most sensitive technique and is the one used in our work. Conceptually simple, this method remains, however, technically difficult.

More recently we have developed a variable angle reflectometer allowing a precise determination of the angular dependence of the reflected beam intensity, for "s" (polarization orthogonal to the incidence plane) or "p" (parallel polarization) incident waves. The possibilities and limits of this optical technique, applied to the study of diffuse interfacial layers, have been theoretically analyzed (20). We concluded that, in order to measure the thickness of typical adsorbed layers (from a few nm to 100 nm), studies in the vicinity of the Brewster angle, for "p" waves, are appropriate. As long as the ratio between the layer thickness and the radiation wavelength remains small, i.e. less than 0.1, the characteristic parameters obtained are essentially the same as in classical ellipsometry; that is, one measures the thickness and the refractive index of an equivalent homogeneous and uniform interfacial layer. If we know how the refractive index varies with concentration, in the high concentration domain, then the adsorbed amount may also be computed. Thus, like ellipsometry, reflectometry measures essentially the zeroth and first moments of the monomer concentration distribution away from the surface. In contrast, Varoqui and

Déjardin showed that, for a polymer adsorbed in a θ solvent, the hydrodynamic technique measures a thickness ℓ_H which greatly over-estimates the mean distance of the monomers to the surface (21). De Gennes (22) improved and extended this analysis to good solvent conditions and showed that the hydrodynamic method provides an esti-mate of the cut-off distance characterising the loops protruding into the solution, away from the surface.

We measured the thickness of fibrinogen layers adsorbed onto glass (hydrodynamic method (23)) or silica (optical method, unpu-blished results). These adsorbents show both similarities and diffe-rences. The results are summarized in Figure 1, and suggest the following comments: (i) in the concentration range explored, the interfacial concentration, measured on glass by radiolabeling (6), increases steadily. This is corroborated by the "optical" data; (ii) hydrodynamic thicknesses are systematically higher than the corres-ponding "optical" values; a value close to 120 Å is in reasonable agreement with the lateral dimensions of the dense parts of the molecule; the higher hydrodynamic thickness may be due to the presence of the dangling "α" chains (24). (iii) in both sets of measurements, the thickness reaches an intermediate plateau value for bulk concentrations varying between 5×10^{-4} and 10^{-2} % by weight; the layer of side-on adsorbed molecules fills up progressively, as the surface concentration increases (iv) at a bulk concentration close to 10^{-2} %, there is a change in the layer structure, as molecules continue to adsorb. The exact nature of this "transition" remains conjectural. With hydraulic permeability data, we notice a sudden increase in the layer thickness, and we suggested that at this point the molecules begin to adsorb with a mean orientation orthogonal to the surface (23). In reflectometry, analysis of the results is still being pursued. If we retain the model of an homo-geneous and uniform layer, then the thickness increases progres-sively, as seen in Figure 1, while the mean refractive index decreases abruptly, as the bulk concentration is raised. The model of a bilayer, i.e. the introduction of a concentration profile away from the surface, may provide a better picture of this complex structural reorganisation. If we couple these observations with the above discussion regarding adsorption isotherms, we come to the tentative conclusion that an elongated, and to some extent flexible, protein molecule like fibrinogen displays a "bimodal" adsorption mechanism which remains to be completely elucidated.

With polyacrylamide adsorbed onto modified glass beads, a "hydrodynamic isotherm" (variation of the thickness ℓ_H with the bulk concentration) could not be obtained through measurements of the suspension viscosity, since at low bulk concentration, adsorption induces bead aggregation (25). Table I presents the plateau values of ℓ_H for different samples, and they consistently overestimate the radius of gyration value by about 70 %. It is therefore clear that, because of strong lateral excluded volume interactions, the chain loops are stretched away from the surface. Moreover, Table I also shows that the ratio between the mean monomer density at the surface and in solution (inside the coil) varies between 20 and 30. Thus, despite the excluded volume interactions, one observes that the

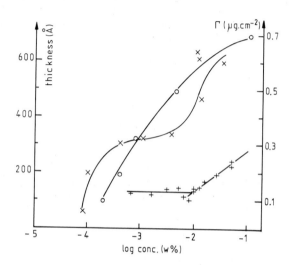

Figure 1: Variations of the surface concentration Γ (o), of the hydrodynamic thickness (x) and of the thickness measured by reflectometry (+) against the bulk equilibrium concentration, expressed in weight percentage, on a decimal logarithmic scale, for fibrinogen adsorbed onto glass (o,x) or silica (+).

chains are literally compressed on the surface, by the interfacial force field. We therefore conclude that macromolecular adsorption induces structural alterations which are usually much more pronounced for flexible polymers than for more or less rigid proteins.

<u>Interfacial conformational transitions.</u> Once adsorption equilibrium is achieved, replacement of the equilibrium bulk solution with pure solvent (or buffer solution) does not generally lead to much desorption, especially with high molecular weight macromolecules. The interfacial layer then remains in a metastable state where it is possible to test its structural response to external pertubations such as variations of pH, temperature, ionic strength or ionic composition. However, if the solvent power is significantly increased with respect to adsorption conditions, significant desorption may occur, and the thickness variations then reflect variations of both molecular interfacial dimensions and concentration.

We analysed the interfacial expansion/contraction of several systems with the hydrodynamic method in a porous medium: with a classical polyelectrolyte, contraction of the layer occurred with increasing salt concentration (<u>26</u>) or calcium ion concentration (<u>27</u>), because of screening or ion binding effects; with polystyrene, Déjardin showed that the thickness of an interfacial layer does not vary with temperature in a good solvent like benzene (<u>28</u>), while an interfacial "pre-collapse" was detected in t-decalin, at the Flory or θ temperature (<u>29</u>); adsorbed poly(glutamic) acid undergoes, as in solution, a reversible helix-coil transition, as a function of pH (<u>30</u>). In other investigations, we also showed that the membrane permeabilities or electrokinetic parameters (membrane potential, transport numbers...) characterizing porous substrates, were also significantly affected (<u>30</u>,<u>31</u>).

In the field of protein adsorption, interfacial transconformations have been published by several authors. We observed an expansion of an adsorbed fibrinogen layer upon increasing the pH. Also thermal denaturation of the end-nodules, detected in solution between 50 and 60 °C (<u>24</u>), was observed by hydrodynamic (<u>23</u>) as well as optical reflectometry measurements (unpublished results).

<u>Kinetic Properties.</u>

Analysis of the dynamic aspects of polymer interactions with an interface is a relatively recent development. Most of the techniques allowing real-time and *in situ* measurements were not used or even available ten years ago. Yet, as will be seen, simple devices can be designed and applied to this problem.

<u>Adsorption kinetics.</u> Kinetically, the adsorption of macromolecules is a complex process in which several steps, occurring almost simultaneously, may be conceptually identified: (i) transport to the interface by diffusion or diffusion/convection, (ii) adsorption/desorption at the interface, described by an interfacial chemical reaction and its related kinetic adsorption and desorption mechanisms, (iii) structural alterations of molecules in contact with the

surface and, at higher occupancy, interactions with other adsorbed molecules, (iv) adsorption competition between molecules of diffe-rent nature or molecular weight.

In the following discussion, we will consider only steps (i) and (ii). When an adsorbing surface is placed in contact with a polymer solution, the molecules which are located in the close vici-nity of the interface become trapped, at a given rate, in the inter-facial potential well. Thus, if x is the distance to the surface, the concentration of the adsorbing species is a function of both time and distance, and is denoted $\Phi(x,t)$ (dimensionless volume frac-tion). If we adopt simple Langmuir chemical kinetics related to equation (1), the adsorption rate may be written:

$$\frac{d\theta}{dt} = k_a \phi [1-\theta] - k_d \theta \qquad d\theta/dt = k_a \Phi(0,t)(1-\theta) - k_d\theta \qquad (4)$$

where k_a and k_d are the adsorption and desorption rate constants, respectively. Depletion in the interfacial layer generates diffusion towards the interface, and we see that there are at least three relaxation times involved in the problem of adsorption rates in the presence of a quiescent solution. These are denoted τ_a (adsorption time), τ_d (desorption time) and τ_D (diffusion time), and are respectively defined by:

$$\tau_a = (\Phi k_a)^{-1} \; ; \; \tau_d = k_d^{-1} \; ; \; \tau_D = L^2/D \qquad (5)$$

L is a characteristic dimension of the macromolecule and D is its diffusion coefficient. For proteins, the characteristic diffusion time varies typically between 0.1 and 100 μs. Two limiting cases must then be considered to describe the initial adsorption rate on a bare surface. If $\tau_D \ll \tau_a$, the rate of the process is controlled by chemical kinetics, and diffusion is fast enough to maintain the interfacial concentration $\Phi(0,t)$ at its bulk value Φ. With experi-mental conditions designed to maintain Φ constant (small surface to volume ratio), the solution to equation (4) is straightforward:

$$\theta(t) = \theta_{eq}\left(1 - \exp[-(k_a\Phi + k_d)t]\right) \; ; \qquad \theta_{eq} = \frac{k_a\Phi}{k_a\Phi + k_d} \qquad (6)$$

This is what we call the Langmuir limit. It should be noted that it is in principle always possible to work experimentally in the vici-nity of this limiting situation, by choosing bulk concentrations which are sufficiently low to have $\tau_a \gg \tau_D$ (see eqn. 5). We have taken advantage of this possibility in the design of experiments using radiolabeling. Conversely, if $\tau_D \gtrsim \tau_a$, diffusion becomes the limiting step: complete depletion in the layer in direct contact with the surface is almost instantaneous. The condition $\Phi(0,t)=0$ has indeed been used by Smoluchowski (32), together with the diffusion equation in the fluid phase, to derive the well-known adsorption rate law, entirely diffusion controlled:

$$\theta(t) \sim \Phi(Dt)^{1/2} \qquad (7)$$

In many cases, the real situation lies between these extre-mes. To provide better insight into this complex process, we present in Figure 2 the variation of the reduced interfacial concentration $\Phi(0,\tau)/\Phi$, as a function of time $\tau=t/\tau_D$, for different values of the coupling ratio $r_a=\tau_D/\tau_a$. These data were obtained by Schaaf and Déjardin ($\underline{33}$), who solved numerically the coupled diffusion and chemical rate equations. As the adsorption relaxation time increases from values close to τ_D, one passes gradually from the diffusion-controlled limit ($\Phi(0,\theta)=0$, $\theta>0$) to the Langmuir limit ($\Phi(0,\theta)=\Phi$, $\theta>0$), reached approximately when $\tau_a>10^5\,\tau_D$.

Varoqui and Pefferkorn have recently published analytic solutions (in the variable $t^{1/2}$) to equation (4), in the limits of low and high surface coverage ($\underline{34}$). They also discussed in detail the assumption of "local equilibrium" proposed in earlier work ($\underline{35}$). This analysis is currently being pursued in our group, by checking different boundary conditions and comparing analytic and numerical solutions. Analysis of the data obtained, under quiescent solution conditions by reflectometry, is also underway.

The limit of chemical control. To eliminate the coupling between transport and chemical kinetics, Pefferkorn suggested the use of the simple "chemical reactor" schematized in Figure 3. A solution of known concentration Φ_a is injected with a speed-variable syringe pump [1] into a cell [3] containing a well-stirred suspension of adsorbing beads. At the outlet, the solution is either collected as fractions at regular time intervals, in the case of β-labeling, or circulated directly past the detector of a γ-scintillation counter [4] coupled to a multichannel analyser and a microcomputer. Thus, the concentration inside the cell, $\Phi(t)$, is recorded as a function of time. Without adsorption (i.e. with no beads or non-adsorbing beads), $\Phi(t)$ relaxes exponentially towards the constant value Φ_a, with a known and controlled relaxation time of the order of several hours. When adsorption takes place, $\Phi(t)$ enters the following mass-balance equation:

$$J_U\Phi_a = V \frac{d\Phi}{dt} + S \frac{d\theta}{dt} + J_U\Phi \qquad (8)$$

where J_U, V and S are the volume flux, the cell volume and the total surface of the adsorbent, respectively. Equation (8) may be coupled to equation (4) if one assumes that the Langmuir model is valid and that there is negligible interfacial depletion. Rapid transport to the interface is enhanced by efficient stirring (see next section) and the possible coupling with transport has been thoroughly checked ($\underline{25},\underline{36}$), and found to be negligible. Comparison of the relaxation times τ_a and τ_D corroborates these arguments.

Significant experimental results have been obtained by Carroy ($\underline{25}$), regarding the adsorption rate of polyacrylamide at the surface of modified glass beads. Under the experimental conditions chosen, desorption was negligible. The influence of both pH and temperature, i.e. of parameters leading to significant variation in the plateau adsorbance, has been checked. Three important results summarized in Table II emerge from this investigation: (1) The adsorption rate curves representing $\theta(t)$ against t could not be

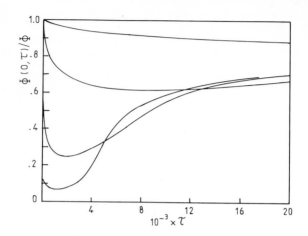

Figure 2: Variations of the reduced interfacial bulk concentration $\Phi(0,\tau)/\Phi$, as a function of the reduced time $\tau=t/\tau_D$, for four different values of the coupling ratio r_a (see text): 1.5×10^{-2}, 1.5×10^{-3}, 1.5×10^{-4}, 1.5×10^{-5}, from the lower to the upper curve, respectively.

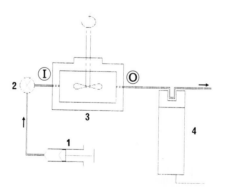

Figure 3: Schematic drawing of the cell designed to work in the limit of chemical control. Symbols are defined in text except: 2: three-way valve; I: inlet; O: outlet.

fitted with one single rate constant K_a throughout the range $0<\theta<1$. Therefore, the values K_a^1 and K_a^2 correspond to low and high surface occupancies, respectively, within ranges of θ values which are given in Table II (note the difference in concentration units, with respect to the constant k_a defined in Eqn.4). (ii) The value K_a^2 is much lower than K_a^1, which means that the probability that a molecule hitting the surface is adsorbed is profoundly reduced, because of a simple entropic barrier exerted by the polymers already adsorbed. (iii) Moreover, a good correlation is observed between the variations of Γ_{max} and of the ratio K_a^1/K_a^2, with the solution pH; in other words, the maximum reduction in the adsorption rate appears at the pH which induces the highest plateau adsorption, that is, the most efficient entropic barrier. Finally, we note that in these experiments, the injection concentration was close to 1.7×10^{-12} mole.cm^{-3}, so that a lower bound of the adsorption relaxation time τ_a is around 10 s, i.e. still higher than $10^5\tau_D$: the Langmuir limit appears to hold throughout the experiment.

Table II

Variation with the pH of the equilibrium solutions, of the kinetic adsorption constants K_a^1 (low surface coverage) and K_a^2 (high surface coverage), for polyacrylamide ($M_w=1.2\times10^6$ g.mole^{-1}) adsorbing on modified glass beads. Γ_{max} is the plateau adsorbance

pH	$\Gamma_{max} \times 10^{13}$ moles.cm^{-2}	$K_a^1 \times 10^{-9}$ cm^3.mole^{-1}.s^{-1}	$K_a^2 \times 10^{-9}$ cm^3.mole^{-1}.s^{-1}	K_a^1/K_a^2
3.0	0.52	9.0 $(0<\theta<0.30)$	1.8 $(0.55<\theta<1.0)$	5.0
4.0	1.11	6.0 $(0<\theta<0.68)$	0.7 $(0.85<\theta<1.0)$	8.6
4.5	1.67	5.5 $(0<\theta<0.75)$	0.55 $(0.84<\theta<1.0)$	10.0
5.5	0.74	5.5 $(0<\theta<0.65)$	0.9 $(0.83<\theta<1.0)$	6.1
7.0	0.51	4.5 $(0<\theta<0.40)$	0.8 $(0.63<\theta<1.0)$	5.6

With radioiodine-labeled proteins, and the possibility of continuous recording, this technique is even more promising. Aptel and Voegel have obtained preliminary results on the adsorption of fibrinogen onto glass (37). A few data are presented in Table III. Since the bulk concentrations injected at the inlet do not correspond to plateau adsorbances, it is possible to estimate, in these experiments, values of the desorption relaxation times which turn out to be of the order of hours, as will be seen later. To fit the adsorption data $\theta(t)$ satisfactorily, three pairs of K_a/K_d values had to be chosen on the adsorption curve, which means that the validity of the simple Langmuir description should be questioned, especially with a complex molecule like fibrinogen. Nevertheless, more experimental evidence on different systems, needs to be obtained to reach firm conclusions. Two further comments are in order. First, the adsorption rates for fibrinogen are more than two orders of magnitude lower than those for polyacrylamide. We already noted a similar difference in the affinity constants $K=K_a/K_d$. Since, as will be seen later, the desorption rates are of the same order of magnitude, these results seem to be coherent. Second, the lower bound of adsorption relaxation times is again close to 10 s, i.e. in the limit of the chemical control.

Table III

Variation of the kinetic adsorption constant K_a and desorption constant K_d describing the adsorption/desorption process of fibrinogen onto glass, for increasing surface coverage and for two different bulk equilibrium concentrations

Protein concentration		0.004 %	0.091 %
Domain 1	Theta	0 ⟶ 0.12	0 ⟶ 0.17
	K_a (cm^3.mol^{-1}.s^{-1})	9.9×10^7	6.7×10^7
	K_d (s^{-1})	9.5×10^{-5}	1.4×10^{-4}
Domain 2	Theta	0.13 ⟶ 0.25	0.21 ⟶ 0.31
	K_a (cm^3.mol^{-1}.s^{-1})	5.0×10^5	3.4×10^5
	K_d (s^{-1})	5.1×10^{-4}	8.4×10^{-4}
Domain 3	Theta	0.26 ⟶ 0.32	0.33 ⟶ 0.45
	K_a (cm^3.mol^{-1}.s^{-1})	5.7×10^5	6.1×10^5
	K_d (s^{-1})	1.3×10^{-4}	1.6×10^{-4}

The Smoluchowski limit. According to the Schaaf-Déjardin numerical analysis, this limit of negligible interfacial bulk concentration during the first part of the adsorption process should be reached as early as $\tau_a < 10^2 \tau_D$. A priori, if we look at the orders of magnitude mentioned previously, such a limit should not apply to a great number of systems. Various experiments performed at the liquid-air interface, seem to be described by the $t^{1/2}$ law ([1],[38]).

However, as far as we are aware, the most convincing set of experimental results has been obtained by Robertson and coworkers ([39],[40]), using the TIRF technique, under flowing conditions. It seems now to be well established that the Lévêque description ([41]) is valid for plasma proteins in the presence of hydrophobic interfaces, in a wide range of concentrations and velocity gradients. Using a protein labeling method, we were also able to observe in situ the variations of the concentration against time, inside a thin silica tube ([42]). The data were not of high precision, but we showed that the power laws relating the adsorption rate to bulk concentration and interfacial velocity gradient were in reasonable agreement with those predicted by the Lévêque solution. However, the absolute values of the interfacial fluxes were lower than those calculated by Lévêque. A possible explanation is that the interfacial concentration does not drop to zero. Results obtained by Mulvihill on the adhesion rate of platelets on different model interfaces, have been interpreted in terms of non-zero interfacial concentration (43): steady-state transport occurs as long as $\Phi(0,t)$ remains close to its minimum value, a value characteristic of the adhesion affinity.

<u>Desorption and Exchange dynamics.</u> The equilibrium between the surface of an adsorbent and a polymer solution may be perturbed by various treatments applied to the system. The simplest is rinsing; i.e. replacing the macromolecular solution by the solvent. This procedure requires a characteristic solution renewal time which may be reduced in some cases to a few seconds. What is then observed, especially with radiolabeling, is a slow, relatively small decrease of the amount adsorbed: this is the origin of the so-called "irreversible adsorption". With heparin-like materials and glass, we indeed observed only slight desorption of fibrinogen, while albumin desorption was significantly greater (44). If we refer to equation (4), with $\Phi=0$, desorption should vary exponentially, with a relaxation time τ_d. This is indeed observed with albumin and the measured relaxation times are of the order of hours. However, if the Langmuir equation is valid, desorption of all the molecules should occur. On the time scale of experiments performed, this is not the case, and there is therefore a population of non-desorbable molecules. At short times, there may be a population of loosely attached and fast desorbing molecules, but only techniques observing *in situ* the structure and composition of the adsorbed layer, with little or no interference of the solution, can provide the appropriate answers. Optical techniques (TIRF, ellipsometry, reflectometry...) seem to be appropriate (45).

Partial desorption of proteins is also induced by changing the solvent power, through temperature, pH or solvent changes. Systematic studies have been published on this subject (46). The desorption rates and amounts measured are of the same order of magnitude as those mentioned previously. These are well-known phenomena in the field of liquid chromatography.

Since an adsorbed layer exists, in the presence of pure solvent, in a metastable state, some authors have concluded that the equilibrium between solution and surface is not a "dynamic" equilibrium: once the plateau surface amount is attained, there should be negligible exchange between surface and solution as a function of time. Thus, the premises of the Langmuir model, or any other model involving a chemical reaction would not be met, thereby casting some doubt on the usual thermodynamic framework. The stimulating approach initiated by Brash and coworkers (14,47,48) shows definitely that partial exchange occurs between solution and interfacial phases. Our recent work (44) on fibrinogen and albumin exchange at the surface of heparin-like adsorbents essentially confirms the earlier results, leading to conclusions which then acquire some generality: two (sometimes three) populations of slowly (sometimes rapidly) exchangeable and of almost non-exchangeable molecules may be identified in the interfacial layer. This remains a crude description, and there is perhaps a whole spectrum of relaxation times which is needed to describe this complex exchange between surface and solution. In this regard a simple chemical equation like equation (1) fails to take account of the evolution of a molecule in the adsorbed state. Contact with the adsorbent surface and, at higher surface coverage, with other adsorbed molecules, may induce some denaturation, especially on hydrophobic materials. It should be remembered that macromolecular concentration polarization at the surface of filters also leads to the introduction of similar concepts and conjectures, regarding "gel-like" layers (49).

Similar, yet significantly different results have been published on the interaction of polyacrylamide with modified glass beads (5). First, desorption is negligible, indeed almost undetectable, once the surface has been rinsed. Second, complete exchange between labeled and non-labeled molecules occurs over a period of a day. All the experiments have been performed with saturated layers, and to explain the long-term exchange, a bimolecular mechanism of exchange, which can be formally deduced from equation (1), has been postulated

$$B + A^* \rightleftharpoons A + B^* \qquad (9)$$

where A^* and B^* represent labeled molecules respectively in the surface and bulk phases. An exponential decrease in the surface radioactivity has been observed for contact times with a solution of non-labeled molecules lasting from 7 to 18 hours. For times less than 7 hours, the exchange rate is faster, but not easily analysed. Thus, the adsorbed molecules do not all have the same dynamics in the interfacial layer.

Conclusion

We have tried to present a unified picture of results obtained on the adsorption/desorption phenomena of both synthetic and biological polymers. It appears that proteins like albumin or fibrinogen do adsorb on many surfaces with little structural alterations at the global molecular level, while the structure of flexible polymers is radically altered once the molecule is trapped in the interfacial force field. Conversely, the slow turnover existing between surface and solution macromolecules was shown to be total with polyacrylamide, while only partial with proteins, especially with fibrinogen. Thus, flexibility and reversibility are preserved in the adsorbed state, for synthetic polymers, even if the time-scale of the molecular dynamics is considerably different from that in solution. For adsorbed proteins, however, the existence of a population of non-exchangeable molecules remains to be fully explained.

Acknowledgments

The authors acknowledge the financial support of the Institut Français du Pétrole and of the CNRS, and the joint project "GRECO Polymères Hémocompatibles".

Literature Cited

1. MacRitchie, F. Adv. Protein Chem. 1978, 32, 283–326.
2. Biomaterials: Interfacial Phenomena and Applications; Cooper, S.L.; Peppas, N.A., Eds.; Adv. in Chemistry Series No. 199; American Chemical Society: Washington, D.C. 1982.
3. Fleer, G.J.; Lyklema, J. In Adsorption from Solution at the Solid-Liquid Interface; Parfitt, C.D.; Rochester, C.A., Eds.; Academic Press: New York, 1983, ch.4, p.153–220.
4. Schmitt, A., in Colloïdes et Interfaces; Cazabat, A.M.; Veyssie, M., Eds.; Les Editions de Physique: Paris, 1983, p.245–288.
5. Pefferkorn, E.; Carroy, A.; Varoqui, R. J. Polymer Sci. Polymer Phys. Ed. 1985, 23, 1997–2008.
6. Schmitt, A.; Varoqui, R.; Uniyal, S.; Brash, J.L.; Pusineri, C. J. Colloid Interface Sci. 1983, 92, 25–34.
7. Norde, W.; Lyklema, J. J. Colloid Interface Sci. 1978, 66, 257–302.
8. Van Dulm, P.; Norde, W.; Lyklema, J. J. Colloid Interface Sci. 1981, 82, 77–82.
9. Koutsoukos, P.G.; Norde, W.; Lyklema, J. J. Colloid Interface Sci. 1983, 95, 385–397.
10. de Baillou, N.; Voegel, J.C.; Schmitt, A. Colloids and Surfaces 1985, 16, 271–288.
11. Pefferkorn, E.; Nabzar, L.; Carroy, A. J. Colloid Interface Sci. 1985, 106, 94–103.
12. de Gennes, P.G.; Pincus, P. J. Physique lett. 1983, 44, L241–L246
13. Eisenriegler, E. J. Chem. Phys. 1983, 79, 1052–1064
14. Brash, J.L.; Samak, Q.M. J. Colloid Interface Sci. 1978, 65, 495–504
15. Chuang, H.Y.K.; King, W.F.; Mason, R.G. J. Lab. Clin. Med. 1978, 92, 483–496.
16. Soderquist, M.E.; Walton, A.G. J. Colloid Interface Sci. 1980, 75, 386–397.
17. Fair, B.D.; Jamieson, A.M. J. Colloid Interface Sci. 1980, 77, 525–534
18. Van Dulm, P.; Norde, W. J. Colloid Interface Sci. 1983, 91, 248–255.
19. Oreskes, I.; Singer, J.M. J. Immunol. 1961, 86, 338–344.
20. Schaaf, P.; Déjardin, P.; Schmitt, A. Revue Phys. Appl. 1985, 20, 631–640.
21. Varoqui, R.; Déjardin, P. J. Chem. Phys. 1977, 66, 4395–4399.
22. de Gennes, P.G. Macromolecules 1981, 14, 1637–1644.
23. de Baillou, N.; Déjardin, P.; Schmitt, A.; Brash, J.L. J. Colloid Interface Sci. 1984, 100, 167–174.
24. Doolittle, R.F. Scientific Amer. 1981, 245, 126–140.
25. Carroy, A. Thèse de Doctorat d'Etat, Université Louis Pasteur, Strasbourg, 1986.
26. Pefferkorn, E.; Déjardin, P.; Varoqui, R. J. Colloid Interface Sci. 1978, 63, 353–361.
27. Déjardin, P.; Toledo, C.; Pefferkorn, E.; Varoqui, R. in Ultra-filtration Membranes, Cooper, A. Ed.; Polymer Sci. and Technol.; Plenum Press, 1980, 13, 203–215.

28. Déjardin, P. Thèse de Doctorat d'Etat, Université Louis Pasteur, Strasbourg, 1981.
29. Déjardin, P. J. Physique 1983, 44, 537-542.
30. Pefferkorn, E.; Schmitt, A.; Varoqui, R. Biopolymers 1982, 21, 1451-1463.
31. Pefferkorn, E.; Schmitt, A.; Varoqui, R. J. Membrane Sci 1978, 4, 17-34.
32. Smoluchowski, M.V. Z. Phys. Chem. 1916, 92, 129-168.
33. Schaaf, P.; Déjardin, P. submitted to Colloids and Surfaces.
34. Varoqui, R.; Pefferkorn, E. J. Colloid Interface Sci. 1986, 109, 520-526.
35. MacCoy, B.J. Colloid Polym. Sci. 1983, 261, 535-539.
36. Pefferkorn, E.; Carroy, A.; Varoqui, R. Macromolecules, 1985, 18, 2252-2258.
37. Aptel, J.D.; Voegel, J.C. submitted to Biophys. J..
38. De Feijter, J.A.; Benjamins, J.; Veer, F.A. Biopolymers 1978, 17, 1759-1772
39. Lok, B.K.; Cheng, Y.L.; Robertson, C.R. J. Colloid Interface Sci. 1983, 91, 104-116.
40. Darst, S.A.; Robertson, C.R.; Berzofsky, J.A. J. Colloid Interface Sci. 1986, 111, 466-474.
41. Lévêque, M. Ann. Mines 1928, 13, 201-409 (especially p.284).
42. Voegel, J.C.; de Baillou, N.; Sturm, J.; Schmitt, A. Colloids and Surfaces 1984, 10, 9-19.
43. Mulvihill, J.N. Thèse de Doctorat d'Etat, Université Louis Pasteur, Strasbourg, 1984.
44. Voegel, J.C.; de Baillou, N.; Schmitt, A. Colloids and Surfaces 1985, 16, 289-299.
45. Lee, J.J.; Fuller, G.G. J. Colloid Interface Sci. 1985, 103, 569-577.
46. Chan, B.M.C.; Brash, J.L. J. Colloid Interface Sci. 1981, 84, 263-265.
47. Brash, J.L.; Uniyal, S.; Samak, Q. Trans. Am. Soc. Artif. Intern. Organs 1974, 20, 69-76.
48. Brash, J.L.; Uniyal, S.; Pusineri, C.; Schmitt, A. J. Colloid Interface Sci. 1983, 95, 28-36.
49. Lonsdale, H.K. J. Membrane Sci. 1982, 10, 81-181.

RECEIVED January 28, 1987

Chapter 16

Adsorption to Biomaterials from Protein Mixtures

Thomas A. Horbett

Department of Chemical Engineering, BF-10, and Center for Bioengineering, University of Washington, Seattle, WA 98195

Protein adsorption from relatively complex mixtures is involved in many different applications of biomaterials. A summary of all studies involving protein adsorption that were done in the author's laboratory is given, largely in the form of an annotated table. Certain aspects of the methods, results, and conclusions of studies that pertain especially to adsorption from mixtures are also discussed.

Protein adsorption to implanted biomaterials is believed to play an important role in determining their biocompatibility with various biological systems and tissues. In the cardiovascular system this is particularly clear because the intrinsic clotting system consists of a series of proteins which act at interfaces to induce fibrin clots. In addition, platelet thrombus formation is strongly influenced by the adsorption of proteins to interfaces. Extravascular implants in soft tissue induce the characteristic foreign body giant cell and fibrous capsule formation by cellular mechanisms. These cellular responses are also strongly affected by proteins at the interface, although this is less well documented. Implants in hard tissues such as bone are also very likely to be influenced by adsorption of proteins to the interface. In this type of implant, adsorbed proteins act through direct effects on the adhesion and growth of osteoblasts, and also through indirect mechanisms such as the surprising ability of proteins to affect the corrosion of metallic implants.

In all these applications of biomaterials, adsorption occurs from relatively complex solutions containing many different proteins. For this reason, much of the effort in my laboratories has focused on the behavior of protein adsorption as it occurs from protein mixtures. In this paper, a summary of these studies is provided with the

purpose of illustrating some of the mechanisms by which
complex adsorption processes appear to occur.

The reference list includes all publications from my
laboratory that have involved or been closely related to
protein adsorption studies (1-29). These publications and
their principal results are summarized and listed in
Table 1 in their order of appearance in the literature,
which approximates the sequence in which they were
actually done. Portions of this work have already been
reviewed in more detail elsewhere (26,27). In what
follows, certain aspects of the methods, results, and
conclusions of these studies are discussed.

Methods

The most frequently used method in studies of protein
adsorption in this lab is the ^{125}I protein technique in
which the prelabelled protein of interest is added to a
mixture of other proteins. An *in situ* radioiodination
method was also developed and extensively used in several
attempts at characterizing the entire "spectra" of
proteins adsorbed to polymers from mixtures. Technical
aspects of both methods have been described in detail
(26).

The use of prelabelled proteins in adsorption studies
appears to be a virtual necessity when complicated protein
mixtures such as blood plasma are employed. None of the
physical methods for measuring protein adsorption have the
required degree of specificity, although it is possible
that Fourier transform infrared spectroscopy may someday
prove to be able to discriminate proteins in some
mixtures. Among biochemical methods, the measurement of
antibody uptake by the adsorbed protein might at first
appear to be better than the prelabelled protein method in
that labelling artifacts cannot influence the adsorption
process. However, it is likely that non-specific
adsorption of antibody by the surface, and changes in the
reactivity of the antigenic "epitope" in the adsorbed
protein due to orientational or conformational effects,
will both influence antibody uptake. These problems limit
the usefulness of the antibody method in obtaining
quantitative measurements. Furthermore, the antibody
methods are intrinsically more difficult than prelabelling
methods due to the additional requirements to produce and
purify a specific antibody and establish a calibration
curve with known amounts of adsorbed antigen.

The principal disadvantage of the prelabelled protein
method is the fact that labelled proteins may not adsorb
the same as the unlabelled protein, thus leading to
potentially large errors. In our studies, we have relied
exclusively on ^{125}I proteins, almost always made with a
chemically very mild ICl method. Preferential adsorption
of ^{125}I proteins made in this way does not seem to occur,
at least as judged by the absence of any effect of changes
in the ratio of labelled to unlabelled protein on the

Table 1

Protein Adsorption to Biomaterials:
A Summary of Studies in the Author's Laboratories

Title: **Cell Adhesion to Polymeric Materials: Implications with Respect to Biocompatibility**

Ref. #	Proteins studied	Surfaces	Adsorption from	Techniques
1	Albumin, γ-globulin and fibrinogen	Radiation grafted poly(2-hydroxyethyl methacrylate) and poly (N-vinyl-2 -pyrrolidone) hydrogels on silicone rubber; silicone rubber	Buffer	125I Preadsorption Elution

Results & conclusions

Chick embryo muscle cell plating efficiency was low on albumin and γ-globulin coated surfaces but the fibrinogen and gelatin coated surfaces were strongly cell adherent.

Adsorbed proteins were retained (79% or more) after contact with serum containing medium or after cell attachment.

Title: **Bovine Plasma Protein Adsorption onto Radiation-Grafted Hydrogels Based on Hydroxyethyl Methacrylate and N-Vinyl-Pyrrolidone**

Ref. #	Proteins studied	Surfaces	Adsorption from	Techniques
2	Fibrinogen, γ-globulin, and albumin	hydroxyethylmethacrylate (HEMA), N-vinyl-pyrrolidone (NVP), and methacrylic acid, (MAAc) radiation grafted onto silicone rubber	water; physiologic buffered saline; blood plasma	Ninhydrin Assay 125I

Results & conclusions

At low ionic strength, small amounts of MAAc in HEMA gels caused great increases in γ-globulin adsorption and, among the proteins studied, adsorption increased in the order of the isoelectric pH of the proteins. At physiologic concentrations of salt, adsorption of all the proteins onto MAAc-HEMA hydrogels was low.

Continued on next page

Fibrinogen adsorption onto the most highly grafted poly(HEMA)/Silastic films (5.8 mg/cm²) was the same as onto films having only about one-fifth the graft, but films with grafts much below this point (1 mg/cm²) showed increased adsorption characteristic of the underlying Silastic.

Adsorption isotherms for Silastic, poly(NVP)/Silastic, and poly(HEMA)/Silastic obtained after 20 hrs of adsorption were approximately "Langmuirian" but a Freundlich fit is also possible.

The saturation level of fibrinogen adsorption from buffer varied by about a factor of four in the order poly(HEMA)/Silastic < Silastic ≅ poly(NVP)/Silastic. The NVP graft was intermingled with Silastic so adsorption to pure NVP was not obtained.

Fibrinogen adsorption from citrated blood plasma was in the order Silastic < poly(HEMA)/Silastic < poly(NVP)/Silastic. The different order from buffer adsorption may be due to lipoprotein adsorption from plasma to Silastic.

Albumin and γ-globulin vs. fibrinogen competition in binary mixtures indicated that these proteins may be quantitatively dominant factors in modifying surface adsorption of fibrinogen from plasma.

Title: *Interactions of Blood and Blood Components at Hydrogel Interfaces*

Ref. #	Proteins studied	Surfaces	Adsorption from	Techniques
3	Fibrinogen	Radiation-grafted HEMA/EMA mixtures on Silastic films	Buffer	125I

Results & conclusions

Fibrinogen adsorption decreased with increasing HEMA/EMA ratio.

Title: *A New Method for Analysis of the Adsorbed Plasma Protein Layer on Biomaterial Surfaces*

Ref. #	Proteins studied	Surfaces	Adsorption from	Techniques
4	Fibrinogen, Albumin, IgG	Silicone Rubber, HEMA-Silicone Rubber, EMA-Silicone Rubber, Acrylamide-Silicone Rubber, Polyurethane, Teflon	Buffer; Plasma	In situ radioiodination; 125I; Elution; SDS gel electrophoresis

The presence of many different proteins in the adsorbed layer was detected due to high sensitivity of this newly developed method.

In a plasma recalcification experiment, several unidentified proteins were adsorbed to HEMA but not to Teflon.®

Title: Solution Stability of Bovine Fibrinogen

Ref. #	Proteins studied	Surfaces	Adsorption from	Techniques
5	Fibrinogen	...	Not done	Clottability; SDS gel electrophoresis

Results & conclusions

The stability of specially purified fibrinogen is adequate for adsorption studies, unlike commercial preparations.

Title: Fibrinogen Adsorption to Surfaces of Varying Hydrophilicity

Ref. #	Proteins studied	Surfaces	Adsorption from	Techniques
6; see also 7	Fibrinogen	Radiation grafted HEMA-EMA-PE	Buffer	ESCA; SEM; 125I SDS gel electrophoresis

Results & conclusions

Adsorption isotherms at 37°C on all surfaces fit the Freundlich equation. The Freundlich parameters varied uniformly with copolymer composition.

Adsorption appeared to be essentially irreversible since both desorption into buffer and the rate of exchange between adsorbed and dissolved protein were of the order of weeks.

Desorption was much more rapid on HEMA than on EMA, indicating that weaker surface-protein interactions occur at more hydrophilic surfaces.

Continued on next page

Gel electrophoresis of SDS eluates from surfaces adsorbed at low concentrations revealed normal fibrinogen subunit distribution but additional altered forms were observed on surfaces adsorbed at high concentrations.

Surface HEMA-EMA ratios were found to be the same as the HEMA-EMA ratio in the monomer solutions used for grafting. Micron sized "lumps" were observed on all grafted surfaces.

Title: The Preferential Adsorption of Hemoglobin to Polyethylene

Ref. #	Proteins studied	Surfaces	Adsorption from	Techniques
8	Hemoglobin, fibrinogen, albumin or γ-globulin	Polyethylene, polystyrene, glass	Plasma; red cell suspensions; buffer;binary mixtures	In situ radioiodination; 125 I

Results & conclusions

In situ radioiodinations, isotherms, binary competitions, and plasma adsorption measurements all indicate that plasma protein affinity for polyethylene varies as follows:

Hemoglobin>>fibrinogen>albumin ≅ γ-globulin.

Title: Hemoglobin Adsorption to Three Polymer Surfaces

Ref. #	Proteins studied	Surfaces	Adsorption from	Techniques
9	Hemoglobin, albumin, fibrinogen and γ-globulin	Polyethylene, polyhydroxyethylmethacrylate	Plasma	125I; SDS gel electrophoresis; In situ radioiodination

Results & conclusions

The adsorbed protein layer formed on these surfaces in plasma was found to be greatly enriched in hemoglobin relative to the bulk plasma composition.

Title: **The Kinetics of Adsorption of Plasma Proteins to a Series of Hydrophilic-Hydrophobic Copolymers**

Ref. #	Proteins studied	Surfaces	Adsorption from	Techniques
10,11	Hemoglobin, immunoglobulin G, fibrinogen, albumin	PE, PEMA/PE, PHEMA/PE, P(HEMA-EMA)/PE	Plasma	125I

Results & conclusions

The adsorption of hemoglobin and immunoglobulin G to each polymer increased with time, although the amount of adsorption depended on the polymer.

Fibrinogen and albumin adsorption on PHEMA/PE and P(HEMA-EMA)/PE also increased with time, but on both PE and PEMA/PE adsorption decreased with time following an initially higher adsorption.

Title: **Analysis of the Organization of Protein Films on Solid Surfaces by ESCA**

Ref. #	Proteins studied	Surfaces	Adsorption from	Techniques
12	Hemoglobin	PTFE	Buffer	Variable angle ESCA on prefrozen, in situ sublimed films; 125I

Results & conclusions

On PTFE, hemoglobin appeared to be localized into islands approximately 50 Å in thickness, with relatively little coverage between the islands. In contrast, however, adsorbed hemoglobin appeared to completely and uniformly cover the platinum surface.

Continued on next page

Title: **Adsorption of Proteins from Plasma to a Series of Hydrophilic-hydrophobic Copolymers. I. Analysis with the in situ radioiodination technique**

Ref. #	Proteins studied	Surfaces	Adsorption from	Techniques
13	Plasma, fibrinogen	Copolymers of hydroxyethyl methacrylate and ethyl methacrylate grafted to PE by the radiation technique	Plasma	In situ radioiodination; Elution; SDS gel electrophoresis

Results & conclusions

Nine or more protein peaks were observed after gel electrophoresis of the SDS eluates from the surfaces.

The fraction of total radioactivity associated with adsorbed proteins was found to vary markedly and systematically among the surfaces.

The spectra of proteins adsorbed at 0.5 min was very different than after 2 hours of adsorption.

Iodination of protein mixtures in solutions revealed selectivity for albumin to be low, probably explaining its low labeling in the adsorbed layer.

Title: **Adsorption of Proteins from Plasma to a Series of Hydrophilic-hydrophobic Copolymers. II. Compositional Analysis with the Prelabeled Protein Technique**

Ref. #	Proteins studied	Surfaces	Adsorption from	Techniques
14	Fibrinogen, immunoglobulin G, albumin, and hemoglobin	HEMA-EMA random copolymers radiation grafted to polyethylene	Plasma	$125I$

Results & conclusions

The adsorption of each protein varied in a characteristic way with copolymer composition, probably reflecting a different affinity of the proteins for the various copolymers.

Results & conclusions (cont)

Fibrinogen adsorption was maximal to the 50 HEMA - 50 EMA copolymer, where albumin adsorption was minimal. IgG and hemoglobin adsorption generally increased with EMA content. Adsorption of all proteins was minimal on the 80 HEMA - 20 EMA surface.

Surface area variations among the copolymers, preferential adsorption of 125I proteins, and the possibility of structural degradation of 125I proteins in plasma were investigated but did not appear to influence the adsorption results.

Title: **Surface Enrichment of Plasma Proteins on Polyhydroxyethylmethacrylate-Ethylmethacrylate Polymers**

Ref.#	Proteins studied	Surfaces	Adsorption from	Techniques
15	Fibrinogen, immunoglobulin G, albumin, and hemoglobin	HEMA-EMA copolymers radiation grafted to PE	Plasma	125I

Results & conclusions

Surface enrichment of the proteins, calculated as the ratio of the weight fraction of each protein in the surface and bulk, varied with copolymer composition, indicating substantial differences in the composition of the surface and bulk phases.

Surface enrichment for the proteins were: fibrinogen (0.8 - 3.4), immunoglobulin G (0.8 - 1.5), albumin (0.3 - 0.8) and hemoglobin (150 - 400), depending on the polymer.

Continued on next page

Title: *Thrombotic Events on Grafted Polyacrylamide-Silastic Surfaces as Studied in a Baboon*

Ref. #	Proteins studied	Surfaces	Adsorption from	Techniques
16	Plasma	pAAm/Silastic, others	Plasma	Silver staining; Elution; SDS gel electrophoresis

Results & conclusions

Large differences in protein adsorption patterns were observed on various polymers. The silver staining method was found to be marginally sensitive and quite variable.

Title: *Adsorption of Proteins from Artificial Tear Solutions to Poly(methyl methacrylate-2-hydroxyethyl methacrylate) Copolymers*

Ref. #	Proteins studied	Surfaces	Adsorption from	Techniques
17	Lysozyme, albumin, and immunoglobulins	Copolymers of poly(methyl methacrylate, poly(2-hydroxyethyl methacrylate), polymerized as slabs	Artificial tear solution	125I

Results & conclusions

Albumin and lysozyme adsorption were minimal on the 35% HEMA / 65% EMA copolymer but IgG adsorption was at a maximum on this surface.

Albumin and lysozyme (but not IgG) uptake were apparently much higher on HEMA rich gels, but we now know this was an artifact due to uptake of unbound iodide by permeable, HEMA-rich polymers.

A linear variation in material composition did not result in a linear variation in the composition of the absorbed layer protein.

Title: **Protein Adsorption on Biomaterials**

Ref. #	Proteins studied	Surfaces	Adsorption from	Techniques
18	Many	Many	Varied	Varied

<u>Results & conclusions</u>

A review of the literature revealed that:

Adsorbed proteins influence and are important in a variety of biological processes, as illustrated by many examples.

Many major questions concerning adsorbed protein remain unanswered, e.g., why certain proteins are more surface active than others.

General properties of proteins, including amphoteric nature, large size, and internalization of hydrophobic groups, are probably the most important sources of surface activity of proteins.

The local concentration of adsorbed proteins is extremely high (ca. 1000 mg/ml), possibly enhancing reaction rates and cellular responses.

The organization of the adsorbed protein layer is influenced by differences in the surface activity of proteins, competition in mixtures, and complex displacement kinetics.

The structure of adsorbed proteins is not well understood, including whether proteins are denatured at the solid/liquid interface.

Cellular response to implanted polymers is the result of specific interactions between components of the adsorbed protein layer on the polymer and the cell periphery. Cellular interactions with foreign materials are probably controlled by the presence of specfic proteins on the foreign surface at sufficiently high surface density and degree of reactivity to elicit a strong cellular response.

Continued on next page

Title: von Willebrand Factor/Factor VIII Adsorption to Surfaces from Human Plasma

Ref. #	Proteins studied	Surfaces	Adsorption from	Techniques
19	von Willebrand factor/factor VIII (vWF/ VIII)	HEMA-EMA-glass, PE, pAAm/Silastic	Plasma	$125I$ vWF/VIII; $125I$ antibody to vWF/VIII

Results & conclusions

VWF/VIII adsorption to the various surfaces differed but was generally quite low (0.24 to 6.5 ng/cm²)

Title: **Mass Action Effects on Competitive Adsorption of Fibrinogen from Hemoglobin Solutions and from Plasma**

Ref. #	Proteins studied	Surfaces	Adsorption from	Techniques
20	Hemoglobin, fibrinogen	Polytetrafluoroethylene, polyethylene, glass	Plasma; Binary mixtures	$125I$

Results & conclusions

The ability of hemoglobin to inhibit fibrinogen adsorption (competitive effectiveness) depends not only on the ratio of the proteins, but is also strongly dependent on the total concentration at which the competition is done.

Fibrinogen adsorption from plasma was found to be maximal at intermediate plasma concentrations, and was considerably lower from 70-100% plasma or from 0.01% plasma than from 0.1 or 1% plasma, an observation called the Vroman effect.

The maximum adsorption occurred at 0.1% plasma for glass, 1% for PE, and about 10% for PTFE.

Both time and concentration effects on competitive adsorption are consistent with a mass action model for competitive effectiveness.

Title: *XPS Studies on the Organization of Adsorbed Protein Films on Fluoropolymers*

Ref. #	*Proteins studied*	*Surfaces*	*Adsorption from*	*Techniques*
21	Hemoglobin, fibronectin	Poly(vinyl fluoride), poly(vinylidene fluoride), and poly(tetrafluoroethylene)	Buffer	ESCA; 125 I

Results & conclusions

The data, when compared with a model, suggested that bare areas of polymer exist after adsorption (i.e., the protein film forms in "islands") and that the degree of coverage of the polymer surface with adsorbed protein increases with increasing polymer critical surface tension.

Fibronectin was found to cover a larger fraction of each polymer surface than hemoglobin.

Title: *Hydrophilic–Hydrophobic Copolymers as Cell Substrates: Effect on 3T3 Cell Growth Rates*

Ref. #	*Proteins studied*	*Surfaces*	*Adsorption from*	*Techniques*
22	Fibronectin, albumin	HEMA-EMA-glass	Serum; Buffer	125 I

Results & conclusions

Fibronectin adsorption from serum to the copolymers varied from 0.1 to 1.5 ng/cm² and was similar to the variation in 3T3 cell spreading after 2 hours of contact of the cells with the surface.

3T3 cell spreading was very slow or did not occur on surfaces with low fibronectin adsorption.

Fibronectin adsorption from serum controls the initial spreading of 3T3 cells on HEMA-EMA copolymers but later spreading is probably mediated by deposition of fibronectin by the cells.

Continued on next page

Title: Rat Peritoneal Macrophage Adhesion to Hydroxyethylmehacrylate-Ethyl Methacrylate Copolymers and Hydroxystyrene-Styrene Copolymers

Ref. #	Proteins studied	Surfaces	Adsorption from	Techniques
23	Albumin, hemoglobin, fibrinogen, fibronectin, IgG	HEMA-EMA-glass	Buffer	Pre-adsorption

Results & conclusions

Fibronectin and IgG preadsorption promoted macrophage attachment while all other proteins markedly decreased or prevented attachment.

Receptors for fibronectin and IgG on macrophages probably are responsible for the effects of these proteins on attachment.

Title: **Changes in Adsorbed Fibrinogen and Albumin Interactions with Polymers Indicated by Decreases in Detergent Elutability**

Ref. #	Proteins studied	Surfaces	Adsorption from	Techniques
24	Fibrinogen, albumin	Silastic, Teflon®, polyethylene, other polymers	Buffer	Detergent elutability

Results & conclusions

The elutability of adsorbed fibrinogen and albumin from polymers decreased slightly over a period of several days of storage in buffer at 4°C but the loss in elutability was much more extensive and proceeded much more rapidly at elevated temperatures.

Proteins apparently may exist in a variety of adsorbed "states". The fraction of proteins in each state will vary with time, temperature and surface, due to differences in the transition rates from one state to another.

Title: **The Kinetics of Baboon Fibrinogen Adsorption to Polymers: In vitro and in vivo Studies**

Ref. #	Proteins studied	Surfaces	Adsorption from	Techniques
25	Fibrinogen	HEMA-EMA grafts, polyethylenes, glass, teflon, several polyurethanes, others	Plasma in vitro; blood in vivo	125I

Results & conclusions:

The adsorption kinetics of fibrinogen to polymers from blood in vivo and from plasma in vitro, and the consumption of platelets in vivo induced by the polymers, all vary with polymer polarity.

Less polar polymers such as polyethylene had high initial fibrinogen adsorption in vitro which rapidly decreased, and had very low and constant fibrinogen adsorption in vivo.

More polar polymers had low initial adsorption in vitro which increased with time but in vivo fibrinogen deposition on these polymers was characterized by a second stage of great increase, inhibitable by heparin. These polymers also caused enhanced platelet consumption.

Polymer surface polarity appears to be a major determinant of fibrinogen adsorption and platelet thrombi formation, possibly because of differences in the state of the adsorbed fibrinogen.

Continued on next page

Title: **Techniques for Protein Adsorption Studies**

Ref. #	*Proteins studied*	*Surfaces*	*Adsorption from*	*Techniques*
26	*Many*	*125 I Prelabelled proteins* *In situ radioiodination*	*Buffer; Binary mixtures; Plasma*	*Many*

Results & conclusions:

This paper is a detailed presentation of the techniques used in my lab for protein adsorption studies.

Typical results obtained with these techniques are also presented.

Title: **Protein Adsorption to Hydrogels**

Ref. #	*Proteins studied*	*Surfaces*	*Adsorption from*	*Techniques*
27	*Many*	*Many*	*Varied*	*Varied*

Results & conclusions

A review of the literature on adsorption to hydrogels revealed:

The data available on protein adsorption to hydrogels are quantitative measurements of the amount of a particular protein adsorbed under a variety of conditions, including exposure to single protein solutions, from simple or complex mixtures, and after various adsorption or desorption times.

The most frequent observation is a quantitative reduction in the amount of a particular protein adsorbed to the hydrogel in comparison to a non-hydrogel, control (typically hydrophobic) surface.

Desorption and exchange studies indicate proteins are less tightly bound to hydrogels than to non-hydrogels.

A qualitative difference in protein adsorption to hydrogel and non-hydrogel surfaces is the apparent preference of certain proteins, especially albumin, for hydrogel surfaces.

It may be possible to produce clinically useful biomaterials with very low protein adsorption and potentially much enhanced biocompatibility using extremely high water content, highly flexible

The Importance of Vascular Graft Surface Composition as Demonstrated by a New Gas Discharge Treatment for Small Diameter Grafts

Ref. #	Proteins studied	Surfaces	Adsorption from	Techniques
28	Fibrinogen	RF plasma, tetrafluoroethylene (TFE) treated Dacron® vascular grafts	Plasma	^{125}I

Results & conclusions

Much lower initial adsorption occurred on RF-plasma TFE treated Dacron® than on Dacron®.

The lower thrombogenicity of the RF plasma/TFE surface may result from insufficient adsorbed fibrinogen to trigger platelet activation.

Title: Platelet Adhesion and Fibrinogen Adsorption: A Re-Examination

Ref. #	Proteins studied	Surfaces	Adsorption from	Techniques
29	Fibrinogen, hemoglobin	Glass	Plasma	Preadsorption

Results & conclusions

Platelet adhesion did not correlate with fibrinogen adsorption to glass pretreated with a series of plasma dilutions since the experiments show the greatest number of adhering platelets did not occur where maximal fibrinogen adsorption occured.

Other plasma proteins besides fibrinogen, or acting in concert with fibrinogen, may be important in influencing platelet adhesion. Alternatively, the state of the adsorbed fibrinogen may be different at different plasma concentrations, i.e., how fibrinogen is adsorbed rather than just how much is adsorbed may be most important.

measured adsorption (7,14). It remains possible, however,
that much more sensitive methods are required to detect
preferential adsorption, e.g., use of depletion methods
requiring high surface area samples. On the other hand, if
the preferential adsorption is so small as to require
detection by a highly sensitive depletion method, it would
seem that the error in the adsorption measurements would
be correspondingly small and therefore not too
significant. Nonetheless, it appears the issue of
preferential adsorption of ^{125}I proteins remains
important, and therefore in need of further work.
Measurement of the kinetics of adsorption from solutions
containing different ratios of labelled to unlabelled
protein might be a better approach to testing for
preferential adsorption than the steady state, single time
point measurements so far employed.

The advantages of the *in situ* radioiodination
technique include extremely high sensitivity and avoidance
of any effects of prelabelling on adsorption behavior. The
in situ radioiodination (or "iodogram") technique has
proven very useful because of its ability to detect the
presence of unsuspected proteins in the adsorbed layer,
including hemoglobin. In addition, this technique was used
to detect variations in the composition of the protein
layer adsorbed from plasma that were induced by either
differences in the chemical properties of the surface or
by differences in the plasma contact time (13). However,
the *in situ* radioiodination method is not quantitative due
to differential and largely unpredictable uptake of iodine
by each of the adsorbed proteins.

Results and Discussion

The results of studies in the author's laboratory
summarized in Table 1 constitute a relatively large amount
of information about how adsorption occurs from the
protein mixtures typically encountered by biomaterials.
Overall, the most important observations made are that
variations in surface chemistry, time of adsorption, and
protein type are major factors in determining the
composition of the adsorbed layer. The adsorbed layer
formed from mixtures thus contains a rather complex and
changeable combination of proteins.

Surprisingly, the enrichment of the surface phase
relative to the bulk phase is not very great for the
plasma proteins so far studied, including fibrinogen,
immunoglobulin G, albumin, or VWF/VIII (14, 15,19).
(Surface enrichment was calculated as the weight fraction
of the adsorbed layer that is due to each protein, divided
by the weight fraction of total protein in the bulk phase
due to this protein). The surface enrichment for these
proteins is in the range of 0.3-3.4 for the HEMA-EMA
series of polymers. Hemoglobin, while not normally
considered a plasma constituent because of its rapid
clearance when released by lysed red cells, is an

exception to this rule in that it is represented in the surface layer at much higher concentrations than one might expect. Nonetheless, as a first approximation, the plasma proteins appear to adsorb in proportion to their bulk concentration, as expected from mass action. Direct verification of this principle has been obtained in studies in which fibrinogen adsorption from various plasmas was shown to increase in proportion to the fibrinogen concentration (S. Slack and T. Horbett, unpublished observations). Strongly adsorbing moieties such as methacrylic acid can, however, lead to greatly enhanced adsorption of protein species of opposite net charge, with which ion exchange can occur (e.g., lysozyme and γ-globulin in the case of methacrylic acid) (2,27). In these cases, the importance of mass concentration is supplanted by the nature and degree of charge on the protein, as characterized by the isoelectric pH.

Fibrinogen adsorption to non-polar surfaces such as polyethylene from plasma after very short contact times constitutes another exception to the mass action generalization. Fibrinogen adsorption from plasma to such surfaces, and, according to others, to more polar surfaces, is initially high but decreases to much lower values as time progresses and the steady state is approached (10,11,25). An analogous displacement effect also occurs if adsorption from a series of diluted plasma solutions is measured. In this case, fibrinogen adsorption reaches a peak at an intermediate plasma concentration that varies with surface type, and is much lower from either more dilute or more concentrated plasma (see (20) and its summary in Table I). These "Vroman effects" are as yet not fully understood, but recent evidence from our laboratories suggest that they may be a characteristic feature of all mixed surfactant systems and be due to differences in the rate of conversion of initially adsorbed molecules to a more adsorptive state (see below). Some workers have reported that the trace clotting factor HMWK appears to be an important factor in the Vroman effect. However, a maximum also occurs in the adsorption of fibronectin to glass and other surfaces from serum diluted to various degrees. Furthermore, we have recently found that the presence of albumin or hemoglobin in binary mixtures with fibrinogen, as well as addition of the synthetic detergent Triton X-100 to a fibrinogen solution, are also able to induce an apparent Vroman effect (S. Slack and T. Horbett, unpublished observations). These latter observations support the generalized mixed surfactant mechanism as an explanation of the Vroman effect.

Adsorbed proteins apparently can undergo some type of transition from a detergent elutable to a detergent non-elutable state (24). This transition occurs very quickly at higher temperatures, is much slower at lower temperatures, and in addition depends on the type of surface used. Thus, the proteins may be becoming more denatured or "spread" on the surfaces. These observations extend many other, less direct observations which indicate that proteins initially adsorb in a different state (e.g., "loosely adsorbed") and then change to a different state (e.g.,"tightly held"; see the chapter by T. Horbett and J. Brash in this book). The transient maximum in fibrinogen adsorption from plasma thus may occur because the competing proteins require more time to completely "spread" and displace fibrinogen. The concept of molecular spreading is new and not yet verified. Considerable experimentation intended to clarify this matter further is presently underway in several laboratories.

The influence of adsorbed proteins on cellular interactions with biomaterials has also been studied, because the response to adsorbed proteins probably constitutes an important mechanism underlying differences in the biocompatibility of various biomaterials. In the case of 3T3 cells, differences in the amount of fibronectin adsorbed to HEMA-EMA copolymers from serum causes differences in initial speading of these cells, and in addition probably dictates whether the attachment step leading to spreading can occur (22). For macrophages, pre-adsorption of certain proteins has been shown to enhance or stabilize attachment (23). In as yet unpublished studies, comparisons of macrophage spreading on the HEMA-EMA copolymer series to the amount of fibronectin or IgG adsorption from serum have shown little or no correlation, despite the presence of receptors for these proteins on macrophages. Similarly, platelet adhesion to glass does not appear to be correlated with the amount of fibrinogen adsorbed from plasma (29). Studies of platelet morphology on protein coated surfaces may prove to be a more definitive approach in determining how adsorbed proteins induce platelet activation. In particular, we are testing the hypothesis that platelet responses may be more sensitive to the way a protein is adsorbed (i.e., its "state") than to the amount of the protein that is adsorbed. In summary, platelet and macrophage responses do not correlate with the amount of adsorption of proteins for which they have receptors, but 3T3 cell spreading and fibronectin adsorption are highly correlated. Thus, more than one mechanism appears to be important in cell interactions with the adsorbed protein layers that form on biomaterials exposed to the biological milieu.

Conclusion

The studies of protein adsorption to biomaterials undertaken so far have shown that the process is understandable in terms of the principles of surface activity, mass action, surface chemistry, and transitions in the structure of proteins. Furthermore, cells with receptors for certain of the adsorbed proteins are likely to respond to surfaces in proportion to the amount of this protein on the surface, although other processes involving

conformational alterations to different states of the
protein may also prove equally important. Since progress
in these problems is important to biomaterials science,
since better methodologies for studying them are rapidly
becoming available, and because a group of trained and
talented investigators has now devoted itself to the
problem, it appears that important strides in clarifying
the nature of protein adsorption and its connection to
cellular responses to materials will be made in the next
few years. This progress should greatly aid the rational
design of improved biomaterials.

Acknowledgments

The generous financial assistance of the National Heart,
Lung, and Blood Institute through grant HL19419 has made
most of this work possible. My colleagues, students,
postdoctoral associates, and technicians all contributed a
great deal to this work and I gratefully acknowledge their
collaboration.

Literature Cited

1. Ratner, B.D.; Horbett, T.A.; Hoffman, A.S.; Hauschka,
    S.D. J. Biomed. Mater. Res. 1975, 9, 407.
2. Horbett, T.A.; Hoffman, A.S. In Applied Chemistry at
    Protein Interfaces, Advances in Chemistry Series ;
    Baier, R.E., Ed.; American Chemical Society:
    Washington, D. C., 1975; Vol. 145, p 230.
3. Hoffman, A.S.; Horbett, T.A.; Ratner, B.D. Ann. N. Y.
    Acad. Sci. 1977, 283, 372.
4. Weathersby, P.K.; Horbett, T.A.; Hoffman, A.S. Trans.
    Am. Soc. Artif. Int. Organs 1976, 22, 242.
5. Weathersby, P.K.; Horbett, T.A.; Hoffman, A.S. Thromb.
    Res. 1977, 10, 245.
6. Weathersby, P.K.; Horbett, T.A.; Hoffman, A.S. J.
    Bioeng. 1977, 1, 395.
7. Weathersby, P.K.; Horbett, T.A.; Hoffman, A.S.; Kelly,
    M.A. J. Bioeng. 1977, 1, 381.
8. Horbett, T.A.; Weathersby, P.K.; Hoffman, A.S. J.
    Bioeng. 1977, 1, 61.
9. Horbett, T.A.; Weathersby, P.K.; Hoffman, A.S. Thromb.
    Res. 1978, 12, 319.
10. Horbett, T.A. In Adhesion and Adsorption of Polymers,
    Part B; Lee, L.H., Ed.; Plenum Publishing Co.: New
    York, 1980; p 677.
11. Horbett, T.A. A.C.S. Org. Coat. Plast. Chem. Prepr.
    1979, 40, 642.
12. Ratner, B.D.; Horbett, T.A.; Shuttleworth, D.; Thomas,
    H.R. J. Coll. Interf. Sci. 1981, 83, 630.
13. Horbett, T.A.; Weathersby, P.K. J. Biomed. Mater. Res.
    1981, 15, 403.
14. Horbett, T.A. J. Biomed. Mater. Res. 1981, 15, 673.

15. Horbett, T.A. In Biomaterials 1980, Advances in
    Biomaterials; Winter, G.D.; Gibbons, D.F.; Plenk, H.,
    Jr, Eds.; John Wiley and Sons, Ltd.: Chichester,
    England, 1982; Vol. 3, p 383.
16. Hoffman, A.S.; Horbett, T.A.; Ratner, B.D.; Hanson,
    S.R.; Harker, L.A.; Reynolds, L.O. In Biomaterials:
    Interfacial Phenomena and Applications, ACS Advances
    in Chemistry Series; Cooper, S.L.; Peppas, N.A., Eds.;
    American Chemical Society: Washington, DC, 1982; Vol.
    199, p 59.
17. Royce, F.H., Jr; Ratner, B.D.; Horbett, T.A. In
    Biomaterials: Interfacial Phenomena and Applications,
    ACS Advances in Chemistry Series; Cooper, S.L.;
    Peppas, N.A., Eds.; American Chemical Society:
    Washington,DC, 1982; Vol. 199, p 453.
18. Horbett, T.A. In Biomaterials: Interfacial Phenomena
    and Applications, ACS Symposium Series; Cooper, S.L.;
    Peppas, N.A., Eds.; American Chemical Society:
    Washington, D. C., 1982; Vol. 199, p 233.
19. Horbett, T.A.; Counts, R.B. Thromb. Res. 1984, 36,
    599.
20. Horbett, T.A. Thromb. Haemostas. 1984, 51, 174.
21. Paynter, R.W.; Ratner, B.D.; Horbett, T.A.; Thomas,
    H.R. J. Coll. Interf. Sci. 1984, 101, 233.
22. Horbett, T.A.; Schway, M.B.; Ratner, B.D. J. Coll.
    Interf. Sci. 1985, 104, 28.
23. Lentz, A.J.; Horbett, T.A.; Hsu, L.; Ratner, B.D. J.
    Biomed. Mater. Res. 1985, 19, 1101.
24. Bohnert, J.L.; Horbett, T.A. J. Coll. Interf. Sci.
    1986, 111, 363.
25. Horbett, T.A.; Cheng, C.M.; Ratner, B.D.; Hoffman,
    A.S.; Hanson, S.R. J. Biomed. Mater. Res. 1986, 20,
    739.
26. Horbett, T.A. In Techniques of Biocompatibility
    Testing; Williams, D.F., Ed.; CRC Press, Inc.: Boca
    Raton, Florida, 1986; Vol. II, p 183.
27. Horbett, T.A. In Hydrogels in Medicine and Pharmacy;
    Peppas, N.L., Ed.; CRC Press, Inc.: Boca Raton, FLA,
    1986; Vol. I, p. 127.
28. Hoffman, A.S.; Ratner, B.D.; Garfinkle, A.M.; Horbett,
    T.A.; Reynolds, L.O.; Hanson, S.R. In Vascular Graft
    Update: Safety and Performance; Kambic, H.E.;
    Kantrowitz, A.; Sung, P., Eds.; American Society for
    Testing and Materials: Philadelphia, 1986; ASTM
    Special Technical Publication 898, p 137.
29. Horbett, T.A.; Mack, K. Trans. Soc. Biomat. 1986, IX,
    45.

RECEIVED January 29, 1987

Chapter 17

Adsorption of Gelatin at Solid–Liquid Interfaces

A. T. Kudish[1] and F. R. Eirich[2]

[1]Department of Chemistry, Ben-Gurion University of the Negev, Beer Sheba, Israel
[2]Department of Chemistry, Polytechnic University, Brooklyn, NY 11201

We studied the adsorption of dilute acid- and alkali precursor gelatins on pyrex and stainless steel surfaces. The adsorptions found, treated as points of Langmuir isotherms, became maximal around the IEP, close to the positions of the minima of the viscosities. We determined also the thicknesses, Δr, of the adsorbed layers by a viscosity method allowing calculation of the volume of the interphases. Comparing molecular concentrations and dimensions with those of the free molecules, we obtained information on the relative compression of the adsorbate. The adsorption remains substantial on the acid side of the IEP, so that the maximal segmental densities occur to the left of the IEP and of the pH of the maximal layer thicknesses. The data indicate also that the state of expansion of the adsorbed molecules, more than the amounts adsorbed or the segmental densities, determine the layer thicknesses.

We wish to report some work on the adsorption of acid and alkali-precursor gelatins, the water-soluble products of collagen. The study of both types allowed us to determine whether the differences exhibited by the two kinds in solution are reflected in the adsorbed state, and offers some insight into the state of flexible molecules at interfaces. The use of two dissimilar adsorbents, glass and stainless steel powders, should yield information on segment-surface interactions during the adsorption process. Some effects of added calcium ions were also studied since the Ca-ions of hydroxyapatite in bone and teeth are intimately related to collagen.

The study of the adsorption of macromolecules requires in

0097-6156/87/0343-0261$06.00/0
© 1987 American Chemical Society

most cases substantial instrumental and theoretical efforts (1-5). The interpretation of the results is satisfactory when the system consists of well defined synthetic or natural polymers that correspond to the theoretical assumptions on the many facets of the adsorption of flexible chains (5). For less well defined polymers, one is reduced to applying empirical equations to obtain parameters which pertain to the essential features of the adsorbed states. Fortunately, many polymers approximate an overall behavior that is described by the simplest type of adsorption theory, that of the Langmuir isotherm (LI) (6). This is neither a coincidence, nor does it mean that polymers fulfill all the assumptions of Langmuir's theory. The approximate agreement is due to the fact that the LI describes the mechanism of monolayer adsorption from dilute solutions with repulsion but no attraction within the adsorbate layer, where the rates of adsorption and desorption depend on the fractions of unoccupied and occupied surface areas, and a point of surface saturation is reached. These happen to be also features of many cases of polymer adsorption from very dilute solutions.

We have thus employed the LI and its formalism to obtain two sets of parameters for the adsorption of gelatin, the initial slopes of the isotherms, the "affinities", A, and the saturation values, a_s which we call the "capacities". In our discussion of the results, we will use the affinity data briefly for a comparison of the tendencies of the gelatins to become adsorbed. The capacities will be used for a description of the fully adsorbed state. We will be helped in this by our results on the hydrodynamic thicknesses, Δr, of the adsorbed layers. The product of the areas occupied per molecule at a_s and Δr, assuming monolayers, yield the volume occupied per molecule in the interphase. For a similar approach, see (7). This may be compared with the molecular volume in solution and allows a description of the relative states of the adsorbed molecules. The resulting data, while approximate and relative, will be seen to be quite informative.

EXPERIMENTAL

MATERIALS: Gelatin. Gelatins are the water-soluble products of the denaturation of collagen. We used derivatives (Courtesy Peter Cooper Corp., Gowanda, N.Y.) of soluble collagens obtained by a treatment which disordered the individual chains of the triple helices but degraded them only little (8). One gelatin used in this study was obtained by acid treatment and extraction of pig skin collagen and designated P-1. It is characterized by an isoelectric range of pH 7-9, after a treatment that allowed most of the asparagine and glutamine residues to remain. Based on experimental evidence, Veis (8) believes that these gelatins still contain some intramolecular links between their 3 unbroken strands, each of approximately 100,000 molecular weight. Our alkali-treated precursor gelatin from calf's skin, designated C-1, had an isoelectric point of 4.5– 5. , since many of the side amide groups were lost through hydrolysis. The C-1 molecules resemble closely the   α–chains of collagen, degraded to a MW of approximately 30,000.

Both gelatins were electrodialyzed, freeze–dried and stored at 4°C.   Amino acid analyses showed them to be representative of their species, see also Table 1. The water used for the solutions was deionized and distilled.   All other chemicals were reagent grade and used without further purification.

Adsorbents.   The adsorption studies were performed on powdered Pyrex glass #7748 (donated by the   Corning Glass Company), and powdered stainless steel, type 316L (donated by the Pall Corp.) both 325 mesh size. The fritted disks were kindly prepared for us by the donating companies. The glass powder and discs were cleaned by heating at 80°C in a 50/50 nitric acid–hydrochloric acid mixture with continuous stirring for 24 hours. The powder was freed from residual acid by repeated slurrying with distilled water and centrifugation. The surface area was found to be 2.1 m^2/g by B.E.T. nitrogen adsorption.

The stainless steel powder and discs were cleaned by immersing in a boiling bath of 15% NaOH for one hour with continuous stirring.   The caustic was decanted and the surfaces neutralized by the addition of hot 10% nitric acid for less than five minutes.   The acid was decanted and the stainless steel powder was rinsed in the same manner as the glass powder and freeze dried.   The surface after such a passivation treatment consists probably of ferric ferrite, Fe(FeO$_2$)$_3$. The surface area was 0.14 m^2/g by B.E.T. nitrogen adsorption.

Adsorption Isotherms.   The solutions were prepared by dissolving the gelatin in water at 40°C to avoid aggregates, adjusting the pH with HCl and NaOH, and equilibrating at 30°C for at least an hour prior to use. The pH's of the solutions were again controlled by the addition of NaOH or HCl.   The resulting ionic strength was 0.01 N.   The adsorption experiments were performed in the usual manner, finding the depletion of the supernatant.   The samples and controls were thermostatted at   30°C ± 0. 5°C, rotated for a minimum of 16 hours, centrifuged at 30°C and the supernatant analyzed for gelatin with a Cary-14 recording spectrophotometer, measuring the absorption at 190 or 195   nm under a nitrogen blanket.   The calibration plots conformed to the Beer–Lambert Law.

The hydrodynamic thicknesses were measured by observing the change in capillary flow due to adsorption on fritted disks prepared by sintering the powders used for adsorption. The results will be reported only for the glass disks. The data from the steel disks were poorly reproducible. For the operational method, see (10) and the dissertation by A. Kudish (11). However, their flow apparatus was modified so that the liquid could flow through the disks in both directions. This was found to reduce the time required to attain equilibrium and compensates for any one-way necks present in the pores.   Poiseuille's equation was used in a modified form (12,13) assuming the disk to be a packet of parallel microcapillary tubes with a tortuosity factor. The ratio of the specific volume flow rates of solution and solvent, becomes:

$$V/V_o = (1/\eta_{rel}) \frac{\Sigma(r_i - \Delta r)^4}{\Sigma r_i^4} \tag{1}$$

where, V, is the specific volume flow rate of the solution, V_o that of the solvent, η_{rel}, the relative viscosity, r_i, the radius of the ith pore and, Δr, the hydrodynamic thickness of the adsorbed layer. The relative viscosity can be measured independently, so that Δr can be calculated from an expansion of Equation (1):

$$(V/V_o)\ \eta_{rel} = 1 - 4 \times [\frac{<r^3>}{[r^3]} - \frac{3}{2} \times \frac{<r^2>}{[r^2]} + \chi^2 \frac{<r>}{[r]} - \frac{1}{4}\chi^3]$$

where: $<r>^a = \frac{1}{N} \Sigma N r_i^a$ \hfill (2)

$$[r] = [\frac{1}{N} \Sigma N r_i^4]^{\frac{1}{4}}$$

$$\chi = \frac{\Delta r}{[r]}$$

The values for $<r>^a/[r]^a$ were obtained from mercury intrusion measurements performed by Rowland (12) using an Aminco-Winslow mercury intrusion porosimeter. The use of equation (2), solved by an IBM 7040, is limited by the requirement that the number of pores, (N), and their size distribution remain the same during calibration and measurements. The radial dimensions of the gelatin molecules were less than 20% of the mean radius of the pores, so that plugging of the pores and changes in size distribution were likely to be minor. However, some plugging can not be excluded so that the values found by us for the adsorbed thicknesses may be on the high side.

The determination of Δr, from the values of χ, requires a knowledge of the 4th root mean radius of the disk pores, $[r]$. This was determined by measuring the volume flow rate of nitrogen gas through the disk, assuming that the average pore size is the same as seen by the liquid, and by applying Poiseuille's equation for compressible fluid flow through a porous plate, described previously, (12).

RESULTS AND DISCUSSION

The adsorption isotherms on glass and stainless steel powders were determined as a function of pH for both gelatins, and also in the presence of 0.1 M $CaCl_2$. The isotherms exhibit in all cases the steep initial rise and the plateaus characteristic of monolayer adsorption of macromolecules. The results are shown, in the form of representative reciprocal plots of some isotherms, Figure 1, and by the plots of the saturation capacities a_s, as a function of pH, see Figures 2 and 3. An example of a plot of affinity constants, also as functions of pH and determined by a least-squares analysis of the adsorption data inserted in the reciprocal LI, Equation (3), is shown in Figure 4. The intrinsic viscosities and the hydrodynamic thicknesses of the adsorbed layers at saturation, as function of pH, and the capacities for comparsion are shown in Figures 5a,b. A summary of the maximal values of the capacities, affinities, and Δr-values is given in Table II.

Our data show, as briefly discussed before that the adsorption date of our gelatin samples approached L.I.'s rather closely. Thus in a general sense, we may interpret the slopes as measures of the affinity between adsorbate and adsorbent, while we use, a_s the values for surface saturation to calculate the spaces occupied per molecule (or monomeric segment, respectively). We compare then the thicknesses of the adsorbed layers, hydrodynamically defined by our capillary flow method, with the dimensions of the free macromolecules as characterized by their viscosities in the equilibrated solution. There is a difference though, between the types of flow. The solvent moves in both cases in simple shear, while the gelatin molecules follow simple shear when freely dispersed, but were stretched in pure shear when adsorbed. The resulting differences are small and do not outweigh the advantage that the molecular dimensions, and therefore the molecular states, are in both cases hydrodynamically defined. An earlier study (14) has indeed shown that the thicknesses of the very same gelatins, calculated from viscosity increases due to adsorption on dispersed latex particles, were within 20% of those reported here, and showed much the same trends with pH. Other studies (10) demonstrated a proportionality between the intrinsic viscosities of free chain molecules and the thicknesses of adsorbed layers when measured by capillary flow. Reference 10 reports also that Δr was proportional to $M^{1/2} \alpha^3$, confirmed by ellipsometry (15). See also (16).

Deriving molecular dimensions in solution from viscosities depends on the model assumed for the conformations of the free molecules. Since any α- or - triple helical sections of our gelatins would be melted at 30°C. we assume near randomness for the chains, and a low ellipticity for the molecular envelopes. Further, the success of Flory's viscosity theory (17) has shown that the hydrodynamically effective volume of randomly coiled (and of many other) chain molecules is not very different from the volume encompassed by the meandering segments. Thus we treated our data as if they pertained to random coil molecules. The measured layer thicknesses then describe the level within the adsorbed interphase below which the segmental density is equal to, or larger, than the effective coil density of the free molecules.

Proceeding on the basis of these arguments, we converted a_s (mg amounts adsorbed per g adsorbate), into the surface areas per

Table I. Composition of charged groups in gelatins P-1 and C-1*

Group	pKa**	IEP 4.5-5.0 Alkali precursor gelatin (C-1) meq./g	IEP 7.-9. Acid precursor, gelatin (P-1) meq./g
Free carboxyl	3.4-4.7	1.22	0.90
Free α amino	7.6-8.4	0.01	0.01
Imidazole	5.6-7.0	0.05	0.05
Free α amino	9.4-10.6	0.42	0.42
Guanidino	11.6-12.6	0.48	0.49
Charges at full ionization		+1.22 −0.96	+0.90 −0.97

*After G. Thomas (1971) (9) .
**The values used for the calculation of the net charges entered in
 Table I were the arithmetic means of the ranges of pKa's shown.

Amino acid analyses courtesy, Drs. I. Listowsky and S. Seifter,
Dept. of Biochem., Albert Einstein School of Medicine, N.Y., N.Y.

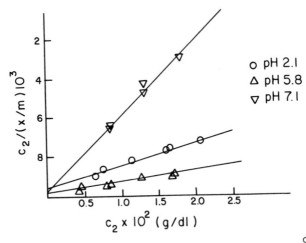

Fig. 1. C-1 gelatin adsorbed on glass powder (30 °C).
Points plotted as reciprocal isotherms in accordance
with Langmuir's equation:

$$c_2/(x/m) = c_2/a_s + 1/Ka_s \qquad (3)$$

where c_2 is the equilibrium concentration in solution, x/m in mg/g are the amounts adsorbed, a_s, the
maximal values of adsorption, and K, the affinity, is the
initial slope of the isotherms.

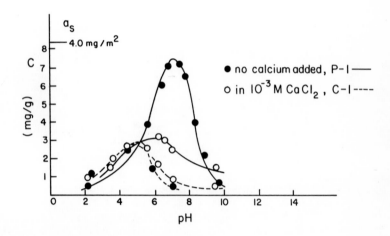

Fig. 2.   Saturation   capacities   of   glass   powder   for   P-1
and   C-1   gelatins   vs.   pH   (30°C), with   and   without
Ca-ions added.

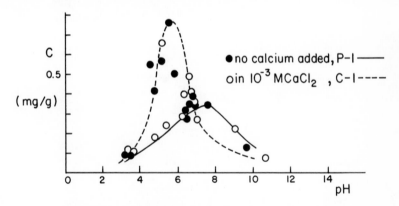

Fig. 3.   Saturation capacities of stainless steel powder   for
P-1 and C-1 gelatins vs. pH (30°C), with and without
Ca-ions added.

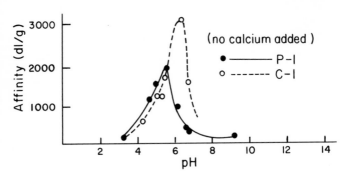

Fig. 4. Slopes of adsorption isotherms (affinities) of P-1
         and C-1 gelatins for stainless steel powder vs. pH,
         (30°C).

Table II. Affinity, Capacity and Layer Thickness Maxima, occurring
           at the given pH values

		A dl/g	pH	C mg/m²	pH	Δr A°	pH
	Glass	4000	5.3	1.5	5.3	500	5.3
	Glass +Ca	2900	4.-	1.5	5.3	480	4.5
C-1	Steel	7500	6.-	5.5	5.5	---	---
	Steel +Ca	(6000)	6.-	5.0	5.5	---	---
	Glass	6000	5.5	3.5	7.4	750	7.6
	Glass +Ca	3000	5.0	1.5	8.-	---	---
P-1	Steel	800	5.5	1.5	8.-	---	---
	Steel +Ca	(600)	5.5	(2.-)	8.-	---	---

adsorbed gelatin molecule. Assigning to an amino acid unit an average molecular weight of 100, we also calculated the area per monomeric unit attached to, but not necessarily lying on the interface. The depth of these anchored, floating, populations, as measured by our hydrodynamically effective thicknesses, Δr, Figures 5a,b, are 700 A and 475 A respectively for P-1 and C-1 at their maxima, i.e., are of the order of 100 times the average thickness of a flat layer of amino acid residues, and about 25% for P-1, and 50% for C-1, of the contour lengths of the chains of the adsorbed gelatins. These results are in line with the assumption that a major portion of the gelatin molecular segments is situated within a layer on the interface that is characterized by Δr (16). The experimental difficulties (1) and the uncertainty to which extent our gelatin fits the models used in the theories of the distribution of loops, tails and surface runs (18,19), deterred us from trying to determine the profile of segmental densities in our interphases. However, since we find the average segmental densities of the interphase to be no higher than five times that of the corresponding average densities of the free molecules, and since the dimensional changes with pH of gelatin as an ampholyte do not differ much from those of the free molecular coils (20), our case may be characterized as one of light to moderately strong adsorption (19).

We will first compare the dimensions of the adsorbed with those of the free P-1 molecules. In Figures 6, we plot, a_s the amounts adsorbed in mg/g, further the areas per adsorbed molecule, and the cross sectional areas. For a comparison we entered our Δr-values for P-1 and the diameters of the free coils in Figure 6b. We then calculated and compared the volumes encompassed and the ratios of the corresponding segmental densities of the P-1 gelatin molecules, when free, or adsorbed, Figure 7. The cross sections and hydrodynamic volumes of the counterpart free molecules were approximated with the help of Fox and Flory's (21) equation for random coils:

$$[\eta] = 6^{3/2} R_G^{\ 3}/M; \quad \Phi = 2.1.10^{21}; \text{ and } R_\eta = 0.85 R_G \text{ (22)}.$$ Thirdly, in Figures 8a,b, we show the Δr's and the densities of the adsorbed layers of P-1 and C-1 versus pH. The density curve of the free coils of P-1 is also shown for comparison.

We wish to emphasize only a few of the many striking features of the diagrams. The curves of area/adsorbed molecules and of Δr of Figures 6, run opposite courses, showing a minimum and a maximum respectively between pH 7 and 8. The areas of the adsorbed molecules drop 5-10 times below those of the free ones. This is equal to a drop per adsorbed monomeric unit below 5 A^2 compared with an average of about 20 A^2 for flat lying amino acid residues, and 40 A^2 within the free coils, and indicates a substantial 2-dimensional pressure. The compaction is not as high as in many other cases of macromolecular adsorption, but it is enough to raise the interphase concentration to where, at lower temperatures, gel formation would occur.

Figures 5 show further the courses of the intrinsic viscosi-

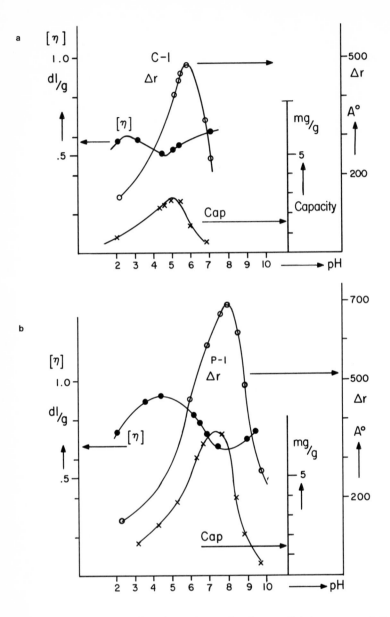

Fig. 5a. The intrinsic viscosity, the layer thickness, and the
          capacity of C-1 gelatin as function of pH.

     b.   The same for P-1 gelatin.

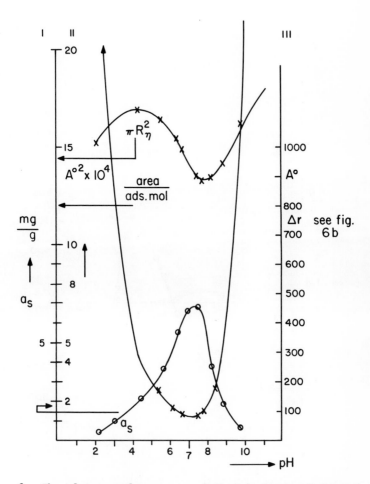

Fig. 6a. The plateau values, a_s, of the isotherm of the P-1 gelatin and the calculated areas occupied per adsorbed gelatin molecule. Shown for comparison is the cross sectional area per molecule in solution as a function of pH.

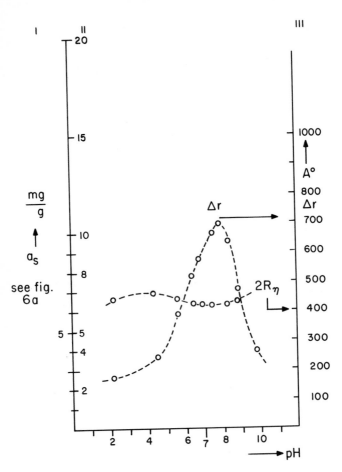

Fig. 6b. For comparison with 6a, the thickness of the layers of adsorbed P-1, and the diameters of the free molecules in solution, are also shown as a function of pH.

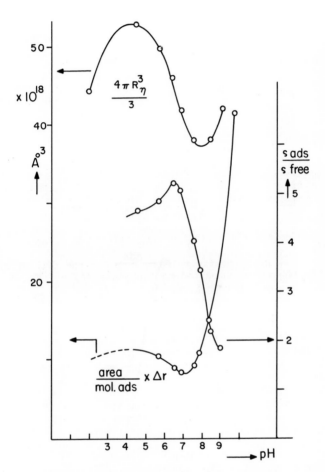

Fig. 7.   The volume of the adsorbed P-1 molecules, their relative densities and, for comparison, the volumes of
free P-1 molecules in solution, as a function of pH.

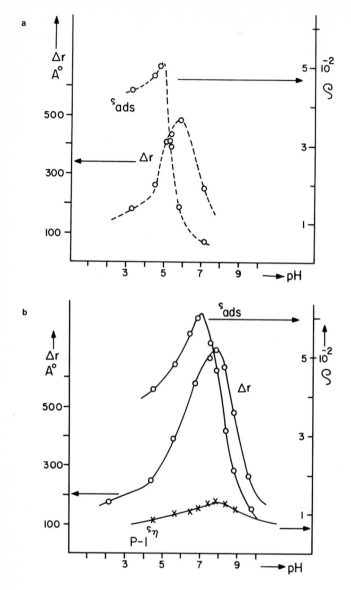

Fig. 8a. The thicknesses of the adsorbed C-1 gelatin layer,
and the absolute densities of the adsorbed molecules
in this layer, as function of pH.

b. The same for P-1 gelatin. For comparison, see also
the absolute densities of the P-1 gelatin segments
in the free molecular coils in solution.

ties, and Figures 6 the adsorbed areas per molecule and the layer thicknesses Δr, of P-1 as function of pH. In the more acid range, the layer thicknesses are smaller than the free coil diameters, but both show only small changes with pH. Going from pH 5 to 6, the Δr values rise strongly due to the rapidly increasing amounts of adsorption which become possible as a consequence of the loss of positive net charge on the gelatin. Because of this increase in Δr, the coils become only about 1/2 as dense as they would according to the reduction of occupied areas. Thus, the volumes pervaded by the chains do not stay constant, nor is the compression of the gelatin molecules isodiametrical. Either coil compression, or interpenetration must occur during adsorption.

The curves for the two cross sections, for Δr and for the free diameters (our Figures 6) are seen to intersect at low and high pH. However, these points of equal Δr and $2R_n$, and of equal area/molecule and πR_n^2, do not coincide in their pH-values, i.e. the molecules do not become adsorbed in the state of their free conformations. Figure 7 shows the courses of the volumes of free vs. adsorbed coils, and of the ratios of the densities of the free coils and in the interphase. The curves for the volumes are very far apart and will become equal only at very low or high pH's, that is, for very small amounts of adsorption.

We saw that over the range of crowding in the interphase, Δr does not increase as much as the area per molecule shrinks on adsorption, but we can not tell whether this is due to coil compaction or interpenetration. Interpenetration of neutral chains has to work against elastic and osmotic resistance without compensating factors; in polyampholytes of near neutral, or neutral net charge, penetration may be accompanied by a decrease in electrical free energy, so that charge pairing per volume helps to overcome the elastic and osmotic resistances. The density maxima in the adsorbed layer lie to the left of the I.E.P., and the densities remain high into the acid side while gelatin and surfaces remain oppositely charged. The affinities stay high in the same range, i.e. segment-surface attractions outweigh intra-segmental repulsion. This, too, would enhance a certain amount of coil interpenetration.

The results for our C-1 gelatin show the smaller values expected for a lower M.W., but by no means as small as the 10-fold difference in MW would lead one to expect. The areas per molecule are about 1/3, and the Δr values about 2/3 of those for P-1. The segmental densities of both gelatins change from being almost equal at the lower pH's, to C-1 becoming about 1/3 as dense for pH's larger than I.E.P., showing that the relation between the opposite tendencies of densification and osmotic pressure are different for both gelatins.

The addition of Ca-ions affects the behavior of C-1 gelatin more than P-1, shifting the adsorption maxima of both gelatins to the acid side. The surface areas occupied per molecule at the I.E.P are twice as large with as without Ca-ions. Since the Δr's stay about the same, Ca-ions attached to the gelatin reduce coil contraction or interpenetration. Surface shielding by adsorbed

Ca-ions is not likely, since the affinities of the gelatins plus Ca-ions remain equal, or turn higher. Possibly, some Ca-ions become adsorbed at the interface while remaining attached to gelatin segments, thus forming strong adsorption sites.

The adsorption on steel powder is smaller for P-1 and larger for C-1. It reflects further the fact that the steel's positive charge extends into the weakly alkaline range, as compared to the switch from a negative to a positive surface charge for glass below pH 2 (Steigman, J., Downstate Medical School, Private Commun., 1969). Thus, adsorption due to opposite charges occurs on steel also to the right of the I.E.P.'s, shown by the less rapidly falling-off of the a_s-values. The greater adsorption on steel of C-1 than P-1 might be attributed to the presence of very low concentrations of Fe^{3+}-ions near the surface that bind strongly to COOH-groups which are more numerous on C-1 than on P-1.

Among the many questions that remain, the most vexing concerns the extent to which Δr may be affected by the rather high rates of shear in the capillaries of our porous disks. We will discuss this in a subsequent paper. Suffice it is to say at this time that, thanks to recent studies (23), we seem justified in our assumption that the rate of shear was not a factor in determining coil shapes, densities, or Δr values.

Conclusion

In conclusion, we believe that combined analyses of amounts and dimensions of adsorbed gelatins, as obtained by evaluating adsorption data as Langmuir isotherms and measuring the hydrodynamic thickness of the adsorbate layers, yield substantial information on the states of the adsorbed molecules. The amounts adsorbed depend largely on the pH, exhibiting maxima around the IEP close to the positions of the minima of the viscosities of the free gelatins. The adsorption remains substantial on the acid side of the IEP, while the surfaces and the adsorbates are oppositely charged. Thus, the densities of the adsorbed layers increase into the acid range and the positions of the maximal segmental densities lie to the left of the IEP and Δr's. To the right of the I.E.P., where the gelatins carry charges of the same sign as the glass interface, the Δr-values decline more slowly than the amounts adsorbed. Based on this and the other observations we find that the state of expansion of the adsorbed molecules as well as the amounts adsorbed and the segmental densities determine the layer thicknesses. This conclusion is supported by our observations on steel surfaces where the adsorptive maxima of both gelatins are shifted to the right, and where calf- skin gelatin becomes more strongly adsorbed than our pigskin gelatin. Lastly, both gelatins become adsorbed even where they carry moderate charges of the same sign as the interfaces, showing that non-electrostatic attraction also plays a role.

Literature Cited

1. Tadros, T. F., "The Effect of Polymers on Dispersion Properties", Academic Press, N.Y., (1983), p.1.
2. Simha, R., Frisch, H. L., and Eirich, F. R., J. Phys. Chem.,
   (1953), 69, 584.
3. Frisch, H. L., and Simha, R., in "Rheology", (F. R. Eirich,
   ed.), Acad. Press, N.Y., (1956), Vol. 1, p. 525.
4. Frisch, H. L., and Simha, R., J. Chem. Phys., (1957), 27, 702.
5. Scheutjens, J. M. H., and Fleer, G. J., "The Effect of
   Polymers on Dispersion Properties", Acad. Press, N.Y.,
   (1981), p. 111.
6. Langmuir, I., JACS, (1918) 40, 1361.
7. Garvey, M. J., Tadros, Th. F., and Vincent, B., J. Colloid
   Interface Sci., (1974), 55, 440.
8. Veis, A., "Macromolecular Chemistry of Gelatin", Acad. Press,
   N.Y., (1964).
9. Thomas, G., Ph.D. Dissertation, Polytechnic Institute of New
   York, (1971), Brooklyn, N.Y.
10. Rowland, F., and Eirich, F. R., J. Polym. Sci., (1965), A1, 4,
    2033, 2401.
11. Kudish, A. T.,Ph.D. Dissertation, Polytechnic Institute of New
    York, (1970), Brooklyn, N. Y.
12. Rowland, F., Ph.D. Dissertation, Polytechnic Institute of New
    York, (1963), Brooklyn, N.Y.
13. Rowland, F., Bulas, R., Rothstein, E., and Eirich, F. R., Ind.
    Eng. Chem., (1965), 57, 46.
14. Chough, E., M. S. Thesis, Polytechnic Institute of New York,
    (1968), Brooklyn, N.Y.
15. Stromberg, R. R., Tutas, D. J., and Passaglia, E., J. Phys.
    Chem., (1965), 69, 3955.
16. Pefferkorn, E., Dejardin, P., Varoqui, R., J. Colloid Interf.
    Sci., (1978), 63, 353.
17. Flory, P. J., "Principles of Polymer Chemistry", Cornell Univ.
    Press, (1953), p. 602.
18. Scheutjens, J. M. H., and Fleer, G. J., J. Phys. Chem.,
    (1979), 83, 1619.
19. Hesselink, F. Th., J. Phys. Chem., (1969), 73, 3488; (1971),
    75, 65.
20. Scheutjens, J. M. H., and Fleer, G. J., J. Colloid Interf.
    Sci., (1986), 111, 446.
21. Fox, T. G. and Flory, P. J., J. Phys. Chem., (1942), 53, 187.
22. Auer, R. L. and C. S. Gardiner, J. Chem. Phys., (1955), 23,
    1546.
23. Lee, J. J., and Fuller, G. G., Macromolecules, (1984), 17, 375.

RECEIVED February 26, 1987

Chapter 18

Microelectrophoretic Study of Calcium Oxalate Monohydrate in Macromolecular Solutions

P. A. Curreri[1], G. Y. Onoda, Jr.[2], and B. Finlayson[3]

[1]Space Science Laboratory, Marshall Space Flight Center, National Aeronautics and Space Administration, Huntsville, AL 35812
[2]Watson Research Center, IBM, Yorktown Heights, NY 10598
[3]Division of Urology, Department of Surgery, College of Medicine, University of Florida, Gainesville, FL 32610

Electrophoretic mobilities were measured for calcium oxalate monohydrate (COM) in solutions containing macromolecules. Two mucopolysaccharides (sodium heparin and chondroitin sulfate) and two proteins (positively charged lysozyme and negatively charged bovine serum albumin) were studied as adsorbates. The effects of pH, calcium oxalate surface charge (varied by calcium or oxalate ion activity), and citrate concentration were investigated.

All four macromolecules showed evidence for adsorption. The macromolecule concentrations needed for reversing the surface charge indicated that the mucopolysacchrides have greater affinity for the COM surface than the proteins. Citrate ions at high concentrations appear to compete effectively with the negative protein for surface sites but show no evidence for competing with the positively charged protein.

The majority of renal stones are predominantly calcium oxalate mono-hydrate and dihydrate, with the former being the most common form present. Interdispersed throughout a stone between the crystalliz-ing phases is a macromolecular substance which is typically around 2.5% (1). It is not known whether the matrix substances play an active or passive role in the formation of urinary stones. However, it is known that several soluble polymeric species and natural macromolecules have a pronounced effect on the kinetics of growth of calcium oxalate crystals (2,3). Such molecules can also affect the manner in which crystalline particles in suspension interact with each other. Adsorbed molecules can help prevent the coagulation (aggregation) of particles in suspension by providing an electrical barrier or steric hindrance (4). When the colloidal particles are only partially covered with polymer, the macromolecules can function as flocculating agents, causing the particles to be bridged together to form large flocs. Coagulation and flocculation phenomena provide

0097-6156/87/0343-0278$06.00/0

one possible mechanism for creating larger units of matter from
finely divided crystals.

The electrical charge residing on the surfaces of calcium oxalate crystals exposed to aqueous solutions should be strongly modified by the adsorption of charged ionic macromolecules. This effect
has been reported for other solid surfaces such as silica (5),
silver iodide (6,7), latexes (8), calcium phosphate (9), etc. That
certain macromolecules adsorb on calcium oxalate has been demonstrated (10), but the effect on surface charge has not been examined
in detail.

In a previous paper (11), we described the effects of small
ionic species on the surface charge of calcium oxalate monohydrate
(COM). The effects were detected by measuring the electrophoretic
mobility of the particles in the aqueous phase. The influences of
the activity of calcium and oxalate ions; monovalent electrolytes;
and sulfate, phosphate, pyrophosphate, and citrate ions on the
electrophoretic mobility were studied. It was found that the results could be accounted for by certain established theories for the
electrical double layer, which is also useful for analyzing the
results of the present work.

In this investigation, we have used bovine serum albumin,
lysozyme, sodium heparin, and chondroitin sulfate as adsorbates. In
addition to their practical interest (12-14), these macromolecules
represent negatively charged proteins, positively charged proteins,
and two distinct types of mucopolysaccharides.

Methods

Commercially available (Nutritional Biochemicals Inc., Cleveland,
Ohio) bovine serum albumin (2X crystallized), lysozyme (muramidase),
sodium heparin, and chondroitin sulfate were used. The serum albumin and lysozyme were globular proteins with isoelectric points at
pH 4.9 and 11, respectively. The sodium heparin and chondroitin
sulfate were negatively charged mucopolysaccharides with random coil
structures. Other chemicals were of reagent grade. The water was
deionized and then distilled in a borosilicate glass still. The
specific conductivity of the water was less than 1.5×10^{-6} (ohm
cm)$^{-1}$. All stock solutions were passed through a 0.22 μm filter to
remove any undissolved impurities.

The COM crystals were precipitated by mixing equimolar concentrated solutions of calcium chloride ($CaCl_2$) and sodium oxalate
($Na_2C_2O_4$). The precipitate was washed with the purified water until
sodium could no longer be detected by atomic adsorption spectroscopy
in the wash. X-ray analysis of dry precipitate confirmed the whe-
wellite form of $CaC_2O_4 \cdot H_2O$. The crystals had a nominal surface area
of 3 m^2/g as measured by gas adsorption.

Suspensions of $CaC_2O_4 \cdot H_2O$ in water were prepared by ultrasonic
dispersion at a concentration of 0.35 g/L. The prepared suspensions
were equilibrated for at least 12 hours. Final suspensions were
made by adding various macromolecular solutions to the dispersion at
a ratio of nine parts suspension to one part solution. For some
experiments the pH was adjusted by adding HCl or NaOH.

Before making electrophoresis measurements, all suspensions
after final compositional adjustments were allowed to equilibrate

for at least 2 hours at 37 °C. Electrophoresis was carried out
using a commercial instrument (Zeta Meter, Inc., New York) in a
constant temperature chamber at 37 °C.

Except for the electronic components, the electrophoretic apparatus was housed in a constant-temperature chamber, held at 37 °C
± 1 °C, with glove ports and windows through which the eyepieces of
the microscope extend outward. Standard techniques discussed elsewhere (15,16) were used to meausre the electrophoretic mobility of
the particle suspensions. Electrophoretic mobilities for each suspension were determined at least in triplicate for a minimum of 20
particles each time to obtain mean values with 95% confidence intervals.

The pH was measured with glass electrodes. For some of the
suspensions, the protein concentrations remaining in solution were
determined by eliminating the solids by filtration through a 0.22 μm
filter and analyzing the filtrate for proteins using solution transmission spectroscopy at 280 nm wavelength.

Results

The changes in the electrophoretic mobility of COM when increasing
concentration of the four macromolecules was present are shown in
Figure 1. In the previous study it was shown that the charge on
calcium oxalate in its own natural saturated solution is positive
and constant throughout a broad pH range (Figure 2). In Figure 1 it
is seen that the two mucopolysaccharides cause a reversal in charge
at relatively low solution concentrations. The serum albumin
appears to have no tendency for change reversal.

Because of the low solid content of the suspensions, there was
not enough surface area present to expect detectable solution depletion as a result of adsorption. This was confirmed in the case
of the proteins by measuring their concentrations in solution before
and after exposure to the solid. To detect the possible precipitation of a second solid phase, we used the following methods. The
protein solutions in equilibrium with calcium oxalate in our various
experiments were filtered after electrophoresis, and the filtrates
were examined with transmission spectrophotometry at 280 nm for
solution depletion of protein against standard solutions. Since the
solids concentration used in these experiments is too small to detect significant solution depletion due to adsorption, any depletion
noted could be attributed to precipitation. The remaining solutions
were prepared to nine-tenths final volume and then were observed
after one-half hour for precipitation before the slurry was added.
Finally, a second solid phase can be recognized by the presence of
particles of two distinct mobilities during electrophoretic measurement. No evidence of precipitation was found in any of the systems
presented in this paper.

The effect of pH on the electrophoretic mobility of COM in systems containing fixed amounts of the four macromolecules is shown in
Figure 2. In general, higher pH values led to more negative values
of the electrophoretic mobility. The most marked effect was with
serum albumin, where a reversal of charge occurred near pH 5.5.

The activity of calcium and oxalate ions in solution has been
shown to strongly affect the electrophoretic mobility of COM (11).
A surface isoelectric point was found at a pCa of 5.2 (corresponding

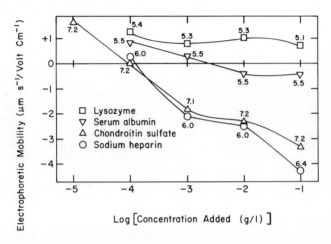

Figure 1. Electrophoretic mobility of calcium oxalate monohydrate vs. macromolecule concentration. The numbers near the data points are the corresponding solution pH values.

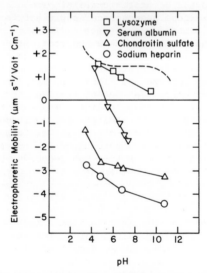

Figure 2. Electrophoretic mobility of calcium oxalate monohydrate with 0.1 g/l of macromolecule as a function of solution pH. The dashed curve without data points represents the mobility vs. pH without macromolecules present.

to a pC_2O_4 of 3.45). It follows that for calcium ion activities
above that of the isoelectric point, the surface charge of COM is
positive, and for oxalate activities above that of the isoelectric
point, the electrokinetic surface charge is negative. The data in
Figure 1 were obtained under conditions where the surface of COM is
normally positively charged. The relatively small effect of lysozyme on surface charge might be attributable to the fact that the
charge on the lysozyme has the same sign as the surface. For the
above reasons, we were interested to see if the lysozyme would adsorb to the COM surface if the surface had been negatively charged
at the beginning. In our previous work, we showed that a negative
surface charge is brought about by increasing the oxalate activity
in solution. Thus, to accomplish this, different strengths of lysozyme were added to COM suspensions that were equilibrated with sodium oxalate. In Figure 3 it is seen that increasing concentartions
of lysozyme reduce the negative mobility of the originally negative
surfaces. The negative mobility is reduced to near zero at high
lysozyme concentations.

For the adsorption of the two proteins, the role of surface
charge due to variations in the concentrations of the potential
determining ions (calcium and oxalate ions) was investigated further. In Figure 4, the mobility of COM suspensions containing
either of the two proteins is given as a function of the calcium and
oxalate concentrations. These concentrations were varied by additions of calcium chloride or sodium oxalate. The change in mobility
for COM suspensions without macromolecules present is shown with the
solid data points.

In a previous paper (17), it was shown that citrate ions adsorb
strongly onto COM. It was of interest to determine how this relatively simple species would perturb the electrokinetic response of
suspensions containing the two proteins. The mobility for suspensions containing fixed concentrations of the two proteins as a function of the concentration of added sodium citrate is given in Figure
5. These are compared with the results found in suspensions containing no proteins. In all three cases, it was found that increasing concentration of sodium citrate resulted in increasing the negative mobility, but this effect is somewhat reduced by the presence
of the proteins.

Discussion

The development of increasingly negative mobilities of COM with
three of the added macromolecules, Figure 1, results from one of two
mechanisms. One is the adsorption of the negatively charged macromolecules. The other is that the macromolecules bind calcium in
solution and caused increased oxalate activity, which, in turn,
would cause the mobility to become more negative. This second
mechanism does not seem plausible, however, because of the increase
in the oxalate activity that would be required to account for the
observed changes in mobility. Based on the previous work, it would
require more than 0.01 molar oxalate ions in solution to bring about
a COM mobility reversal from +1.7 to -1.7 mobility units. The
amount of solids in the suspension is not sufficient, even if all
were to dissolve, to produce 0.002 molar oxalate. In addition, the
magnitude of calcium binding by proteins indicated by values given

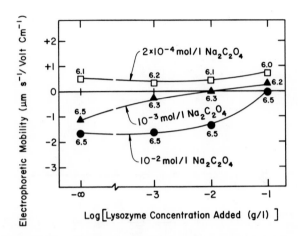

Figure 3.   Electrohoretic mobility of calcium oxalate monohydrate versus lysozyme concentration for different sodium oxalate concentrations. The numbers near the data points are solution pH. The left-most data points are mobilities without lysozyme.

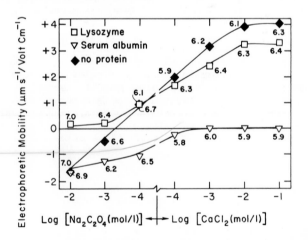

Figure 4.   Electrophoretic mobility of calcium oxalate with 0.1 g/1 macromolecule for various calcium chloride or sodium oxalate additions. The numbers near the data points are solution pH.

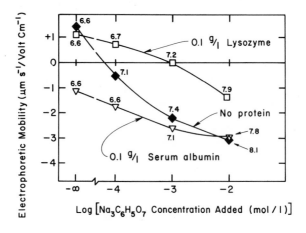

Figure 5. Electrohoretic mobility of calcium oxalate monohydrate with 0.1 g/l proteins vs. citrate concentration. The numbers near the data points are solution pH. The left most data points are zero citrate concentrations.

in the literature (18,19) would not be enough of a depletion of
calcium activity to appreciably affect the COM mobility. Thus, it
appears that the adsorption mechanism is the more likely of the two
alternatives.

We see in Figure 1 that the two mucopolysaccharides adsorb onto
the positively charged solid. The less negatively charged serum
albumin also appears to adsorb, but the positively charged protein
(lysozyme) shows little effect. On an originally negative surface,
as produced by oxalate addition, the lysozyme has a greater effect
on making the surface more positive with increasing concentrations
(Figure 3). Some adsorption tendency is suggested by the small
charge reversal that takes place for the intermediate sodium oxalate
concentration.

It was shown (11) that the mobility of COM remained unaffected
over a pH range of 4 to 10. However, changes in pH are known to
alter the net charge on the macromolecules. If surface coverage is
relatively high and the adsorbed layer is thick, changes in the
charge of adsorbed polymer molecules due to pH variations should be
reflected in the electrophoretic mobility. In fact, this assumption
is often made in the study of charges on macromolecules adsorbed on
glass capillaries (20). The variations in mobility exhibited in
Figure 2 qualitatively follow what is expected for the changes in
the charge of the macromolecules. The isoelectric points based on
mobility can be estimated by extrapolation or interpolation of the
curves. These values agree closely with the known isoelectric
points of the macromolecules (21,22). The values for the isoelectric points are expected to be closer if the charge on the sorbent is close to that on the polymer. This is demonstrated by the
data presented in Figures 2 and 4. In Figure 2, it is shown that
the electrophoretic mobility of positively charged COM particles
covered with lysozyme approaches zero at pH 11; whereas, according
to Figure 4, the i.e.p. of lysozyme at negatively charged COM is
reached at pH 7.

The nature of the adsorption process at high solution concentration of proteins appears to be particularly interesting. From
Figure 4 it can be inferred that when the surface originally has a
high charge (positive or negative), the adsorption of relatively
high concentrations of a protein having an opposite charge from the
surface occurs in a manner that reduces the mobility to zero. When
the activity of calcium ions is high (giving a high positive charge
to COM), the negatively charged protein adsorbs to an extent that
reduces the mobility to zero. Similarly, at high oxalate activity
(a negative surface charge), adsorption of positively charged protein reduces the COM mobility to zero. In simple double layer
theory, it is not expected that chemical adsorption of ions of opposite charge to the surface would produce zero mobility (23). The
mobility should change continuously, and at some point specific
adsorption should cause a reversal in charge. Further study is
necessary to determine if these data result from a change in adsorption mechanism (e.g., from monolayer to multilayer) or if it is
merely a fortuitous phenomenon.

The studies with citrate provide some information on competition
between a strongly adsorbing, small molecule and a macromolecule.
It might be expected that if molecules compete for sites on the
surface, the addition of higher concentrations of citrate should

lead to a reduction in the amount of protein adsorbed. We find in Figure 5 that at low citrate concentrations the presence of the negatively charged protein causes the mobility to be much more negative than if only the citrate is present. However, when the citrate concentration is large, the presence of the protein has no effect. At 10^{-2} mol/l citrate, the ratio of citrate to protein molecules in solution is around 1000; whereas, the same ratio at 10^{-4} mol/l citrate is only 10. It appears that the presence of high concentrations of citrate prevent appreciable adsorption of the negatively charged protein. In contrast, the positively charge protein always has the effect of providing a more positive surface, regardless of the amount of citrate in the system. Citrate apparently does not interfere with the adsorption of the lysozyme, suggesting that they are not competing for the same surface sites.

Conclusions

Two mucopolysaccharides (sodium heparin and chondroitin sulfate) adsorb onto calcium oxalate. A negatively charged protein, bovine serum albumin, also adsorbs weakly onto positively charged calcium oxalate surfaces. A positively charged protein, lysozyme, adsorbs on negatively charged calcium oxalate, as produced by adjustment in the oxalate activity. The adsorption mechanism of proteins appears to depend on the magnitude of the surface charge.

Citrate ions at high concentrations appear to effectively compete with the negative protein for surface sites. They show no evidence for competing with the positively charged protein.

Acknowledgments

This work was supported by National Institutes of Health Grant No. AM20586-01. We wish to thank Brian McKibben for assistance in some of the experimental measurements, Linda Curreri for help in figure preparation, and Art Smith and Lindreth DuBois for technical assistance. This work was also supported in part by Omni Med Corporation.

Literature Cited

1. Boyce, W. H.; Amer. J. Med. 1968, 45, 673.
2. Dent, C. E.; Sutor, D. J. Lancet. 1971, II, 775-78.
3. Nakagawa, Y.; Kaiser, E. T.; Coe, F. L. Biochem. Biophys. es. Commun. 1978, 84, 1038-44.
4. Vincent, B. Adv. Colloid Interface Sci. 1974, 4, 193-277.
5. Dixon, J. K.; La Mer, V. K.; Cassian, Li; Messinger, S.; Linford, H. B. J. Colloid Interface Sci. 1967, 23, 165-73.
6. Vincent, B.; Bijsterbosch, B. H.; Lyklema, J. J. Colloid Interface Sci. 1971, 37, 171-78.
7. Fleer, F. J.; Koopal, L. K.; Lyklema, J. Kolloid-Z. u. Z. Polymer 1972, 250, 689-702.
8. Norde, W.; Lyklema, J. J. Colloid Interface Sci. 1978, 66, 285-94.
9. Healy, T. W.; La Mer, V. K. J. Colloid Interface Sci. 1964, 19, 323-32.
10. Leal, J. J.; Finlayson, B. Invest. Urol. 1977, 14, 278-83.

11.   Curreri, P.; Onoda, G. Y., Jr.; Finlayson, B.   J. Colloid
      Interface Sci. 1979, 69, 170-82.
12.   Boyce, W. H.; Swanson, M.   J. Clin. Invest. 1955, 34, 1581-89.
13.   Maxfield, M.   Ann. Rev. Med. 1963, 14, 99-110.
14.   Kentel, H. L.; King, J. S. Invest. Urol. 1964, 2. 115.
15.   Hunter, R. J.   Zeta Potential in Colloid Science: Principles
      and Applications; Academic Press: New York, 1981.
16.   James, A. M.   In Surface and Colloid Science; Good, R. J.;
      Stromberg, R. R., Eds.; Plenum Press: New York, 1979, Vol. II.
17.   Finlayson, B.; Curreri, P.; Onoda, G.; Brown, C.   Pathogensis
      and Clincal Treatment of the Kidneystone, X, 1984, pp 174-183,
      Vahlensieck, W.; Gasser, G., Eds.; Steinkopff Publishing
      Company: Darmstadt.
18.   Munday, K. A.; Mahy, B.W.J.   Clin. Chem. Acta 1964, 10, 144-51.
19.   Blatt, W. F.; Robinson, S. M.   Analy. Biochem. 1968, 26, 151-
      73.
20.   Shaw, D. J.   Electrophoresis; Academic Press: New York, 1969; p
      97.
21.   Cohn, E. J.; Hudges, W. L.; Weare, J. H.   J. Am. Chem. Soc.
      1947, 69, 1753-60.
22.   Anderson, E. A.; Alberty, R. A.   J. Phys. Chem. 1948, 52, 1345-
      64.
23.   Overbeek, J. Th. G.   In Colloid Science; Kruyt, H. R., Ed.;
      Elsevier: Amsterdam, 1952; Vol. 1, p 229.

RECEIVED January 3, 1987

ADVANCES IN METHODOLOGY

Chapter 19

Human and Hen Lysozyme Adsorption: A Comparative Study Using Total Internal Reflection Fluorescence Spectroscopy and Molecular Graphics

D. Horsley, J. Herron, V. Hlady, and J. D. Andrade

Department of Bioengineering, College of Engineering, University of Utah, Salt Lake City, UT 84112

Total internal reflection intrinsic fluorescence (TIRIF) spectroscopy and molecular graphics have been applied to study the adsorption behavior of two lysozymes on a set of three model surfaces. A recently devised TIRIF quantitation scheme was used to determine adsorption isotherms of both hen egg-white lysozyme (HEWL) and human milk lysozyme on the three model surfaces. This preliminary study suggests that the adsorption properties of the two lysozymes are significantly different, and that further comparative studies of the two lysozymes might prove to be beneficial in understanding how protein structure might influence adsorption properties. Molecular graphics was used to rationalize the adsorption results from TIRIF in terms of the proteins' surface hydrophobic/hydrophillic character.

The understanding and control of the interactions of proteins with solid surfaces is important in a number of areas in biology and medicine. In the last twenty years there has been considerable interest in protein interactions with materials used in medical devices (1-3). One area of particular interest to the contact lens industry is in the interaction of tear proteins with contact lenses. One of the major constituent of protein deposits on lenses is lysozyme. An understanding of human lysozyme interaction with contact lens materials is essential to the minimization and elimination of contact lens deposits.

In view of the importance of tear protein interactions to the long-term efficacy of contact lenses, it is surprising that so few basic studies are

0097-6156/87/0343-0290$06.00/0
© 1987 American Chemical Society

available. Only recently have lysozyme adsorption studies on polymers and contact lens materials begun to appear (4-14).

Our lab is using two rather recently introduced techniques to offer a comparative study of human and hen lysozyme adsorption on three different model surfaces.

The first technique is known as total internal reflection intrinsic fluorescence spectroscopy (TIRIF). This recent technique provides a sensitive, real time, interfacial method for detecting the fluorescence of proteins (labeled or non-labeled) adsorbed to the totally reflecting interface. The technique was recently reviewed (15-17) and has been used by a number of biomaterial research groups to study protein adsorption (15-20). Up to this point, the major problems with using TIRIF have been its lack of reproducability and its inaccuracy in measuring absolute amounts of protein on the surface (due to the fact that changes in the quantum yield of the protein upon adsorption are ignored). A recently devised quantitation scheme developed in our lab has facilitated our acquisition of reproducable adsorption data from TIRIF (21), but its absolute accuracy in quantitating the amount of protein adsorbed on the surface remains to be improved (see discussion below).

The second technique is the use of molecular graphics. Although still in its infancy stage, it appears to have a very promising future. Using molecular graphics we have been able to rationalize our data obtained from TIRIF by examining the surface hydrophobic/hydrophillic characteristics of the two lysozyme molecules.

Experimental

Materials. Amorphous silica microscope slides were obtained from ESCO products. The silanes 3-amino-propyltriethoxysilane (APS) and dimethyldichlorosilane (DDS) were both purchased from Petrarch Systems Inc.. Hen egg-white lysozyme (3X crystalline) and human milk lysozyme (highly purified, salt free powder) were products of Calbiochem. The fluorescence standard, 5-hydroxytryptophan methyl ester hydrochloride (TrpOH) was also a product of Calbiochem. PBS buffer (pH 7.4, $[KH_2PO_4]$=0.013M, $[Na_2HPO_4]$=0.054M, $[NaCl]$=0.1M) made from analytical grade reagents and low conductivity water was used to prepare all protein and fluorescence standard solutions. These solutions were prepared fresh, prior to each experiment.

Preparation of surfaces. Adsorption studies were performed using charged and hydrophobic surfaces. Amorphous silica microscope slides were used as the substrate for all surfaces. Slides were cleaned in hot (80 degrees C) chromic acid for 20-30 minutes, cooled to room temperature, and then rinsed well in ultra-pure water (Milli-Q reagent water system). Slides were then desiccated for 12 hours at 100 degrees C. Cleanliness was confirmed by the measurement of Wilhelmy plate contact angle (no hysteresis with a clean surface).

The amorphous silica microscope slides exhibit an intrinsic negative charge at pH 7.4 (the surface silanol groups have been determined as having a pK_a of 5-7 by several workers (22-24)) and were used without further modification for adsorption of lysozyme onto negatively charged surfaces. The average ζ potential of equivalent silica slides was determined by Van Wagenen et al. (25) to be -65 mV.

Positively-charged surfaces were prepared by reacting the cleaned microscope slides with 3-amino-propyltriethoxysilane (APS). At pH 7.4 the end amino group (pK_a=10-11) of the immobilized APS molecule bears a positive charge. Clean slides were dip cast in a solution of 5% APS (v/v) in ethanol-H_2O (95:5), and allowed to react for 30 minutes. Slides were then rinsed several times in H_2O, followed with ethanol. Non-covalently bound APS was removed by vacuum desiccation at 60 degrees C for 12 hours. The measured contact angles of the APS coated slides exhibited mean values of 70 degrees advancing angle and 20 degrees receding angle with less than 10% variability among several APS coated surfaces. The average ζ potential of similarly prepared APS coated slides was determined by Van Wagenen et al. (25) to be -32 mV.

Hydrophobic surfaces were prepared by dip casting cleaned microscope slides with dimethyldichlorosilane (DDS). A similar protocol was used for DDS as for APS , except that the reaction mixture for DDS was 10% DDS (v/v) in dry toluene, and slides were rinsed with ethanol before rinsing with H_2O in order to remove the toluene. The measured contact angles of the DDS coated slides exhibited mean values of 110 degrees advancing angle and 90 degrees receding angle with less than 10% variability among several DDS coated surfaces.

All of the model surfaces were used within 4 days of preparation.

Figure 1 illustrates the proposed surface structure of the silanized surfaces as well as the untreated silica surface (26,27).

Experimental Apparatus. The TIRIF apparatus used in these experiments has been described in detail elsewhere (28). The incident light totally internally reflects at the quartz-aqueous interface and produces a standing wave normal to the reflecting interface inside the quartz (the optically more dense medium) due to the superposition of the incident and reflected waves. The electric field of the standing wave has a non-zero amplitude (E^0) at the interface which decays exponentially with distance (z) normal to the interface into the aqueous phase (the optically less dense medium), thereby creating a surface evanescent wave that selectively excites molecules within a few thousand angstroms from the surface.

a   Hydrophillic Silica

b   3-Aminopropyltriethoxysilane (APS) silica

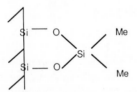

c   Dichlorodimethylsilane (DDS) silica

Figure 1. Proposed surface structures of a) hydrophillic silica, b) APS treated silica, and c) DDS treated silica.

Adsorption Isotherms. All protein adsorption isotherms were determined at room temperature. Fluorescence emission was generated by exciting at 280 nm and collecting at 335 nm (several workers have shown that tyrosine emission at 335 nm of both human and hen lysozymes is insignificant when exciting at 280 nm (29-31)). Slits of 16-nm half-bandwidth were used in both excitation and emission monochromators. For _in situ_ protein adsorption experiments, the TIRIF cell was first primed with buffer and the background fluorescence was taken. The buffer solution was then replaced with external standard (TrpOH) solutions of increasing concentration and the fluorescence intensity recorded (fig. 2). After the fluorescence of the TrpOH standard of the highest concentration was recorded, the cell was again flushed with buffer to return the signal to the background level. At this point, 5 ml of the least concentrated protein solution was injected into the cell at a flow rate of 20 ml/min. after which the flow was stopped and the protein was allowed to adsorb for 50 minutes (30 minutes after no further increase in surface fluorescence was seen). A shutter was used to prevent overexposure to UV light during the adsorption time. At the end of 50 minutes, the shutter was opened and fluorescence signal was recorded (N_{tot}). The cell was then flushed with buffer (20 ml/min.) to remove nonadsorbed proteins and the resulting signal recorded (N_a). The procedure was then repeated for the protein solution of the next higher concentration and so on. In this manner, a step-adsorption experiment was performed as opposed to a single-shot adsorption experiment where the clean surface is exposed only once to each protein solution.

The quantitation scheme developed by Hlady, _et al._ (21) was then used to quantitate the amount of protein adsorbed at each protein concentration from the recorded fluorescence intensities.

Molecular Graphics. Molecular graphics studies of hen and human lysozymes were performed using a Silicon Graphics IRIS-2400 graphics workstation. Computer programs for displaying and manipulating protein models were developed by the computer graphics laboratory at the University of California at San Francisco. The principal program, called MIDAS acts as a display vehicle for several different data structures including atomic coordinates in Protein Data Bank format, van der Waals surfaces, solvent-accessible surfaces, and electrostatic potential surfaces.

Atomic coordinates for both hen egg-white and human lysozymes were based on crystallographic structures determined by Blake _et al._ (32-36) and deposited in the Protein Data Bank (Brookhaven National Laboratory). The distribution of hydrophobic, polar, and charged atoms on the surface of the two proteins was analyzed by calculating

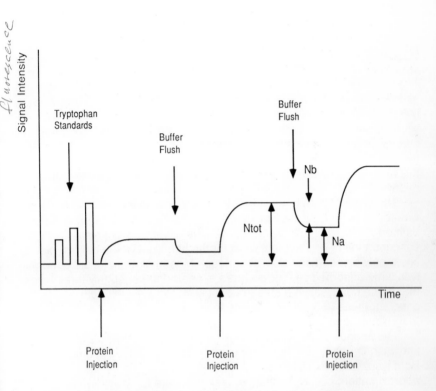

Figure 2. Time course of a typical step-adsorption experiment.
Ntot : Signal due to adsorbed and bulk protein.
Nb : Signal due to bulk protein only.
Na : Signal due to adsorbed protein only.

Corey-Pauling-Kaltun (CPK) surfaces and coloring atoms according to two coloring schemes (see tables I and II). Both the Feldmann scheme (37) and the scheme based on the Eisenberg Atomic Solvation Parameter (ASP) (38), were based on how each residue would be charged at pH 7.4.

Results

The step-adsorption isotherms for both hen and human lysozymes on hydrophobic (DDS), negatively-charged (silica), and positively-charged (APS) surfaces are presented in figure 3. Since the TIRIF quantitation scheme used assumes that the quantum yield of the protein does not change upon adsorption (an assumption we are currently trying to test), the actual amount adsorbed may differ from that presented in the figure if this assumption proves to be invalid for these proteins (see discussion below).

The order of both lysozymes' affinity toward the three model surfaces was found to be DDS > silica > APS. The human lysozyme exhibited well defined plateaus on all three surfaces at relatively low protein concentrations (< 10 mg/ml), while for the hen lysozyme a definite plateau could only be seen on the DDS surface. Moreover, the amount of human lysozyme adsorbed per surface area appeared to be roughly three times that of the hen lysozyme on equivalent surfaces at low protein concentrations (< 5 mg/ml). Again, these results may not be accurate due to the limitations of the TIRIF quantitation scheme.

The images of the two lysozymes from a few different viewpoints are shown in figure 4. The coloring schemes are as outlined in tables I and II. Both lysozymes exhibited a number of similarities. To a first approximation, positively-charged residues were fairly evenly spaced over the entire protein, although the electrostatic surface potentials (ESP) computed for the two proteins showed a slight assymetric distribution of surface positive charge, with the larger lobe showing more positive character than the smaller lobe (39). Both proteins were seen to have long positive side chains extending into the solvent.

There were approximately one-third as many negatively-charged groups as there were positively-charged groups on the surface of both proteins. These negatively-charged residues barely reach the surface and are not nearly as accessible as the positively-charged residues. ESP calculations have shown that the active site cleft shows the greatest concentration of negative surface charge on both proteins (39).

Nonpolar residues formed a hydrophobic patch in the middle of the back side (opposite the active site cleft) of both proteins as can be seen in figure 4 e & f. The hydrophobic patch on the human lysozyme appeared to be slightly larger than that on the hen lysozyme.

Perhaps the most dramatic difference in the surface properties of the

Table I

Coloring of atoms according to their hydrophobicity is based on D. Eisenberg's Atomic Solvation Parameter ($\Delta\sigma$) : $\Delta\sigma = \Delta G/A$

where ΔG = free energy of transfer of the
         atom from n-octanol to water

A = water accessible surface area of
    the atom

Atom	$\Delta\sigma$ (cal $\text{Å}^{-2}\text{mol}^{-1}$)	Color
S	+21 (+/- 10)	1 (green)
C	+16 (+/- 2)	2 (light green)
N/O	-6 (+/- 4)	6 (light blue-green)
O-	-24 (+/- 10)	10 (light blue)
N+	-50 (+/- 9)	15 (blue)

Table II

Feldmann's Functional Color Code

Atom	Color
Carbon	White
Sulfur	Yellow
Oxygen (δ-)	Pink
Oxygen (-)	Red
Nitrogen (δ+)	Light Blue
Nitrogen (+)	Blue

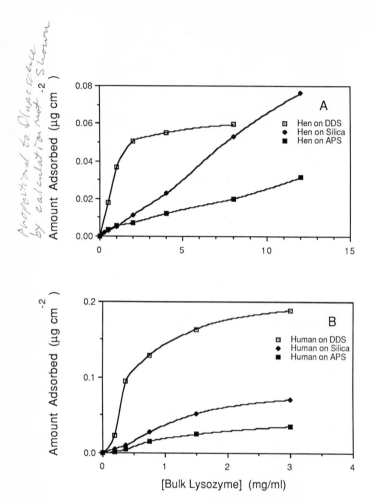

Proportional to fluorescence
by calculation not shown

Figure 3. Comparison of lysozyme adsorption on hydrophobic (DDS), negatively-charged (silica), and positively-charged (APS) surfaces. Step adsorption isotherms for the hen egg-white protein are plotted in Panel A, and for the human milk protein in Panel B.

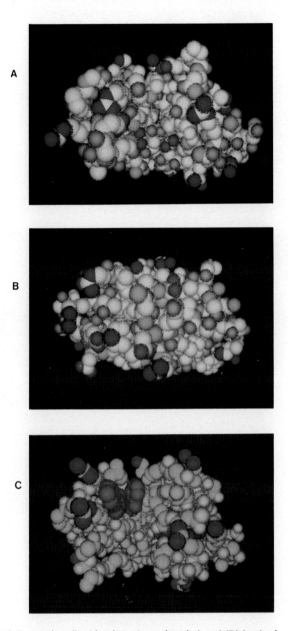

Figure 4. (A) Front view (looking into the active site) and (B) back view of human lysozyme colored according to Feldmann's functional color scheme. Front view of (C) human lysozymes colored according to Eisenberg's atomic solvation parameters. Tryptophan residues are additionally colored red. *Continued on next page.*

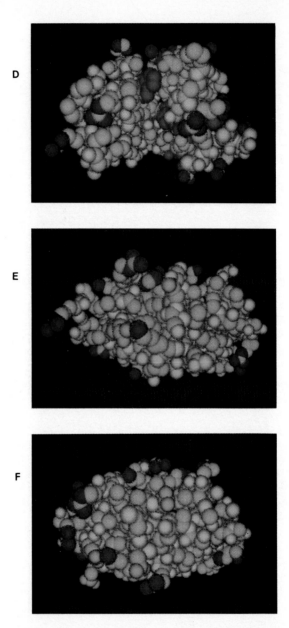

Figure 4.—*Continued.* Front view of (D) hen lysozymes colored according to Eisenberg's atomic solvation parameters. Tryptophan residues are additionally colored red. Back view of (E) human and (F) hen lysozymes colored according to Eisenberg's atomic solvation parameters. Note the hydrophobic patch seen on the back view of both proteins.

two lysozymes was seen by the ESP calculations which showed that the human lysozyme had a significantly greater overall electrostatic surface potential than the hen lysozyme had (39).

Discussion

As stated above, the primary problem with the TIRIF quantitation scheme is the fact that it ignores changes that might occur in the quantum yield of the protein upon adsorption. Hlady, *et al.* (21) used a radiolabeling technique to estimate the change in quantum yield that might occur upon adsorption of BSA and IgG onto hydrophillic silica and found that the quantum yield decreased by a factor of at least two for both proteins. Consequently, the TIRIF isotherms for BSA and IgG were at least two-fold less than the isotherms determined by radiolabeling the proteins. The origin of the decreased quantum yield of adsorbed protein would most likely be due to a change in the conformation of the protein upon adsorption.

Several reports have pointed out that iodination of proteins may affect their adsorption to solid surfaces and their chromatographic behavior (40-43). Consequently, we have chosen to determine the quantum yield of unlabeled, adsorbed lysozyme via fluorescence lifetimes. We are currently in the process of modifying our fluorescence equipment to obtain such fluorescence lifetimes.

The TIRIF adsorption isotherms for the two lysozymes presented in fig. 3 seem to imply that the human protein has a higher affinity for all three surfaces examined than the hen protein does. Until quantum yield determinations are made, nothing quantitative can be said about these differences in isotherms. However, these TIRIF isotherm differences do suggest an interesting point. Namely, if further research shows that there isn't a significant difference between the two lysozymes with respect to the amounts adsorbed, then these TIRIF isotherms clearly indicate that the adsorption process results in a greater change in Φ_a/Φ_b (Φ_a= quantum yield of adsorbed protein, Φ_b= quantum yield of protein in solution) for one of the lysozymes than for the other lysozyme.

Table III is a brief synopsis of some interesting differences between human and hen lysozymes. As can be seen in the table, the two enzymes show little antibody cross-reactivity and different amino acid compositions on their surfaces. From these results, one would expect that the two proteins would exhibit differences in their adsorption characteristics, due to their surface differences.

The fact that the human protein has one less disulfide bond and a greater susceptibility to thermal denaturation than the hen protein does, suggests that the human protein might also be more susceptible to

Table III

Differences Between Hen and Human Lysozymes

Property	Differences	Reference
Amino Acid Composition	40% of positions are different, most of which are on the surface of the protein.	51, 52
Disulfide Bonds	4 in hen; 3 in human.	51
Tryptophans	6 for hen; 5 for human.	51
Secondary Structure CD-far UV	Far UV spectra are similar, suggesting similar secondary structures.	52-55
CD-near UV	Spectra are very different, suggesting different Trp and Tyr environments.	51, 55
Fluorescence	Emission max. is 330 for human; 336 for hen. Quantum yields are different.	51, 52, 56, 57
Denaturation	Human is more susceptible to thermal denaturation.	51
Enzymatic Activity	Human is 3 times more active.	51, 52
Antibody Reactivity	Little to no antibody cross-reactivity, suggesting different surface epitopes.	58, 59
Crystal Group	Different, suggesting surface groups and inter-molecular associations are different.	53, 54
Self-Association	Hen lysozyme dimerizes and oligomerizes at pH > 5 and high protein concentrations. Little is known about the human protein.	49-51

surface denaturation. Iodide quenching studies of adsorbed lysozyme recently performed in our laboratory, have partially confirmed this prediction on the two hydrophillic surfaces (39).

The TIRIF isotherm results, suggesting that both lysozymes exhibited the overall order of affinity, DDS > silica > APS was not surprising. Many studies of lysozyme adsorption have shown that it has a higher affinity to hydrophobic surfaces than to hydrophillic surfaces (44-46). Lysozyme adsorption on hydrophobic surfaces may occur at the site of the hydrophobic patch seen on the back side of both proteins. At pH 7.4, both lysozymes are positively-charged and would be expected to adsorb more strongly to the negatively-charged silica surface than to the positively-charged APS surface. This was seen to be the case for both proteins. As noted above, both lysozymes have long positive side chains extending out into the solvent while the surface negative charges are not nearly as accessible. This point further suggests a preferential adsorption onto negative surfaces. With regard to the effect of electrostatic repulsion between lysozyme and the positively-charged APS surface, it is interesting to note that both Perkins (47) and Salton (48) found that lysozyme activity was inhibited by positively charged groups in the mucopeptide of bacterial cell walls (lysozyme's biological substrate), presumably due to electrostatic repulsion since the activity was increased when the free amino groups in the mucopeptide were neutralized.

The hen lysozyme showed non-saturating adsorption behavior on both charged surfaces. This may suggest that charged surfaces promote the formation of multiple protein layers at the surface in some manner which is not yet clear. Hen lysozyme forms dimers and higher aggregates under conditions of high protein concentration (5-10 mg/ml) and pH > 5 (49-51). This kind of self-association might be going on at the surface where the adsorption process inherently concentrates the proteins.

Acknowledgments

This work was supported in part by The Center for Biopolymers at Interfaces.

References

1. Vroman, L. and Leonard, E. F., eds. (1977) *Behavior of Blood and Its Components at Interfaces, Ann. N.Y. Acad. Sci.* **283**.
2. Baier, R. E., ed., (1975) *Applied Chemistry at Protein Interfaces, Adv. Chem. Ser.* **145**.
3. Cooper, S. L. and Peppas, N. A. (1982) *Biomaterials, Adv. Chem. Ser.* **199**.
4. Holly, F. J. and Refojo, M. F. (1976) In: *Hydrogels for Medical and Related Applications* (J. D. Andrade, ed.), ACS Sym. Series #31, p. 267.

304 PROTEINS AT INTERFACES

5. Ratner, B. D. and Horbett, T. A. (1984) *Trans. Soc. Biomat.* **10**, 76.
6. Gachon, A. M., Bilbaut, T., and Dastugue, B. "Adsorption of Tear Proteins on Soft Contact Lenses", in press.
7. Castillo, E. J., Koenig, J. L., Anderson, J. M., and Lo, J. (1984) *Biomaterials* **5**, 319.
8. Castillo, E. J., Koenig, J. L., and Anderson, J. M. (1985) *Biomaterials* **6**, 338.
9. Castillo, E. J., Koenig, J. L., and Anderson, J. M. (1986) *Biomaterials* **7**, 9.
10. Castillo, E. J., Koenig, J. L., and Anderson, J. M. (1986) *Biomaterials* **7**, 89.
11. Baszkin, A., Proust, J. E., and Boissonnade, M. M. (1984) *Biomaterials* **5**, 175.
12. Proust, J. E., Baszkin, A., Perez, E., and Boissonnade, M. M. (1984) *Colloids and Surfaces* **10**, 43.
13. Chen, J., Dong, D. E., and Andrade, J. D. (1982) *J. Colloid Interface Sci.* **89**, 577.
14. Royce, F., Ratner, B., and Horbett, T. (1980) In: *Biomaterials: Interfacial Phenomonon and Applications* (S. L. Cooper and N. A. Peppas, eds.) ACS Advances in Chemistry Series **199**, p. 453.
15. Axelrod, D., Burghard, T. P., and Thompson, N. T. (1984) *Ann. Rev. Biophys. Bioeng.* **13**, 247.
16. Cheng, Y-L., Lok, B. K., and Robertson, C. R. (1985) In: *Surface and Interfacial Aspects of Biomedical Polymers, Vol 2 Protein Adsorption* (J. D. Andrade, ed.), Plenum Press, p. 121.
17. Darst, S. A., Robertson, C. R. (1986) In: *Spectroscopy in the Biomedical Sciences* (R. M. Gendreau, ed.), CRC Press, p. 175.
18. Beissinger, R. L. and Leonard, E. F. (1980) *ASAIO J.* **3**, 160.
19. Eberhart, et. al. (1984) Abstract, *Devices Tech. Branch Contractor's Conference,* p. 43.
20. Van Wagenen, R. A., Rockhold, S., and Andrade, J. D. (1980) In: *Biomaterials: Interfacial Phenomenon and Applications* (S. L. Cooper and N. A. Peppas, eds.) ACS Advances in Chemistry Series **199**, 351.
21. Hlady, V., Reinecke, D. R., and Andrade, J. D. (1986) *J. Colloid. Interface Sci.* **111**, 555.
22. Van Wagenen, R. A., Andrade, J. D., and Hibbs, J. B. (1976) *J. of the Electrochem. Soc.* **123**, 1438.
23. Hair, M. L., and Hertl, W. J. (1970) *J. Phys. Chem.* **74**, 91.
24. Marshall, K., Ridgewell, G. L., Rochester, C. H., and Simpson, J. (1974) *Chem. Ind. (London)* **19**, 775.
25. Van Wagenen, R. A. et. al. (1981) *J. Colloid. Interface Sci.* **84**, 155.
26. Arkles, B. (1977) *CHEMTECH,* 766.
27. Waddell, T. G., Leyden, D. E., and DeBello, M. T. (1981) *J. Am. Chem. Soc.* **113**, 5303.
28. Hlady, V., Van Wagenen, R. A., and Andrade, J. D. (1985) In: *Surface and Interfacial Aspects of Biomedical Polymers, Vol 2 Protein Adsorption* (J. D. Andrade, ed.), Plenum Press, p. 81.
29. Imoto, T., Forster, L. S., Rupley, J. A., and Tanaka, F. (1971) *Proc. Nat. Acad. Sci.* **69**, 1151.
30. Teichberg, V. I., Plasse, T., Sorell, S., and Sharon, N. (1972) *Biochim. Biophys. Acta* **278**, 250.
31. Lehrer, S. S. and Fasman, G. D. (1966) *Biochem. Biophys. Res. Comm.* **23**, 133.
32. Banyard, S. H., Blake, C. C. F., and Swan, I. D. A. (1974) In: *Lysozyme* (E. F. Osserman, R. E. Canfield, and S. Beychok, eds.) Academic Press, p. 71.
33. Artymiuk, P. J., and Blake, C. C. F. (1981) *J. Mol. Biol.* **152**, 737.
34. Blake, C. C. F., Loenig, D. F., Mair, G. A., North, A. C. T., Phillips, D. C., and Sarma, V. R. (1965) *Nature* **206**, 757.
35. Blake, C. C. F., Johnson, L. N., Mair, G. A., North, A. C. T., Phillips, D. C., and Sarma, V. R. (1967) *Proc. Roy. Soc.* **B167**, 318.

36. Blake, C. C. F., Mair, G. A., North, A. C. T., Phillips, D. C., and Sarma, V. R. (1967) *Proc. Roy. Soc.* **B167**, 365.
37. Feldmann, R. J. and Bing, D. H. *Teaching Aids for Macromolecular Structure*, Taylor-Merchant Corp., New York.
38. Eisenberg, D. and McLachlan, A. D. (1986) *Nature* **319**, 199.
39. Horsley, D. G. M. Sc. thesis, Univ. of Utah (In preparation).
40. Koshland, M. E., Englberger, F. M., Erwin, M. J., and Gaddone, S. M. (1963) *J. Biol. Chem.* **238**, 1343.
41. Greenwood, F. C. (1971) In: *Principle of Competitive Protein Binding Assay* (W. D. Odell and W. H. Daughady, eds.), Lippincott, Philadelphia, p. 288.
42. Van der Scheer, A., Feijen, J., Elhorst, J. K., Krugers-Dagneaux, P. G. L. C., and Smolders, C. A. (1978) *J. Colloid Interface Sci.* **66**, 136.
43. Crandall, R. E., Janatova, J., and Andrade, J. D. (1981) *Prep. Biochem.* **11**, 111.
44. Hansen, J. (1985) M. S. Thesis, University of Utah.
45. Chen, J., Dong, D. E., and Andrade, J. D. (1982) *J. Colloid Interface Sci.* **89**, 577.
46. Halperin, G., Breitenbach, M., Tauber-Finkelstein, and Shaltiel, S. (1981) *J. Chromatogr.* **215**, 211.
47. Perkins, H. R. (1967) *Royal Soc. of Lond. Proc.* **167**, 443.
48. Salton, M. R. J. (1964) *Med. Hyg.* **22**, 985.
49. Bruzzesi, M. R., Chiancone, E., and Antonini, E. (1965) *Biochemistry* **4**, 1796.
50. Deonier, R. C. and Williams, J. W. (1970) *Biochemistry* **9**, 4260.
51. Imoto, T., Johnson, L. N., North, A. C. T., Phillips, D. C., and Rupley, J. A. (1972) In: *The Enzymes*, Vol. 7, 3rd ed. (P. D. Boyer, ed.), Academic Press, p. 665.
52. Mulvey, R. S., Gaultieri, R. J., and Beychok, S. (1974) In: *Lysozyme* (E. F. Osserman, R. E. Canfield, and S. Beychok, eds.) Academic Press, p. 281.
53. Banyard, S. H., Blake, C. C. F., and Swan, I. D. A. (1974) In: *Lysozyme* (E. F. Osserman, R. E. Canfield, and S. Beychok, eds.) Academic Press, p. 71.
54. Artymiuk, P. J., and Blake, C. C. F. (1981) *J. Mol. Biol.* **152**, 737.
55. Halper, J. P., Latovitzki, N., Bernstein, H., and Beychok, S. (1971) *Proc. Natl. Acad. Sci. USA* **68**, 517.
56. Mulvey, R. S., Gaultieri, R. J., and Beychok, S. (1973) *Biochemistry* **12**, 2683.
57. Teichberg, V. I., Plasse, T., Sorell, S., and Sharon, N. (1972) *Biochim. Biophys. Acta* **278**, 250.
58. Atassi, M. Z., and Habeeb, A. F. S. A. (1979) In: *Immunochemistry of Proteins*, Vol. 2 (M. Z. Atassi, ed.), Plenum, p. 177.
59. Arnon, R. (1977) In: *Immunochemistry of Enzymes and Their Antibodies* (M. Salton, ed.), Wiley, p. 1.

RECEIVED March 4, 1987

Chapter 20

Protein Adsorption at Polymer Surfaces: A Study Using Total Internal Reflection Fluorescence

Aron B. Anderson, Seth A. Darst, and Channing R. Robertson

Department of Chemical Engineering, Stanford University, Stanford, CA 94305

The application of total internal reflection fluorescence spectroscopy (TIRF) by this laboratory to the study of protein adsorption at solid-liquid interfaces is reviewed. TIRF has been used to determine adsorption isotherms and adsorption rates from single- and multi-component protein solutions. Initial adsorption rates of BSA can be explained qualitatively by the properties of the adsorbing surface. Most recently, a TIRF study using monoclonal antibodies to probe the conformation of adsorbed sperm whale myoglobin (Mb) elucidated two aspects of the Mb adsorption process: 1) Mb adsorbs in a non-random manner. 2) Conformational changes of adsorbed Mb, if they occur, are minor and confined to local regions of the molecule. Fluorescence energy transfer and proteolytic enzyme techniques, when coupled with TIRF, can characterize, respectively, the conformation and orientation of adsorbed Mb.

Proteins generally adsorb onto solid surfaces from solution. This process is of importance in a number of applications. For instance, the composition, conformation, and orientation of adsorbed proteins are believed to influence cell/substrate interactions (1-3). Also, adsorption of serum proteins onto biomaterials is generally recognized as the initial event in the sequence that culminates in thrombus formation (4,5). Consequently, protein behavior at solid-liquid interfaces has been extensively studied (6-11). Many fundamental questions about the protein adsorption phenomenon, however, remain unanswered (12).

Current understanding of protein adsorption has been synthesized from investigations employing a number of techniques. One of these techniques, total internal reflection fluorescence (TIRF), has emerged as a versatile tool for studying proteins at surfaces. The investigation of protein adsorption at solid-liquid interfaces in this laboratory using TIRF is reviewed. Several reviews of the TIRF technique have appeared recently (13-15).

0097-6156/87/0343-0306$06.00/0

Whereas these reviews encompass general principles, methodologies, and applications of TIRF, this review focuses mainly on protein adsorption characteristics that have been studied in this laboratory.

The Technique of TIRF

Total Internal Reflection Principles.   The theory that encompasses total internal reflection is discussed extensively by Harrick (16) and will be only briefly outlined here. As shown schematically in Figure 1, a beam of light incident on an interface between two transparent media will totally internally reflect if the angle of incidence, α, exceeds the critical angle, $\alpha_c = \sin^{-1}(n_2/n_1)$, where n_1 and n_2 are the indices of refraction for medium 1 and medium 2, respectively. From the point of reflection, a standing wave extends into the rarer medium (here, medium 2). The amplitude of this evanescent field decays exponentially with distance normal to the interface. The depth of penetration depends on the wavelength of the incident light, the indices of refraction n_1 and n_2, and α. In our apparatus (see below) the amplitude of the evanescent wave decreases to e^{-1} of the interfacial value at a distance on the order of 1000 Å (17). Since the wave is surface confined, a fluorescence signal induced by its interaction with fluorescently-labeled protein can be used as a sensitive probe of adsorption.

TIRF Apparatus.   The TIRF apparatus used to study protein adsorption in our laboratory is shown schematically in Figure 2. Important features of the apparatus are described in the Figure caption. In a typical experiment, an aqueous buffer solution of fluorescently-labeled protein (medium 2 of Figure 1) is pumped through the flow cell. The protein diffuses to and adsorbs onto a polymer film (medium 1 of Figure 1) coated on the glass microscope slide that constitutes one wall of the flow cell. The adsorption of protein is monitored continuously as a function of time through the fluorescence induced by the evanescent field.

Interpretation of TIRF Results.   The detected fluorescence signal must be interpreted with caution. The signal may not arise exclusively from surface-adsorbed protein (17). The measured intensity may also include fluorescence from solution protein excited by scattered light or by the evanescent field. Therefore, the measured signal can be interpreted as a true indicator of surface-adsorbed protein only if the fluorescence from the bulk solution protein is negligible. To insure that this is indeed the case, scattered light has been minimized by development of a surface preparation technique that includes meticulous cleaning of the microscope slides prior to coating, depositing polymer films reproducibly with a spin-coating technique (17), and characterizing the polymer surfaces using electron spectroscopy for chemical analysis (17,22). Additionally, Lok et al. (17,18) have determined that fluorescence from solution protein excited by the evanescent field can be rendered negligible by establishing bounds on both the fluorescent labeling ratio (moles label/moles protein) and the bulk solution concentration of protein.

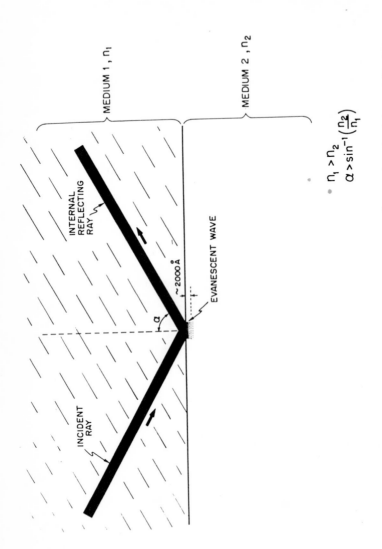

Figure 1. Total internal reflection geometry. (Reproduced with permission from Ref. 17. Copyright 1983 Academic Press.)

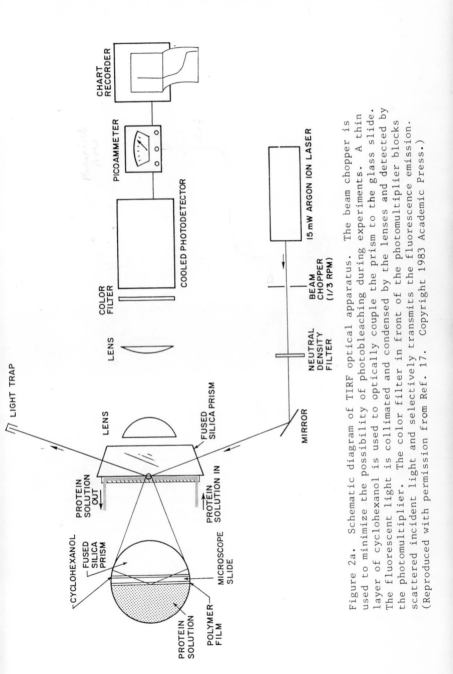

Figure 2a. Schematic diagram of TIRF optical apparatus. The beam chopper is used to minimize the possibility of photobleaching during experiments. A thin layer of cyclohexanol is used to optically couple the prism to the glass slide. The fluorescent light is collimated and condensed by the lenses and detected by the photomultiplier. The color filter in front of the photomultiplier blocks scattered incident light and selectively transmits the fluorescence emission. (Reproduced with permission from Ref. 17. Copyright 1983 Academic Press.)

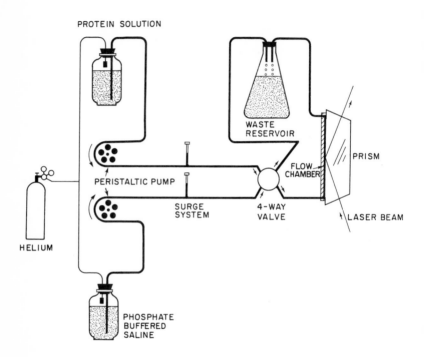

Figure 2b. Schematic diagram of TIRF flow apparatus. The surge system damps out the pulsations of the peristaltic pump. Helium is used to purge the system of fluorescence-quenching oxygen. The 4-way valve is placed close to the flow-cell to reduce priming volumes to 0.1 ml. The flow-cell/prism assembly is mounted vertically so that fluid enters from the bottom and exists at the top of the chamber. (Reproduced with permission from Ref. 17. Copyright 1983 Academic Press.)

Also, the measured fluorescence signal may not be proportional to the surface concentration of protein. This has been observed during adsorption experiments using fluorescently-labeled γ-globulins (17) and sperm whale myoglobin (Mb) (19). Therefore, the TIRF apparatus must be properly calibrated to determine whether fluorescence intensity is indeed proportional to surface concentration. Some early calibration techniques must be viewed with caution because they were not performed under the same conditions as the protein adsorption experiments (20,21). More recently, a reliable calibration procedure that uses a double labeling technique has been developed (17).

Experimental Applications of TIRF

The TIRF technique has been refined in this laboratory to provide reproducible results about protein adsorption phenomena at solid-liquid interfaces (17-19). The investigation of protein adsorption behavior using TIRF in this laboratory is discussed below. The overall goal of this research is a complete, general description of the protein adsorption process.

The TIRF apparatus developed by Watkins and Robertson (20) and subsequently refined by Lok *et al.* (17) can be used to study protein adsorption _in situ_ and noninvasively. The initial work by this laboratory focused on quantifying macroscopic properties of protein adsorption using the plasma proteins albumin, fibrinogen, and γ-globulin together with polymer surfaces possessing a wide range of surface properties. More recently, research has focused on molecular aspects of protein adsorption using the thoroughly characterized protein Mb.

Macroscopic Protein Adsorption Properties

Protein Adsorption Isotherms.

Lok *et al.* (18) reported isotherms for bovine serum albumin (BSA) and bovine fibrinogen adsorbing on polydimethylsiloxane (PDMS). The BSA adsorption isotherm at 37 $^{\circ}C$ is shown in Figure 3. It was determined that these isotherms could not be described by a simple Langmuir isotherm (18). Isotherms are generally believed to represent a dynamic equilibrium between solution and adsorbed species. Given the size and conformational adaptability of proteins, however, this idea may not apply to protein adsorption isotherms. For instance, the rates of adsorption at low bulk protein concentrations, as well as the rates of exchange and desorption, are very slow. If adsorption experiments are terminated before equilibrium is attained, interpretation of the "isotherm" will be ambiguous. It has been observed that adsorption experiments that had apparently reached equilibrium, if allowed to continue for longer periods of time, showed a significant increase in the amount of adsorbed protein (22). Additionally, it is not clear whether protein adsorption can be described as a reversible process. Indeed, conformational changes of the adsorbed protein may occur that affect adsorption reversibility and give rise to topologically different structures. This would invalidate the use of an isotherm description that implies reversibility. It is clear that one must be cautious when

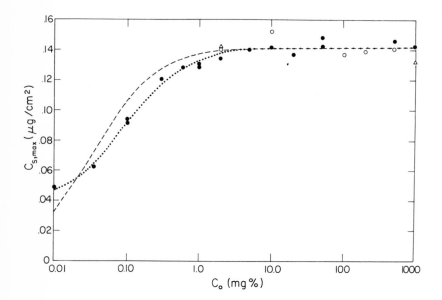

Figure 3.   Adsorption isotherm for BSA on PDMS at 37 °C. Dashed
line is a least squares fit to a Langmuir isotherm. Dotted line
is a least squares fit to a two parameter isotherm. (Reproduced
with permission from Ref. 18.   Copyright 1983 Academic Press.)

interpreting protein adsorption isotherms.   Protein adsorption isotherms have been discussed in detail by Cheng <u>et al.</u> (13).

<u>Protein Adsorption and Desorption Rates and Kinetics</u>.   The TIRF flow cell was designed to investigate protein adsorption under well-defined hydrodynamic conditions.   Therefore, the adsorption process in this apparatus can be described by a mathematical convection-diffusion model (17).   The rate of protein adsorption is determined by both transport of protein to the surface and intrinsic kinetics of adsorption at the surface.   In general, where transport and kinetics are comparable, the model must be solved numerically to yield protein adsorption kinetics.   The solution can be simplified in two limiting cases: 1) In the kinetic limit, the initial rate of protein adsorption is equal to the intrinsic kinetic adsorption rate.   2) In the transport limit, the initial protein adsorption rate, as predicted by Lévêque's analysis (23), is proportional to the wall shear rate raised to the 1/3 power.   In the transport-limited adsorption case, intrinsic protein adsorption kinetics are unobservable.

At present, the protein/surface interactions that determine adsorption kinetics are unclear. To clarify these interactions, the effects of polymer surface properties on protein adsorption and desorption rates have been investigated.   BSA adsorption from a 1 mg% solution (1 mg% = 1 mg/100 mL) was studied using several polymers chosen for their wide range of surface properties and functionalities (22, Cheng, Y.L. <u>et al.</u>, J. Coll. Int. Sci., in press).   The polymers and their surface properties (under the conditions of the BSA adsorption experiments) are listed in Table I.

TABLE I

General Surface Properties of Polymers Used for BSA Adsorption

Surface	Hydrophobicity	Charge	H-Bonding Capability
Polydimethylsiloxane (PDMS)	Strongly Hydrophobic	Uncharged	None
Polydiphenylsiloxane (PDϕS)	Strongly Hydrophobic	Uncharged	None
Polycyanopropylmethylsiloxane (PCPMS)	Hydrophilic	Uncharged	Yes
Polystyrenesulfonate (PSS)	Hydrophilic	Negative	Yes
Polymethylmethacrylate (PMMA)	Slightly Hydrophobic	Uncharged	Weak
Polyethyleneoxide (PEO)	Hydrophilic	Uncharged	None

The initial BSA adsorption rate onto PDMS, PDϕS and PCPMS is transport-limited for wall shear rates ranging from 25 to 4000 s^{-1}. BSA adsorption on PSS is transport-limited below 70 s^{-1} but becomes kinetically-limited at increasing shear rates. No adsorption (on the time scale of the experiment) occurs on PEO. Only on PMMA is BSA adsorption kinetically limited over the entire range of shear rates investigated. The initial adsorption of BSA on PMMA can be described by a kinetic rate expression that is first order in BSA concentration.

The BSA desorption rates are much slower than the adsorption rates. Desorption of BSA from each polymer surface is kinetically-limited for all shear rates studied. The desorption rates are also insensitive to bulk protein concentration.

This study indicates that BSA adsorbs most rapidly on the strongly hydrophobic surfaces (PDMS and PDϕS) and the uncharged, hydrogen-bonding surface (PCPMS), whereas no BSA adsorption occurs on PEO, which is hydrophilic, uncharged, and offers no hydrogen-bonding capability. Thus, the initial adsorption rate of BSA can be explained qualitatively in terms of the general properties and chemical functionalities of each surface.

Adsorption rates of other protein/surface pairs have been investigated. The initial adsorption of bovine fibrinogen from a 1 mg% solution onto PDMS is transport-limited up to a wall shear rate of 410 s^{-1} (18). The initial adsorption of Mb onto PDMS is also transport-limited over the same range of shear rates (19). At higher bulk concentrations, where the time-independent assumption inherent in the Lévêque analysis is no longer valid and therefore a numerical solution must be applied, the initial adsorption of BSA and bovine fibrinogen onto PDMS is also transport-limited (18). These results imply that small proteins, by virtue of their higher diffusion coefficients, will initially adsorb faster than large proteins during competitive adsorption from multi-protein solutions. The TIRF technique is amenable to study of competitive adsorption. One application is discussed in the next section.

Competitive Adsorption of Fibrinogen and Albumin.   When a foreign surface contacts blood it encounters a complex mixture of plasma proteins. The adsorption rates and surface coverages determined for proteins individually will undoubtedly differ when several proteins challenge the surface simultaneously. Therefore, a study of the adsorption characteristics of multi-component protein solutions has been conducted using TIRF. When extrinsic labeling is employed, TIRF is particularly suitable for studying competitive adsorption.

Lok et al. (18) have reported initial adsorption rates and surface coverages for the binary system of BSA and bovine fibrinogen adsorbing onto PDMS. Under the conditions of the experiments (using approximately the physiologic concentration ratio of albumin to fibrinogen) albumin initially adsorbs faster than fibrinogen due to its higher diffusion coefficient, as predicted above. At long times, however, fibrinogen becomes preferentially adsorbed (18). Other investigators have reported similar results (24,25). Apparently the faster diffusing BSA will only delay, not halt, the subsequent adsorption of fibrinogen (which is more thrombogenic than albumin (26)) onto PDMS.

Molecular Aspects of Protein Adsorption Using Mb

Most recently, we have undertaken TIRF experiments using site specific monoclonal antibodies to directly probe the conformational states of a protein antigen, Mb, adsorbed onto PDMS. The goal of these experiments is to define, at the molecular level, the physical state of the adsorbed protein in terms of its conformation and orientation.

The Mb protein and the antibodies used in this study have been extremely well characterized (see below). This allows the results of experiments to be interpreted more completely than would be possible with other protein/antibody systems. This work has been performed in collaboration with Dr. J.A. Berzofsky, Metabolism Branch, National Cancer Institute, National Institutes of Health.

Introduction. Several monoclonal antibodies (designated as clones 1, 2, 3.4, 4, and 5) against the protein Mb have been isolated and characterized (27,28). The binding of each of the antibodies has been shown to be sensitive to the conformation of the Mb antigen, indicating that each antibody recognizes a topographic, as opposed to a linear, epitope on the Mb molecule. In addition, for three of these antibodies (clones 1, 3.4, and 5), amino acid residues on the Mb molecule essential for the binding of the antibody have been identified. Thus, the antibody most likely interacts directly with these amino acid residues. The locations of the residues indicate the regions of the Mb molecule that are recognized by each antibody. Figure 4 shows a computer-generated model of the three-dimensional structure of Mb taken directly from the x-ray crystallographic coordinates (29). The location of the amino acid residues essential for the binding of the clone 1, clone 3.4, and clone 5 antibodies are indicated (see Figure caption for details). These antibodies can be used as conformational probes of the adsorbed Mb. Also, certain regions of an adsorbed Mb molecule may be sterically blocked from access by an antibody if the Mb surface binding site includes that region. Thus, the antibodies with known recognition sites on the Mb molecule can be used to obtain information concerning the orientation of the adsorbed Mb.

Prior to performing the experiments designed to observe the binding of the Mb-specific antibodies to surface adsorbed Mb, it was first necessary to characterize the adsorption of Mb onto PDMS. The work characterizing the adsorption of Mb onto PDMS has been described elsewhere (19) and will not be discussed in this review.

Experimental Protocol. For all of the experiments discussed hereafter, unless otherwise specified, the following protocol was used:

- Adsorb Mb onto a PDMS film under the following conditions:

  - Mb Concentration - 0.85 mg%
  - Buffer   - 10 mM phosphate, 150 mM NaCl, 3 mM NaN_3, pH 7.40
  - Wall Shear Rate   - 94 s^{-1}
  - Temperature   - 37 $^\circ$C

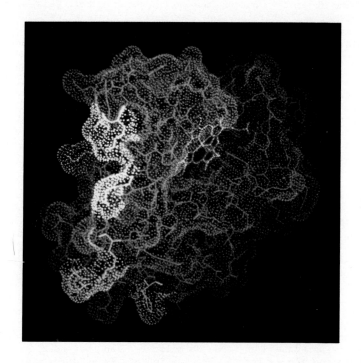

Figure 4. Computer generated model of the three-dimensional
structure of Mb. The amino acids essential for the binding of
three of the monoclonal antibodies are shown. The amino acids
and the antibodies correspond as follows: Green, clone 3.4;
orange, clone 1; purple, clone 5. The heme group is shown in
red.

- After 10 hrs, replace the Mb solution with a solution of FITC-antibody. Use TIRF to monitor antibody binding to adsorbed Mb.

Comparison of IgG and Fab Binding of the Same Antibody, Clone 4.

All of the antibodies used in this study are of the IgG class. IgG antibodies have a molecular weight of about 150,000 and consist of three units: two identical Fab units (for antibody binding fragment) containing the binding sites of the antibody, and one Fc unit (for complement fragment). Each Fab and Fc is a globular unit with a molecular weight of approximately 50,000. In the IgG molecule, the three fragments form a "Y" shape, with the binding sites at the ends of the two Fab units. Digestion of IgG with the proteolytic enzyme papain results in the cleavage of the molecule into its three fragments. The active Fab fragments can be purified from the inactive Fc fragment, resulting in effectively univalent antibodies.

Isotherms for the binding of the clone 4 IgG and Fab antibodies to the previously adsorbed Mb were determined by measuring the antibody binding over a range of antibody solution concentrations. It can be demonstrated that a reversible equilibrium exists between the antibody in solution and the surface, Mb-bound antibody. The results are plotted in the form of a Scatchard plot in Figure 5. For both the clone 4 IgG and Fab, the adsorption appears to reach a point which is limited by steric exclusion. Both IgG and Fab molecules are larger than the Mb molecule. For that reason, the maximum fractional surface coverages cannot be unity. The maximum surface coverages shown in Figure 5 are close to values that can be estimated by treating the adsorbing molecules as randomly adsorbing spheres (30) with radii equal to the hydrodynamic radii calculated from the known diffusion coefficients.

The equilibrium constant for the univalent Fab-adsorbed Mb reaction is 5×10^8 M^{-1}. This compares very favorably with the value for the clone 4 equilibrium constant in solution, 7.1×10^8 M^{-1} (27). That the IgG isotherm is significantly steeper than the Fab isotherm indicates that bivalent binding of the IgG molecules occurs.

Comparison of Fab Binding for Each Antibody.

Isotherms were determined as above for the Fab fragments of each antibody binding to the previously adsorbed Mb. The results for clones 1 and 2 were similar to the clone 4 Fab binding already discussed. The binding of clone 3.4 and 5 Fab fragments to the adsorbed Mb, however, was significantly perturbed by the adsorption process. Both clone 3.4 and clone 5 Fab fragments recognize Mb in solution as determined by a solution radioimmunoassay.

An additional control has been performed for the clone 5 antibody. It is known that the clone 4 and clone 5 antibodies can bind to the Mb molecule simultaneously (27). If the clone 4 antibody is adsorbed to the PDMS surface initially, and then Mb is introduced into the solution, some of the clone 4 antibody adsorbed to the PDMS surface will remain active and will bind Mb from solution. The bound Mb is thus surface-confined without interacting with the PDMS surface and in an orientation that the clone 5

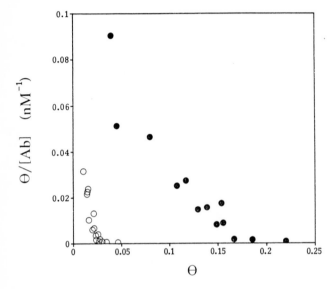

Figure 5. Scatchard plot for the clone 4 IgG (○) and clone 4 Fab (●) binding to Mb adsorbed under the conditions outlined in the text. The surface coverage, θ, is defined in the following manner:

$$\theta = \text{(moles of adsorbed Ab)/(moles of adsorbed Mb)}$$

[A] is the Antibody (Ab) solution concentration in nM.

antibody should be able to recognize and interact in an antibody-Mb-antibody sandwich. Thus, possible conformational changes to the clone 5 antigenic site due to Mb adsorption on PDMS will not occur and steric blocking of the Clone 5 antigenic site will also not occur. The clone 5 Fab binds rapidly to the clone 4-Mb surface. The amount that binds is also very close to the steric limit discussed above. Thus, this and the other controls show that the diminished binding affinities of the clone 3.4 and clone 5 antibodies to Mb adsorbed directly onto PDMS are due to an effect of the adsorption process on the Mb molecule itself.

Table II compares the equilibrium constant for the reaction with adsorbed Mb and with Mb in solution for the antibodies studied. For the clone 1, 2, and 4 antibodies, the equilibrium constant for the reaction with adsorbed Mb compares favorably with the equilibrium constant for the solution reaction, whereas the clone 3.4 and clone 5 antibody affinities are at least 2 orders of magnitude less for the adsorbed Mb.

TABLE II

Summary of Adsorbed Mb/Antibody Interactions

Antibody	Affinity for Mb $(M^{-1} \times 10^9)$	
	Solution[a]	Surface
Clone 1	1.9	0.95
Clone 2	2.2	1.3
Clone 3.4	0.2	$< 10^{-3}$
Clone 4	0.71	0.53
Clone 5	1.6	$< 10^{-2}$

[a]Determined in reference 27 except for clone 3.4, which was determined in reference 28

Summary of Mb Studies.   The binding of five Mb-specific monoclonal antibodies to previously-adsorbed Mb has been investigated. Three of the antibodies bound to the adsorbed Mb with affinities close to their normal affinities for Mb in solution, but the affinities of two of the antibodies for the adsorbed Mb were diminished by at least two orders of magnitude. At present, we are unable to determine whether the diminished antibody binding is due to adsorbed-Mb conformational changes or simply to steric blocking of the antibody site. Experiments described below should allow us to make this distinction. Nevertheless, two important conclusions can be made:

   i.   The Mb molecules are not adsorbed in a random fashion. If the Mb molecules were adsorbed randomly, with no preferred mode of adsorption, then every antibody would be expected to bind to the adsorbed protein with the same affinity relative to that for Mb in solution. This preferred mode of adsorption can either be a) a preferred orientation; b) random orientation but with conformational changes to the clone 3.4 and clone 5 antigenic sites (preferred conformational changes); c) a combination of a and b.

ii. Conformational changes of the Mb molecule due to the adsorption process, if occurring at all, are confined to local regions of the adsorbed Mb. Although conformational changes of the antigenic sites of the clone 3.4 and 5 antibodies may result in diminished binding of these antibodies, the binding of the clone 1, 2, and 4 antibodies are essentially unaffected by the adsorption process, indicating that the topography of the antigenic sites for these antibodies is not significantly altered by adsorption related phenomena. Due to the compact structure of the Mb molecule, this implies that any conformational changes are minor since the effects of gross unfolding of portions of the molecule would be expected to be propagated to other portions of the molecule.

Further Studies of Mb Adsorption

The recent investigation of Mb adsorption on PDMS (above) has probed the conformation and orientation of a layer of protein adsorbed on a hydrophobic surface. Our present goal is further characterization of the conformation and orientation of Mb on PDMS. Two techniques, fluorescence energy transfer and proteolytic enzyme cleavage, can be used to achieve this goal.

Fluorescence Energy Transfer Studies. The technique of fluorescence energy transfer has been shown to be a sensitive probe of intramolecular distances in macromolecules (31). The technique has been widely used to estimate distances within macromolecules in solution. Recently, energy transfer has been employed to investigate conformational changes of adsorbed BSA (32). The energy transfer measurements of adsorbed BSA were unable to reveal specific conformational changes, because randomly-labeled BSA was used and the exact three-dimensional structure of BSA is undetermined at present. By using specifically-labeled Mb, however, the energy transfer technique can give direct information concerning the conformation of adsorbed Mb.

If appreciable deformation of adsorbed Mb occurs there should be a concomitant change of distances between specific sites in the Mb molecule. Fluorescent labels can be introduced at specific amino acid residues (33,34) and the heme pocket in Mb (35). The efficiency of singlet-singlet energy transfer between two such fluorescent dipoles can be used to estimate the distance between the labels. The three-dimensional structure of Mb is known in great detail (29). Therefore any distance changes observed for adsorbed Mb can be directly related to conformational changes of the Mb molecule. By performing energy transfer experiments with fluorescent labels at various well-defined locations in the Mb molecule, a detailed picture of any deformation of adsorbed Mb can be constructed.

Proteolysis Studies. Mb in solution is known to be susceptible to proteolytic enzyme cleavage at well-defined locations in its amino acid sequence (36,37). The reaction rates of these enzymes depend on Mb conformation. Bonds that are highly exposed are cleaved quickly, whereas bonds that reside in the interior of the molecule are cleaved much more slowly or not at all. Proteolytic enzyme cleavage studies described below can be used to determine regions of adsorbed Mb that are exposed to solution. Regions of contact between Mb and the surface can then be inferred from these results.

As mentioned above, fluorescent labels can be attached to specific amino acid residues in Mb. Mb possessing a label at a single site can be adsorbed to PDMS in the TIRF test cell and challenged by solutions of proteolytic enzymes. The rate of decay of the fluorescence signal will indicate the degree of exposure of the labeled amino acid residue, because as bonds near the label are cleaved by the enzyme, the peptide with the fluorescent label will be released and subsequently washed out of the test cell. Since the position of the label will be known the regions of the Mb molecule exposed to solution can be determined.

<u>Summary of Further Studies</u>. The experiments presented in the previous two sections afford a direct investigation of the conformation and orientation of Mb adsorbed on PDMS. These experiments can be used to clarify the results of the antibody experiments concerning the orientation and the deformation of adsorbed Mb. The experiments may also provide insight about certain as yet unexplained observations of Mb adsorption (19).

Conclusions

TIRF occupies a unique niche, providing a noninvasive method for studying protein adsorption <u>in situ</u> and in real time. The TIRF technique, as it is applied in our laboratory, has been used successfully to study both macroscopic and molecular aspects of protein adsorption. The present goal of this laboratory is to elucidate the interactions occurring when a protein adsorbs to a solid surface using the TIRF technique. It is hoped that, eventually, a complete, general description of the protein adsorption process will be attained.

Acknowledgments

Research in this laboratory was funded by the National Institutes of Health under grant NIH-2-R01-HL-27187. S.A.D. was the recipient of a Kodak fellowship.

Literature Cited

1. Grinnell, F.; and Feld, M.K. <u>J. Biomed. Mater. Res.</u> 1981, <u>15</u>, 363.
2. Grinnell, F.; and Feld, M.K. <u>J. Biol. Chem.</u> 1982, <u>257</u>, 4888.
3. Grinnell, F.; and Phan, T.V. <u>Thrombosis Res.</u> 1985, <u>39</u>, 165.
4. Baier, R.E.; and Dutton, R.C. <u>J. Biomed. Mater. Res.</u> 1969, <u>3</u>, 191.

5. Packham, M.A.; Evans, G.; Glynn, M.P.; and Mustard, J.F. J. Lab. Clin. Med. 1969, 73, 686.
6. Andrade, J.D., Surface and Interfacial Aspects of Biomedical Polymers Vol. 2 Protein Adsorption; Andrade, J.D., Ed.; Plenum Press: New York, 1985
7. Cooper, S.L., and Peppas, N.A., Eds., Biomaterials: Interfacial Phenomena and Applications, Adv. Chem.Ser. No. 199; American Chemical Society: Washington DC, 1982.
8. Norde, W., In Adhesion and Adsorption of Polymers Vol. 2, Lee, L-H. Y., Ed.; Plenum Press: New York, 1980, p.801.
9. MacRitchie, F. Adv. Protein Chem. 1978, 32, 283.
10. Morrissey, B.W., Ann. N.Y. Acad. Sci. 1977, 288, 50.
11. Baier, R.E., Ed., Applied Chemistry at Protein Interfaces, Adv. Chem. Ser. No. 145; American Chemical Society: Washington, DC, 1975
12. Andrade, J.D. In Surface and Interfacial Aspects of Biomedical Polymers Vol. 2 Protein Adsorption; Andrade, J.D., Ed.; Plenum Press: New York, 1985; p. 1
13. Cheng, Y.-L.; Lok, B.K.; and Robertson, C.R. In Surface and Interfacial Aspects of Biomedical Polymers Vol. 2; Andrade, J.D., Ed.; Plenum Press: New York, 1985; p. 121.
14. Darst, S.A.; and Robertson, C.R. In Spectroscopy in the Biomedical Sciences; Gendreau, R.M., Ed.; CRC Press: West Palm Beach, Fl, 1986; p. 175.
15. Axelrod, D.; Burghardt, T.P.; and Thompson, N.L. Ann. Rev. Biophys. Bioeng. 1984, 13, 347.
16. Harrick, N.J. "Internal Reflection Spectroscopy", John Wiley and Sons Inc.: New York, 1967.
17. Lok, B.K.; Cheng, Y.-L.; and Robertson, C.R. J. Coll. Int. Sci. 1983, 91, 87.
18. Lok, B.K.; Cheng, Y.-L.; and Robertson, C.R. J. Coll. Int. Sci. 1983, 91, 104.
19. Darst, S.A.; Robertson, C.R.; and Berzofsky, J.A. J. Coll. Int. Sci. 1986, 111, 466.
20. Watkins, R.W.; and Robertson, C.R. J. Biomed. Mater. Res. 1977, 11, 915.
21. Burghardt, T.P.; and Axelrod, D. Biophys. J. 1981, 33, 455.
22. Cheng, Y.-L. Ph.D Dissertation, Stanford University, Stanford, CA, 1983.
23. Lévêque, M. Ann. Mines 1928, 13, 284.
24. Brash, J.L.; and Uniyal, S. J. Polym. Sci. C. 1979, 66, 377.
25. Horbett, T.A.; and Hoffman, A.J. In Applied Chemistry at Protein Interfaces, Baier, R.E., Ed., ACS Advances in Chemistry Series, No. 145, American Chemical Society, Washington, DC; 1975,p. 230.
26. Salsman, E.W. The Chemistry of Biosurfaces, Hair, M.L., Ed.; Marcel Dekker: New York; 1972, p. 489.
27. Berzofsky, J.A.; Hicks, G.; Fedorko, J.; and Minna, J. J. Biol. Chem. 1980, 255, 11188.
28. Berzofsky, J.A.; Buckenmeyer, G.K.; Hicks, G.; Gurd, F.R.N.; Feldmann, R.J.; and Minna, J. J. Biol. Chem. 1982, 257, 3189.
29. Phillips, S.E.V. J. Mol. Biol. 1980, 142, 531.
30. Feder, J.; Giaever, I. J. Colloid Interface Sci. 1980, 78, 144.
31. Stryer, L. Ann. Rev. Biochem. 1978, 47, 819.

32. Burghardt, T.P.; and Axelrod, D. <u>Biochemistry</u> 1983, <u>22</u>, 979.
33. Zukin, R.S.; Hartig, P.R.; and Koshland, D.E., Jr. <u>Proc. Natl. Acad. Sci. USA</u> 1977, <u>74</u>, 1932.
34. Iwanij, V. <u>Eur. J. Biochem.</u> 1977, <u>80</u>, 359.
35. Boxer, S.G.; Kuki, A.; Wright, K.A.; Katz, B.A.; and Xuong, N.H. <u>Proc. Natl. Acad. Sci. USA</u> 1982, <u>79</u>, 1121.
36. Edmundson, A.B. <u>Nature</u> 1963, <u>198</u>, 354.
37. Mihalyi, E. <u>Application of Proteolytic Enzymes to Protein Structure Studies</u>, 2nd Edition; CRC Press: West Palm Beach, Fl; 1978.

RECEIVED January 3, 1987

Chapter 21

Adsorption of Fibronectin to Polyurethane Surfaces: Fourier Transform Infrared Spectroscopic Studies

W. G. Pitt, S. H. Spiegelberg, and S. L. Cooper

Department of Chemical Engineering, University of Wisconsin, Madison, WI 53706

The infrared spectra of plasma fibronectin adsorbed to three polyurethanes shows evidence of structural change upon adsorption. These block copolymers have identical hard segment chemistry, but they differ in soft segment composition and surface energy as measured by contact angle. On the more hydrophobic surfaces, the amount of adsobed fibronectin and the extent of spectral changes were greater than on the more hydrophilic surface. When compared to the spectrum of FN in solution, the spectra of the protein which adsorbs first appear to have more extensive spectral changes than protein adsorbing at later times. On all surfaces, increasing the concentration of protein in solution increased the amount of adsorbed protein.

Plasma fibronectin (FN) or cold insoluble globulin is a high-molecular-weight globular glycoprotein which is thought to mediate cell adhesion and growth processes on artificial surfaces. Recent reviews of the structure and function of FN have been published (1,2). The role of FN in mediating platelet adhesion and thrombus formation on polymer surfaces exposed to non-anticoagulated whole blood has been previously studied in this laboratory using an ex vivo A-V femoral shunt in canines (3-6). These studies have shown that when FN is pre-adsorbed to various polymers prior to implantation, the amount of platelet and fibrinogen deposition on the polymers is increased by an order of magnitude, suggesting that the adsorbed FN had retained its ability to bind platelets. Other in vitro studies, however, have indicated that when FN is adsorbed, its native conformation is changed and it loses some biological activity toward antibody binding (7-10). These types of conformational changes were observed to a greater extent on hydrophobic surfaces than on hydrophilic surfaces, suggesting that the substrate surface energy affects the conformation of the adsorbed protein. Also, FN was observed to adsorb to a greater extent on the hydrophobic surfaces than on the more polar surfaces (7-9).

0097-6156/87/0343-0324$06.00/0

Although the surface energy is probably not the only parameter which determines the state of adsorbed FN, it does appear to be important. An understanding of how surface energetics effect the interaction of FN with polymer surfaces would be an important contribution in the areas of cell growth and the interaction of blood proteins with synthetic polymers.

This study addresses the question of how bulk polymer chemistry and surface energy affect the amount and the conformation of FN adsorbed to a series of polyurethaneureas. The technique of Fourier transform infrared spectroscopy (FTIR) coupled with attenuated total reflectance (ATR) optics was used to continuously and non-invasively measure the kinetics of FN adsorption, as well as to monitor conformational changes occuring during adsorption.

Materials and Methods

Protein Purification.   Canine plasma fibronectin was used in this study in order to correlate these in vitro studies with canine ex vivo experiments involving preadsorbed canine proteins. Canine FN was isolated from citrated canine plasma using the methods of Ruoslahti (11). The FN was suspended in phosphate buffered saline (PBS) containing 0.02% NaN_3, and then snapfrozen and stored at -70°C until less than 24 hours before use. The protein was then snapthawed at 40°C, filtered (0.22 μm Millex GV, Millipore, Bedford, MA), and diluted to concentrations of 0.07 or 0.21 mg/ml as determined by UV absorbance at 280 nm ($\varepsilon_1^{1 \text{ mg/ml}}{}_{cm} = 1.28$). The purity and homogeneity of the thawed FN were verified by polyacrylamide gel electrophoresis in sodium dodecyl sulfate (SDS-PAGE). A transmission FTIR spectrum of 5.02 mg/ml FN in PBS between CaF_2 windows with a path length of 3 μm was collected at 8 cm^{-1} resolution.

Polymer Surface Preparation and Characterization.   Three polyurethane ureas were prepared as previously described (12). These contain a methylene bis(p-phenyldiisocyanate) (MDI) hard segment, an ethylene diamine chain extender, and a polyether soft segment in mole ratios of 2/1/1 respectively. The soft segment materials were polyethyleneoxide (PEO) and polytetramethyleneoxide (PTMO), both of 1000 molecular weight, and polydimethylsiloxane (PDMS) of 2000 molecular weight. In this paper, these polymers will be refered to as PEO-PEUU, PTMO-PEUU and PDMS-PEUU respectively. These were dissolved in N,N-dimethyl acetamide (DMA) to make a 0.1 wt.% solution.

Germanium internal reflection elements (IRE, 50x20x3, 45° aperture) were polished twice with 0.3 um alumina polish, rinsed with distilled water, rinsed with ethanol, and then cleaned in a radio frequency plasma discharge. The IRE's were pulled vertically (dip coated) from the polymer solutions at 3mm/min. Each coated IRE was dried in a vacuum oven at 60°C for at least 4 hours, and was stored under vacuum until use.

Underwater contact angles of air and octane in double distilled deionized water were determined as previously described (13). The harmonic mean equation (14) was used to estimate the surface energy parameters of the polymer coated IRE's.

The thickness of the polymer coatings was determined as follows. Polymer films were spin cast (15) from a MDI/ED/PTMO polyurethaneurea of known composition. The film thickness, measured by a micrometer, was found to correlate linearly with the height of the 1600 cm^{-1} peak (ν(C=C) benzene ring), thus allowing determination of a Beer's law extinction coeffecient for the mass fraction of benzene rings in these polymers. Transmission spectra of the dip coated IRE's were obtained, and the polymer thickness determined assuming that the Beer's law extinction coefficient was unchanged on these thin films.

Measurement of Protein Adsorption Using FTIR/ATR. The study of protein adsorption using FTIR/ATR is presented in detail elsewhere (13,16-21), and are only briefly reviewed here. Infrared radiation from the spectrometer source enters and then reflects internally along the length of an internal reflection element (IRE) mounted in a flow cell. Each reflection of the infrared beam on the IRE surface produces an "evanescent" wave which decays exponentially as it extends into the protein solution. Polymer, buffer and protein within this evanescent wave absorb some of the infrared energy and decrease the intensity of the infrared radiation exiting the IRE. This attenuation is detected and processed by the FTIR, producing an infrared absorbance spectrum of the molecules within the evanescent wave. By subtracting the spectra of polymer and buffer (and water vapor, when present), one obtains the spectrum of the protein near the polymer-solution interface. This spectrum contains contributions from both the adsorbed and the non-adsorbed (or "bulk" solution) protein within the evanescent wave (19,20). The details of processing the spectra have been reported previously (13,16).

The infrared spectrum of a protein provides information on both the amount (the amide II band) and the conformation (the amide I and III bands) of the protein. Previous studies have used both the height (17) and area (22) of the amide II band to quantitate the amount of protein on a surface.

The polyurethaneureas were exposed to the FN solutions in a flow cell contained inside a constant temperature (39°C) compartment built into a Nicolet 170SX FTIR equipped with a MCT detector (Nicolet, Madison, WI). The polymer surfaces were exposed to PBS buffer in the flow cell for at least 40 minutes prior to introduction of the protein in order to attain some equilibration between the polymer and buffer. The buffer was displaced by injecting at 1 ml/sec at least 4.5 ml of FN solution from a syringe. During injection the wall shear rate was 400 sec^{-1}, and after injection the solution was static until termination of the experiment. The volume of solution injected was shown by residence time distribution experiments (data not shown) to remove 97% of the buffer. From the time of injection of the FN into the flow cell, spectra were collected continuously at 8 cm^{-1} resolution for at least 2 hr, although a few experiments were continued for up to 18 hr. During data collection, the number of coadded scans increased from 4 at the first time point to 2000 for times after 30 min. Upon completion of an adsorption experiment, the non-adsorbed protein was displaced by flowing buffer gently through the flow cell, and a final spectra of the remaining adsorbed protein was collected. Two adsorption experiments were done on each polymer at each protein concentration.

Quantitation of Adsorbed Fibronectin.   Grinnell has shown that ^{125}I-labeling of FN causes loss of some of its biological activity (7).   Instead of using radiolabeled FN, the area of the amide II absorbance band (22) was used to quantitate the amount of adsorption in this study.   Known amounts of FN were dried from a 0.1 M NaCl solution onto bare IRE's which were then placed in the same optical configuration as in the adsorption experiments.   Spectra of the dried protein were obtained, and the areas of the amide II from 1590 to 1474 cm^{-1} band were measured.   The area of the amide II absorption did not change significantly as the protein was dried (±5% standard deviation).   The linear correlation between the amide II area and the surface concentration of FN, shown in Figure 1, was used to determine the amount of FN adsorbed in the experiments.

Results and Discussion

Surface Characterization.   Table I presents the contact angles (through the water phase), interfacial energies, and polymer thicknesses of the polymers coated onto the Germanium crystal.   The contact angles for the polymers decrease in the order PTMO-PEUU, PDMS-PEUU and PEO-PEUU, indicating that the polymer surfaces become more polar in that order.   The γ_{sw} also decreases in this order, although these values should only be taken as approximate values considering that the relationship between contact angles and surface energy on polar solids in not well established (23).   That these thin films have lower surface energy than thicker films of the same polymers (12) may be a result of their being applied as a very thin coating which could disrupt the bulk polymer morphology that exists in thicker films.   Scanning electron microscopic examination of the surfaces did not reveal evidence of holes or breaks in the films.

Table I.   Properties of Dip Coated Polyurethane Surfaces

Polymer	air-water contact angle (degrees)	octane water contact angle (degrees)	γ_{sw} (dyn/cm)	thickness (Å)
PEO-PEUU	35 ± 3	59 ± 1	8.7	-
PDMS-PEUU	38 ± 3	68 ± 1	17.8	173
PTMO-PEUU	59 ± 5	87 ± 2	19.6	137

The thickness of the dip coated PEO-PEUU could not be determined.   The polymer thicknesses reported here should be taken as only approximate values, due to the cumulative errors in the spectral measurements associated with the very low signal/noise of these thin films.   The thicknesses of these polymers are about 1/20 of the depth of penetration of the infrared evanescent wave (about 400 nm at 1550 cm^{-1}) (21).   Thus there is adequate space within the evanescent wave for adsorption of the 4x60 nm FN molecules to be observed (24).

Adsorption Kinetics. Figures 2A and 2B show the FN adsorption
kinetics on the three surfaces from 0.07 and 0.21 mg/ml FN solutions respectively. Each line is the average of two experiments on
a given polymer. At each protein concentration, the initial rate
of adsorption is independent of the type of polymer substrate, and
adsorption from 0.21 mg/ml is nearly 3 times faster than from 0.07
mg/ml FN. The initial adsorption rates are linear in time$^{1/2}$ until
adsorption exceeds 0.06 µg/cm^2 on PEO-PEUU and 0.10 µg/cm^2 on the
other polymers (data not shown). This suggests that the adsorption
is diffusion controlled up to the above surface concentrations,
after which point the adsorption rate decreases and becomes dependent upon the polymer surface chemistry. The amount of FN adsorbed
does not reach a plateau within 120 minutes, nor does it reach a
plateau when adsorption continues for 18 hours (data not shown).
At both concentrations, the lowest adsorption occurs on PEO-PEUU,
supporting previous observations that hydrophilic surfaces adsorb
less FN than more hydrophobic surfaces (7-9). Also at both concentrations, PDMS-PEUU adsorbs slightly more FN than the PTMO-PEUU,
even though the contact angle data indicate that PDMS-PEUU is
slightly more polar, suggesting that the surface chemistry as well
as surface energy influence the amount of adsorption.

Conformational Changes in Adsorbed FN. One of the major advantages
of FTIR spectroscopy is its potential for elucidating protein
structural changes. In this study, changes in the FN spectra were
observed. The transmission FTIR spectrum of FN in PBS is shown in
Figure 3 with the amide III region expanded by a factor of 4. The
amide I, II and III bands are centered at 1642, 1549, and 1247
cm^{-1} respectively, characteristic of a protein containing some β-
sheet structure. Figure 3 also shows typical FN spectra after 2 hr
of adsorption onto the three polymers from the 0.07 mg/ml solution.
     Several differences between the solution and adsorbed spectra
are obvious. First there is a change in the small absorbance band
in the 1740-1720 cm^{-1} region. On PEO-PEUU, this band is at the
same frequency, and is about the same magnitude as observed for FN
in solution. However, on the PDMS-PEUU and PTMO-PEUU, the band has
shifted 20 cm^{-1} and has increased in magnitude. This band could be
assigned to the carbonyl stretching vibration of the COOH group,
the protonated form of the carboxylic acid which produces the band
at 1400 cm^{-1}. In FN, carboxylic acids are found in some protein
residues (Asp and Glu), at the carboxyl end of the polypeptide, and
in the acidic carbohydrates (sialic acid). These have COOH vibrations at 1735-1720 cm^{-1}, 1755-1720 cm^{-1} and 1748-1724 cm^{-1} respectively (25). A similar peak at 1735 cm^{-1} was observed in a study
of lysozyme adsorption on contact lenses, and was assigned to
interaction of Glu or Asp residues with the polymer surface (26).
     Formation of a protonated carboxylic acid occurs when the
pK$_a$ of the particular COOH group is near or greater than the local
pH of the solution, which is not often the case for proteins at
physiologic pH. The pK$_a$ of the α-carboxyl ranges about 3.1-3.5
(27), and that of sialic acid is near 2.6 (28). The pK$_a$ of the
acidic protein residues is usually between 3 and 5, although a
pK$_a$ of 6.5 has been reported for a Asp residue in a nonpolar
region of lysozyme (29).
     The polymer surface chemistry appears to influence the

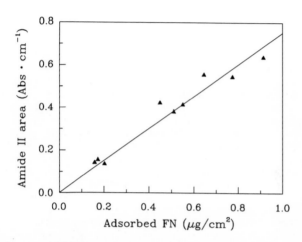

Figure 1.   Correlation of the amide II area with the mass of FN dried on a Germanium IRE.

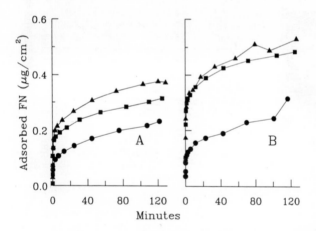

Figure 2.   FN adsorption onto PEO-PEUU (●), PTMO-PEUU (■) and PDMS-PEUU (▲) from 0.07 mg/ml (A) and from 0.21 mg/ml (B).

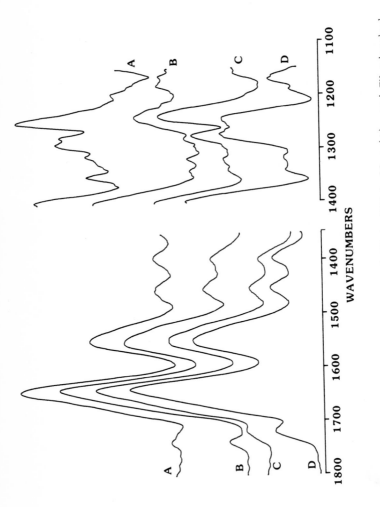

Figure 3. FTIR absorbance spectra of FN in PBS buffer (A), and FN adsorbed from 0.07 mg/ml solution for 2 h to PEO-PEUU (B), PDMS-PEUU (C) and PTMO-PEUU (D). The region from 1400 to 1100 cm^{-1} has been expanded by 4x.

vibration of the COOH group since the band appears at different
frequencies and adsorption times on each polymer surface. On
PEO-PEUU, the band begins to appear at 1740 cm^{-1} after about
20 minutes of adsorption and then increases with time as Figure 4
illustrates. On the PDMS-PEUU surface (Figure 5), a larger and
broader band appears at 1725 cm^{-1} from the time of first contact of
the protein with the polymer. It increases for about 5 minutes and
then remains constant although the other peaks in the protein
spectra increase in magnitude. On the PTMO-PEUU surface
(Figure 6), a band also appears at 1720 cm^{-1} from the time of first
contact and increases with time for about 15 minutes, after which
time it remains constant. On all three polymers, the COOH band is
relatively larger (with respect to the 1550 cm^{-1} band) for adsorption from the 0.07 mg/ml solution than from the 0.21 mg/ml solution
(data not shown).

The appearance of the COOH peak at 1720 cm^{-1} on the more
hydrophobic polymers suggests that these polymers may interact more
strongly with the adsorbed FN. The decrease in vibrational frequency on PTMO-PEUU and PDMS-PEUU could be attributed to many factors, one of which could be hydrogen bonding of the carbonyl group
in COOH which would decrease its vibrational frequency (25). The
increased magnitude of the COOH vibration on the hydrophobic polymers indicates that more COO^{-} groups have been protonated to the
COOH form. Although one cannot rule out the possibility of a major
pK$_a$ change due to protein denaturation, it is more likely that the
carboxyl groups become protonated as they approach (on the order of
Angstroms) the polymer surface. Here the proximity of a surface
with a low dielectric constant would shift the carboxyl group
dissociation equilibrium (COOH = COO^{-} + H^{+}) toward the neutral protonated species (30). Whether protonation occurs from denaturation
or from the proximity of the surface, this effect is greater on the
PDMS-PEUU and PTMO-PEUU surfaces than on the PEO-PEUU surface.

Temporal Changes in Amide I Absorbance. The absorbance of the
amide I band was observed to change on all three surfaces during
the first few minutes of adsorption as Figures 4-6 illustrate. On
all polymer surfaces and at both protein concentrations, the first
seconds of protein adsorption showed an amide I peak growing near
1669 cm^{-1}. Very quickly, usually within two minutes of adsorption,
the 1669 cm^{-1} peak ceased increasing, and an absorbance centered at
1638 cm^{-1} began growing. The 1638 cm^{-1} band continued to grow as
the amount of adsorbed protein increased, enveloping the 1669 cm^{-1}
band which was observable only as a shoulder after about three
minutes.

In protein solutions the absorbance of a band near 1669 cm^{-1}
is usually assigned to the antiparallel-β-sheet conformation.
However, both experimental observations and theoretical calculations indicate that β-sheet amide I vibrations have at least two
bands, a strong absorption near 1635 cm^{-1}, and a smaller one near
1680 cm^{-1} (25,31-33). Thus the assignment of this initial 1669
cm^{-1} band in the absence of a 1635 cm^{-1} band to an β-sheet conformation may be incorrect. Other possibilities for this band assignment are an unordered conformation (1656-1658 cm^{-1}) (25) or a
β-turn conformation (1680 cm^{-1}) (31), although the latter is less
likely since this β-turn vibration has been observed to disappear

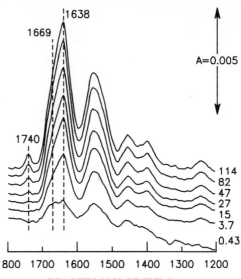

Figure 4. Absorbance spectra of FN adsorbing to the PEO-PEUU polymer from a 0.07 mg/ml solution. The adsorption time in minutes is indicated on each spectrum.

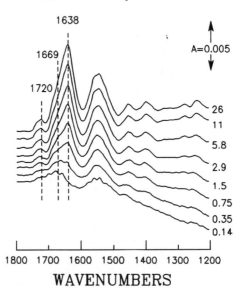

Figure 5. Absorbance spectra of FN adsorbing to the PDMS-PEUU polymer from a 0.07 mg/ml solution. The adsorption time in minutes is indicated on each spectrum.

upon denaturation. At present the 1669 cm^{-1} band which appears at short adsorption times remains unassigned. A similar amide I absorbance has been observed for lysozyme adsorption on contact lens material (26). In that study, Castillo observed an absorption at 1672 cm^{-1} without a concommitant band near 1635 cm^{-1} only at the shortest adsorption time measured.

The amide I vibrations producing the 1638 cm^{-1} peak and the 1669 cm^{-1} shoulder (or peak at early times) were studied by fitting the amide I region with gaussian curves using a non-linear least squares curve fitting routine. The spectrum of FN is solution was fit by 2 curves at 1638 and 1672 cm^{-1} with full widths at half height (FWHH) of 40 and 37 cm^{-1} respectively. All the spectra of adsorbed FN were fit very well by two gaussian peaks at 1634±1 and 1669±3 cm^{-1} with FWHH's of 40 and 45 cm^{-1} respectively.

This curve fitting analysis indicated that at early times, the 1669 cm^{-1} band is much larger than the 1634 cm^{-1} band. But this latter band begins growing at later times and eventually surpasses the 1669 cm^{-1} band. As the amount of adsorbed FN increased, the ratio of the 1634 to 1669 cm^{-1} bands increased from near zero to a plateau value nearer to the same ratio determined from curve fitting the spectra of FN in solution (see Figure 7). Similarly, Castillo observed that the FTIR spectra of adsorbed lysozyme, mucin, albumin, and γ-globulin become more like the solution spectra as the adsorption time increases (22,26,32,33). This does not necessarily imply that the protein which contacts the surface first is denatured but then regains its native conformation with time. Rather, the protein adsorbed first, and in most direct contact with the surface may retain its denatured form, but protein with a more native conformation (and spectra) continue to adsorb. Using the dimensions of 4x60 nm for the FN molecule, the mass of a random packed side-on adsorbed monolayer is less than 0.36 µg/cm^2. Thus the protein which adsorbs first and produces the strong 1669 cm^{-1} band would be less than a monolayer coverage of randomly oriented side-on adsorbed molecules. Also of note in Figure 7 is that the spectra of FN adsorbed on the PEO-PEUU polymer attains a 1634 cm^{-1}/1669 cm^{-1} ratio similar to solution FN (dashed line in Figure 7) at lower total surface concentrations than does PTMO-PEUU and PDMS-PEUU. This suggests that the PEO-PEUU surface has adsorbed less protein with a conformation which produces the strong 1669 cm^{-1} amide I band observed at early times.

It is noteworthy that the amide I absorbance at 1669 cm^{-1} as well as the COOH peak at 1720 cm^{-1} are of greater magnitude on the PDMS-PEUU and PTMO-PEUU polymers than on the PEO-PEUU polymer. In both spectral features, protein adsorbed on the PEO-PEUU surface appears most like the native FN in solution. This supports the observation of others that the FN-polymer interactions are less on the more hydrophilic surface than on the more hydrophobic surfaces. For example, Iwamoto observed changes in the fluorescence spectra of adsorbed FN which indicated that the protein is denatured more on a hydrophobic than on a hydrophilic surface (7). Grinnell and others have shown that the interaction of antibodies with FN adsorbed on hydrophilic surfaces is greater than with FN adsorbed on hydrophobic surfaces, suggesting that the FN adsorbs in a more native conformation on hydrophilic surfaces (8-10).

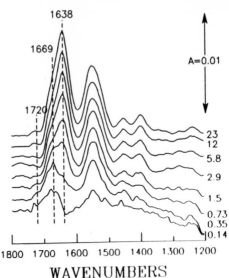

Figure 6. Absorbance spectra of FN adsorbing to the PTMO-PEUU polymer from a 0.21 mg/ml solution. The adsorption time in minutes is indicated on each spectrum.

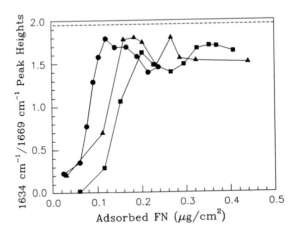

Figure 7. Ratio of the 1634 cm^{-1} peak to the 1669 cm^{-1} peak from the curve fitting analysis. The polymer substrates are PEO-PEUU (●), PTMO-PEUU (■) and PDMS-PEUU (▲). The dashed line is the same ratio for FN in PBS buffer. This data is the average of all adsorption experiments at 0.07 and 0.21 mg/ml concentrations on a given polymer.

Changes in the Amide III Region. In addition to the amide I band, the amide III vibrations are sensitive to the protein conformation. However, the low signal/noise ratio of the amide III region makes spectral interpretation difficult. In these experiments, the amide III region was not identical in each experiment due to random noise and fluctuations in the baseline. However, large changes in spectral features were consistent on a given polymer. In Figure 8, the amide III spectra for 4 spectra at 2 hr of adsorption on a given polymer have been added, thus enhancing the consistent spectral features, and minimizing random noise and baseline fluctuations. The amide III region of the solution FN has peaks at 1247, 1275 and 1290 cm^{-1}, all of which are altered in the spectra of the adsorbed FN. For example, the solution FN peak at 1247 cm^{-1} is absent in the adsorbed spectra, and there is a new broader peak of lower frequency centered near 1240 cm^{-1}. Where the solution FN has peaks at 1275 and 1290 cm^{-1}, the adsorbed spectra have a broad peak centered near 1280 cm^{-1} on the PTMO-PEUU and PDMS-PEUU polymers, while on the PEO-PEUU surface, peaks in this region are absent or very small. These changes further indicate that the conformation of FN is altered as it adsorbs to these polymers.

Previous infrared studies indicate that proteins with a large amount of β-sheet structure absorb near 1240 cm^{-1} and those with α-helix stucture absorb near 1280 cm^{-1} ([22,32]). Absorbances for denatured albumin, reportedly containing random and β-sheet conformations, are found at 1240 and 1260 cm^{-1} ([22]). These assignments correlate with the more studied Raman spectroscopy of the amide III region which has vibrations at 1230-1250 cm^{-1} for β-sheet structure, at 1260-1290 cm^{-1} for α-helix structure, and at 1240-1265 cm^{-1} for unstructured polypeptide ([34,35]).

Applying these vibrational assignments to the spectra of Figure 8 would suggest that solution FN contains some α-helix structure. However, this is probably an incorrect assignment since the amide I band does not have a strong absorbance at 1646-1650 cm^{-1} characteristic of α-helix, and circular dicroism (CD) studies indicate little or no α-helical content for FN ([1,36-39]). It is also doubtful that the peak at 1280 cm^{-1} in the adsorbed FN is due to α-helix structure since the amide I band does not suggest significant α-helix structure. At present, the peaks in the region from 1290 to 1275 cm^{-1} remained unassigned.

Both the amide I peak at 1638 cm^{-1} and the amide III peak at 1247 cm^{-1} support CD observations of the presence of some β-sheet structure. The shift of the latter peak to a lower frequency around 1240 cm^{-1} in the adsorbed state suggest that an increase in β-sheet structure may occur upon adsorption. Also the shift in the amide I band from 1642 cm^{-1} in solution to 1638 cm^{-1} when adsorbed further substantiates the hypothesis that the β-sheet content of FN increases upon adsorption. An increase in β-sheet structure upon adsorption has been previously reported in FTIR/ATR studies of protein adsorption on contact lens materials. Specifically, an α-helix to random and β-sheet transition has been observed for adsorbed albumin and lysozyme, as well as a random to β-sheet transition for mucin ([22,26,32]). However, a decrease in β-sheet structure has also been observed for adsorbed γ-globulin which contains a high content of β-sheet structure in its native form ([33]).

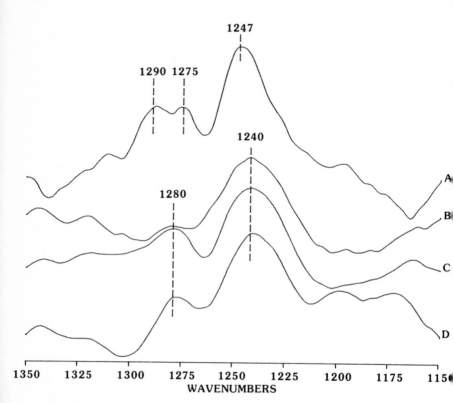

Figure 8. Absorbance spectra of the amide III region of solution FN (A), and FN adsorbed on PEO-PEUU (B), PDMS-PEUU (C) and
PTMO-PEUU (D). Each spectrum is not from a single experiment,
but is the co-addition of 4 spectra on a given polymer at
2 hours of adsorption.

Conclusions

The FTIR/ATR studies of FN adsorption indicate that the polymer surface plays an important role in determining both the amount and the conformation of adsorbed FN. Comparison of the PEO-PEUU surface with the PDMS-PEUU and PTMO-PEUU surfaces show that on the more hydrophilic PEO-PEUU polymer, less protein adsorbs, and the interactions between the protein and polymer take place more slowly and are less intense as shown by the late appearance of the COOH vibration and the smaller 1669 cm^{-1} amide I peak observed at early adsorption times. The protein which adsorbs first appears to interact most strongly with the surface. On all polymers, changes in the amide I and the amide III region of adsorbed FN suggests that the amount of β-sheet structure in FN increases upon adsorption.

Acknowledgments

The authors wish to acknowledge partial support of this work through the National Instututes of Health grants HL-21001 and HL-24046, and fellowship support for WGP from the W. R. Grace Company. The helpful discussions with Dr. Frank Wasacs of Mattson Instruments and Dr. Deane F. Mosher of the University of Wisconsin, Department of Medicine were greatly appreciated.

Literature Cited

1. Mosher, D. F. Prog. Haemo. Thromb. 1980, 5, 111-151.
2. Hynes, R. O. Cell Surf. Rev. 1982, 7, 97-136.
3. Ihlenfeld, J. V.; Mathis, T. R.; Barber, T. A.; Mosher, D. F.; Riddle, L. M.; Hart, A. P.; Updike, S. J.; Cooper, S. L. Trans. Amer. Soc. Artif. Intern. Organs 1978, 24, 727-730.
4. Barber, T. A.; Lambrecht, L. K.; Mosher, D. F.; Cooper, S. L. Scanning Elect. Microsc. 1979, III, 881-890.
5. Young, B. R. Ph.D. Thesis, Univ. of Wisconsin, Madison, Wis., 1984.
6. Lambrecht, L. K.; Young, B. R.; Stafford, R. E.; Park, K.; Albrecht, R. M.; Mosher, D. F.; Cooper, S. L. Thromb. Res. 1985, 41, 99-117.
7. Iwamoto, G. K.; Winterton, L. C.; Stoker, R. S.; Van Wagenen, R. A.; Andrade, J. D.; Mosher, D. F. J. Colloid Interface Sci. 1985, 106, 459-464.
8. Grinnell, F.; Feld, M. K. J. Biomed. Mater. Res. 1981, 15, 363-381.
9. Jonsson, U.; Ivarsson, B.; Lundstrom, I.; Berghem, L. J. Colloid Interface Sci. 1982, 90, 148-163.
10. Grinnell, F.; Feld, M. K. J. Biol. Chem. 1982, 257, 4888-4893.
11. Ruoslahti, E.; Hayman, E. G.; Pirschbacher, M; Engvall, E. Meth. Enzymology 1983, 82, 803.
12. Grasel, T. G.; Cooper, S. L. Biomaterials 1986, 7, 315-328.
13. Pitt, W. G.; Cooper, S. L. Biomaterials 1986, 7, 340-347.
14. Andrade, J. D.; Ma, S. M.; King, R. N.; Gregonis, D. E. J. Colloid Interface Sci. 1979, 72, 488-494.
15. Koberstein, J. T.; Cooper, S. L.; Shen, M. C. Rev. Sci. Instrum. 1975, 46, 1639-1641.

16. Pitt, W. G.; Park, K.; Cooper, S. L. J. Colloid Interface Sci. 1986, 111, 343-362.
17. Gendreau, R. M.; Leininger, R. I.; Winters, S.; Jakobsen, R. J. In Biomaterials: Interfacial Phenomena and Applications; Cooper, S. L.; Peppas, N. A., Eds.; Adv. Chem. Series No. 199; American Chemical Society: Washington DC, 1982; pp 371-394.
18. Gendreau, R. M.; Jakobsen, R. J. J. Biomed. Mater. Res. 1979, 13, 893-906.
19. Fink, D. J.; Gendreau, R. M. Anal. Biochem 1984, 139, 140-148.
20. Chittur, K. K.; Fink, D. J.; Leininger, R. I.; Hutson, T. B. J. Colloid Interface Sci. 1986, 111, 419-433.
21. Harrick, N. J. Internal Reflection Spectroscopy; Harrick Scientific Corporation: Ossining, New York, 1979; Chapter 2.
22. Castillo, E. J.; Koenig, J. L.; Anderson, J. M.; Lo, J. Biomaterials 1984, 5, 319-325.
23. Andrade, J. D.; Smith, L. M.; Gregonis, D. E. In Surface and Interfacial Aspects of Biomedical Polymers; Andrade, J. D. Ed.; Plenum Press, New York, 1985 pp. 249-292.
24. Williams, E. C.; Janmey, P. A.; Ferry, J. D.; Mosher, D. F. J. Biol. Chem. 1982, 257, 14973-14978.
25. Parker, F. S. Applications of Infrared Spectroscopy in Biochemistry, Biology, and Medicine; Plenum Press: New York, 1971; Chapter 6, 10.
26. Castillo, E. J.; Koenig, J. L.; Anderson, J. M. Biomaterials 1985, 6, 338-345.
27. Stryer, L. Biochemistry; W. H. Freeman: San Francisco, 1981; Chapter 2, 4.
28. Spiro, R. G. In Adv. Protein Chem. Afinsen, C. B.; Edsall, J. T.; Richards, F. M., Eds.; Vol. 27, Academic: New York, 1973; pp 349-467.
29. Timasheff, S. N.; Rupley, J. A. Arch. Biochem. Biophys. 1972 150, 318-323.
30. Mukerjee, P.; Cardinal, J. R.; Desai, N. R. In Micellization, Solubilization, and Microemulsions; Mittal, K. L., Ed.; Vol. 1, Plenum: New York, 1977; pp 241-261.
31. Bandekar, J.; Krimm, S. Biopolymers 1980, 19, 31-36.
32. Castillo, E. J.; Koenig, J. L.; Anderson, J. M.; Jentoft, N. Biomaterials 1986, 7, 9-15.
33. Castillo, E. J.; Koenig, J. L.; Anderson, J. M. Biomaterials 1986, 7, 89-96.
34. Lord, R. C. Appl. Spectrosc. 1977, 31, 187-194.
35. Mark, J.; Hudry-Clergeon, G.; Capet-Antonini, F.; Bernard, L. Biochim. Biophys. Acta 1979, 578, 107-115.
36. Osterlund, E.; Eronen, I.; Osterlund, K.; Vuento, M. Biochem. 1985, 24, 2661-2667.
37. Alexander, S. S., Jr.; Colonna, G.; Edelhoch, H. J. Biol. Chem. 1979, 254, 1501-1505.
38. Koteliansky, V. E.; Glukhova, M. A.; Bejanian, M. V.; Smirnov, V. N.; Filimonov, V. V.; Zalite, O. M.; Venyaminov, S. Yu. Eur. J. Biochem. 1981, 119, 619-624.
39. Tooney, N. M.; Amrani, D. L.; Homandberg, G. A.; McDonald, J. A.; Mosesson, M. W. Biochem. Biophys. Res. Commun. 1982, 3, 1085-1091.

RECEIVED January 29, 1987

Chapter 22

Effects of the Environment on the Structure of Adsorbed Proteins: Fourier Transform Infrared Spectroscopic Studies

R. J. Jakobsen and F. M. Wasacz[1]

Mattson Institute for Spectroscopic Research, Ohio State University Research Park, Columbus, OH 43212

The protein backbone vibrations have been assigned for a group of proteins in aqueous solution using deconvoluted spectra for the assignments. These assignments were related to the secondary structure of the proteins and show the effect of one secondary structure on another. Changes in the environment of dissolved and adsorbed proteins have been followed by infrared spectroscopy and these spectral changes have been used to verify the vibrational assignments and determine the secondary structures.

Very few of the infrared studies of proteins have been carried out on aqueous solutions of the proteins. Except for the work of Koenig and Tabb (1), the few aqueous IR studies have been on single proteins. Correspondingly, most of the assignments of the backbone vibrations (the so-called Amide I, II, III, etc. vibrations) have been based on either Raman spectra of aqueous solutions (2) or on infrared spectra of proteins in the solid state (3). Where infrared solution spectra have been obtained, it has mostly been on D_2O solutions (4) - not H_2O solutions. Since these Amide I, II, III, etc. vibrations involve motion of the protein backbone, they are sensitive to the secondary structure of the protein and thus valid assignments are necessary in order to use infrared spectroscopy for determining the conformations of proteins.

The Raman frequencies of the protein backbone vibrations are often different than the infrared frequencies and thus infrared assignments are needed for structure determinations from infrared spectra. Such structural information is needed from the

[1]Current address: Mattson Instruments, Inc., 1001 Fourier Court, Madison, WI 53717

infrared spectra. Albumin (1,5) is an example of a protein for which structural infrared band assignments are needed. Since the structure of proteins can differ between solid and solution states, infrared assignments in the solution state are necessary, especially since the solid state is often not the natural physiological state of the protein. These IR solution assignments have to come from aqueous solutions since D_2O solutions can cause frequency shifts in the protein backbone vibrations and can cause changes in the structure of the protein.

In the last 5-10 years, FT-IR instrumentation has developed to the point where routine IR measurements of proteins dissolved in water can now be obtained (1,6,7). This includes the region of the Amide I vibration of proteins (1625-1655 cm^{-1}) which lies directly under the very strong OH bending mode of water near 1640 cm^{-1}. This strong water absorption band can be computer subtracted permitting reliable measurements of the Amide I vibration of the protein. Automated computer programs for water subtraction have been developed including one specifically designed for aqueous solutions of proteins (8).

In the past, the assignments of the protein backbone vibrations to secondary structures were often made from the spectrum of one compound and no attempts were made to support the assignments by changing the structure of the proteins. In fact structures were often determined from the frequencies and contours of one band. Only recently (4,9,10,11), have resolution enhancement or deconvolution (12) techniques being applied to the infrared spectra of proteins. While these deconvolution techniques appear to be essential for a valid interpretation of protein spectra, it is also necessary to use more than one infrared band and to substantiate assignments.

In order to verify the spectra-structure correlations with the secondary structure of protein, one can take advantage of the sensitivity of proteins to the nature of the surrounding environment. By changing this environment (by altering pH, temperature, pressure, solvent, etc.) one can induce changes in the structure of the protein. Often these environmental effects have been studied by other techniques and the induced structural changes are known. In these cases the infrared spectral changes can be related to the structural changes and spectra-structure correlations can be obtained which will help verify the assignments of the infrared frequencies. Once the assignments have been substantiated, the converse argument become valid, i.e., the effects of the environment can be used to deduce the secondary structure of proteins from the infrared spectra. Infrared techniques have been developed (13,14) permitting variation of pH, temperature, pressure, and ionic strength for either dissolved proteins or adsorbed protein films and also for variation of solvents on adsorbed protein films.

Therefore, the purpose of this paper is to first present new work on the assignment of the protein backbone vibrations for a group of proteins in aqueous solution using the deconvoluted spectra for the assignments. These are the first spectra-structure correlations for a group of proteins in aqueous solutions and using deconvoluted spectra. These assignments also offer the first evidence that one secondary structure affects the vibrations of another secondary structure and that the protein backbone vibrations cannot always be assigned simply as an α-helix vibration or a β-sheet vibration.

The second part of the paper reviews the authors past work on studying the effects of the environment on the structure of dissolved and adsorbed proteins and shows that infrared spectroscopy can follow these induced structural changes. These spectral changes can then be used to either verify vibrational assignments or conversely to determine structures.

The paper ends with preliminary results on the comparison of the structure of adsorbed proteins with the structure of proteins in solution and shows that the structure of some adsorbed proteins change with time.

SPECTRA-STRUCTURE CORRELATIONS.

Table I lists a group of 11 proteins along with the percentages of α-helix and β-sheet secondary structure for each protein. For most of the proteins in Table I the percentages of secondary structure stem from the calculations of Levitt and Greer (15). The values for albumin are taken from Schechter and Blout (16) while those for IgG come from Amzel and Poljak (17). It can be seen in Table I, that based on the percentages of secondary structure, the proteins can be (and have been) placed in five groups according to the amounts of α-helix and β-sheet structure.

Transmission infrared spectra of saline solutions of these 11 proteins were obtained where the solutions were adjusted to pH 7.4 and held at a temperature near 30°C. The protein solutions were made to a 5 wt. percent concentration except for IgG which was at 2 wt. percent. After subtraction of a spectrum of saline, the resultant protein spectrum was deconvoluted (11) which has the effect of resolving into components the broad infrared bands of proteins.

The Amide I and II region of the deconvoluted spectra of the " α" proteins is shown in Figure 1 while the Amide III region (deconvoluted and scale expanded) of these proteins is shown in Figure 2. The Amide I region of " α" proteins is characterized by a strong band near 1655 cm^{-1} with a weak band on each side of the strong 1655 cm^{-1} absorption band. While the weak bands may be components of Amide I vibrations or due to amino acid side chain vibrations, it is only the behavior of the strong Amide I band near 1655 cm^{-1} that is pertinent to this discussion. In the Amide II spectral region there is a strong component near 1550 cm^{-1} with shoulders on each side of the strong band. It is important to note that all the " α" proteins have very similar infrared spectra in the Amide I and Amide II spectral regions. Similarities between the α-helix proteins can also be observed in the Amide III spectral region (1330-1230 cm^{-1}). Note the most intense band in this region is near 1300 cm^{-1} with a weaker component seen at about 1245 cm^{-1}. Thus the spectra of α-helix proteins are dominated by characteristic α-Amide I (1655 cm^{-1}) and Amide III (1300 cm^{-1}) bands. There is a characteristic Amide II frequency

TABLE 1. SECONDARY STRUCTURE CONTENT OF VARIOUS PROTEINS

| PROTEIN | TYPE | SECONDARY STRUCTURE (%) | | |
		α	β	OTHER
MYOGLOBIN		88	--	12
HEMOGLOBIN	α	86	--	14
ALBUMIN		55	--	45
CONCANAVALIN A	β	--	64	36
IgG		--	75	25
CHYMOTRYPSIN	β + Low α	7	55	38
CHYMOTRYPSINOGEN		12	49	39
LYSOZYME	α + Low β	46	19	35
CYTOCHROME C		46	15	39
RIBONUCLEASE A	α + β	23	46	31
β -LACTOGLOBULIN		20	30	50

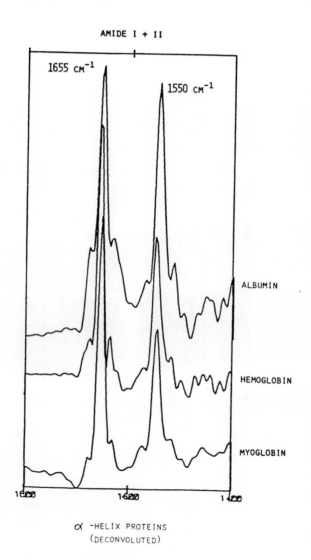

Figure 1.   Deconvoluted aqueous solution spectra of α-helix proteins in the Amide I and Amide II spectral regions.   Top: albumin; middle: hemoglobin; bottom: myoglobin.

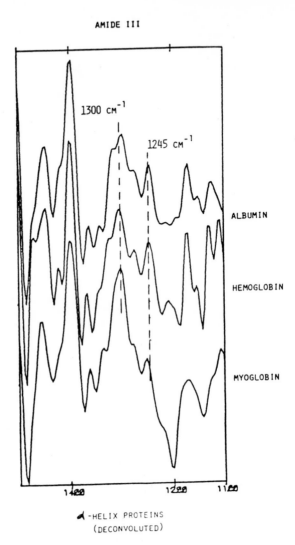

Figure 2.   Deconvoluted aqueous solution spectra of α-helix proteins in the Amide III spectral region.   Top: albumin; middle: hemoglobin; bottom: myoglobin.

near 1550 cm^{-1}, but as will be discussed later, the Amide II
vibrations cannot, as yet, be used to differentiate structures.

Figures 3 and 4 show the deconvoluted spectra of " β "
proteins in the Amide I and II regions and in the Amide III (scale
expanded) region, respectively. For each " β " protein there is a
strong Amide I band near 1638 cm^{-1} with shoulders at higher
frequencies. The Amide II region shows two bands, one on each
side of 1550 cm^{-1}, but the intensity ratio of the two Amide II
bands differs for each " β " protein. The Amide III region has the
most intense band at 1235 cm^{-1} and this band has a high frequency
shoulder. Note the series of four bands marked in Figure 4
between 1370 and 1270 cm^{-1}. The four bands of each compound also
have the same intensity ratios to each other, i.e. weak-strong-
strong-weak. Thus the β-sheet proteins give very similar spectra
in the Amide I and Amide III regions and these spectra are
dominated by a characteristic β-Amide I (1638 cm^{-1}) and Amide III
(1235 cm^{-1}) along with a characteristic pattern of four bands
between 1370 and 1270 cm^{-1}. Proteins which contain only α-helix
or contain only β-sheet structure can be readily identified or
differentiated by the Amide I frequency (1655 cm^{-1} for α vs. 1638
cm^{-1} for β).

In order to use a few figures as possible, the Amide I region
of " β + low α " proteins is not shown since the deconvoluted
spectra are virtually identical to those of " β " proteins (Fig.
3). The Amide III region for these proteins as shown in Figure 5.
The strong β-Amide III vibration seen in the spectra of " β "
proteins (Fig. 4) at 1235 cm^{-1} is clearly split into two
components (1254 and 1236 cm^{-1}) for " β + low α " proteins. Also
the intensity ratios of the four bands between 1370 and 1270 cm^{-1}
has changed to a weak-strong-weak-strong ratio (from high to low
frequency). So for " β + low α " proteins there is a characteristic
β-Amide I (1636 cm^{-1}) band, but the β-Amide III is split into
two components (1254 and 1236 cm^{-1}) and the intensity ratios of
the four bands between 1370 and 1270 cm^{-1} has changed (as compared
to " β " proteins). Thus the small α-helix content of these
proteins was not directly observed - only the effect of the
helix content on the β-vibrations in the Amide III region.

The spectra in the Amide I region of " α + low β " proteins is
also not presented since the deconvoluted spectra are very close
to those of " β " proteins (Fig. 1) except that the average
frequency of the strong Amide I band is lowered to near 1652 cm^{-1}.
Figure 6 shows the Amide III region of the spectra for these " α +
low β " proteins. For the three previous groups of proteins and
for the Amide I region of this group, the spectra within a group
were very similar. For the last group differences in the
Amide III region of the spectra can be observed as well as
similarities. Both proteins in this group have moderately intense
bands near 1315 and 1238 cm^{-1}, but each also has a moderately
strong or strong band between these frequencies (1275 cm^{-1} for
cytochrome C and 1258 cm^{-1} for lysozyme). The 1315 cm^{-1} and the
1238 cm^{-1} frequencies go along with the α- and β- content of
these proteins, but it is not known at the present time if the
1275-1258 cm^{-1} bands are Amide III vibrations or vibrations of
amino acid side chains. Thus " α + low β " proteins have a

AMIDE I + II

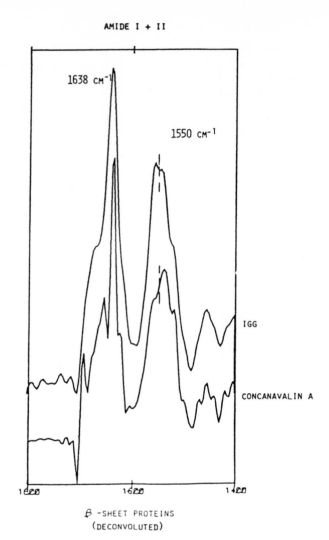

β -SHEET PROTEINS
(DECONVOLUTED)

Figure 3. Deconvoluted aqueous solution spectra of β-sheet proteins in the Amide I and Amide II spectral regions. Top: IgG; bottom: concanavalin A.

AMIDE III

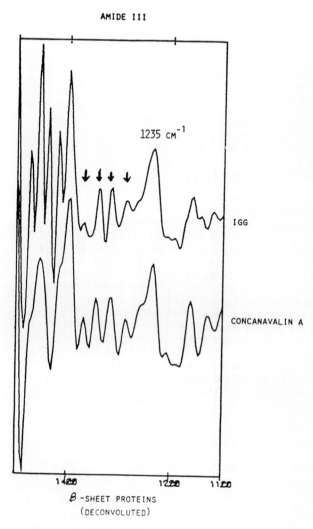

β-SHEET PROTEINS
(DECONVOLUTED)

Figure 4. Deconvoluted aqueous solution spectra of β-sheet proteins in the Amide III spectral region. Top: IgG; bottom: concanavalin A.

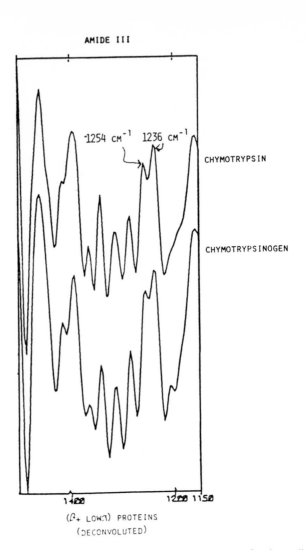

Figure 5. Deconvoluted aqueous solution spectra of "β+ low α" proteins in the Amide III spectral region. Top: chymotrypoin; bottom, chymotrypsinogen.

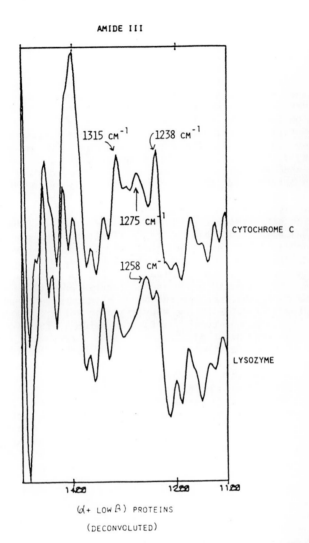

Figure 6. Deconvoluted aqueous solution spectra of "α+ low β" proteins in the Amide III spectral region.   Top: cytochrome C; bottom: lysozyme.

characteristic α–Amide I (1652 cm^{-1}) band with medium intensity
α–Amide III (1315 cm^{-1}) and medium intensity β–Amide III
(1238 cm^{-1}) bands.

Figures 7 and 8 give the spectra of " α+ low β" proteins in
the Amide I spectral region and the Amide III spectral region,
respectively. In the Amide I region (Fig. 7) the strongest band
is near 1645 cm^{-1}, but while ribonuclease has a strong high
frequency shoulder and β–lactoglobulin a weak high frequency
shoulder, only β–lactoglobulin has a low frequency shoulder. Yet
both compounds have a frequency near 1645 cm^{-1} and this frequency
is not typical of either α – or β–secondary structure. In the
previous four groups of proteins the Amide I frequency indicated
the major structure. In this group where there is not a large
difference between the α–helix and β–sheet content the Amide I
has an intermediate frequency which possibly reflects the
influence of one secondary structure on the vibrations of the
other. In the Amide III region of the spectra (Fig. 8), both
proteins exhibit a 1315(α) and a 1237(β) cm^{-1}. Ribonuclease
also has a 1291 cm^{-1} band which is likely another component of the
α –Amide III vibration. Thus " α+ low β" proteins have a non-
characteristic Amide I (1645 cm^{-1}) band along with an α–Amide III
(1315–1293 cm^{-1}) and a β–Amide III (1237 cm^{-1}) band.

From the spectra shown in the figures and following the
discussion in the text the secondary structure assignments for the
five groups of proteins are summarized in Table 2. From this
table is can be observed that for the Amide I vibration the major
structural components dictates the Amide I frequency, but it also
indicates that one structure influences the frequencies of the
other. For example, the α–Amide I frequency falls at 1656 cm^{-1},
but as more and more β–structure is present the frequency shifts
from 1656 cm^{-1} (α) to 1652 cm^{-1} (α+ low β) to 1644 cm^{-1} (α + β
). In the Amide III region the α–helix frequency shifts from
1298 cm^{-1} to 1315 cm^{-1} whenever some β–sheet structure is
present. Similarly, the β–sheet Amide III shifts from 1235 cm^{-1}
to 1241 cm^{-1} when enough α–helix structure is present.

The data in Table 2 lead to the following conclusions
concerning spectra-structure correlations for proteins:

1. The frequencies of the infrared peptide backbone vibrations
   (the so-called Amide vibrations) of proteins can be used to
   differentiate secondary structures (conformations) in proteins
   in aqueous solutions.
A. The Amide I and Amide III vibrations are particularly useful
   for distinguishing conformations. As yet, the frequencies of
   the Amide II vibrations are not especially helpful.
B. Frequencies of one band alone (either Amide I or Amide III) do
   not yield clear-cut distinctive information on conformations.
C. The frequencies due to one type of secondary structure are
   influenced by the presence of other structures and therefore,
   the frequencies can be used to yield information about the
   relative ratios of structures.

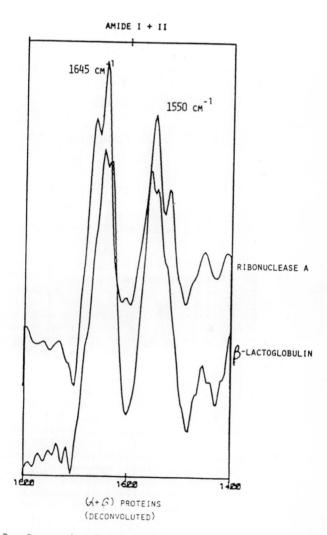

Figure 7. Deconvoluted aqueous solution spectra of "α & β" proteins in the Amide I and Amide II spectral regions. Top: ribonuclease A; bottom: β-lactoglobulin.

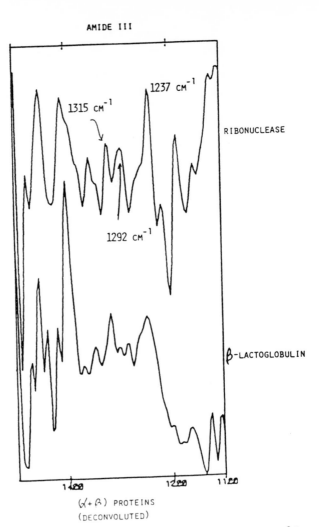

Figure 8. Deconvoluted aqueous solution spectra of "α & β" proteins in the Amide III spectral region. Top: ribonulease A; bottom: β-lactoglobulin.

TABLE 2.   SECONDARY STRUCTURE ASSIGNMENTS (cm^{-1})

TYPE OF PROTEIN	AMIDE I	AMIDE II	AMIDE III
α	1656 ± 2(s)	1548 ± 2(s)	1298 ± 4(m)
			1245 ± 3(w)
β	1637 ± 2(s)	1546 ± 8(s)	1235 ± 1(m)
β +Low α	1636 ± 2(s)	1548 ± 4(s)	1254 ± 1(m)
			1236 ± 1(m)
α +Low β	1652(s)	1547 ± 3(s)	1315(m)
			1266 ± 8(w)
			1238(m)
α + β	1644 ± 2(s)	1549 ± 7(s)	1315(w)
			1241 ± 4(m)

s=strong, m=medium, w=weak

2. For those proteins that posses a high degree of α-helix
   structure (no β-sheet) or a high degree of β-sheet structure
   (no α-helix), there are easily distinguishable Amide I and
   Amide III frequencies.

3. If there are unequal amounts of α-helix and β-sheet
   structure:
A. The frequency of the strong Amide I component indicates the
   major structure.
B. The frequency of the Amide III bands always indicates the
   presence of β-sheet and only indicates the presence of
   α-helix if that is a major component.

4. When there are roughly equal amounts of α- and β- structures:
A. The Amide I frequency falls at an intermediate position
   (1644 cm^{-1})
B. The Amide III frequency generally emphasizes the β-component
   although α-Amide III bands can be observed.

EFFECTS OF THE ENVIRONMENT ON THE STRUCTURE OF ADSORBED PROTEINS.

In the introduction section of this paper, it was discussed that
the structure of proteins could be altered by changing the
environment surrounding the proteins. These protein structure
changes could then be used to validate spectra-structure
correlations or assignments. This section reviews prior work in
which pH changes, pressure changes, and solvent variation have
been used to effect changes in the structure of albumin and IgG.
The induced structural changes in the proteins were reflected in
changes in the infrared spectra and these spectral changes could
be interpreted in terms of the structure changes.
    The experimental techniques used in these studies have been
detailed in the original papers (13, 14) and will only be briefly
discussed here. Studies of pressure and pH variation were carried
out on dissolved proteins while studies of solvent and pH
variation were done on adsorbed protein films. The adsorbed film
studies utilized a liquid attenuated total reflection (ATR) cell
with the proteins adsorbed on the ATR crystal (Ge). In this
arrangement the pH of the saline or the solvent itself could
easily be varied. As in the previous section, virtually all of
the spectra were obtained in aqueous media with the infrared
spectrum of water removed from the final spectrum by computer
substraction. Where non-aqueous solvents were used (14), the
spectrum of the solvent was also removed by spectral subtraction.

Albumin. After albumin as adsorbed onto a germanium ATR crystal
(from a saline solution), the adsorbed film was then exposed to
pure saline followed by either methanol or ethylene glycol
(13,14). After subtraction of the solvent, the spectrum of the
adsorbed albumin film exposed to saline was compared to the
spectrum of the adsorbed albumin film exposed to either of the
non-aqueous solvents. Three major differences were observed in
the films exposed to a non-aqueous solvent:

(1)   The Amide I band (1655 cm^{-1}) and the Amide III complex (around 1300 cm^{-1}) increased in both intensity and band area as compared to the spectrum of the film exposed to saline.   (2) The bandwidth of the Amide I vibration decreased when exposed to a non-aqueous solvent and (3) The frequency of the Amide I vibration increases.

The increase in intensity and band area of the Amide I and Amide III vibrations of albumin has been attributed (14) to an increase in the α -helix content of this protein.   This interpretation is supported by both the frequencies of the Amide I and III vibrations (see previous section) and by the fact that methanol and ethylene glycol have increased the helix structure in other proteins (18).

The interpretation (13) of the Amide I band narrowing and frequency shift was that since most of the bound water molecules of proteins were attached to polar groups located on the surface of the protein (19), some or all of this bound water would be removed by the non-aqueous solvents.   The frequency shift indicated that as the bound water was removed from the protein a more ordered (13) helix structure was produced.   The narrowing of the Amide I bandwidth is also consistent with an increase in order.   Three other types of experiments support this interpretation of the relationships of the bandwidth and order of and Amide I vibration of albumin with the amount of bound water present in the protein (13).   Two of these involve changes in pH or changes in pressure of albumin molecules dissolved in saline. For the pH change study, infrared spectra were obtained on the solutions of albumin at various pH's and the solution were also analyzed by thermal analytical techniques.   Figure 9 shows a plot of the percentage of bound water at various pH's calculated from the thermal analysis results as compared to a plot of the Amide I bandwidth at the same pH's.   Here it can be seen that the shapes of the curves are very similar and have a minimum at the same pH indicating a direct relationship between the Amide I bandwidth of albumin in solution and the amount of bound water.

This bandwidth-bound water relationship also holds true for adsorbed albumin and for solutions of albumin subjected to pressure.   Thus, the four types of infrared experiments demonstrated that for albumin:

(1) The use of the non-aqueous solvents increased the amount of helix secondary structure, (2) There is a direct relationship between the Amide I bandwidth and the amount of bound water, (3) A more ordered form of helix secondary structure is produced as bound water is removed from the protein.

Gamma Globulins.   In experiments (14) similar to those previously described for albumin, adsorbed IgG films were exposed to saline and then to either methanol or ethylene glycol and infrared spectra were obtained.   In addition, spectra of aqueous solutions of IgG at various pH's were obtained as well as spectra of IgG as it adsorbed onto the ATR crystal from a very dilute solution of IgG in ethylene glycol.   In the spectra of the adsorbed IgG exposed to methanol or to ethylene glycol there was increased

ALBUMIN - DISSOLVED

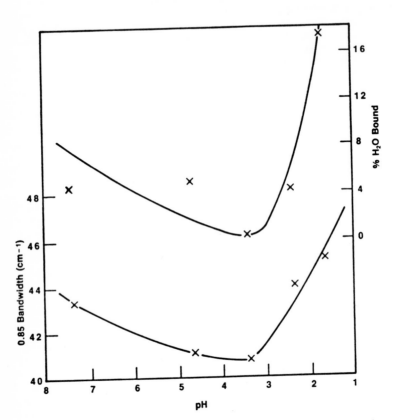

Figure 9.   Top: Percentage of bound water (from TGA analysis) versus pH of albumin solutions.   Bottom: Amide I band width versus pH of albumin solutions.

infrared adsorption near both 1660 and 1300 cm^{-1} indicating an increase in the α-helix structure due to exposure to these solvents. However, for IgG exposed to ethylene glycol there was no β-sheet Amide I vibration observed near 1638 cm^{-1}. Instead, this strong band was replaced by an equally strong band at 1629 cm^{-1} while the 1238 cm^{-1} Amide III band (exposed to saline) appeared at 1242 cm^{-1} in the ethylene glycol case. For the dissolved IgG at low pH, no helix vibrations (1660 and 1300 cm^{-1}) were observed, no β-sheet vibration at 1638 cm^{-1} were seen, but both the 1629 and the 1241 cm^{-1} bands were seen. In the study of the time behavior of the adsorption of IgG dissolved in ethylene glycol, the 1629 appeared as a weak band early in the adsorption process and increased in intensity with time of adsorption. It should be noted that circular dichroism measurement (20) of IgG dissolved in saline at the same pH's as for the infrared measurement indicated no change in the total amount of β-sheet structure, but indicated the formation of a more disordered β-sheet structure. This would indicate that the 1629 cm^{-1} infrared band indicates a more disordered β-sheet structure (as supported by frequency) and since the adsorption time studies indicate some unfolding takes place, it is postulated that this more disordered β-sheet structure is intermolecular in nature. From these results, the following conclusions have been postulated for IgG:

(1) Exposure to methanol produces α-helix structure, but does not affect the β-sheet structure. (2) Exposure to ethylene glycol produces helix structure and alters the conventional ordered, intramolecular β-sheet structure to a less ordered, intermolecular β-sheet structure. (3) Varying the pH does not produce α-helix structure, but does alter the β-sheet structure as described in (2). (4) A mechanism for the changes in (2) can be proposed which involves rapid formation of both α-helix and disordered β-sheet structure from both disordered (random) segments and from the ordered β-sheet structure. The disordered β-sheet structure then increases at the expense of the ordered β-sheet structure until very little of the original β-sheet structure remains.

Adsorbed Proteins. In a previous section, we discussed spectra-structure correlations for proteins in solution and presented a table of assignments for some of the vibrations of these proteins. Some proteins carry out their functions when dissolved, while others function as adsorbed proteins. Because of this predilection of proteins for adsorption, it is important to determine if the assignments for proteins in solution also hold for adsorbed proteins. In other words, is the structure of dissolved and adsorbed proteins the same? Since adsorbed proteins are often used for the solvent and pH variation studies which substantiate the spectra-structure correlations, it is important to know how these assignments compare to those of dissolved proteins. Therefore we have initiated a program to compare the frequencies from the spectra of adsorbed proteins to the frequencies obtained from spectra of dissolved proteins and eventually to study effects of solvent and pH variation on these adsorbed proteins. The initial results of this program are noted in the following paragraphs.

Figure 10 shows spectra (in the Amide III spectral region) of albumin dissolved in saline (top) and adsorbed on an ATR crystal (bottom). While the Amide III bands (1310-1245 cm^{-1}) of the adsorbed film appear to be slightly better resolved, in general, the spectra appear very similar. Thus, the conformation of albumin adsorbed on a germanium surface appears to be identical to the conformation of the protein in solution. Spectra of the adsorbed albumin after being on the surface for over three hours did not show any changes in the conformational structure. IgG behaved in a manner analogous to the behavior of albumin. The spectra showed that the structure of adsorbed IgG was identical to the structure of IgG solution and the adsorbed IgG did not change with time of adsorption.

Spectra of hemoglobin are shown in Figure 11. Here there are distinct differences between the spectrum (top) of dissolved hemoglobin and the spectrum (bottom) of adsorbed hemoglobin. In the Amide III region, the helix Amide III band (near 1300 cm^{-1}) in the spectrum of the dissolved hemoglobin is decreased in intensity (relative to the 1245 cm^{-1} band) in the spectrum of the adsorbed hemoglobin. Thus the conformational structure of the dissolved and adsorbed hemoglobin is different.

When an aqueous solution of hemoglobin is placed in contact with an ATR crystal, the spectra of the adsorbed hemoglobin at various adsorption times are shown in Figure 12. Comparing the top spectrum of Figure 12 to the top spectrum of Figure 11, it can be seen that there are similar (but not identical) 1300 cm^{-1}/ 1254 cm^{-1} intensity ratios between the two spectra. Thus, after three minutes of adsorption, the helix structure is still maintained although it may already have begun to change since the 1300 cm^{-1}/1245 cm^{-1} intensity ratio is higher for the dissolved hemoglobin (Figure 11, top) than for the adsorbed hemoglobin (Figure 12, top). As the adsorption time increases, the structure of the adsorbed film changes and after about an hour, the adsorbed film has achieved a stable structure.

Thus, the structure of adsorbed hemoglobin is initially very similar to the structure of hemoglobin in solution, but with increases in adsorption time, the structure of hemoglobin changes or rearranges. The stable adsorbed structure is not the same as the structure in solution.

Thus, our very preliminary results would indicate that adsorption on a germanium surface produces a protein structure similar to the structure in solution, i.e. it is the solution structure that first adsorbs. However, this initial adsorbed structure may not be the most stable structure and thus, rearrangement to a more stable structure may occur for some proteins. It is important to note that these results apply to germanium surfaces. It has already been noted (7) that the structure of adsorbed albumin and adsorbed IgG changes with the nature of the adsorption surface.

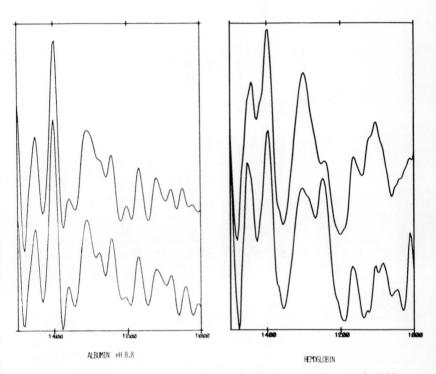

ALBUMIN pH 8.8

HEMOGLOBIN

Figure 10 (left).   Infrared spectra of albumin in the Amide III
spectral region.   Top: Albumin dissolved in saline at pH 8.8.
Bottom: Adsorbed albumin film exposed to saline at pH 8.8.   (Reproduced with permission from Ref. 13.   Copyright 1986 Wiley.)

Figure 11 (right).   Infrared spectra of hemoglobin in the Amide III
spectral region.   Top: Hemoglobin dissolved in saline.   Bottom:
Adsorbed hemoglobin exposed to saline.   (Reproduced with permission
from Ref. 13.   Copyright 1986 Wiley.)

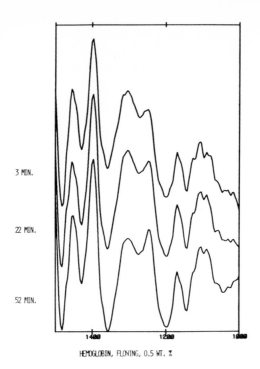

3 MIN.

22 MIN.

52 MIN.

HEMOGLOBIN, FLOWING, 0.5 WT. %

Figure 12. Infrared spectra of adsorbed hemoglobin in the Amide
III spectral region versus time of adsorption from a flowing, 0.5
wt. percent hemoglobin solution. Top: 3 minutes adsorption time;
middle: 22 minutes; bottom: 52 minutes. (Reproduced with permission
from Ref. 13. Copyright 1986 Wiley.)

Acknowledgments

Much of the data reported in this publication was obtained while
the authors were with Battelle Columbus Laboratories. Support for
the work was provided by Mattson Instruments, Inc. and by the
National Institutes of Health under Grant No. DRR-01367. The
authors would like to thank Katrina B. Smith of Ross Laboratories
for her assistance with parts of the experimental work.

Literature Cited

1. Koenig, J.L.; Tabb, D.L. Analytical Applications of FT-IR to
    Molecular and Biological Systems; Durig, J.R., Ed.; D. Reidel
    Publishing Co.: New York. 1980; pp. 241-255.
2. Carey, P.R.; Biochemical Applications of Raman and Resonance
    Raman Spectroscopies; Academic Press: New York, 1982.
3. Parker, F.S.; Applications of Infrared, Raman and Resonance
    Raman Spectroscopy in Biochemistry; Plenum Press: New York,
    1983.

4. Purcell, J.M.; Susi, H; J. Biochem. and Biophys. Methods; 1984, 9, 193.
5. Lin, V.J.; Koenig, J.L.; Biopolymers; 1976, 15, 203.
6. Gendreau, R.M.; Leininger, R.I.; Winters, S.; Jakobsen, R.J.; Biomaterials: Interfacial Phenomena and Applications; Cooper, S.L.; Peppas, N.A., Eds.; ACS Advances in Chemistry Series 199; American Chemical Society: Washington, D.C., 1982; pp. 371-394.
7. Jakobsen, R.J.; Wasacz, F.M.; Smith, K.B.; Chemical, Biological and Industrial Application of Infrared Spectroscopy; Durig, J.R., Ed.; J. Wiley and Sons, Ltd.: London, 1985; pp. 199-213.
8. Powell, J.R.; Wasacz, F.M.; Jakobsen, R.J.; Appl. Spectroscopy; 1986, 40, 339.
9. Susi, H. Byler, D.M.; Biochem. Biophys. Res. Comm.; 1983, 115, 391.
10. Dev. S.B.; Chokyun, R.; Walder, F.; J. Biomol. Structure and Dynamics; 1984, 2, 431.
11. Kauppinen, J.K.; Mofatt, D.J.; Mantsch, H.M.; Cameron, D.G.; Appl. Spectroscopy; 1981, 35, 271.
12. Yang, W.J.; Griffiths, P.R., Byler, D.M.; Susi, H.; Appl. Spectroscopy; 1985, 39, 282.
13. Jakobsen, R.J.; Wasacz, F.M.; Brasch, J.W.; Smith, K.B.; Biopolymers; 1986, 25, 639.
14. Wasacz, F.M.; Olinger, J.M.; Jakobsen,R.J.; Biochemistry; 1987, 26, 1464-1470.
15. Levitt, M.; Greer, J.; J. Mol. Biol.; 1977, 114, 181.
16. Schechter, E.; Blout, E.R.; Proc. Nat. Acad. Sci.; 1964, 51, 695.
17. Amzel, L.M.; Poljak, R.J.; Ann. Rev. Biochem.; 1979, 48.
18. Singer, S.J.; Advances in Protein Chemistry; Anfinsen, C.B.; Anson, J.L.; Bailey, K.; Edsall, J.T., eds.; Academic Press; New York, 1962, 17, 1-68.
19. Ghellis, C.; Yon, J.; Protein Folding; Academic Press; New York, 1982, 117-118.
20. Doi, E.; Jirgenson, B.; Biochemistry; 1970, 5, 1066.

RECEIVED March 4, 1987

Chapter 23

Fourier Transform Infrared Spectroscopic and Attenuated Total Reflectance Studies of Protein Adsorption in Flowing Systems

K. K. Chittur, D. J. Fink[1], T. B. Hutson, R. M. Gendreau, R. J. Jakobsen[2], and R. I. Leininger[3]

National Center for Biomedical Infrared Spectroscopy, Battelle Memorial Institute, Columbus, OH 43201

This paper summarizes a series of experiments directed to the development of Fourier transform infrared spectroscopic (FT-IR) techniques for monitoring the events that occur when blood contacts the surface of a biomedical device. Special emphasis is placed on the methodology used for quantification and compositional analysis of protein adsorption from complex protein mixtures in aqueous solutions.

Most researchers now agree that the early molecular interactions that occur when blood contacts a surface play an important, but incompletely understood, role in the course of thrombogenesis on surfaces.

The primary objective of this research has always been the monitoring of the adsorption of proteins from blood onto surfaces of biomedical interest, with the intent of correlating the molecular events occurring with the surface chemistry and hemocompatibility. This is a most ambitious objective, due to the complexity of the competitive adsorption of proteins that occurs when a surface is exposed to blood. FT-IR methods were developed for studying protein adsorption because we felt that this technology has the highest potential for meaningful study of the blood-surface interface. This technology is still in the early stages of development, but the power of the methodology has been demonstrated in our laboratories, as well as in others around the world.

[1]Current address: Bio-Integration, Inc., 1989 West Fifth Avenue, Suite #11, Columbus, OH 43212
[2]Current address: Mattson Institute for Spectroscopic Research, Ohio State University Research Park, Columbus, OH 43212
[3]Current address: 1973 Milden Road, Columbus, OH 43221

ADVANTAGES OF FT-IR

FT-IR instruments employ a Michelson interferometer to produce an interferogram of transmitted or reflected light, the Fourier transform of which is the infrared spectrum of the sample. This technology has been greatly advanced in the past decade with the increase in processing speed of minicomputers, software for fast Fourier transforms, and improved mercury-cadmium-telluride detectors, interfero-meters and sample cells. Commercially available FT-IR instruments are now capable of very rapid spectral collection rates, with high signal-to-noise ratios and with surprisingly simple sample preparation requirements (1). The sensitivity of the new instrumentation has permitted the analysis of biological macromolecules in aqueous media, which has historically been hindered by the interference of the strong absorption of the OH-bending vibration of water (centered at 1640 cm^{-1}) that overlaps the region of the strongest absorption bands of proteins (the Amide I band at 1630-1660 cm^{-1}). The currently available FT-IR instruments, and their associated minicomputers, have the capability to subtract digitized spectra, permitting the separation of the contributions of the solvent, solutes and polymer in the combined IR spectrum. A series of other numerical manipulations, such as spectral deconvolution or differentiation, can then be used to reveal the subtleties of the protein IR features. When this FT-IR technology is combined with flow cells built around Attenuated Total Reflectance light guides (2), a very powerful system for investigating protein adsorption phenomena is produced.

PROBLEMS WITH FT-IR/ATR

Most of the observations summarized here are qualitative or semi-quantitative, owing to the difficulties involved in the analysis of complex IR spectral features for which little or no previous interpretive background data exists. The primary technical problems that make quantitative analysis of protein spectra difficult include:

    1.   The objective subtraction of the absorption due to water from the spectra of protein solutions;

    2.   The complexity of the FT-IR spectra collected in the Attenuated Total Reflectance (ATR) mode which can include contributions from (a) the adsorbed protein layer or (b) the soluble proteins in the liquid layer adjacent to the ATR surface but within the "depth of penetration" of the evanescent field of the IR beam (this so-called "bulk" effect is especially significant when near-physiological protein concentrations are studied);

    3.   The inherent "noise" of the FT-IR optics/electronics system that can result in variable spectral features and often in poor baseline stability;

    4.   The general similarity of protein IR spectra, which requires protein spectral recognition based on weaker secondary bands or slight variations in band intensities or frequencies rather than on strong bands that are unique for each species;

5. The sensitivity of the IR signals of proteins to microenvironmental factors such as pH, ionic strength, and the presence of other solutes or surfaces; this can result in subtle spectral changes from "reference" protein spectra taken under different conditions;

6. The inherent complexity of the plasma, including the presence of low concentrations of proteins that might have an important effect on the nature of the adsorbed film. This complexity makes analyses based on standard reference proteins suspect to oversimplification.

Most of our efforts during this program have been directed to assessing the magnitude of and solving these problems.

OVERVIEW OF FT-IR STUDIES

The study of protein adsorption by FT-IR has not been a single project; instead it has involved a series of research steps that are generally outlined in flowsheet format in Figure 1. This perspective demonstrates the need to solve certain technical problems before more advanced, and more technically interesting, experiments can be performed and/or interpreted. For example, just as it is necessary to develop appropriate flow cells before kinetic data can be acquired, so also must the approaches to the analysis of protein mixtures from their infrared spectra be learned before software can be optimized for multicomponent analysis.

The flow chart emphasizes the two types of information that can be derived from FT-IR investigations:

1. The rates and amounts of adsorption of specific classes of proteins --this is the <u>quantitative</u> aspect of the adsorption process that most studies of this problem have been directed towards; and

2. Conformational changes in the adsorbing species that might indicate a transition of the proteins from "normal" to "foreign" in terms of recognition by the cascade of thrombosis events--this is the more subtle <u>structural</u> issue, and the one most analytical techniques are unable to address

The quantitative and structural aspects of protein adsorption are of importance in surface-induced thrombogenesis, although neither role has been completely described for even a simple model system. The infrared spectrum is rich in information related to both, making the development of these methods of fundamental importance to these investigations. The status of each of the aspects of research of this methodology are summarized in the following sections, except for the investigations of protein structure that are described in a separate paper at this meeting. (Jakobsen et al).

FLOW CELL SYSTEM

A dual-channel flow cell has been designed and built for use in protein adsorption and <u>ex vivo</u> shunt studies. The current cell incorporates several design modifications intended to minimize the poor liquid flow and difficult operational characteristics of two previous generations of flow cells <u>(3-6)</u>.

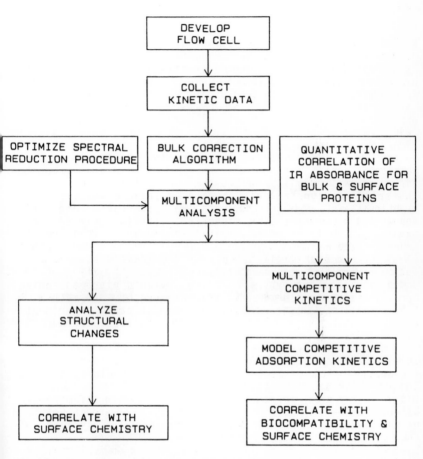

Figure 1. Flow sheet describing the evolution of FT–IR/ATR experiments for the study of protein adsorption onto solid surfaces.

Flow Cell Hydrodynamics. The flow cell contains two identical channels with dimensions of approximately 8 mm wide x 1 mm deep x 8 cm long, through which protein solutions can be pumped at up to 100 ml/min producing wall shear rates of up to 2100 sec^{-1}, thus spanning the range of arterial shear rates. The hydrodynamics of this cell were characterized by a series of experiments with non-adsorbing dextran having an average molecular weight of 82,000 daltons. Plots of the dextran IR band at 1160 cm^{-1} as a function of time and wall shear rate indicated that the convective diffusion model developed by Lok et al (7) can be used to describe solute transport mechanisms occurring in this cell.

Experimental Set Up. In the normal mode of operation, the flow cell is placed in the FT-IR system as shown schematically in Figure 2. A combination of peristaltic pumps, tubing and four-way valves, with two complete flow circuits per channel, permit continuous perfusion of the cell with saline or protein solutions, and rapid switchover from one to the other. Flow cells of this or an earlier design have also been used in a series of ex vivo experiments with dogs (4) and sheep (8). Exteriorized arterio-venous shunts were connected directly to the flow cell tubing to provide a supply of fresh, non-anticoagulated blood from the unanestheized animals. Spectra are acquired using a Model FTS-15 FT-IR system from Digilab, Inc. (Cambridge, Massachusetts) equipped with fast scan capabilities, a Hycomp 32 array processor and a high-sensitivity, narrow-range, liquid nitrogen-cooled, mercury- cadmium-telluride (MCT) detector. A less sensitive Model FTS-10 FT-IR system was used in many of our early experiments. Attenuated total internal reflectance (ATR) optics were supplied by Harrick Scientific (Ossining, New York).

Germanium ATR crystals were coated with thin polymer films by spin-coating techniques. A commercially available spin coater (Headway Research, Inc., Model EC-101) with a specially designed Teflon chuck was used to hold the Ge crystal. For example, Biomer--the medical polyurethane supplied by Ethicon (Somerville, NJ)--was applied in three to four coats in a 1.5 percent solution of cyclohexanone to produce stable films with thickness less than the depth of penetration of the IR field.

Kinetics Data Acquisition. Using the flow cell and FT-IR spectrometer described above, the capability to acquire up to four co-added spectral scans in 0.8 seconds (9) has been demonstrated in flowing protein experiments, and appears to be sufficiently fast to capture the rates of appearance and adsorption of the plasma proteins studied to date (10). In normal runs, data are collected at 0.8 second intervals for the first 2 minutes using the Digilab GCDATACOL software package, then every 10 seconds for the next 3 minutes, and finally every 15 minutes until the end of the experiment using the Digilab IMX software. In this way, more data are collected early in the experiment when the greatest rates of spectral change occur. Water subtraction remains a procedure that is performed manually on all spectra. The primary criteria for this subtraction are the attainment of a horizontal baseline between 2000 and 2500 cm^{-1} along with a "normal" ratio of the Amide II (1550 cm^{-1}) to Amide I (1650 cm^{-1}) bands, and with no

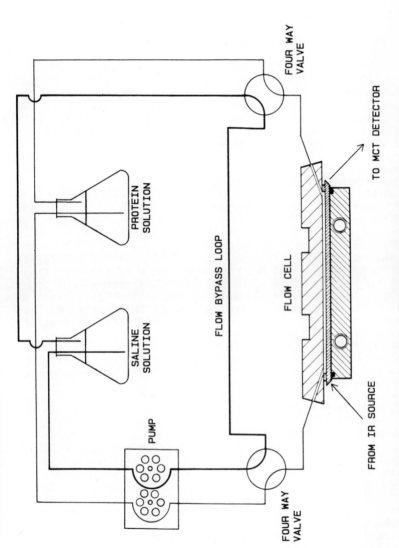

Figure 2. Flow cell and plumbing setup for the protein adsorption experiments.

obvious dispersion bands in the vicinity of the Amide I and II bands. More objective criteria for performing the water subtraction have recently been proposed (11).

Bulk Correction Algorithms. In ATR experiments performed with physiological concentrations of proteins, the IR signal is composed of contributions from adsorbed proteins, which may continue to build up for hours, and from soluble proteins in the "bulk" solution, which tend to stabilize in 30-60 seconds. A procedure for estimating the magnitude of the bulk contribution and correcting for this effect, based on comparative transmission spectra of the bulk solution has been described (6). This procedure is, however, not valid in the early stages of the adsorption process. Therefore a "two-cycle" approach for separating the adsorbing protein features from those of the proteins still in the liquid layer near the surface of the IRE was investigated (10). In this approach, a clean surface is exposed to a flowing protein solution for a period of 1-2 hours to produce a stable protein film on the IRE surface; the cell is then washed with buffer to clear the protein in the supernatant and to yield an IR spectrum of the adsorbed film. In the second cycle, more protein is perfused through the same cell under conditions identical to those of cycle 1. Subtraction of the spectrum of the adsorbed protein film from each spectrum of cycle 2 yields a series of spectra that describes the filling of the flow cell in the absence of protein adsorption to the IRE surface--and that can be used to approximate the development of the "bulk" even when adsorption does occur. Subtraction of this series of "bulk" spectra from those in cycle 1, at the corresponding times after introduction of protein to the flow cell, results in a series of spectra that approximate the build-up of the adsorbed protein film. This procedure has been performed to date for single and binary (albumin/ gamma-globulins) protein solutions adsorbing onto germanium and Biomer-coated surfaces (10).

PROTEIN QUANTITATION

Unlike simple organic molecules that generally have unique bands that can be used for their characterization and quantification, protein infrared spectra cannot be easily distinguished by using peak heights or areas above standard baselines. Protein IR spectra in aqueous solutions are complex combinations of bands arising from primary, secondary or tertiary structural bonds that are manifested in broad spectral features composed of many overlapping bands. Therefore, the development of characteristic spectral "fingerprints" for specific proteins or classes of proteins must be based on the shapes of the more intense bands (Amide I and II) and/or on the intensities of the weaker Amide III complex. Qualitative methods for estimating the composition of protein films were described in many of our earlier publications (12,13).
          The most useful quantitative methods for estimating the composition of mixtures of proteins from their IR spectra appear to be based on matrix algebra, as described in a series of papers by Brown et al. (14,15). While the use of such techniques is not

new to IR spectroscopy, the application to aqueous mixtures of proteins has not been reported previously. The basic elements of this work will be reported here only briefly but will be reported in more detail elsewhere (Chittur, K.K. etal, Manuscript in preparation).

Mathematics of P and K Matrix Methods. Two basic forms of the matrix approach are in use, the so-called "K" and "P" matrix methods. Both methods involve using mixtures of known compositions as standards, then estimating concentrations in unknown mixtures using a set of calibration coefficients generated from the standard spectra. At each wavelength of the spectrum and for each calibration mixture, an extended form of Beer's law can be written as follows:

$$A_j = k_{1j} c_1 + k_{2j} c_2 + \ldots \ldots + k_{ij} c_i$$

where A_j is the absorbance at the j-th IR frequency, c_i is the concentration of the i-th component and k_{ij} is the proportionality coefficient at the j-th frequency for the i-th component. The equation can be reduced to the "K" matrix notation as:

$$A = K C$$

where A and C are the vectors of absorbances and concentrations, respectively. In the calibration procedure, a "K" matrix is calculated from a set of spectra derived from mixtures with known concentrations. It is normal practice to have more mixtures and more wavelengths than components. This over-determination results in a "K" matrix that is best in the least-square sense.

In the "P" matrix form, concentration is written as a function of absorbance:

$$C = P A$$

where A and C have the same meaning as above but the proportionality matrix is termed P. The "P" matrix is established in the least-square sense by over-determination of the number of spectra with respect to the number of components. Unlike the "K" matrix, however, the number of frequencies must be less than or equal to the number of calibration spectra.

This simple-looking inversion of the concentration-absorbance relationship has rather important consequences. It is possible, by use of the "P" matrix approach, to quantify the composition of mixtures in the presence of variable amounts of "impurities", without being able to specify the composition or concentration of those impurities. The only requirement is that the impurities must be present in the calibration standards in amounts that bracket the concentrations of these impurities in the unknown samples. This property of the "P" matrix method makes it feasible to generate calibration spectra of plasma or whole blood containing known amounts of the proteins of interest (albumin, gamma-globulins, fibrinogen, fibronectin, transferrin . . .) and the remainder of the blood components can be treated as impurities.

Application of both these techniques to extensive sets of binary and ternary acqueous solutions of albumin, IgG and fibrinogen indicated that the "P" matrix estimates of unknown compositions were more accurate than those made using the "K" matrix method. The "P' matrix also allowed quantitation inspite of the uncertainty in water subtraction and the definition of baselines.

The analysis of protein mixtures by both these approaches are limited by the nature of the calibration spectra. For example to analyze protein mixtures where one or many of the individual protein(s) may have changed conformations, the calibration set of spectra must include spectra of the protein(s) in different conformational states. This can be achieved for example by obtaining the respective spectra by changing the pH or the solvent type. Analysis of protein mixtures without the correct calibration set of spectra may result in incorrect predictions of the concentrations in unknown spectra.

A rather surprising result of the examination of the "P" matrix spectral reduction procedure is that the best approach identified to date for quantifying ternary mixtures of albumin, gamma-globulins and fibrinogen is based only on the intensities of the Amide II band (1480-1600 cm^{-1}), not on the more unique--but less intense--Amide III and 1400/1450 complexes as we have described previously[11]. This finding is probably due to the greater intensity of the Amide II band (Figure 3a) and to the substantial differences among the three proteins in this region of the spectrum as revealed by the partially resolved bands (Figure 3b) following deconvolution by the Kauppinen technique (16). Based on this observation, it may be possible that researchers with less sensitive instruments than the one we use can still perform multicomponent spectral analysis on protein mixture spectra.

Much of our recent effort on this project has been devoted to the analysis of complex mixtures of proteins because it seems clear that these methods are central to accomplishing the objective of monitoring the protein adsorption process from whole blood.

Multicomponent Analysis of Kinetic Data. In preliminary work described in a recent paper (10), the kinetics of adsorption of albumin and gamma-globulins and their mixtures on germanium surfaces using the two-cycle bulk correction procedure has been determined. In single-component experiments on germanium, for albumin adsorbing from 30 mg/ml solutions, the initial IR signal was almost entirely due to the appearance of albumin in the evanescent layer--very little irreversible adsorption occurred in the first 3-4 minutes. In contrast, for gamma-globulins at 20 mg/ml, almost half of the initial signal came from proteins that had irreversibly adsorbed. The development of the evanescent field in these experiments confirmed that albumin filled the liquid layer near the surface more rapidly than the gamma-globulins, as expected from its higher concentration and diffusion coefficient. In a similar experiment with competitive adsorption of these proteins on germanium, essentially the same processes occurred at the same rates. Although the evanescent field was

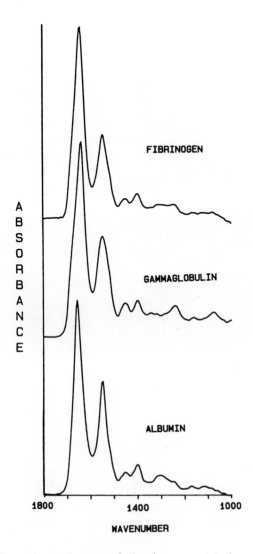

Figure 3a. Water subtracted spectra of albumin, gammaglobulin, and fibrinogen.

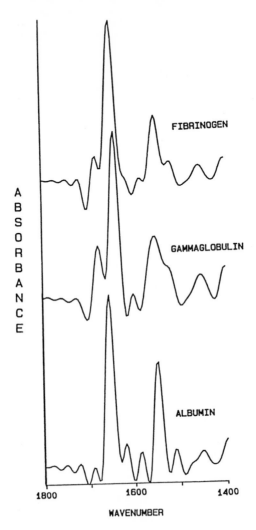

Figure 3b. Water subtracted spectra of albumin, gammaglobulin, and fibrinogen after deconvolution.

filled first with albumin, the gamma-globulins began to adsorb
almost instantaneously. Thus, while the appearance of protein
very near the surface is necessary for the adsorption process to
occur, it is not sufficient to insure that the protein adsorbs
irreversibly. More recent experiments have shown that albumin
adsorbs much more rapidly onto Biomer-coated crystals than it does
onto uncoated germanium. Thus two forms of these proteins can
exist on the surface-- a loosely associated (reversible) form that
is converted to a stabilized (irreversible) form--and that the
rate of this stabilization process may play a critical role in the
eventual biocompatibility of the material. Therefore, the
competitive adsorption model described and investigated by
Beissinger and Leonard (17) appears appropriate for fitting the
kinetics of this protein adsorption process.

CORRELATION OF IR ABSORBANCE WITH QUANTITY OF PROTEIN

An essential element in the quantification of protein mixtures by
FT-IR is the correlation of observed spectral features with the
quantity of protein in the field, which must be performed by a
second method for determining protein concentration. For protein
in the bulk solution, this process is relatively easy --the
protein is permitted to adsorb until a stable film has developed,
then a protein solution of known concentration is introduced into
the cell to determine the effective extinction coefficient for the
ATR cell based on the increase in absorbance. This process has
been demonstrated with albumin (Chittur, K.K., etal, Manuscript in
preparation) and in addition we have shown that the response
follows Beer's law up to protein concentrations of at least 60
mg/ml. In the case of adsorbed protein films, however, this
correlation is not so straightforward to achieve. An extensive
series of experiments was conducted in which[125]I-labeled proteins
have been (1) adsorbed to IRE surfaces, (2) washed with buffer to
produce a stable protein film, of which the FT-IR spectrum is
taken, and (3) desorbed into a 2 percent detergent solution and
counted to yield an estimate of the surface concentration of the
protein in the stable film. This procedure has resulted in an
excellent Beer's law correlation for IgG adsorbed onto Ge crystals
(5), which appeared to be valid for the three strongest bands in
the protein spectrum(1640, 1550 and 1400 cm^{-1}). Tests run in this
way using solutions of albumin or fibrinogen have not been as
successful, however. (Chittur, K.K. etal, Manuscript in
preparation). The correlation of peak height vs. surface
concentration for albumin is much more scattered than are the IgG
data. But these data appear to have approximately the correct
relationship to IgG that is expected based on the ratio of
extinction coefficients computed from transmission spectra. The
adsorbed fibrinogen data, although less scattered, appear to have
a much lower extinction coefficient (only approximately 25-30
percent) than that estimated from transmission data. The cause of
this apparent discrepancy will be the focus of continued study in
the future.
     The analysis of the protein adsorption process will be
complicated by any changes that occur in the protein spectra
during this process. Very little is known about this effect at

this time because it will depend on a variety of factors including
the structure of each protein, the surface properties, the nature
of competing proteins and other solutes, and other
microenvironmental factors. Preliminary observations have been
made of spectral changes upon adsorption of fibronectin (Cooper,
S.L. University of Wisconsin-Madison, personal communication),
fibrinogen and albumin (Chittur, K.K. Unpublished observations).

Ex Vivo SHEEP EXPERIMENTS

Thirty-eight ex vivo sheep blood experiments have been performed
to date using the protocol described above, including five each on
Biomer- and Silastic-coated and three on PVC-coated germanium
crystals. Clot formation during these experiments prevented the
separation of bulk and surface events by the two-cycle procedure.
Therefore the following qualitative conclusions, based on this
series of experiments, illustrate that these studies are possible
to perform but not completely interpretable in a quantitative
sense with the analytical software currently available.

1. Up to 60 percent of the total protein signal from whole
blood runs can be due to proteins in the bulk (supernatant)
solution. It is also possible that some loosely bound protein
might be desorbed when the flowing blood is replaced by saline.
2. Based on these preliminary experiments with plasma and
whole blood, we conclude that (a) hemoglobin was not detected in
any significant quantities on the surface or within the evanescent
field and (b) the presence or absence of red cells did not appear
to alter the protein adsorption pattern observed when bagged and
heparinized blood are employed. More recent observations by
Horbett (18) and Brash (19) suggest that hemoglobin adsorption may
play an important role in biocompatibility of surfaces.
3. Excellent spectral reproducibility between channels was
obtained under all conditions tested; therefore, the two
channels can be used to compare protein adsorption on different
surfaces.
4. When experiments were performed within 3-4 weeks after
shunt implantation, fairly good reproducibility was obtained from
studies run on the same sheep on different days --however, results
were not as reproducible as the comparison between channels run on
the same day.
5. Reasonable qualitative and quantitative reproducibility
was obtained between the spectra from different sheep adsorbing
onto the same surface, but the small number of these experiments
made under identical conditions limits the validity of this
comparison.
6. Two differences were observed between blood taken from
heparinized or non-heparinized sheep: (a) an infrared band near
1035 cm^{-1} was observed only in the heparinized sheep runs, perhaps
caused by heparin adsorption onto the ATR element; and (b) total
adsorbed protein was decreased in the heparin runs.
7. Differences were observed between the infrared spectra of
proteins from bagged blood and unmodified blood taken directly
from a live animal. For example, (a) more of the 1365-cm^{-1}
component adsorbed from the live blood, and (b) another unknown

component absorbing at 1290 cm^{-1} appeared in both the adsorbed and bulk spectra of the live blood but not the bagged blood.

8.   Protein adsorption onto germanium and Biomer-coated surfaces were compared on the same day using live, unmodified sheep blood in the dual-channel ATR cell. Qualitatively, most of the same events occurred at similar rates on the two surfaces, but clear differences were observed. The Biomer surface (a) adsorbed more albumin and total protein than the Ge surface, (b) adsorbed small amounts of a component having a 1280-cm^{-1} band not seen on the Ge surface; (c) lost more of the 1365-cm^{-1} and 1080-cm^{-1} species and more albumin during desorption with saline, and (d) retained more albumin and fibinogen/gamma globulin than the Ge surface, while most of the 1280-cm^{-1} species ended up in the desorbed film.

OBSERVATION OF ADSORPTION OF NON-PROTEIN SPECIES

Observations made during the ex vivo sheep blood experiments suggested that the adsorption of proteins is not the first event that occurs in the blood-surface interaction sequence. Several bands were observed in the first few seconds of contact that are not normally associated with proteins, the strongest being two with bands at 1365 and 1080 cm^{-1}. These observations were subsequently also made with bagged blood and plasma. A comparison of spectra from flowing plasma adsorption experiments to transmission spectra of plasma indicated that the amounts of these components might be concentrated relative to the water band at 1640 cm^{-1} in the adsorption studies. After the first few seconds of adsorption, the intensity of the two bands appear to decrease relative to other absorbing species, indicating that the early-adsorbing species are either displaced by later-arriving solutes, or their presence is simply masked by these species.

Whole sheep plasma was dialyzed, then the permeate was ultrafiltered to obtain a $<$ 1000 dalton fraction that contained the 1365-cm^{-1} and 1080-cm^{-1} components, as well as another material(s) absorbing at 1400 cm^{-1}. No Amide I or II bands, characteristic of proteins or peptides, were observed in these spectra. Efforts to separate these materials into distinct fractions by using HPLC on C$_{18}$-reverse phase columns or by solvent extraction from basic and acidic solutions of the $<$ 1000 dalton plasma fraction were unsuccessful. Some fractionation was obtained by gel permeation chromatography on Bio-Gel P-2, which led to fractions containing predominately the 1365-cm^{-1} and the 1080-cm^{-1} components, and another fraction with a combination of the 1365-cm^{-1} and 1400-cm^{-1} material(s). However, more than one component probably exists that is responsible for each of these bands. The possiblility also exists that these species may represent either degradation products or aggregates of major plasma constituents.

In adsorption experiments with the $<$ 1000 dalton plasma fraction alone, it appeared that some of the 1080-cm^{-1} and 1400-cm^{-1} components bound tightly to the surface (i.e., they did not come off in a saline wash), while the 1365-cm^{-1} component(s) were more loosely associated with the surface and were removed during the saline rinse (Hutson, T.B. Unpublished observations).

The possible adsorption of low molecular weight components from sheep plasma onto surfaces carries with it the implication of some possible interaction of these components with plasma proteins or the surface itself, which might be as significant as the proteins themselves in determining the thrombogenicity of the surface.

CONCLUSIONS

We have been investigating Fourier transform infrared spectroscopic methods for monitoring blood-surface interactions for over six years, during which period the following capabilities of the FT-IR technique for this application have been demonstrated:

1.   Currently available FT-IR spectrometers can produce spectra of plasma proteins with sufficient speed, sensitivity and reproducibility, and at high protein concentrations, so that the protein adsorption process can be followed under physiological conditions.

2.   FT-IR spectra can be obtained for proteins in aqueous solution or adsorbed onto surfaces by a combination of transmission and attenuated total reflectance techniques.

3.   A dual-channel ATR flow cell has been fabricated and tested in a variety of protein adsorption/desorption studies, permitting the direct comparison of two surfaces or two flowing solutions under otherwise identical conditions.

4.   Spin-coating methods have been demonstrated for applying thin polymer films to the germanium ATR crystals so that protein deposition can be investigated on surfaces of practical interest.

5.   Proteins can be characterized and quantified on the basis of their intrinsic IR spectra, without the need for extrinsic labels such as fluorescent or radioactive tags.

6.   Structural changes in proteins, e.g., resulting from microenvironmental factors such as pH or from protein-protein interactions, can be observed in the IR (especially in the conformation-sensitive Amide III region), also without requiring added probes of the protein structure.

7.   An ex vivo shunt system has been demonstrated with sheep such that fresh, unmodified blood can be tested in the ATR flow system.

8.   The IR spectra obtained by these techniques contain spectral information about all molecular species in the IR field, providing opportunities for observing unanticipated participants in the adsorption process such as the unidentified plasma components we have frequently observed to concentrate near the surface before proteins arrive.

9.   Computer software is now available or nearing completion to permit rapid collection of FT-IR spectra and a variety of digital manipulations including spectral subtractions, differentiation, deconvolution, baseline straightening, and multicomponent spectral analysis.

ACKNOWLEDGMENTS

This research was funded by Grant Number HL-24015 from the National Heart, Lung and Blood Institute of the National Institutes of Health (NIH). The facilities of the National Center for Biomedical Infrared Spectroscopy, funded by Grant Number RR-01367 from the NIH's Division of Research Resources, were used for a major portion of the experimental work.

LITERATURE CITED

1. Griffiths, P.R.; deHaseth, J.A. Fourier Transform Infrared Spectroscopy ; J. Wiley & Sons, New York, 1986.
2. Harrick, N.J. Internal Reflection Spectroscopy; Interscience, New York, 1967.
3. Gendreau, R.M.; Leininger, R.I.; Jakobsen, R.J. In: Biomaterials; Winter, G.D., Gibbons, D.F., Plenk, H.Jr. Eds.; John Wiley & Sons: New York, 1982; pp415-421.
4. Gendreau, R.M.; Winters, S; Leininger, R.I.; Fink, D.J.; Hassler, C.R.; Jakobsen, R.J. Applied Spectroscopy 1981, 35, 353-57.
5. Leininger, R.I.; Fink, D.J.; Gendreau, R.M.; Hutson T.B.; Jakobsen, R.J. Transactions Amer.Soc.Artif.Inter.Organs. 1983, 29, 152-57.
6. Fink, D.J.; Gendreau, R.M. Analytical Biochemistry 1984, 139, 140-48 .
7. Lok, B.K.; Cheng, Y.L.; Robertson, C.R. J.Colloid Interface Sci. 1983, 91, 87-103.
8. Winters, S.; Gendreau, R.M.; Leininger, R.I.; Jakobsen, R.J. Applied Spectroscopy 1982, 36, 404-9.
9. Gendreau, R.M. Applied Spectroscopy 1982, 36, 47-49.
10. Chittur, K.K.; Fink, D.J.; Leininger, R.I.; T. B. Hutson. J.Colloid Interface Sci. 1986, 111, 419-33.
11. Powell, J.R.; Wasacz, F.M.; Jakobsen, R.J. Applied Spectroscopy 1986, 40, 339-44.
12. Gendreau, R.M.; Jakobsen, R.J. J. Biomed. Mater. Res., 1979, 13, 893-906.
13. Gendreau, R.M.; Leininger, R.I.; Winters, S.; Jakobsen, R.J. In: Adv.Chem.Ser. Biomaterials. Interfacial Phenomena and Applications; Cooper, S.L.; Pappas, N.A. Eds.; American Chemical Society, Washington DC, 1982, Vol. 199, pp 371-394.
14. Brown, C.W.; Lynch, P.F.; Obremski, R.J.; Lavery, D.S. Analytical Chemistry 1982, 54, 1472.
15. Brown, C.W. Spectroscopy 1986, 1, 32-37.
16. Kauppinen, J.K.; Moffatt, D.J.; Mantsch, H.H.; Cameron, D.G. Applied Spectroscopy 1981, 53, 1454.
17. Beissinger, R.L.; Leonard, E.F. J. Amer. Soc. Artif. Inter. Org. 1980, 3, 160.
18. Horbett, T. A. J. Biomed. Mater. Res. 1981, 15, 673-95.
19. Uniyal, S., Brash, J. L., Degterev, I.A. In: Adv.Chem.Ser. Biomaterials.Interfacial Phenomena and Applications; Cooper, S.L.; Pappas, N.A. Eds.; American Chemical Society, Washington DC, 1982, Vol 199, pp 277-92.

RECEIVED January 29, 1987

Chapter 24

Organization of Albumin on Polymer Surfaces

Robert C. Eberhart, Mark S. Munro, Jack R. Frautschi, and Viktor I. Sevastianov[1]

Department of Surgery and Biomedical Engineering Program, University of Texas Health Science Center, Dallas, TX 75235

Our concern has been to understand the initial organization of plasma proteins on polymer surfaces and to devise methods to inhibit the activation of those proteins implicated in host responses. Visualization of adsorbed proteins suggests the adsorption pattern varies; fibrinogen, fibronectin and gammaglobulin each form reticulated films with protein-protein binding, while albumin forms globular deposits, desorbed by fluid shear or competition with other proteins. Adsorbed albumin exists in two conformations: a loosely held layer of native and partially unfolded molecules and an irreversibly adsorbed component of molecules with substantial chain unfolding. We developed a surface alkylation technique to increase polymer albumin affinity. Treated polymers exposed to whole blood are rapidly covered by a dense albumin layer which inhibits fibrinogen binding. Red and white thrombus formation have been strongly inhibited by this means.

Our studies of proteins at interfaces began 15 years ago as an outgrowth of experience gained in clinical extracorporeal membrane oxygenation (ECMO) for adult acute respiratory failure. Significant thromboembolic episodes were observed in patient perfusions for periods ranging from three to 20 days with an ECMO circuit containing 12 square meters of blood-polymer contacting material. Proteinaceous material was observed on all blood contacting surfaces at disassembly. The amount of material depended upon the local blood flow rate and the general hematologic state of the patient. Beyond these general observations, it was not possible to predict the thromboembolic potential for the patient/ECMO system (1). Baier and Dutton's protein preconditioning hypothesis was emerging at this time (2) and we were struck by the logic of this approach. Thus we initiated studies to allow direct observation of protein deposits on

[1]Current address: Institute of Transplantology and Artificial Organs, Moscow 123436, Union of Soviet Socialist Republics

polymer surfaces, reasoning that identification of specific proteins
and the influences governing their adsorption patterns might help to
explain the wide variation in protein adsorption we had observed in
the clinical ECMO study. The ramifications of that study have carried us far from ECMO. However, it now seems that our findings may
well yield the thromboresistant, biocompatible surface treatment of
ECMO circuits we have sought since 1972.

Protein Adsorption

Partial Gold Decoration Transmission Electron Microscopy.

We bor-
rowed a technique from the metallurgy literature, originally developed to decorate grain boundaries, in which a small amount of gold
or other noble metal thermally evaporated onto a surface under vacuum migrates on that surface to preferential binding sites and provides stable nuclei at those sites. Not only defects in polymer
surface structure, but also a measure of the surface free energy
distribution may be deduced by this technique. The extent of surface migration of the metal atoms on the polymer is related to the
surface binding forces, and can be quantified from the size and
spacing of metal nuclei (3). Moreover, the gold nucleus density is
significantly higher on adsorbed biological coatings than on the
polymer substrate. Thus it is possible to visualize, with high
resolution, protein coats on a number of polymers.

We observed differences in the organization of various proteins adsorbed from solution (4). Albumin exhibited a bulk concentration-dependent, characteristic globular deposition pattern
with low coverage of the surface for a number of hydrophobic polymers includin Teflon (Fig. 1), with similar results for some hydrophilic polymers. Air drying had a denaturing effect on the albumin
adsorbate, leading to formation of aggregates, as observed in
Figure 1. The sparse coverage feature was preserved, even with a
gentler sample preparation with sequential ethanol dehydration and
CO_2 critical point drying; however, aggregation was not as extensive. Fluid shear rate influenced the extent of surface coverage,
especially at low wall shear rate (Fig. 2); surface material and
roughness had less influence on albumin deposits.

In contrast, the glycoproteins fibrinogen, gammaglobulin and
plasma fibronectin, evaluated in single protein species adsorption
studies, all exhibited more extensive coating of the surfaces, with
a characteristic reticulated pattern (Fig. 3a) (5). The extent of
surface coverage depended on bulk protein concentration and surface
roughness; preferential binding of fibrinogen, with possibly some
fibrinogen polymerization, was observed by stereo pair TEM in surface cracks of the order of 1 μm. These data confirm the observation of Vroman, et al. (6) and support the common finding that
smooth surfaces are less thrombogenic than rough ones. Shear rate
had less influence on surface coverage by fibrinogen (Fig. 2). Surface preparation also influenced the observed degree of coverage;
air-dried protein films had larger pores than those prepared by the
gentler critical point drying technique (Fig. 3b). However, the
reticulated pattern was seen for all three glycoproteins, which
suggested that the organization of glycoprotein coats adsorbed from
solution is influenced by the protein-protein binding capacity of
the protein as well as the other listed parameters. Under high mag-

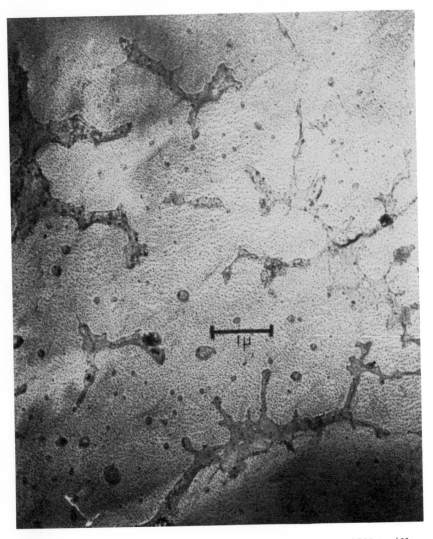

Figure 1. Teflon, exposed to human albumin solution, 2500 mg/dl,
for 1 hr at room temperature. Air drying and partial gold decoration TEM technique were employed. (Reproduced with permission
from Ref. 4. Copyright 1977 American Society for Artificial
Internal Organs.)

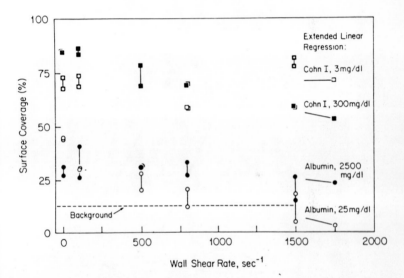

Figure 2. Fluid shear dependence of protein sorption on Teflon. Critical-point drying and partial gold decoration TEM techniques were employed. (Reproduced with permission from Ref. 5. Copyright 1982 American Chemical Society.)

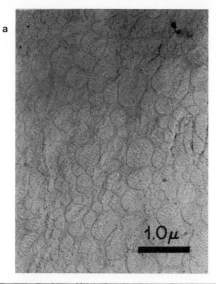

Figure 3.   Cohn I human fibrinogen, 3 mg/dl, adsorption on Teflon; gold decoration TEM technique.   (a) air dried; (b) critical-point dried with bar equal to 0.1 µm.   (Reproduced with permission from Ref. 5.   Copyright 1982 American Chemical Society.)

nification it was possible to observe the characteristic Hall-Slayter trilobed dumbbell form of fibrinogen molecules on Teflon and other hydrophobic polymer surfaces. At low bulk concentration, we observed not only the side-on binding configuration but also saw alignment of the lobes side-to-side and end-to-end (Fig. 4). This finding has gained significance in view of the recent understanding of the activity of the fibrinogen D region (7). In the end-on configuration this region is apparently more reactive, and prone to interact with cell membrane receptors. Fibrinogen's side-on configuration, including fibrinogen-fibrinogen binding, observed for these hydrophobic surfaces suggests less reactivity, at least under these in vitro conditions. On pyrolitic carbon, a more reactive material, we have observed denser and thicker coatings, using radiolabel PGDTEM methods, with fewer pores in the protein coat.

Sequential and concurrent albumin-fibrinogen adsorption studies were carried out, utilizing this new visualization technique, and it is noteworthy that albumin pretreatment, but not concurrent treatment, significantly reduced the total amount of protein (albumin plus fibrinogen) visualized on the surface (Fig. 5) (8). These findings supported the initial observation of Chang (9). However, as others have shown, albumin precoating is not universally successful, being desorbed by fluid shear as well as interaction with competing species at the polymer substrate. This led us to search for methods to improve the albumin affinity of the polymer.

Pulse Intrinsic Fluorescence Fluorimetry. The literature provides ample evidence that surface contact of proteins might unmask groups which could activate various host defenses (10-12). We reasoned that the protein-protein binding observed in the side-on configuration of fibrinogen and other glycoproteins, and the albumin aggregatory phenomenon might be a consequence of the protein conformational alteration. Thus we sought, especially for albumin, since it had an apparent passivating tendency, a measure of protein conformational alteration upon adsorption. Building on the work of Robertson, et al. (13) and Andrade, et al.(14), we developed a fluorescence decay measurement technique which allowed probing of only those albumin molecules at or in the immediate vicinity of the surface. The tryptophan and, to a lesser extent, tyrosine residues exhibit intrinsic fluorescence and thus provide a measure of structural alterations in their microenvironment. The single tryptophan residue in one of the central alpha-helices of albumin (15) suggests that it would be a sensitive indicator of conformational alteration involving the internal hydrophobic grooves.

Utilizing time resolved internal reflection spectroscopic technique (Fig. 6), we were able to isolate the tryptophan intrinsic fluorescence and observe its ≃20 ns fluorescence lifetime for albumin in bulk and in the surface microenvironment of a hydrophilic quartz material. The pH dependence of bulk albumin fluorescence lifetime served to "calibrate" albumin in terms of native (≃7 ns time constant) protein at pH 7.2 and unfolded (≃4 ns) protein at the isoelectric pH 3.8. The fluorescence lifetime data (Tables I,II) supported the hypothesis that the adsorbed albumin exists in two forms on a hydrophilic quartz surface, each with a possibly different structure (16). A loosely held "layer," consisting of microaggregates, native and partially unfolded albumin molecules with

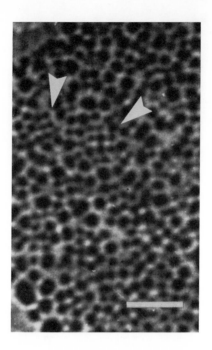

Figure 4.   Cohn I human fibrinogen, 30 mg/dl, adsorption on
Teflon; partial gold decoration TEM technique.   Bar equal to
1000 Å.

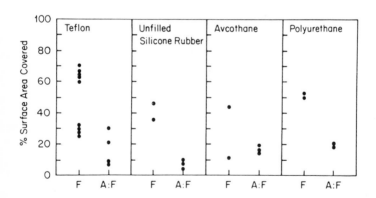

Figure 5.   Effect of albumin preadsorption on fibrinogen adsorption.   Air drying and partial gold decoration TEM were used for
all samples.   Key: F, 30 mg/dl fibrinogen (Cohn I) and A:F,
2500 mg/dl albumin followed by 30 mg/dl fibrinogen.   (Reproduced
with permission from Ref. 5.   Copyright 1982 American Chemical
Society.)

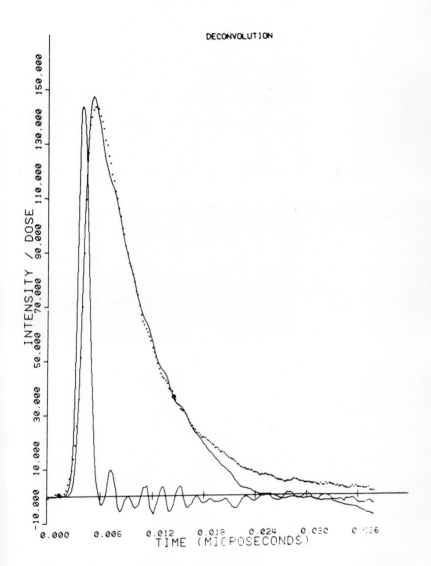

Figure 6. Monoexponential decay curve fit for bovine serum albumin, pH 3.8, bulk solution, 150 mg/dl, time constant 4.1 ns. First curve is the system response to the laser pulse, and is deconvolved from the raw decay curve.

TABLE I: FLUORSCENCE LIFETIMES (ns) OF BSA IN BULK SOLUTION (CUVETTE)

| Bulk Solution | pH 3.8 | | pH 7.2 | | Heat Treated | |
	Fraction V	Monomer	Fraction V	Monomer	Fraction V	Monomer
15 mg%	4.0	4.1	6.6	7.1	4.6	5.5
150 mg%	4.3	4.1	6.8	7.1	4.9	5.6

24. EBERHART ET AL. *Organization of Albumin on Polymer Surfaces* 387

TABLE II: FLUORESCENCE LIFETIMES (ns) OF ADSORBED BSA (TIRF CELL) [a]

	pH 3.8		pH 7.2		Heat Treated	
	Fraction V	Monomer	Fraction V	Monomer	Fraction V	Monomer
Protein in TIRF Cell						
15 mg%	5.4	4.6	5.8	6.3	4.7	5.2
150 mg%	5.4	5.0	6.5	6.8	4.6	5.5
After Gentle Wash						
15 mg%	4.1	4.4	5.0	4.7	3.6	4.4
150 mg%	4.0	3.8	5.1	5.1	4.1	4.7
After Vigorous Wash						
15 Mg%	3.8	4.2	4.1	4.1	3.0	3.0
150 mg%	3.9	3.5	4.1	4.0	3.5	3.3

time constants of the order of 7 ns appeared to exist on the undisturbed surface. A gentle wash removed some, if not most of this
native and partially unfolded material, reducing the time constant
of retained material to the order of 5 ns. Vigorous washing removed
all but the most tightly held protein, which appeared to exist, at
time constants of 4 ns., in a different configuration. Additional
technical details remain to be worked out (independent calibration
of quantum yield, operation with interposed polymer sheet, etc.).
However, the intrinsic fluorescence decay method appears well suited
to serve as a quantitative probe of the kinetics of protein conformational change upon adsorption, and thus to observe the structure
of surface-interacting albumin.

Alylation Improves the Albumin Affinity of Polymers

In Vitro Studies. Prealbuminated polymers, treated by a number of
means, had been shown by many to have a modest thromboresistive
effect of short duration (9,17). Munro, in our laboratory, proposed
a method to increase polymer albumin affinity which overcame many of
the problems associated with earlier efforts (18). His method takes
advantage of the internal binding sites, the hydrophobic grooves,
created by the arrangement of nonpolar residues in the alpha-helical
region of the molecule. Binding of saturated circulating free fatty
acids is efficiently carried out at these sites (Table III). Munro
reasoned that synthetic aliphatic chains might be grafted to the
polymer and, in an appropriate orientation, mimic the circulating
free fatty acids. It later turned out that analogous techniques are
used for sample purification by affinity chromatography (19).

A number of surface functionalization chemistries were explored
which yielded substantially improved affinity of albumin for a number of polymer substrates (18,20,21). Albumin binding was shown to
increase as the degree of surface C_{18} alkylation was increased (22).
Albumin bound rapidly (within 30 sec) in an apparent substrate specific manner (Fig. 7). While these results demonstrated that, given
enough time, more albumin would bind to untreated surfaces, it was
also found that the hydrophobic interaction at the alkyl site gave a
more tenacious bond, resisting both fluid shear desorption (Fig. 8)
and elution with sodium dodecyl sulfate and other protein denatur-
ants (23,24).

The distribution of albumin on the alkylated polymer surface
was visualized for a polyether polyurethane. Using a double antibody technique, it was possible to uncover the dense, homogenous
nature of the albumin adsorbate on the alkylated surface, and on
controls, the patchy coverage with apparent albumin aggregates
(Fig. 9) (23). Incubation of polyurethane (Biomer) C_{18} alkylated
to different degrees with a mixed albumin (15 mg/dl), fibrinogen
(15 mg/dl) solution showed that increasing albumin binding markedly
reduced binding of fibrinogen to the surface (Fig. 10) suggesting
that the increased affinity of polymer for albumin was responsible
for reduction in fibrinogen binding. Thus it appeared that the enhanced albumin affinity function might: (1) block binding of other
proteins at surface active sites, or (2) complex albumin with other
surface-bound proteins so that triggering of host defenses and organization of more extensive surface coverage could be inhibited.
In either case, the albumin coat would serve as a biological buffer

TABLE III: COMPOUNDS WHICH BIND TO SERUM ALBUMIN

Anion	K_A (M^{-1})	Conditions pH	Conditions Temp. (°C)	Est. % bound to albumin in normal plasma
Oleate	2.6×10^8	7.4	37	99.9
Palmitate	6.2×10^7	7.4	37	99.9
Bilirubin	1×10^8	7.4	38	99.8
	7×10^7	7.4	25	
Hematin	5×10^7	7.5	23	
l-Thyroxine	1.6×10^6	7.4	24	10
l-Tryptophan	1.6×10^4	7.4	2	75
Estradiol	1×10^3	7.4	5	
Progesterone	3.7×10^4	7.4	5	
Cortisol	5×10^3	7.4	37	30
Corticosterone	1.3×10^4	7.4	5	30
Aldosterone	5×10^3	7.4	5	60
Testosterone	4.2×10^4	7.4	25	6
Prostaglandin	7×10^4	7.5	37	
Urate	3×10^2	7.4	37	15

K_A is the affinity coefficient for human albumin

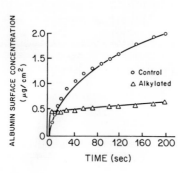

Figure 7. Albumin adsorption kinetic on polyetherpolyurethane using human serum spiked with ^{125}I albumin (Sevastianov's data). Curves are calculated according to Sevastianov's adsorption-desorption kinetics model.

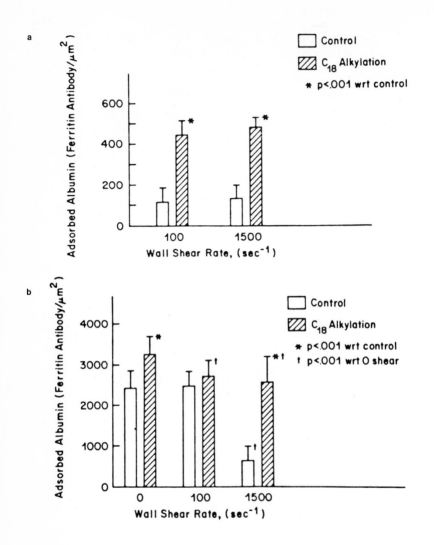

Figure 8.   (a) Albumin binding from solution (25 mg/dl) on polyurethane at indicated wall shear rate applied for 1 hr.   (b) Same
measurement following 1 hr static incubation and 1 hr saline
exposure at indicated wall shear rate.   Immunoferritin TEM technique.   (Reproduced with permission from Ref. 23.   Copyright 1985
American Society for Artificial Internal Organs.)

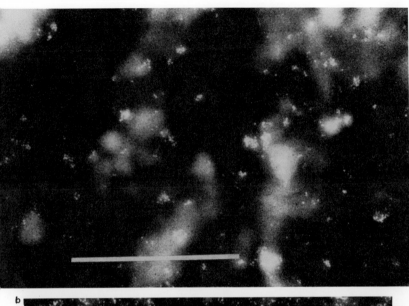

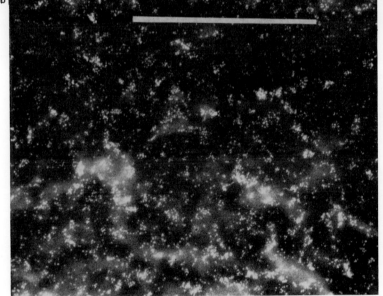

Figure 9.   (a) Albumin adsorption on polyurethane, 25 mg/dl,
1 min static exposure. Ferritin conjugated anti-IgG complex with
antialbumin treated sample. Ferritin particles register as white
dots. Line is 1 µm.   (b) Albumin adsorption on C_{18} alkylated
polyurethane. Same incubation and ferritin labeling conditions.
(Reproduced with Permission from Ref. 23. Copyright 1985 American
Society for Artificial Internal Organs.)

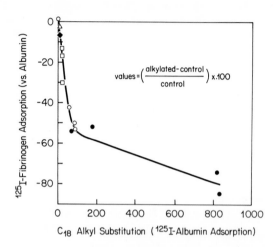

Figure 10. Reduction of fibrinogen adsorption (simultaneous albumin incubation) with C_{18} alkylated polyurethanes. Abscissa: enhancement of albumin adsorption on a duplicate treated sample, using identical methodology. (Reproduced with permission from Ref. 25. Copyright 1983 American Society for Artificial Internal Organs.)

layer which might inhibit the surface-induced activation of other plasma proteins.

The buffering potential of the alkylated surface was supported by in vitro study of the complement activating potential of C_{16} alkylated Cuprophan dialysis membrane. Using a poly-4-vinyl pyridine (P4VP) graft intermediate, it was shown that more albumin bound to C_{16}-P4VP treated Cuprophan than to controls, and that complement protein C3 activation for C_{16}-P4VP was significantly less than for controls (24). The albumin bound to the alkylated surface passivated that surface more effectively than spontaneously bound albumin.

Considering all the in vitro data, we hypothesize that the albumin held by hydrophobic interaction to the C_{16} or C_{18} derivatized surfaces is maintained in a form akin to its tertiary structure in bulk. It may be that the straight chain hydrocarbons further stabilize the hydrophobic strips on the alpha-helices, adding to the stabilization normally provided by disulfide bridges. Receptor sites which might otherwise be exposed to contribute to the contact activation of serum proteins or other blood constituents, are thus unavailable. We further hypothesize that in time the albumin molecules denature, possibly by disulfide bridge rupture, and the alignment of the alpha-helices is lost. The hydrophobic groove is thus lost, removing the affinity of the albumin for the surface aliphatic chain and leading to desorption of the albumin molecule. However, the alkyl chain remains, since it is covalently bound to substrate. If not internalized, it is available for hydrophobic interaction with other functional albumin molecules. Thus a continuing renewal of the albumin layer at the surface may be maintained for an indefinite period, and the biocompatibility of the surface may be enhanced for prolonged periods of time.

<u>In Vivo Studies</u>. The first in vivo studies tested the ability of a C_{18} derivatized polymer to bind albumin sufficiently rapidly to inhibit blood coagulation and platelet aggregation. Four mm I.D. wire reinforced polyurethane tubes were derivatized by a one-step method employing octadecylisocyanate. This surface treatment attaches the alkyl group at the urethane nigrogen atom. Control and derivatized 4-5 cm long tubes were implanted bilaterally, end-to-end, at femoral arterial sites in the dog. Tubes were filled with saline; no albumin precoating, systemic heparinization or other anticoagulant treatment was employed. Following 30 min of exposure to flowing whole blood, the samples were removed, gently flushed with saline, noting any thrombotic material which emanated from the tube, and fixed in gluteraldehyde solution. Gross and electron microscopic examination showed a dramatic improvement in the surface characteristics of the implanted tubes (Fig. 11). No fibrin or platelet aggregate was seen on the alkylated surfaces whereas controls were coated with a complete mural red and white thrombus. Identical results were obtained in all five experiments. Similar results to these 30 min studies were obtained in 2- and 4-hr implantations. Radiolabeled fibrinogen was infused in a 20-hr implant series and periodically scanned. In support of the in vitro study, fibrinogen binding to the alkylated polymer surface was significantly reduced (Fig. 12). Platelet adhesion and aggregation were also markedly reduced. However, an increasing fibrinogen (fibrin) buildup was observed at the suture line, which we attribute to coagulation

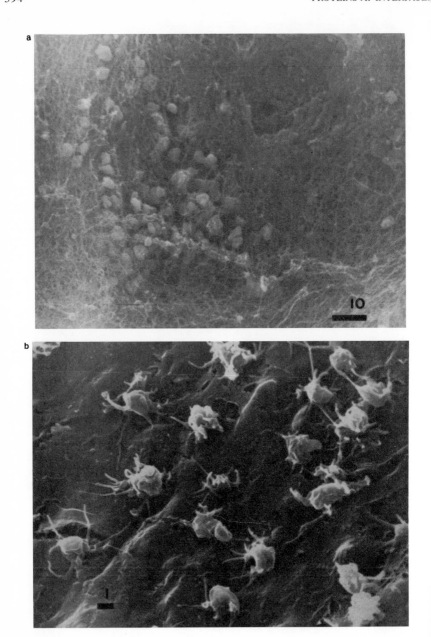

Figure 11.   (a) SEM of midsection of control polyurethane tube implanted in canine femoral artery, 30 min exposure.   (b) SEM of midsection of alkylated polyurethane tube implanted at the bilateral position.   (Reproduced with permission from Ref. 25.   Copyright 1983 American Society for Artificial Internal Organs.)

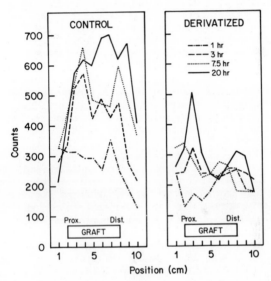

Figure 12.   Canine ^{125}I fibrinogen uptake on control and C_{18} alkylated polyurethane tubes at bilateral femoral arterial sites. Fibrinogen first appears at the anastomoses, then propagates through the tubes.   (Reproduced with permission from Ref. 25. Copyright 1983 American Society for Artificial Internal Organs.)

initiated at the interface of the host vessel with the graft. While these results suggested that additional treatment would be necessary to provide a viable small vessel prosthesis, they did not detract from the observed significant improvement in thromboresistance in the body of the grafts away from the host vessel interface (25).

Albumin turnover was evaluated in a series of in vivo experiments, employing the same femoral arterial canine model. Radiolabeled canine albumin solution was preincubated in situ, in crossclamped derivatized and control grafts for 30 min, followed by saline flush and exposure to flowing blood. Radiolabeled albumin rapidly desorbed from both control and derivatized surfaces. However, statistically significant differences in the amounts of labeled albumin retained on the derivatized and control surfaces were observed during the 2-hr course of the whole blood exposure (Fig. 13). Following a second cross-clamping, washout of the residual whole blood and reincubation with radiolabeled albumin solution, the experiment was repeated, with similar results. However, on the second 2-hr blood pass the statistical significance between albumin binding on derivatized and control surfaces was lost. This suggested that competing materials were acting to organize the surface, possibly a hydrophobic interaction between substrate and immunoglobulin, with subsequent binding of albumin. This view was supported by an additional experimental series in which whole blood was incubated with the control surface prior to radiolabeled albumin exposure. More albumin bound to the blood preincubated control surface than to one not preexposed to blood. Perhaps the most important finding of this series was that when whole blood was preexposed to C_{18} alkylated surfaces, less labeled albumin was subsequently bound to these than to controls (Fig. 14). The alkyl binding sites were apparently blocked by some substance adsorbed from whole blood. The most obvious candidate, considering the protein makeup of plasma and the in vitro kinetics studies in plasma (Fig. 7) would be endogenous albumin. If we adopt this interpretation, a partial confirmation of the underlying albumination hypothesis is obtained.

Conclusion

Plasma proteins organize on polymer substrates in different ways. Adsorbates are influenced by substrate physicochemical properties and by environmental factors, especially fluid shear and bulk protein distribution. Different types of binding interactions and more than one conformation for adsorbed protein are observed. In the case of albumin, the irreversibly adsorbed conformation, as measured by pulse intrinsic fluorescence, appears to be substantially altered from that of bulk albumin. Microaggregated albumin and undenatured forms are seen at the polymer interface, which are readily desorbed by viscous drag.

A method to substantially improve the albumin affinity of a number of polymers was developed, based on surface C_{16}-C_{18} alkylation. In vitro studies suggest the albumin is bound in a stable form, possibly involving one or more hydrophobic grooves in the alpha-helical region. The bound albumin may be presented to contact-activating proteins as if it were in bulk form, but one which is not easily desorbed. This treatment is demonstrated, in vivo, to inhibit contact activated coagulation and platelet aggregation.

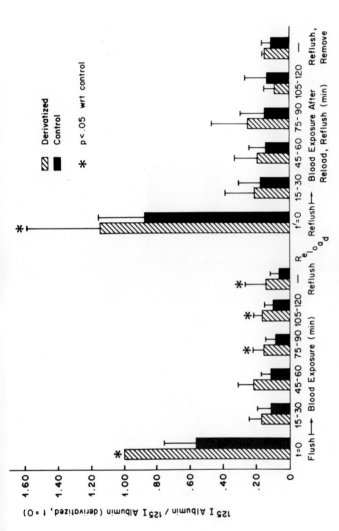

Figure 13. Albumin retention on C_{18} alkylated and control polyurethane tubes at bilateral femoral arterial sites for two sequential blood exposure periods, with reincubation with albumin following saline flush at 2 h. Values are compared with 125I albumin uptake on derivatized graft following 30 min. incubation. (Reproduced with permission from Artificial Organs [manuscript in press]. Copyright 1987 Raven Press.)

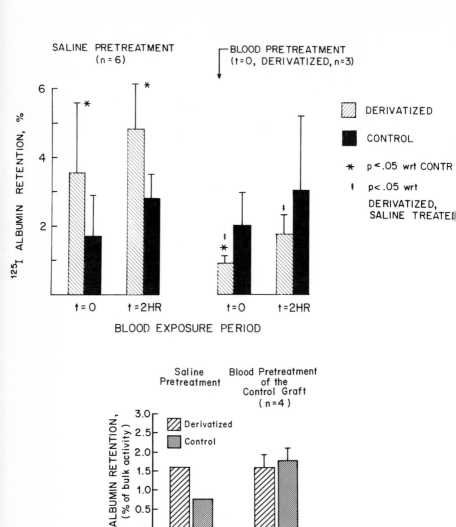

Figure 14. Effects of saline and saline/blood pretreatment on uptake of canine ^{125}I albumin on polyurethane tubes implanted at bilateral femoral arterial sites. The blood pretreatment step was carried out only for derivatized tube at t = 0. (b) Same conditions as (a) but blood pretreatment of control tube. (Reproduced with permission from Artificial Organs [manuscript in press]. Copyright 1986 Raven Press.)

Acknowledgments

This work was supported by grants from the USPHS (R01: HL 19173 and HL 28690) and the American Heart Association, Texas Affiliate. Graduate students John Wissinger, Samuel Riccitelli, Michael Lynch, Kevin Tingey and Mark Rainbow contributed significantly to the research.

References

1. Rodvien, R. Artif. Organs 1978, 2, 12-18.
2. Baier, R. E.; Dutton, R. C. J. Biomed. Mater. Res. 1969, 3, 191.
3. Bilge, F. H. Ph.D. Thesis, University of Texas Health Science Center at Dallas, Dallas, TX, 1982.
4. Eberhart, R. C.; Prokop, L. D.; Wissinger, J. G.; Wilkov, M. Trans. Am. Soc. Artif. Intern. Organs 1977, 23, 134-140.
5. Eberhart, R. C.; Lynch, M. E.; Bilge, R. H.; Wissinger, J. F.; Munro, M. S.; Ellsworth, S. R.; Quattrone, A. J. In Biomaterials: Interfacial Phenomena and Applications; Cooper, S. L.; Peppas, N. A., Eds; Adv. Chem. Ser. 199; American Chemical Society: Washington, DC; 1982; pp 293-316.
6. Vroman, L.; Adams, A. L.; Fischer, G. C.; Munos, P. C.; Standford, M. In Biomaterials: Interfacial Phenomena and Applications; Cooper, S. L.; Peppas, N. A., Eds; Adv. Chem. Ser. 199; American Chemical Society: Washington, DC; 1982, pp 265-92.
7. Lindon, J. N.; McManama, G.; Kushner, L.; Ware, J. A.; Merrill, E. W.; Salzman, E. W. In Surface Chemistry in Biology, Medicine and Dentistry: Proteins at Interfaces; Horbett, T. A.; Brash, J., Eds.; American Chemical Society Symposium Series; May, 1987.
8. Eberhart, R. C.; Wissinger, J. G.; White, W. A.; Wilkov, T. Trans. Soc. Biomater. 1978, 2, 136.
9. Chang, T. M. S. Can. J. Physiol. Pharmacol. 1969, 4, 1043.
10. Lee, E. S.; Kim, S. W. Trans. Am. Soc. Artif. Intern. Organs 1979, 25, 124-131.
11. Colman, R. W. J. Clin. Invest. 1984, 73, 1249-53.
12. Chenoweth, D. E.; Cooper, S. W.; Hugli, T. E.; Stewart, R. W.; Blackstone, E. H.; Kirklin, J. W. N. Engl. J. Med. 1981, 304, 497-502.
13. Watkins, R. W.; Robertson, C. R. J. Biomed. Mater. Res. 1977, 11, 915.
14. Hlady, V.; Van Wagenen, R. A.; Andrade, J. A. In Surface and Interfacial Aspects of Biomedical Polymers; Andrade, J. A., Ed.; Plenum Press: New York; 1985; p 81.
15. Brown, J. R. In Albumin Structure, Biosynthesis, Function; Peters, T.; Sjoholm, I., Eds; 11th Meeting FEBS; Pergamon: New York, 1977; Vol. 50, p 1.
16. Rainbow, M. R.; Atherton, S. J.; Rodgers, M. J.; Eberhart, R. C. Trans. Soc. Biomater. 1985, 8, 2.
17. Lyman, D. J.; Knutson, K.; McNeil, B.; Shibatani, K. Trans. Am. Soc. Artif. Intern. Organs 1975, 21, 49.

18. Munro, M. S.; Quattrone, A. J.; Ellsworth, S. R.; Kulkarni, P.;
     Eberhart, R. C. Trans. Am. Soc. Artif. Intern. Organs 1981,
     27, 499.

19. Parikh, I.; Cuatrecasas, P. Chem. & Eng. News; Aug. 26, 1985.

20. Tingey, K. G.; Frautschi, J. R.; Lloyd, J. R.; Eberhart, R. C.
     Trans. Soc. Biomater. 1986, 9, 125.

21. Frautschi, J. R.; Munro, M. S.; Lloyd, D. R.; Eberhart, R. C.
     Trans. Am. Soc. Artif. Intern. Organs 1983, 29, 242.

22. Munro, M. S. Ph.D. Thesis; University of Texas Health Science
     Center at Dallas, Dallas, TX, 1986.

23. Riccitelli, S. D.; Schlatterer, R. G.; Hendrix, J. A.; Williams,
     G. B.; Eberhart, R. C. Trans. Am. Soc. Artif. Intern. Organs
     1985, 31, 250.

24. Frautschi, J. R.; Eberhart, R. C. Proc. 5th South. Biomed.
     Eng. Conf. 1986, 425.

25. Munro, M. S.; Eberhart, R. C.; Maki, N. J.; Brink, B. E.; Fry,
     W. J. asaio J. 1983, 6, 65-75.

RECEIVED January 21, 1987

Chapter 25

Interactions of Proteins at Solid–Liquid Interfaces: Contact Angle, Adsorption, and Sedimentation Volume Measurements

D. R. Absolom[1,2,3], **W. Zingg**[1,3], and **A. W. Neumann**[1,2,3]

[1]**Research Institute, The Hospital for Sick Children, Toronto, Ontario M5G 1X8, Canada**
[2]**Department of Mechanical Engineering, University of Toronto, Toronto, Ontario M5S 1A4, Canada**
[3]**Institute of Biomedical Engineering, University of Toronto, Toronto, Ontario M5S 1A4, Canada**

The hydrophobicity of proteins can be assessed quantitatively through contact angle measurements on thick films of proteins generated by means of ultrafiltration with membrane filters. The surface tension values of proteins obtained from such contact angle measurements may be used as input data for a thermodynamic model for protein adsorption. This model predicts an increase of protein adsorption with decreasing surface tension of the substrate for high surface tensions of the suspending liquid; the opposite pattern is predicted for low-surface-tension liquid phases. Similar correlations between the extent of protein adsorption and surface tensions exist for the interaction of different proteins with one and the same substrate. The experimental observations follow closely the pattern predicted by the model. The detailed analysis of the adsorption data allows determination of the surface tension of the proteins independent of contact angle measurements. On the other hand, the sedimentation volume of a fixed mass of small particles in a liquid assumes an extremum when the surface tension of the particles is equal to the surface tension of the suspending liquid. This technique may be used to study properties of the adsorbed protein films. For particles coated with low-bulk concentrations of the protein, surface tensions of the adsorbed protein film are observed that are lower than those of fully hydrated proteins. This is possibly due to substrate-induced conformational changes of the adsorbed proteins. At high bulk concentrations of the coating protein solution, surface tensions for the protein-coated particles are found that are in excellent agreement with the results obtained from contact angles as well as protein adsorption. Blood plasma proteins in their native state are hydrophilic, with surface tensions typically in the range of 60–70 ergs/cm^2.

0097-6156/87/0343-0401$06.25/0
© 1987 American Chemical Society

Protein adsorption at an interface is of importance for many
reasons. Foremost amongst these is that the surface properties of
the substrate material are inevitably altered as a result of
contact with a protein-containing solution. The substrate may be
either a solid or liquid. There are many areas in which such
phenomena have direct technological applications including emulsion
stabilization, hydrophobic chromatography, biomedical devices,
enzyme immobilization and immunology.

During the past several years we have attempted to address the
following questions: 1. What causes a protein to adsorb to a
polymer substrate?; and 2. What effect does protein adsorption
have on the surface properties of the "naked" polymer substrate?
These questions will be the central theme of this article.

In blood contact situations protein adsorption is believed to
be an extremely rapid process occurring within a few seconds of
exposure (1-3). Protein adsorption has at least two major consequences. The adsorbed protein molecules alter the substrate surface
properties significantly. This will influence dramatically subsequent events such as the extent of platelet and leukocyte adhesion
and hence thrombus formation on those surfaces. The level of cell
adhesion may be reduced (4-8) as a result of the protein-coated
material exhibiting markedly different surface properties as compared to the uncoated material. Alternatively the level of cell
adhesion may increase significantly due to the presence of specific
receptor-ligand interactions between the adhering cells and the
adsorbed protein molecules. One example in this regard is the
increase in platelet adhesion on fibrinogen-coated polymers (9,10)
which is believed to be due to the presence of fibrinogen receptors
on the platelet membrane (11). Depending on both the nature of the
adsorbing protein molecules and the substrate materials the resultant protein-coat may vary considerably giving rise to surfaces
which may induce markedly different secondary responses, such as
the degree of cell spreading (12,13) and secondary granule release
(14), in the adherent cellular material. All-in-all protein
adsorption is a key event in blood surface interactions and needs
to be understood more fully in order to understand and control the
mechanisms of surface-induced thrombosis.

Surface Characterization of Proteins

Several authors (15-18) have related the extent of protein adsorption (or cell adhesion) to the hydrophobicity of both the "naked"
substrate material (or the protein-coated substrate) and the
adhering entities themselves. In this regard the hydrophobicity of
the substrate material has generally been assessed by means of
contact angle measurements. In the case of proteins the hydrophobicity has been determined in a number of different ways including
two-phase partition studies (19), alcohol precipitation (20), cis-
paranaric acid binding (21,22) and also by means of contact angle
measurements (18,23,24). It is generally accepted that in an
aqueous solvent all serum proteins exhibit overall a high degree of
hydrophilicity. This statement does not in any way exclude the
existance of hydrophobic patches or domains in the protein molecule
but simply implies that in their native state proteins exhibit an

overall hydrophilic character. This is illustrated for several
serum proteins in Figure 1 in which contact angle data on relatively
thick hydrated layers of the proteins are reported. The protein
layers were formed by means of ultra-filtration of protein solutions
through anisotropic cellulose acetate membranes. As illustrated
the contact angles are measured as a function of time. The plateau
contact angle values, which are stable for approximately 30–60
minutes, have been identifed as the relevant contact angle (25).
These values are independent of experimental variables such as relative humidity, temperature and thickness of the ultra-filtered
layer. These factors change all features of the contact angle
versus time curves such as the initial slope, the duration of the
plateau, the end values, etc., with the striking exception of the
contact angle value as given by the plateau (18,25). The observed
plateau values reflect solely the properties of the hydrated
protein layer and are unique for each protein investigated. These
contact angle values can for purposes of a more ready comparison,
be used to derive the surface tension of the protein layer. For
this purpose we use an equation-of-state approach (26). The validity of this approach has been discussed in detail elsewhere (27).
The surface tensions of the highly hydrated protein layers obtained
in this manner are summarized in Table 1. It is readily seen from
this table that the proteins under these circumstances all exhibit
a high surface tension value, i.e. they are hydrophilic. Also
given in Table 1 for comparison purposes are the surface tension
values of the same proteins obtained by other independent means.
These results confirm the hydrophilic nature of the native proteins.

TABLE I.   Comparison of the Surface Tension of Fully Hydrated,
           Non-Denatured Proteins, Determined by
           Three Independent Methods

| Protein | Surface tension (ergs/cm^2) determined from: | | |
	Contact Angles	Adsorption	Sedimentation Volume
α_2M	70.9	71.0	–
HSA	70.3	70.2	69.7
IgM	69.5	69.4	69.4
IgG	67.3	67.6	67.7
Fibrinogen	63.5	–	63.2

Under certain circumstances however the same proteins can
exhibit a markedly hydrophobic character (23,24). This is illustrated in Table 2 in which contact angle measurements have been
performed on thin layers of proteins adsorbed onto various polymer
surfaces and subsequently exposed to an air interface. This
results, due to reorientation and/or structural changes, in a much
more hydrophobic tertiary configuration. Thus, when referring to
the relative hydrophobicity/hydrophilicity of proteins it is
important to bear in mind the environmental conditions under which
the experimental data are generated.

TABLE II.   Contact Angles and Corresponding Surface Tensions
Obtained on Air-Dried Layers of Human Serum Albumin Coated
on Three Different Polymer Surfaces

Polymer	Contact Angle (degrees)		Surface Tension (ergs/cm^2)	
	Saline[a]	Glycerol[b]	Saline	Glycerol
PTFE (17.6 ergs/cm^2)	51	36	52.6	53.1
Low Density Polyethylene (32.5 ergs/cm^2)	50	38	53.2	52.4
Sulphonated Polystyrene (66.7 ergs/cm^2)	52	36	51.8	53.1

[a]γ_{LV} = 72.8 ergs/cm^2
[b]γ_{LV} = 63.1 ergs/cm^2

Protein Adsorption to Polymer Surfaces

The effect of protein hydrophobicity on the extent of adsorption to
a hydrophobic substrate, poly-tetrafluorethylene (PTFE) is shown in
Figure 2. The four proteins are dissolved at various bulk concentrations in phosphate buffered saline, pH 7.3, with a surface
tension of 72.9 ergs/cm^2. It is clear from this figure that an
adsorption plateau occurs for each protein. The height of the
plateau is unique for each protein and corresponds to the hydrophobicity of the proteins: increased surface concentration corresponding to increased protein hydrophobicity (cf. Table 1). Similar
experiments have been performed for these proteins adsorbing onto a
variety of other substrate materials exhibiting a wide range of
surface tensions. The plateau level of adsorption ($\mu g/cm^2$) for
each protein on each of the substrates has been plotted as a
function of substrate surface tension, γ_{SV}, in Figure 3. It is
clear from this figure that for these experimental conditions that
the plateau level of adsorption decreases with increasing substrate
surface tension, γ_{SV}, for each of the proteins examined. The
differences in the slopes of the curves is significant as discussed
below. Thus both the substrate surface tension as well as protein
hydrophobicity are important parameters in determining the extent
of protein adsorption.
    Several authors have reported that protein adsorption is a
maximum on hydrophobic substrates whereas others claim that it is
more pronounced on hydrophilic surfaces (15-17). In an attempt to
clarify the situation and to elucidate the fundamental mechanisms
involved in bioadhesion we have examined the adhesion of various
biological cells to a number of polymer surfaces (28-33). The

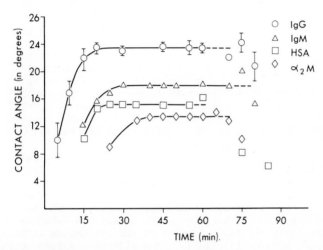

Figure 1:   Contact angle on four serum proteins deposited by
ultrafiltration on anisotropic cellulose acetate membranes.
Errors indicated are 90% confidence limits. For graphical
reasons errors are shown only in some cases; errors are similar
in all cases. o————o, IgG; Δ————Δ, IgM; □————□,
human serum albumin (HSA); ◊————◊, α_2-macroglobulin.
(Reproduced with permission from Ref. 18. Copyright 1981,
Elsevier/North Holland Press).

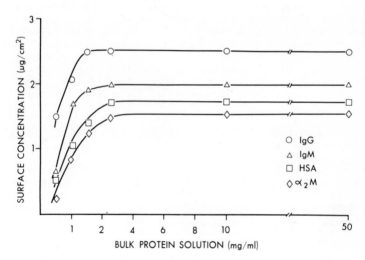

Figure 2:   Adsorption isotherms of four serum proteins to
poly-tetrafluorethylene. Proteins dissolved at various bulk
concentrations in phosphate buffered saline, pH 7.3; surface
tension = 72.0 ergs/cm^2. Proteins as indicated. (Reproduced
with permission from Ref. 18. Copyright 1981, Elsevier/North
Holland Press).

results have shown that the extent of bioadhesion can be predicted
by means of a thermodynamic model. The question was then raised as
to whether this model could also be used to predict the extent of
protein adsorption to various polymer surfaces.

The model, in terms of protein adsorption, is based on surface
thermodynamic considerations: Consider a protein molecule (P)
initially suspended in a buffer solution (L) adsorbing to a solid
surface (S) which is also immersed in the same buffer as illustrated schematically in Figure 4. In the absence of specific
interactions of the receptor-ligand type, the change in the Helmhotz
free energy (ΔF^{ads}) due to the process of adsorption is:

$$\Delta F^{ads} = \gamma_{PS} - \gamma_{PL} - \gamma_{SL} \qquad [1]$$

where γ_{PS}, γ_{PL} and γ_{SL} are the protein-solid, protein-liquid
and solid-liquid interfacial tensions, respectively. The validity
of this model has been discussed elsewhere (27).

It is apparent from Eq. [1] that the free energy of adsorption
of a protein onto a surface should depend not only on the surface
tension of the adhering protein molecules and the substrate material but also on the surface tension of the suspending liquid. As
an illustration shown in Figure 5 is a plot of the free energy of
adsorption (ΔF^{ads}) as a function of substrate surface tension,
γ_{SV}, for immunoglobulin M (IgM) suspended in liquids of two
different surface tensions, γ_{LV}. The input data required for
the development of such a plot are the respective surface tensions
of the adsorbing proteins (γ_{PV}), the polymer substrates (γ_{SV})
and the suspending liquid medium (γ_{LV}). The γ_{LV} and γ_{SV}
values are determined using any of a variety of procedures (34).
The surface tension, γ_{PV}, of the hydrated protein molecules is
most easily determined by means of contact angle measurements on
thick layers of proteins as discussed earlier.

Consideration of such theoretical calculations as illustrated
in Figure 5 leads to a distinction between two situations. For

$$\gamma_{LV} < \gamma_{PV} \qquad [2]$$

ΔF^{ads} decreases with increasing γ_{SV}, predicting increasing
protein adsorption with increasing substrate surface tension,
γ_{SV}, over a comparatively wide range of γ_{SV} values. On the
other hand when

$$\gamma_{LV} > \gamma_{PV} \qquad [3]$$

the opposite pattern of behaviour is predicted. For the limiting
case of the equality

$$\gamma_{LV} = \gamma_{PV} \qquad [4]$$

ΔF^{ads} becomes equal to zero independently of the value of γ_{SV}.
In this limiting case protein adsorption should not depend on the
surface tension of the substratum and, in principle, should be zero
if no other effects, such as electrostatic interactions, come into
play (28-31). Thus the model suggests that the surface tension of
not only the substrates and the adsorbing protein but also of the
suspending liquid medium is an important parameter in determining
protein adsorption; yet this important physico-chemical property is
seldom reported or even considered.

We have evaluated the role of γ_{LV} and tested the predictions
of protein adsorption experimentally (18). Shown in Figure 6 are
the experimental data for IgM suspended in various liquids of
different surface tensions. (A phosphate buffer containing varying
small amounts of dimethyl-sulfoxide (DMSO) was used. The pH of the

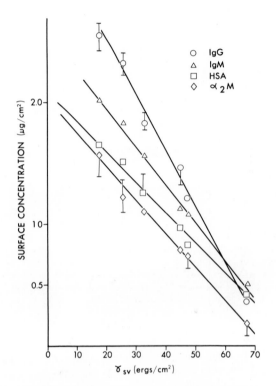

Figure 3: Plateau level of protein adsorption to a wide range of substrates plotted as a function of substrate surface tension, γ_{SV}. (γ_{LV} = 72.9 ergs/cm^2). Proteins as indicated. Errors indicated are 95% confidence limits. (Reproduced with permission from Ref. 18. Copyright 1981, Elsevier/North Holland Press).

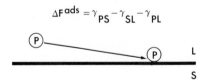

Figure 4: Schematic representation of the process of protein adsorption. P, protein; L, liquid; S, substrate.

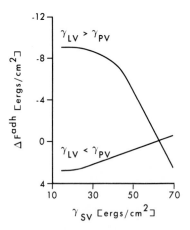

Figure 5: The free energy of adsorption (ΔF^{ads}) as a function of substrate surface tension, γ_{SV}.

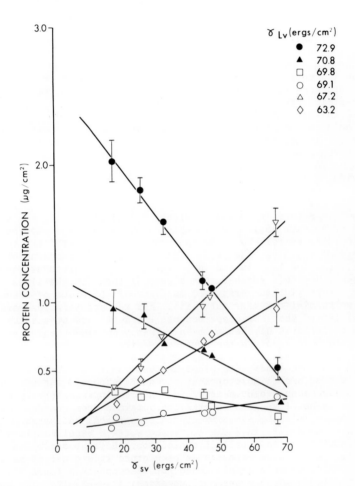

Figure 6: Immunoglobulin M (IgM) adsorption to various polymers with a wide range of surface tensions, γ_{SV}. IgM used at a bulk concentration of 2.5 mg/ml corresponding to the plateau region in Figure 2. Protein dissolved in phosphate buffered saline, pH 7.3, containing varying amounts of DMSO and hence having different liquid surface tensions, γ_{LV}: •, 72.9 ergs/cm^2; ▲, 70.8; □, 69.8; o, 69.1; ▽, 67.2; ◊, 63.2. Errors indicated are 95% confidence limits. For graphical reasons errors are given only in some cases; errors are similar in all cases. (Reproduced with permission from Ref. 18. Copyright 1981, Elsevier/North Holland Press).

solution was maintained at pH 7.3). The theoretical predictions inherent in Figure 5 and their implications are substantiated experimentally. At the lowest DMSO concentration, corresponding to the highest surface tension γ_{LV} of the suspending medium, protein adsorption decreases with increasing substrate surface tension, γ_{SV}. As the DMSO concentration is increased and the surface tension γ_{LV} correspondingly lowered, the change in the degree of protein adsorption with increasing γ_{SV} becomes less pronounced. At a certain intermediate γ_{LV} value the extent of IgM adsorption becomes independent of γ_{SV} and finally, at yet lower values of the liquid surface tension, γ_{LV}, adsorption increases with increasing γ_{SV}. Similar adsorption studies have also been performed for human serum albumin (HSA), immunoglobulin G (IgG) and $\alpha_2$macroglobulin (18).

Aside from the intrinsic interest of these data there are two further points to be made. First the thermodynamic model underlying Eq. [1] describes the qualitative features of protein adsorption well. Second, the data of Figure 6 lend strong support to the method of contact angle measurements on thick layers of hydrated protein (35), as follows. The thermodynamic model predicts that in the case of $\gamma_{LV} = \gamma_{PV}$ (Eq.[4]), ΔF^{ads} should be independent of γ_{SV}, a situation that is indeed contained in the curves of Figure 6. To investigate this further, the slopes of the straight lines in Figure 6 were plotted versus γ_{LV} in Figure 7, by means of a second-order polynomial computer curve fit. For each protein investigated it is inferred that the slope becomes equal to zero at a value of γ_{LV} characteristic of that protein. This γ_{LV} value according to the model is equal to the surface tension, γ_{PV}, of the adsorbing species. For example, consideration of Figure 7 reveals for IgM that the adsorption slope becomes equal to zero when $\gamma_{LV} = 69.4$ ergs/cm^2 implying that the surface tension of this protein is also equal to 69.4 erg/cm^2. This is in excellent agreement with the value of $\gamma_{PV} = 69.5$ ergs/cm^2 obtained from contact angle measurements, via the equation-of-state approach, on layers of the protein. Summarized in Table 1 are the surface tension values obtained with the adsorption technique for the four serum proteins. Comparison with the data obtained from other techniques shows good agreement; in all cases the discrepancy between the various methods is less than 0.5 ergs/cm^2.

The above discussion assumes naively that protein molecules act as small particles having a distinct surface tension. It should be noted, however, that agreement between thermodynamic model predictions and experimental results does not stipulate that the individual macromolecules have a surface tension. The fact is that, since there is a close correlation between the thermodynamic free energy of adhesion and van der Waals interactions (36,37), the results can be understood as manifestations of the latter.

Characterization of the Surface Properties of Protein-coated Polymers

In many situations it is desirable to characterize the surface properties of protein-coated polymer surfaces. This is not a trivial task and most available methods employ either exposure of the proteinated surface to an air interface or other denaturing pretreatment (23,24,38,39). For example, as discussed earlier in this article, contact angle measurements performed in air on layers of proteins adsorbed onto polymer surfaces show that the adsorbed protein is hydrophobic (23,24). It is believed that this observation is an artifact due to the exposure to air during the measurement and does not reflect the in situ condition. (These comments do not apply to contact angle measurements on thick layers of hydrated proteins which indicate that the proteins in their native state are hydrophilic.) Recently, however, a novel strategy, called the sedimentation volume method, has been developed which permits the determination of the surface tension of protein–coated polymer particles in situ under non–denaturing conditions (40).

In the sedimentation volume technique a fixed mass of polymer particles (less than 100 μm in size) is exposed to a known volume of the protein solution for a certain length of time. Thereafter the particles are rinsed by means of a dilution/displacement method in order to remove non–adherent protein molecules. The protein–coated particles are then resuspended in liquid mixtures of various surface tensions. Such liquids are prepared by incorporating small volumes of a surface tension lowering additive such as propanol or DMSO into the buffer. The pH of the buffer remains fixed at pH 7.2. The surface tension of the solution is measured immediately before use. The particles are then quantitatively transferred to graduated Wintrobe[R] tubes, and allowed to settle for sixteen hours under the influence of gravity. (After this time no further changes in the sedimentation volume is observed.) Thereafter the sedimentation height in each of the tubes is determined. A photo–micrograph of one such experiment for Nylon–6,6 (N–6,6) particles coated with HSA is given in Figure 8. The sedimentation volume of the particles in each tube is then plotted as a function of the surface tension, γ_{LV}, of the suspending liquid medium in each tube as shown in Figure 9. It is apparent from these figures that the final sedimentation volume, V_{sed}, changes as a function of the composition and hence the surface tension, γ_{LV}, of the liquid mixtures such that an extremum (here: a maximum) occurs at a certain γ_{LV} value. From previous work with pure polymer systems (41) it is known that this maximum occurs where the surface tension of the suspending liquid, γ_{LV}, equals the surface tension, γ_{PV}, of the particles. Since the surface tension, γ_{LV}, of the liquids is readily measured the sedimentation volume method provides a direct means for determining the surface tension of the protein–coated particles.

Shown in Figure 10 are the experimental data for three polymer systems coated with HSA at different bulk concentrations. It is clear that each bulk concentration gives rise to a unique V_{sed} extremum. For example in the case of HSA used at bulk concentrations of 25 mg/ml and 5 mg/ml to coat PTFE particles, V_{sed} maxima occur at γ_{LV} values of 69.7 and 57.0 ergs/cm^2, respectively. These results imply that the surface tension of PTFE coated with these concentrations of HSA is 69.7 and 57.0 ergs/cm^2, respectively. Thus the V_{sed} technique is able to discriminate between the surface properties of one and the same polymer coated with different bulk concentrations of the same protein.

Shown in Figures 11 and 12 are the data for the same three polymers coated with different bulk concentrations of IgG and

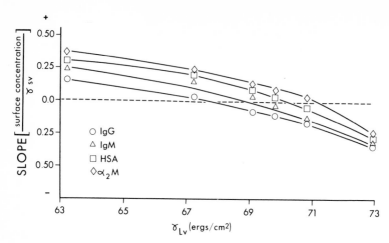

Figure 7: Slopes of the straight lines of the type indicated in Figure 6 _versus_ γ_{LV}. The slope is zero for $\gamma_{LV} = \gamma_{PV}$. The points are computer curve fitted to a second order polynomial and the intercepts are taken from that curve fit. (Reproduced with permission Ref. 18. Copyright 1981, Elsevier/North Holland Press).

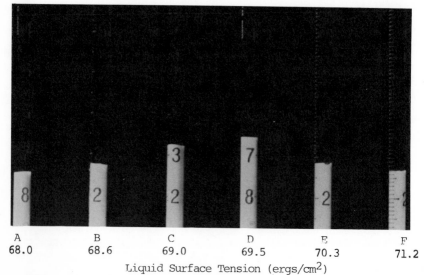

Figure 8: Sedimentation of Nylon-6,6 particles coated with a 25 mg/ml bulk concentration of Human Serum Albumin. The particles in each tube are suspended in binary mixtures of Hanks Balanced Salt Solution (HBSS) containing different amounts of dimethylsulfoxide, pH 7.2. (Reproduced with permission from Ref. 40. Copyright 1986, Academic Press Inc.)

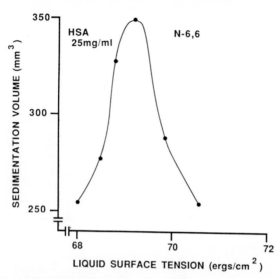

Figure 9: Sedimentation volume of Nylon–6,6 particles given in Figure 8 plotted as function of the surface tension, γ_{LV}, of the suspending liquid in each tube. (Reproduced with permission from Ref. 40. Copyright 1986, Academic Press Inc.)

fibrinogen. Each of the polymers coated with the same bulk concentration of HSA, IgG or fibrinogen give rise to V_{sed} extrema which are different for each protein. There are several points to be made about the data contained in Figures 10–12.

The first is that protein adsorption, even out of dilute solutions, changes the apparent surface tension of the uncoated polymer material very markedly. As an example, a bulk concentration of 1 mg/ml of HSA, IgG or fibrinogen gives rise to a surface tension of 53.0, 58.8 and 59.0 ergs/cm², respectively, on PTFE. These values represent an increase in surface tension of more than 30 ergs/cm² over the value of uncoated PTFE particles (18).

Next, it is clear that the position at which the V_{sed} maxima occur changes with varying bulk protein concentrations. The extent of these changes is related to the nature of both the substrate material and to the adsorbed protein itself. For example as shown in Figure 10 for HSA, it is noted that for lower bulk concentrations the difference is more pronounced on the more hydrophobic PTFE material than on the other substrates. By way of illustration of this point it is seen that for an HSA bulk concentration of 10 mg/ml that Nylon–6,6, which has a surface tension of 44.1 ergs/cm², gives rise to a sedimentation maximum occurring at 68.8 ergs/cm², whereas polyvinyl chloride (32.1) and PTFE (19.8) gives rise to V_{sed} maxima occurring at 63.3 and 61.5 ergs/cm², respectively. Similar trends are also observed for both IgG and fibrinogen.

The influence of the substrate surface tension is further revealed through a consideration of the range in V_{sed} maxima for the various bulk concentrations of each protein adsorbed onto the different surfaces. It is clear from the data contained in Figures

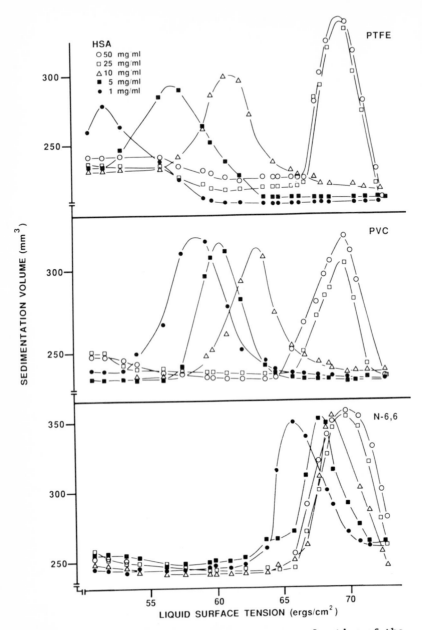

Figure 10: Sedimentation volume, V_{sed}, as a function of the surface tension, γ_{LV}, of the suspending liquid (propanol/HBSS) for albumin-coated polymer particles. Concentration of the coating protein solution is indicated. Polymers: Poly-tetrafluoethylene (PTFE); Polyvinylchloride (PVC); Nylon-6,6 (N-6,6). (Reproduced with permission from Ref. 40. Copyright 1986, Academic Press Inc.)

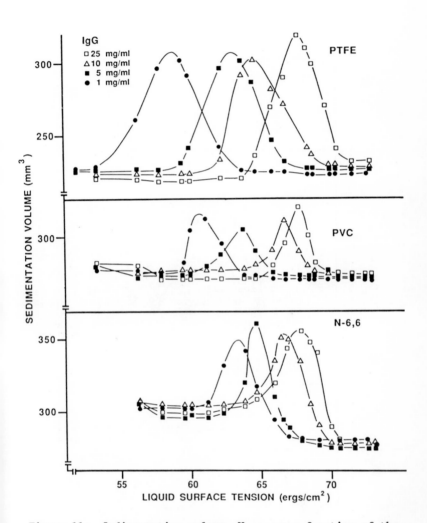

Figure 11: Sedimentation volume, V_{sed}, as a function of the surface tension, γ_{LV}, of the suspending liquid (propanol/HBSS) for IgG-coated polymer particles. Concentration of the coating protein solution is indicated. For identification of the polymers see Figure 10. (Reproduced with permission from Ref. 40. Copyright 1986, Academic Press Inc.)

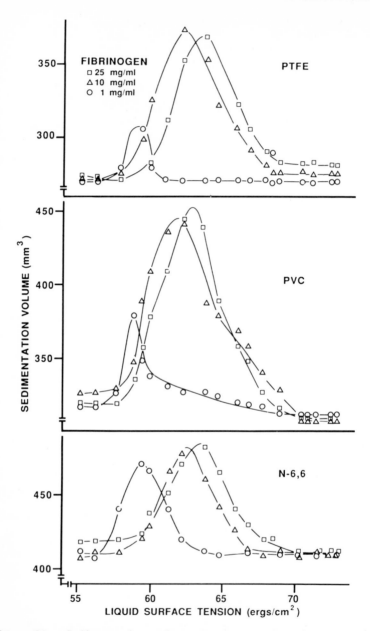

Figure 12: Sedimentation volume, V_{sed}, as a function of the
surface tension, γ_{LV}, of the suspending liquid
(propanol/HBSS) for fibrinogen-coated polymer particles.
Concentration of the coating protein solution is indicated.
For identification of the polymers see Figure 10. (Reproduced
with permission from Ref. 40. Copyright 1986, Academic Press
Inc.)

10-12 that the "spread" in the V_{sed} maxima values increases with decreasing substrate surface tension. For example, in the case of HSA over the bulk concentration range of 25 to 1 mg/ml the range in V_{sed} maxima is approximately 4 ergs/cm^2 for N-6,6, 11 ergs/cm^2 for PVC and 17 ergs/cm^2 for PTFE.

The range in V_{sed} maxima also appears to be due, in part, to the nature of the adsorbed protein. As shown in Figure 13 at bulk concentrations equal to or greater than 25 mg/ml it is noted that HSA gives rise to the highest surface tension (69.7 ergs/cm^2) followed by IgG (67.8) and then fibrinogen (63.5). These values are independent of the surface properties of the underlying substrate material. This order of hydrophobicity, however, is not consistently maintained as the bulk concentration is decreased. HSA is the protein which gives rise to the largest range in V_{sed} maxima followed by IgG and fibrinogen. For example PTFE coated with a HSA bulk concentration of 25 mg/ml yields a V_{sed} maximum at approximately 70 ergs/cm^2 whereas a 1 mg/ml bulk concentration yields a V_{sed} maximum at 53 erg/cm^2; i.e. a change in surface tension of approximately 17 erg/cm^2 over this range of bulk concentrations. For the same bulk protein concentrations of IgG, the corresponding PTFE values are 68 and 58 ergs/cm^2 respectively: a decrease of 10 ergs/cm^2. For the most hydrophobic protein studied, fibrinogen, at the same bulk concentrations the observed PTFE values are approximately 63 and 59 ergs/cm^2, respectively. This corresponds to a decrease of only 4 ergs/cm^2. One possible explanation for these observations is that HSA undergoes more extensive substrate induced conformational changes than the more rigid IgG and fibrinogen molecules.

In contrast to the pattern at low bulk protein concentrations, there is a unique limiting maximum at high bulk concentrations for each of the proteins, even though the underlying substrate surface tensions are markedly different. This is illustrated in Figure 13 in which it is shown that the polymer particles when coated with protein at a concentration of 25 mg/ml or larger gives rise to a V_{sed} maximum which is characteristic of each protein and is independent of the hydrophobicity of the underlying substrate material. These limiting values agree well with the surface tension values of the proteins determined from saline contact angle measurements on relatively thick hydrated protein layers (18,35), hydrophobic interaction chromatography (42) and adsorptivity measurements (18). These values are compared in Table 1.

There is one more striking observation in connection with the sedimentation volume of fibrinogen coated particles. As illustrated in Figure 13 for all bulk concentrations examined, fibrinogen, when adsorbed onto any of the three polymers, results in markedly larger sedimentation volumes than either HSA- or IgG-coated particles. This is particularly noticeable at high bulk concentrations and on the high energy surface, Nylon-6,6. The reason for these large values in V_{sed} is not known at this time. However, it may be related to steric considerations since even in phosphate buffer without any additive, the V_{sed} values for fibrinogen-coated particles are about 50% larger than for the same polymers coated with either HSA or IgG at the same bulk concentration. In this regard it should be noted that fibrinogen molecules are the largest among the three proteins studied having a

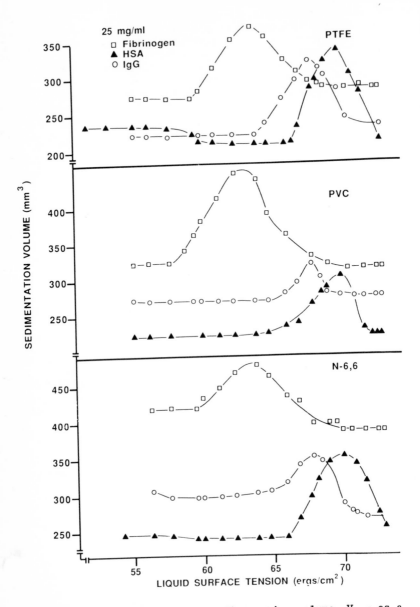

Figure 13: Comparison of the sedimentation volume, V_{sed} as a function of the surface tension, γ_{LV}, of the suspending liquid (propanol/HBSS) for different polymers coated with three serum proteins at a bulk concentration of 25 mg/ml. For identification of the polymers see Figure 10. (Reproduced with permission from Ref. 40. Copyright 1986, Academic Press Inc.)

molecular weight 340,000 and an axial ratio of 1:7. The large
V_{sed} values for fibrinogen coated particles cannot, however, be
entirely ascribed to protein molecular dimensions alone since HSA
(67,000; 1:3) and IgG (169,000; 1:5) in virtually all cases give
rise to similar V_{sed} values.

The data discussed here do not exclude the possibility that the
adsorbed layer might be comprised of more than a single layer of
protein molecules. Evidence for such multilayer formation has been
given elsewhere for both fibrinogen and HSA (43,44).

Conclusions

In their native state most serum proteins are quite hydrophilic.
Characterization of the surface properties of the native proteins
is best achieved by time dependent contact angle measurements on
relatively thick protein layers or by the adsorption method. With
high bulk concentrations of the coating solution the V_{sed} method
may also be employed.

The relative extent of protein adsorption onto polymer surfaces
is influenced by the surface tensions of the substrate material, of
the suspending liquid and of the proteins themselves. For one and
the same substrate material the extent of protein adsorption
depends on the relative hydrophobicity of the proteins. For the
situation where $\gamma_{LV} > \gamma_{PV}$ the more hydrophobic proteins
will adsorb to the largest extent.

Protein adsorption changes the surface properties of the
"naked" substrate markedly. These changes may be characterized by
means of the sedimentation volume method. Most polymers are
rendered significantly more hydrophilic as a result of protein
adsorption. The extent of the change of substrate properties is
dependent on the nature of the adsorbing protein, the bulk
concentration of the coating protein solution and the substrate
surface properties. At low bulk concentrations substrate surface
properties are important in that they influence the subsequent
surface tension of the protein-coated polymer particles possibly
through substrate-induced conformational changes in the adsorbed
molecules. At high bulk concentrations the surface tension of the
protein coated polymer is independent of the underlying substrate
surface properties. Under these conditions of high bulk concentration each protein gives rise to a unique surface tension which is
characteristic of the coating protein.

Acknowledgments

Supported by research grants from the Medical Research Council of
Canada (No. MT-5462, MT-8024, MA-9114), the Natural Sciences and
Engineering Research Council of Canada (No. A-8278, UO-493) and the
Heart and Stroke Foundation of Ontario (No. 4-12, AN-402). One of
us (D.R.A.) acknowledges receipt of a Senior Research Fellowship
from the latter agency.

Literature Cited

 1. Gendreau, R.M.; Winters, S.; Leininger, R.I.; Fink, D.; Hassler
    C.R.; Jakobsen, R.J. Appld. Spectroscopy 1981, 35, 353-7.

2. Vroman, L.; Adams, A.L.; Klings, M.; Fisher, G.C.; Munoz P.C.; Solensky, R.P. Annals N.Y. Acad. Sci. 1977, 283, 65-76.
3. R.A. Wagenen, S. Rockhold and J.D. Andrade. Amer. Chem. Soc. Adv. Chem. Ser. 1982, 199, 351-70.
4. Zingg, W.; Hum, O.S.; Absolom, D.R.; Neumann, A.W. Thromb. Res. 1981, 23, 247-53.
5. Neumann, A.W.; Moscarello, M.A.; Zingg, W.; Hum O.S.; Chang, S.K. J. Polymer Sci. 1979, 66, 391.
6. Steinberg, J.; Zingg, W.; Neumann A.W.; Absolom, D.R. . Proc. 12th Ann. Mtg. Amer. Soc. Biomaterials. Abstract #8, Minneaplois-St. Paul, MN., June, 1986.
7. Mason, R.G.; Shermer R.W.; Zucker, W.H. . Am. J. Pathol. 1973, 73, 183-200.
8. Stoner, G.E. Biomed. Med. Dev. Artif. Organs, 1973, 1, 155-62
9. Kim, S.W.; Wisniewski, S.M.; Lee E.S.; Winn M.L. . J. Biomed. Mater. Res. 1977, 8, 23-31.
10. Ward, C.A.; Stanga, D. J. Colloid Interface Sci. (In Press).
11. Packham, M.A.; Evans, G.; Glynn M.F.; Mustard.J.F. J. Lab. Clin. Med. 1969, 73, 686-97.
12. Absolom, D.R.; Hawthorn L.A.; Chang, G.E. J. Biomed. Mater. Res. (Submitted).
13. van Wachem, P.B.; Beugeling, T.; Feijen, J.; Bantjes, A.; Demeters J.P.; van Aken, W.G. Biomaterials 1983, 6, 403-8.
14. Wicher, S.J.; Brash, J.L. J. Biomed. Mater. Res. 1978, 12, 181-207.
15. Brash, J.L.; Lyman, D.J. J. Biomed. Mater. Res. 1969 3, 175-89.
16. Baszkin, A.; Lyman D.J. J. Biomed. Mater. Res. 1980, 14, 393-403.
17. Chuang, H.Y.; Kung W.F.; Mason, R.G. J. Lab. Clin. Med., 1978, 92, 483-95.
18. van OSS, C.J.; Absolom, D.R.; Neumann, A.W.; Zingg, W. Biochim. Biophys. Acta 1981, 670, 64-73.
19. Albertsson, P.A. Adv. Prot. Chem. 1979, 24, 309-41.
20. Schultz, H.E.; Heremans, J.F. Molecular Biology of Human Proteins Vol. 1, Elsevier/North Holland, 1966.
21. Keshavarz, E.; Nakai, S. Biochim. Biophys. Acta 1979, 576, 269-79.
22. Kato, A.; Nakai, S. Biochim. Biophys. Acta 1980, 624, 13-20.
23. Lee, R.G.; Adamson, C.; Kim S.W.; Lyman, D.J. Thromb. Res. 1973, 3, 87-90.
24. Absolom, D.R.; van OSS, C.J.; Neumann, A.W.; Zingg, W. Biochim. Biophys. Acta 1981, 670, 74-8.
25. Absolom, D.R.; Zingg W.; Neumann, A.W. J. Colloid Interface Sci. 1986, 112, 599-601.
26. Neumann, A.W.; Good, R.J.; Hope C.J.; Sejpal, M. J. Colloid Interface Sci. 1974, 49, 286.
27. Smith, R.P.; Absolom, D.R.; Spelt J.K.; Neumann, A.W. J. Colloid Interface Sci. 1986, 110, 521-31.
28. Neumann, A.W.; Absolom, D.R.; Zingg W.; van OSS, C.J. Cell Biophysics 1979, 1, 79-92.
29. Absolom, D.R.; van OSS, C.J.; Genco R.J.; Neumann, A.W. Cell Biophysics 1980, 2, 113-26.
30. Absolom, D.R.; Zingg, W.; van OSS C.J.; Neumann, A.W. J. Colloid Interface Sci. 1985, 104, 51-9.

31. Absolom, D.R.; Thomson, C.; Kruzyk, G.; Zingg W.; Neumann, A.W. Colloids and Surfaces (In Press).
32. Absolom, D.R.; Lamberti, F.V.; Policova, Z.; Zingg, W.; van OSS, C.J.; Neumann, A.W. J. Appld. Environ. Microbiol. 1983, 46, 90–97.
33. Absolom, D.R.; Snyder, E.L. J. Dispersion. Sci. Technol. 1985, 6, 37–54.
34. Absolom, D.R. Methods in Enzymology 132, G. Di Sabato and R. Evers, eds., Academic Press, New York, 1987 (In Press).
35. van OSS, C.J.; Gilman, C.F.; Neumann, A.W. Phagocytic Engulfment and Cell Adhesiveness as Cellular Surface Phenomena. Marcel Dekker, New York, 1978.
36. Neumann, A.W., Omenyi S.N., van OSS, C.J. Colloid and Polymer Sci. 1979, 257, 413.
37. Neumann, A.W.; Omenyi S.N.; van OSS, C.J. J. Phys. Chem. 1982, 86, 1267.
38. Cuypers, P.A.; Heremans W.T.; Hemeker, H.C. Annals N.Y. Acad. Sci. 1977, 283, 77–87.
39. Morrissey, B.W. Annals N.Y. Acad. Sci. 1977, 283, 50–64.
40. Absolom, D.R.; Policova, Z.; Bruck, T.; Thomson, C.; Zingg W.; Neumann, A.W. J. Colloid Interface Sci. (Accepted for publication).
41. Vargha-Butler, E.I.; Zubovitz, T.; Hamza H.A.; Neumann, A.W. J. Dispersion Sci. Technol., 1985, 6, 357–79.
42. van OSS, C.J.; Absolom D.R.; Neumann, A.W. Separ. Sci. Technol. 1979, 14, 305–17.
43. De Baillou, N.; Dejardin, P.; Schmitt A.; Brash, J.L. J. Colloid Interface Sci. 1984, 100, 167–74.
44. Penners, G.; Priel Z.; Silberberg, A. J. Colloid Interface Sci. 1981, 80, 437–44.

RECEIVED January 13, 1987

Chapter 26

Electron Tunneling Used as a Probe of Protein Adsorption at Interfaces

J. A. Panitz

Department of Cell Biology, University of New Mexico School of Medicine, Albuquerque, NM 87131

The ability of electron tunneling to detect protein adsorption on a metal surface has been investigated. Tunneling at a vacuum-metal interface is discussed. Field-electron emission tunneling experiments are reviewed; they suggest a fundamental limit on the ability of an electron tunneling microscope to probe protein adsorption at an interface, or to image the structure of a biological macromolecule.

Tunneling is a ubiquitous phenomenon. It is observed in biological systems ($\underline{1}$), and in electrochemical cells ($\underline{2}$). Alpha particle disintegration ($\underline{3}$), the Stark effect ($\underline{4}$), superconductivity in thin films ($\underline{5}$), field-electron emission ($\underline{6}$), and field-ionization ($\underline{7}$) are tunneling phenomena. Even the disappearance of a *black hole* (or the fate of a multi-dimensional universe) may depend on tunneling, but on a cosmological scale ($\underline{8}$-$\underline{9}$).

Classical physics dictates that a particle constrained by an energy barrier can become free only if it acquires an energy greater than the height of the barrier. In quantum mechanics, this restriction is eased. For example, quantum mechanics allows an electron to escape from the interior of a metal by *tunneling* through the potential barrier that confines it. The height of this barrier is called the work function of the metal (Φ). The work function is a property of a metal surface which can be locally modified by the presence of an adsorbate. For a clean metal surface, Φ = 1-6 eV

When a potential difference is applied to two metal electrodes in high vacuum, two types of tunneling can be observed: *metal-vacuum-metal* tunneling when the electrodes are separated by 1-2nm, or *field-electron emission* tunneling when the gap is much larger. In practice, it is difficult to measure a tunneling current in a vacuum gap when the gap is very small because the electrode spacing must be maintained without electrode contact. For this reason, the first successful tunneling experiments between closely spaced electrodes used a thin, insulating layer to define the electrode gap, and fix the electrode separation ($\underline{10}$).

0097–6156/87/0343–0422$06.00/0

Electron Tunneling Phenomena

If all sources of conduction current in a vacuum gap are eliminated (for example, the current that would flow through an an asperity that might span the gap), a tunneling current can be observed (11-12). At small electrode separations, the tunneling current depends exponentially on both the separation of the electrodes and the work function of the cathode surface, and linearly on the voltage applied between them. A simplified, one-dimensional picture of the tunneling barrier is shown schematically in Figure 1.

If an adsorbate is placed in the tunneling gap, the tunneling current will be modified by the local change in work function that the adsorbate produces. To observe a tunneling current, electrons must tunnel from one electrode (the cathode) into the adsorbate, and then conduct through the adsorbate to the other electrode (the anode). Alternately, electrons could tunnel completely through the adsorbate, but this process becomes more improbable as the thickness of the adsorbate increases. As the adsorbate thickness increases, the electrode gap that contains it must also increase. If the adsorbate is a protein molecule, the gap must be increased to tens or hundreds of nanometers. At these distances, a tunneling current could normally not be measured.

The dimensional stability of the tunneling gap is of primary importance when proteins are placed in the gap. If the electrode separation increases, the proteins may not completely fill the gap; if the separation decreases, the proteins may be deformed or destroyed. Field-electron emission provides an alternative way to probe the tunneling properties of proteins without the difficulties imposed by small, random changes in the separation of the tunneling electrodes. Unlike metal-vacuum-metal tunneling, *field-electron emission tunneling does not explicitly depend on the separation of the electrodes in a tunneling apparatus.* As a result, large protein molecules can be placed on the cathode, and a tunneling current measured, independent of the anode position.

Field-electron emission tunneling depends exponentially on the work function of the cathode, and exponentially on the electric field strength at its surface (13). At a field strength of a few volts per nanometer, the width of the tunneling barrier will be reduced and electrons will tunnel with high probability from the cathode surface. If protein molecules are placed on the surface, the local tunneling probability will reflect their presence. Field-emitted electrons emerge as free particles in the vacuum gap, and accelerate to the anode through the potential difference that is applied across the electrodes. To avoid electrical breakdown, the electrode separation must be large, and the potential difference must be small. An easy way to generate the required field strength under these conditions is to enhance the electric field at the cathode surface by using a highly curved, needle-like cathode known as a field-emitter *tip* (See Figure 2).

A tip with the required shape and size can be prepared from fine wire by standard electropolishing techniques (14). The highly curved apex of the tip can be made smooth on an atomic scale by annealing the tip in high vacuum close to its melting point (15). Tips with an apex radius of curvature of 10-1000nm can be easily fabricated by these techniques. It is important to realize that on the scale of a single protein molecule, the highly curved apex of a large radius field-emitter tip looks like a flat surface of infinite extent.

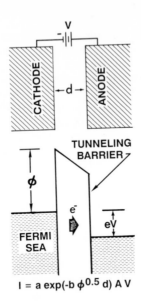

$$I = a \exp(-b \phi^{0.5} d)\ A\ V$$

Figure 1. Metal-vacuum-metal tunneling.

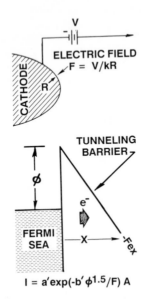

$$I = a'\exp(-b' \phi^{1.5}/F)\ A$$

Figure 2. Field-electron emission tunneling.

Protein deposition on field-emitter tips

Reproducible deposition of protein molecules on the apex of a field-emitter tip can present formidable problems. Unlike a small organic adsorbate that can be sublimed directly onto the apex of a field-emitter tip in high vacuum, a large protein molecule must be deposited onto the tip apex from an aqueous environment, and then dried without introducing artifacts. Surface tension forces during the drying process can rearrange or distort the structure of the protein molecules adsorbed on the tip apex; proteins can even be removed from the apex as an air-liquid interface is traversed. Fortunately, the deposition problem has been solved. A surprisingly simple protocol can be used to deposit protein molecules (and most other species of biological interest) onto the apex of a field-emitter tip in a reproducible fashion (16-17).

The success of a particular deposition procedure can be determined from a series of control experiments in which the coverage of a biological species on the tip surface is determined by imaging the tip profile in the transmission electron microscope (TEM). Isolated species on the tip surface can be visualized if they are stained with uranyl acetate, or coated with a thin layer of tungsten prior to imaging (18). For example, figure 3 shows tobacco mosaic virus particles deposited from aqueous solution onto a large radius, tungsten field-emitter tip. The enzyme-cleaved virus particles were rotary shadowed with tungsten prior to imaging the tip apex in the TEM (19).

Field-electron emission Microscopy

A field-emitter tip has a unique advantage when used as the cathode in a tunneling apparatus: the electron tunneling probability at the tip apex can be directly visualized in the *field-electron emission microscope* (20). Electrons that tunnel from the apex of a field-emitter tip emerge as free particles in vacuum, and are accelerated along electric field lines that rapidly diverge into space. In the field-electron emission microscope (FEEM), the anode of a tunneling apparatus is coated with a suitable phosphor and placed far from the tip apex. The tunneling electrons that strike the phosphor form a highly magnified image that reflects their point of origin at the tip apex. Bright regions in the image reflect regions of increased electron tunneling; dark regions reflect a decrease in the electron tunneling probability.

The magnification of an FEEM image is determined by the radius of the tip apex and the distance between the tip and the phosphor-coated anode that displays the image. In practice, a magnification of several hundred thousand times is easily achieved (21). The resolution of an FEEM image is about 2nm; it is limited by the lateral velocity component of the tunneling electrons as they emerge from the tunneling barrier, and by the Heisenberg uncertainty principle that ultimately obscures their precise point of origin on the tip apex (22). Unlike other electron microscopes, the FEEM is a very simple device. Image quality is not affected by external vibrations, and high contrast images are stable for long periods of time. It has been noted that *in the absence of lenses, illuminating devices, and automatic controls a field-emission microscope is less of an apparatus and more of a direct aid to the eye and brain* (23).

Figure 3. Transmission electron microscopy of enzyme-cleaved, tobacco mosaic virus particles on the apex of a tungsten field-emitter tip (imaged at 200kV). TMV sample kindly supplied by P. J. Butler, the MRC, Cambridge, England.

FEEM imaging of small organic molecules

Many small organic molecules can be conveniently imaged in the FEEM because they can be directly sublimed onto the apex of a field-emitter tip without sacrificing the high vacuum environment of the microscope. The first attempt to image such molecules in the FEEM was made in 1950 (24-25). Two planar molecules were studied: copper-phthalocyanine (a four-fold symmetric molecule), and flaventhrene (a two-fold symmetric molecule). Figure 4A is an FEEM image that is characteristic of a clean, (110)-oriented tungsten field-emitter tip. The symmetry of the image reflects the symmetry of the tip apex about the axis of the wire from which it was made (a result of the electropolishing technique mentioned above). Figure 4B shows the result of subliming copper phthalocyanine molecules onto the tip apex. Figure 4C shows an FEEM image of another tungsten tip exposed to the same flux of molecules for a greater time (resulting in an increased coverage of molecules on the tip apex). Figure 4D shows the result of subliming flaventhrene onto a different tungsten tip. The bright features that appear after sublimation reflect a decrease in the local work function of the surface at each adsorption site. Although these regions seem to reflect the known symmetry of each adorbate, the correspondence may be fortuitous: three-fold symmetric adsorbates (and other non-symmetric molecules) also produce two-fold and four-fold symmetric FEEM images, and other unique shapes have also been reported (26).

The symmetry of a phthalocyanine or a flaventhrene image feature is thought to reflect a complex scattering phenomenon within the molecule. The potential well defined by the molecule may tend to open a window, or *aperture*, in the tunneling barrier at the cathode surface, increasing the tunneling current at the adsorption site (27). Tunneling electrons, elastically (and inelastically) scattered from the aperture, then reemitted into space, could produce the patterns that are observed (28). Careful experiments have demonstrated that an FEEM image can accurately reflect the adsorption of a single organic molecule on the tip apex, but will not necessarily reflect its true shape or size (29).

The size of a molecular image feature in Figure 4 is about an order of magnitude larger than the size of the molecule that produced it. The increase in local image magnification has been explained by assuming that a molecule acts like a small metallic protrusion on the tip apex (30). A small metallic protrusion will distort the trajectories of the tunneling electrons in its vicinity, causing them to diverge more rapidly into space. If a molecule contains a number of quasi-free (i.e. pi) electrons, the electric field in the vicinity of the molecule will tend to be excluded from its interior, and the molecule will act like a small metallic protrusion. Although some molecules may behave in this way (e.g. semiconducting phthalocyanines), others may behave more like an insulator than a metal. The electric field will penetrate almost completely into the interior of an insulating (dielectric) protrusion. Electron trajectories in the vicinity of the protrusion will be relatively undisturbed by its presence, and the local magnification at the site of the protrusion will not change by an appreciable amount.

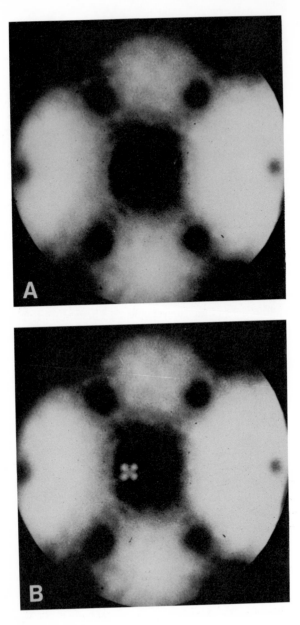

Figure 4. Field-electron emission microscopy of small, organic adsorbates. (A) 110-oriented tungsten tip. (B) With copper-phthalocyanine adsorption.

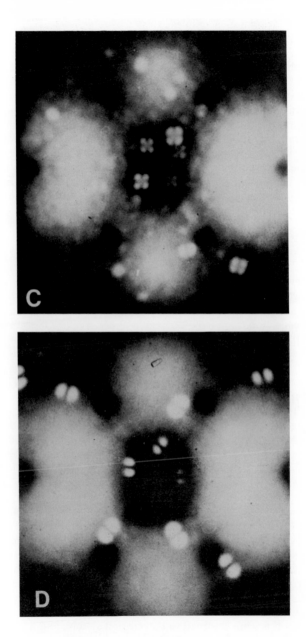

Figure 4.—*Continued.* Field-electron emission microscopy of small, organic adsorbates.
(C) With increased copper-phthalocyanine. (D) With flaventhrene adsorption.
(Courtesy of Dr. A. J. Melmed, The National Bureau of Standards, Gaithersburg, MD.)

FEEM imaging of Immune Complexes

The tunneling characteristics of protein molecules and virus particles have been studied by observing how they affect the appearance of an FEEM image. These experiments highlight the difficulty in handling biological species that must be removed from an aqueous environment for examination in an FEEM under ultra-high vacuum conditions. To insure some semblance of statistical reliability, many tips must be examined under reasonably identical conditions. With this in mind, thirty or forty tips are usually examined, divided into groups, with two tips in each group. One tip in each group is called the *active* tip; it is exposed to buffer containing the immune complex. The other tip is called the *control* tip; it is transferred with the active tip, in and out of the FEEM, during each stage of the imaging protocol. The control tip is used to assess the effect of tip contamination by adsorbed gas or impurities from laboratory ambient (31).

It is instructive to review the imaging protocol that was developed for studying the tunneling characteristics of ferritin/goat anti-rabbit IgG conjugate because this protocol illustrates the type of control that is required for examining any biological species in the FEEM:

(1) Two tips were cleaned by repeated heating in vacuum to 2100C. The heating schedule was designed to remove contaminant species from the tip apex by thermal desorption.

(2) An FEEM image of each tip was taken without breaking vacuum to record the field-electron emission pattern of the clean tip surface (Figure 5A and 5D).

(3) Both tips were transferred into laboratory ambient. The *active* tip was placed for 180s into an aqueous solution of 20mM Tris-Cl buffer containing 150mM NaCl at pH 7.6. The tip was rinsed in distilled water, transferred wet into a mixture of 90% ethanol in water for fifteen seconds, and then dried in air. The *control* tip remained in laboratory ambient during this time.

(4) Both tips were returned to the vacuum system and an FEEM image was taken of each tip after a 12 hour pumpdown (Figure 5B and Figure 5E). The bright features in Figure 5B are characteristic of exposing a tip to an aqueous solution of buffer (that does not contain protein molecules) as described above.

(5) Both tips were transferred into laboratory ambient. The *active* tip was placed in buffer containing ferritin/IgG conjugate at a concentration of about 12.5 micrograms/ml. After three minutes the *active* tip was rinsed as described in (3), above. Previous TEM images confirmed that this procedure resulted in a saturation coverage of the immune complex on the tip surface. The *control* tip remained in laboratory ambient during the deposition procedure.

(6) Both tips were returned to the vacuum system, and an FEEM image was taken of each tip. The image of the *active* tip (Figure 5C) reflects the adsorption of gas phase contaminants during tip transfer in air, the adsorption of buffer molecules, and the adsorption of ferritin/IgG complexes from solution. The total tunneling current from a tip exposed to the protein complex is greatly reduced (or eliminated) when compared to the tunneling current from a clean tip, or a control tip (Figure 5F).

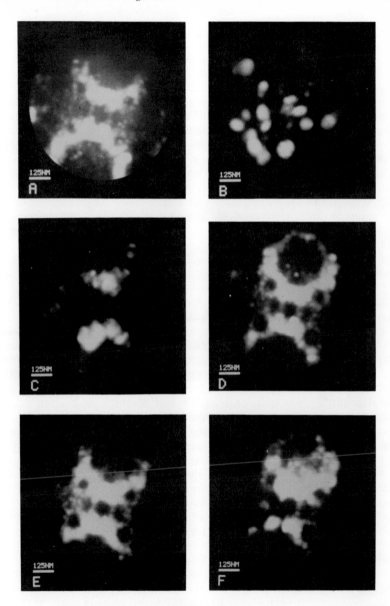

Figure 5. Field-electron emission microscopy of protein adsorbates. (A) Active tip prior to deposition. (B) Deposition in buffer without ferritin-IgG. (C) Deposition in buffer with ferritin-IgG. (D) Control tip prior to air exposure. (E) After exposure to laboratory ambient. (F) After subsequent exposure to laboratory ambient. (Reproduced with permission from Ref. 31. Copyright 1984 The American Institute of Physics.)

Conclusions

As a result of the FEEM imaging experiment described above, and other (unpublished) FEEM experiments, the following general conclusions have been reached:

(1) Repeated exposure of a tungsten tip to laboratory ambient does not seem to appreciably alter its tunneling characteristics. Since the tip apex must be covered with a monolayer of gas phase contaminants (as a result of exposure to laboratory ambient), the adsorbed species must either not affect FEEM image contrast, or the adsorbates must desorb from the tip surface during the pumpdown cycle prior to imaging. The latter effect is probably responsible for the image contrast that is observed. Localized regions of increased image brightness are occasionally seen in an FEEM image after exposing a tip to laboratory ambient (indicating the presence of adsorbed contaminants), but these features are short lived and do not survive minor increases in field strength (of the order of 0.5%).

(2) If a clean tip is exposed to buffer (the type of buffer does not seem to matter), a characteristic FEEM image is recorded in vacuum. Image features consist mainly of bright, circular regions of increased contrast, often superimposed on a weak background image that looks similar to an FEEM image of a clean tip. Unlike the bright regions that are occasionally observed in a control tip image after air exposure, these bright regions are stable, even if the imaging field is increased by several percent. The increased emission has been correlated with the presence of salt in the buffer solution.

(3) The FEEM image of a tip exposed to an aqueous solution of buffer containing a protein (the exact protein appears to be unimportant) shows characteristically less emission than the FEEM image of a clean tip, or a control tip. Electron tunneling from large regions of the tip surface is suppressed, apparently by the presence of protein molecules in these regions. The reduction in the total emitting area is in qualitative agreement with the coverage of protein on the tip apex as judged by subsequent imaging in the TEM (unpublished). We interpret these observations by suggesting that a protein molecule behaves like a thick, insulating protrusion on the tip apex. Tunneling seems to occur with high probability only from the regions of the tip apex that are not covered with protein. Precise, *probe-hole* measurements of the tunneling current are needed to quantify this effect (<u>32</u>).

Implications for STM imaging

The scanning tunneling microscope (STM) is a high resolution, non-contacting, surface profilometer (<u>33</u>). Contrast is generated in an STM image by mapping the tunneling probability of electrons across a surface scanned by a field-emitter tip. A tip is used to limit and define the tunneling region of the surface. As the tip is rastered within 1nm of the surface (by piezoelectric crystals), a metal-vacuum-metal tunneling current is recorded. The tunneling current is kept constant as the tip scans the surface by allowing the tip to move vertically with respect to the surface below. A plot of raster position verses tip elevation records the surface profile in three dimensions. STM images of semiconductor surfaces show structure at the atomic level, but the appearance of an image depends on the bias voltage that is applied between the tip and the surface (<u>34-35</u>).   The STM has also been used to image unstained virus particles in laboratory ambient, but the images have not been reproduced and are unconvincing when compared to their TEM counterparts (<u>36</u>). STM images of fatty acid bilayers deposited by the Langmuir-Blodgett technique and imaged in air have also been reported (<u>37</u>).

Unstained protein molecules, and unstained virus particles (unpublished results), do not image in the FEEM. Tunneling appears to be negligible or absent at the adsorption site of these species. STM images reflect a reasonably large tunneling probability for these species; FEEM images do not, and the dichotomy is puzzling. Field-electron emission images are consistent with a picture of a protein molecule (or a virus particle) that behaves like a large, insulating species while STM images suggest that these species are at least quasi-conductors for the tunneling electrons. Perhaps the different degree of hydration of the species that result from the two imaging techniques may account for the different tunneling characteristics that have been reported (FEEM images of biological species must be produced in a high vacuum environment while STM images can be taken in laboratory ambient).   More complete studies of electron tunneling through protein molecules under a variety of deposition and imaging conditions will be needed to resolve the fundamental questions that have been raised by these two types of tunneling experiments.

Acknowledgment

The author wishes to acknowledge the Defense Advanced Research Projects Agency (Advanced Biochemical Technology Program) for supporting this research under ARPA contract 4597.

Literature Cited

1.   Frauenfelder, H.   In *Tunneling in Biological Systems*; Chance, B., Ed.; Academic Press: New York, 1979; p 627.
2.   Gurney, R. W. *Proc. R. Soc. London.* 1932, **A134**, 137.
3.   Gamow, G. *Z. Phys.* 1928, **51**, 204.
4.   Lanczos, C. Z. *Z. Phys.* 1931, **68**, 204.
5.   Giaever, I.; In *Tunneling Phenomena in Solids*; Plenum: New York, 1969; Chapter 19.
6.   Eyring, C. F.; Mackeown, S. S.; Millikan R. A. *Phys. Rev.* 1928, **31**, 900-09.
7.   Inghram, M. G.; Gomer, R. *J. Chem. Phys.* 1954, **22**, 1279-82
8.   Page, D. N. *Nature.* 1986, **321**, 111.
9.   M. J. Perry. *Nature.* 1986, **320**, 679.
10.   Giaever, I.; *J. Appl. Phys.* 1961, **32**, 172-77.
11.   Teague, E. C. *Bull. Am. Phys. Soc.* (March) 1978, **23**, 290
12.   Binnig, G.; Rohrer, H.; Gerber, Ch.; Weibel, E. *Appl. Phys. Lett.* 1982, **40**, 178-80.
13.   Fowler, R. H.; Nordheim, L. W.   *Proc. R. Soc. London.* 1928, **A119**, 173.

14.   Muller, E. W.; Tsong, T. T.   *Field-Ion Microscopy: Principles and Applications*. Elsevier: New York, 1969; p 119-27.

15.   Boling, J. L.; Dolan, W. W. *J. Appl. Phys*. 1958, **2**, 556-59.

16.   Panitz, J. A.; Andrews, C. L.; Bear D. G. *J. Elec. Micros. Techn*. 1985, **2**, 285-92.

17.   Panitz J. A. *Rev. Sci. Instrum*. 1985, **56**, 572-74.

18.   Panitz, J. A. In *Science of Biological Specimen Preparation*; Muller, M; Becker, R. P.; Boyde, A.; Wolosewick, J. J., Eds.; SEM Inc.: AMF O'Hare (Chicago), 1985; p 283.

19.   Panitz, J. A.; Bear, D. G. *J. Micros. (Oxford)*. 1985, **138**, 107-10.

20.   E. W. Muller. *Z. Physik*. 1937, **106**, 541-50.

21.   Panitz, J. A. In *Methods of Experimental Physics*; Park, R. L.; Lagally, M, Eds.; Academic: New York, 1985; Vol. 22, Chapter 7.

22.   Gomer, R.   *Field Emission and Field Ionization*;   Harvard Press: Cambridge, 1961; Chapter 2.

23.   Rochow, T. G.; Rochow, E. G. *An Introduction to Microscopy by Means of Light, Electrons, X-rays, or Ultrasound*; Plenum: New York, 1978; p 35.

24.   E. W. Muller. *Naturwissenschaften*. 1950, **14**, 333.

25.   E. W. Muller. *Life*. (June) 1950, **28**, 67.

26.   E. W. Muller. *Ergebnisse d. exakt. Naturwiss*. 1953, **27**, 290-360.

27.   Gomer, R.   *Field Emission and Field Ionization*;   Harvard Press: Cambridge, 1961; Chapter 5.

28.   Gadzuk, J. W.; Plummer, E. W. *Rev. Mod. Phys*. 1973, **45**, 487-548.

29.   Melmed, A. J.; Muller, E. W. *J. Chem. Phys*. 1958, **186**, 1037.

30.   D. J. Rose. *J. Appl. Phys*. 1956, **27**, 215-20.

31.   Panitz, J. A. *J. Appl. Phys*. 1984, **56**, 3319-23.

32.   Muller, E. W. *Z. Phys*. 1943, **120**, 261, 270.

33.   Quate, C. F. *Physics Today*. (August) 1986, **86**, 26-33.

34.   Golovchenko, J. A. *Science*. 1986, **232**, 48-53.

35.   Hamers, R. J.; Tromp, R. M.; Demuth, J. E.   *Phys. Rev. Letts*. 1986, **56**, 1972-75.

36.   Baro, A. M.; Miranda, R.; Alaman, J.; Garcia, N.; Binnig, G.; Rohrer, H.; Gerber, Ch.; Carrascosa, J. L. *Nature*. 1985, **315**, 253-54.

37.   Rabe, J.; Gerber, Ch.; Swalen, J. D.; Smith, D. P. E.; Bryant, A.; Quate, C. F. *Bull. Am. Phys. Soc*. (March) 1986, **31**, 289.

RECEIVED February 18, 1987

Chapter 27

Characterization of the Acquired Biofilms on Materials Exposed to Human Saliva

H. J. Mueller

Council on Dental Materials, American Dental Association, Chicago, IL 60611

A variety of dental alloys were submitted to adsorption experiments with human saliva. FT-IR and SIMS were used to analyze the surface films. IEF compared the protein patterns from surface extracts and salivas used in protein adsorptions to those from unexposed saliva controls. Results support both selective and nonselective adsorption processes. The SIMS spectrum showed variabilities in elemental intensities between substrates of different compositions, while IEF patterns of surface extracts from eleven different compositions of powder all appeared to contain the same acidic protein bands. FT-IR spectrum showed variabilities in the protein to carbohydrate intensity ratios at different sites on the same alloy surface, and suggested that other factors besides substrate material may be important in protein adsorption.

Surfaces coming into contact with saliva become adsorbed in a short time with a thin film of organic matter.(1-4) Much interest has been generated in characterizing this film for purposes of elucidating, (i) demineralization-mineralization processes of enamel,(5) (ii) interactions of enamel with fluoridation treatments,(6) (iii) its role as precursor to the attachment of microorganisms and the formation of plaque, caries, and periodontal disease (7), and its role in tarnishing and corrosion of dental alloys.(8) The role of adsorbed salivary proteins, especially mucins, in protecting the oral tissues against environmental insult, potential pathogens, and in lubricating has been taken as routine biological functions.(9)

Saliva - Enamel Interactions. Enamel becomes adsorbed with a bacteria-free film almost instantaneously after contacting saliva,(10) and is constantly renewable if

0097-6156/87/0343-0435$06.00/0
© 1987 American Chemical Society

lost or abraded.([11]) Langmuir's adsorption isotherm has been used with some success especially at lower protein concentrations in following the adsorption.([12,13]) Thicknesses of the order of 10-20 nm after the first 1-2 hours of saliva exposure have been detected.([1-4]) although much thicker films of the order of microns have also been demonstrated.([11]) The films in contrast to enamel are acid insoluble, although an acid soluble fraction also occurs,([14]) and act as diffusion barriers against acids, ([15]) thereby reducing the acid solubility of enamel and inhibiting the adherence of organisms. It has become customary to refer to the initial bacteria-free integument as the acquired pellicle.([14,15]). Aged pellicles contain in addition to the adsorbed proteins, microorganisms, plaque, mineralized products, and other debris.([7,11,16])

Analysis of extracted two hr in-vivo enamel pellicle showed it to be negatively charged and containing both lower and higher MW proteins. Glycine, glutamic acid, and serine were in abundance and with an amino acid content similar to a reported salivary phosphoprotein.([17]) In addition, the pellicles contain carbohydrate, which includes up to 70 % glucose([4]) and lipid.([2]) New(1-2 hr) pellicles contain 30% of proline-rich proteins. Their degradation begins after about 24 hrs. The proline-rich protein content in aged pellicles is less than 0.1 %.([18]) Anionic disc gel electrophoresis of extracted two hr pellicles indicated four major bands and with three bands indicating multiple sub-bands.([19])

The adsorption of salivary proteins to enamel may include exchange reactions in which the protein phosphate groups replace surface phosphate in the enamel hydroxyapatite.([20,21])

Salivary Binding Proteins. Salivary proteins that bind to hydroxyapatite and which may be important in pellicle formation include,([22]) (i) mucous glycoproteins (MW=3-5 x 10^5, pI=2, 70% carbohydrate), (ii) proline-rich basic glycoprotein (MW=3.5 x 10^5, pI=9.5, 40% CHO), (iii) proline-rich acidic protein (MW=6-12 x 10^3, pI=4-4.7), (iv) tyrosine-rich protein known as statherin (MW=5.2 x 10^3, pI=4.2), v) histidine-rich protein (MW=4.5 x 10^3, pI=7), and (vi) calcium glycoprotein of mixed saliva (MW= 6.2 x 10^4, pI=4.7, 15% CHO). Proteins in part iii, iv, and vi also bind calcium. A calcium precipitable glycoprotein of submaxillary saliva (MW=12 x 10^3, 5% CHO), while not binding to hydroxyapatite does bind calcium and is phosphorylated. Proteins in parts iii, and v are also phosphorylated while the proteins in parts i, and ii are sulfated.

Saliva - Dental Materials Interactions. Besides enamel and other tissues, surfaces from metallic, polymeric, and ceramic dental materials are capable of becomimg adsorbed with organic films. Germanium and silica infrared spectrometer prisms formed oral films at high speeds and were

stable over time. Detection of protein, carbohydrate, and
lipid was made.(2) The amino acid content of films formed
on several plastics and glass varied and was different
from that formed on enamel. It was concluded that the
chemical compositoion of the substrate surface has an
important influence on the type of proteins which they
retain.(23) For pellicles on dentures, specific proteins
seemed to be precursors in forming the films.(10) For the
pellicles from different restorative materials, C, N, and
O predominated the compositions. The thickness of the film
formed on gold alloy was about 10 x, 4 x, and 1.25 x the
thicknesses that formed on amalgam, enamel, and composite
resin, respectively. No film occurred on silicate cement.
The release of F^- may have competed for binding sites on
the anionic adsorbing proteins. Copper was found in film
on gold alloy, while tin was found in film on amalgam.
The cation may be a factor in the formation and adhesion
of the films, and the attachment of microorganisms.(3)
Besides C, N, and O, the in-vivo tarnished films on gold
alloys contain Cl, S, Ca, Ni, Mg, Si, Sn, Fe, K, Na, Al
and P. Copper and Zn were the only alloying elements in
the films.(8)

Objective

Salivary proteins-dental materials interactions have not
been fully addressed. The few in number of reports have
been inconclusive regarding important issues pertaining
to adsorption. Further results are required to elucidate
in greater detail the adsorption of salivary components
to dental materials surfaces. Whether the adsorption
processes on all surfaces are specific to a few proteins
or whether the adsorbed proteins depend upon the underlying substrate composition is very much of interest.
It was the goal of this project to investigate the
effect of dental material composition upon the adsorbed
film characteristics. What elements become adsorbed and
what compounds are formed or adsorbed were investigated.
Included was a comparison of the adsorbed proteins to
those occurring in saliva. Since the films are usually
only nm in thickness, the appropriate analytical methods
were required for their analysis. It was the purpose of
this poroject to utilize both fourier transform infrared
spectrometry(FT-IR) and secondary ion mass spectrometry
(SIMS), two highly surface orientated techniques, as well
as isoelectric focusing. FT-IR is capable of detecting
organic structures, while SIMS is capable of detecting
most of the elements within several of the outermost
monolayers. Isoelectric focusing is an electrophoretic
technique made in pH gradient. The resulting bands along
the pH gradient correspond to the isoelectric points of
included proteins.

Materials and Methods

Saliva and Alloy Surface Preparations.   Unstimulated

whole saliva was collected into ice chilled polyproplyene
beakers from one donor at the beginning of each of the
different experiments. The samples were centrifuged in
polyproplyene tubes at 1600 x g for 30 minutes. The supernatants were decanted and used in adsorption tests with
various alloy surfaces or powders to be described.

Alloy surfaces for FT-IR and SIMS were initially
ground to a no. 600 grit finish on silicon carbide
strips, followed by 10 um and 5 um grinding on Struer's
rotating SiC discs. The surfaces were polished with 3 and
1 um diamond pastes on nap cloths. Samples were immersed
ultrasonically in deteregent, deionized water rinsed, and
degreased by immersions in acetone followed by CCl_4.

Fourier Transform Infrared Spectrometry (FT-IR) Four
crown and bridge alloys, alloys A, B. C, and E in Table I,

Table I. Alloy Surfaces Analyzed by FT-IR or SIMS

Alloy	Manufacturer	Composition(wt %)				
		Au	Pd	Ag	Cu	other
A Szabo	Heraeus	77.4	2	12.5	8.1	
B Midas	Jelenko	46.0	6	39.5	7.5	
C Albacast	Jelenko	-	25	70.0	5.0	
D Tytin	S.S. White	-	-	34.5	7.5	16Sn, 42Hg
E MS	Monarch	-	-	-	72.0	20Al Fe,Ni,Mn
F Biobond	Dentsply	76Ni, 12Cr, 3Mo, Sn,Nb,Si,B bal				

were cast into square shapes 8 x 8 mm and 1.5 mm thick.
The polished samples were exposed for a few days to the
oral environment. The samples with holes drilled length-
wise were attached via orthodontic elastic thread to
plastic brackets cemented to the buccal surfaces of upper
molar or bicuspid teeth.(24) After removal, the samples
were rinsed with distilled H_2O and air dried. The samples
were analyzed in the reflective mode on a Digilab FT-IR.

In another series, gold alloy A(Table I) was exposed
in-vitro for 1 hr to the supernatant of saliva. After
distilled water rinsing and air drying, the surface was
viewed under the FT-IR microscope and surface scrapings
made with a scapel tip at different locations across the
surface. Organic material removed at the different sites
was obtained as thin films on KBr discs. An Analect FT-
IR was used to obtain the spectrum

Secondary Ion Mass Spectrometry. Polished alloys, A, C,
D, E, and F in Table I were exposed to the supernatant of
human saliva, water rinsed, air dryed, and analyzed by
SIMS. A Cambridge Stereoscan scanning electron microscope
with an attached argon beam gun and quadrapole mass
analyzer (Kratos SIMS unit) was used. Both the positive
and negative spectrum were taken. After SIMS analysis,
the samples were reground and repolished and analyzed
again to obtain SIMS spectrum of the as-polished surfaces.
Adsorption Methods for Isoelectric Focusing. Five ml

of the supernatant were mixed with 50 mg of different
dental materials powders contained in polystyrene tubes.
Table II characterizes the powders. Initially the tube

Table II. Alloy Powders for Protein Adsorption with IEF

	Powder	Source	Particle Size
1	Human Enamel	ground tooth	-200 mesh
2	Hydroxyapatite	Sigma	-200 mesh
3	Porcelain	Dentsply	-200 mesh
4	PMMA resin	General Dental	-150 mesh
5	Palladium	Alfa	0.25-0.55 um
6	Silver	Goldsmith	-325 mesh
7	Ag-Cu eutectic	Consolidated Astro	-325 mesh
8	Copper	Sargent-Welch	-150 mesh
9	Tin	Fisher	-325 mesh
10	Amalgam Alloy	Engelhard	-325 mesh
11	Bismuth	Goldsmith	-325 mesh

contents were vigorously shaken with tube mixer followed
by a 4 hr incubation with a moderate linear back and forth
motion. The powders were separated from the saliva by
centrifuging at 1600 x g for 15 minutes. The powders were
washed with 5 ml of distilled H_2O by vigorously shaking
for 5 min with tube mixer and collected by centrifuging.
Two extractions followed, the first with 0.2 M NaH_2PO_4
and the second with 0.2 M $EDTA(Na_4)$. Five ml of the
extraction solution were added to each tube, vigorously
shaken for several minutes and followed by an incubation
for 4 hr. The extracts were collected by centrifuging,
dialyzed for 24 hrs against distilled H_2O with 1000
MWCO membranes. Concentration follwed by dehydration
to near dryness in a 35% solution of poly ethylene glycol.
The solutioins were reconstituted by added 0.15 ml of
distilled water.

Polyacrylamide and Agarose Isoelectric Focusing   IEF
was performed on a LKB Multiphor 2117 electrophoresis unit
with either polyacrylamide or agarase gel plates which
contained ampholine carrier ampholytes for generating the
pH gradient 3.5 to 9.5. With the 1 mm thick polyacrylamide
precast gel plates, 1 M H_3PO_4 and 1M NaOH were added to
the anode and cathode, respectively, while with the 0.5 mm
thick cast agarose plates a 0.5 M acetic acid and 0.5 M
NaOH were used. Cooling of the gels was by 5 deg C water
flowing in the cooling plate. In focusing a plate having
dimensions of 110 x 245 mm, a constant power of 20 W was
applied for 1 1/2 hrs with the acrylamide gels and for 1/2
hr with the agarose gels. The pH gradient across the gel
widths was measured with an LKB surface combination pH
electrode. The gels were fixed in a 11.5% trichloroacetic
acid - 3.5% sulphosalicylic acid solution and rinsed in
95% ethanol to remove ampholine. The polyacrylamide gels
were stained with a solution containing 1.15% coomassie
blue R 250, 8% acetic acid, and 25% ethanol solution.

Destaining was with staining solution without added coomassie blue. The agarose plates were stained with silver
(25) After removing ampholine and drying, the gels were
immersed in a 2% KFeCN for 5 min, rinsed in water, and
stained for 15 min in developing solution which contained
35 % of solution A and 65 % of solution B. Solution A was
composed of 8% Na_2CO_3, while solution B was composed
of 0.19% of NH_4NO_3, 0.2% $AgNO_3$, 1% tungstosilicic acid,
and 7.3% (v/v) formalin(37%).

Results

FT-IR. Figure 1 presents an FT-IR spectra taken from
the surface of alloy A in Table 1 after in-vivo exposure.
The amide I and II protein bands are detected at about
1650 cm^{-1} and 1530 cm^{-1}, respectively, as well as
additional protein bands at 1450 cm^{-1} and 1390 cm^{-1}.
Strong carbohydrate adsorption is detected at 1060 cm^{-1},
as well as moderate lipid content at 1250 cm^{-1}. Additional
adsorption peaks include CH_3 at about 2930 cm^{-1} and CO_2
at 2350 cm^{-1}. Spectrum from the remaining alloys (B, C, E
in Table I) were similar to that from alloy A. The use of
deconvolution methods failed to produce any significant
differences due to alloy composition.

The spectrum from the scrapings for different sites
on the surface of alloy A after only 1 hr of in-vitro
exposures to the supernatant of saliva showed varied
results. A comparison of the ratios of the protein (or
lipid) to carbohydrate peak intensities at four different
surface sites is presented in Table III.

Table III. Protein (or Lipid) to Carbohydrate Peak Ratios
(% T/% T) at Four Surface Sites on Alloy A

Peak(cm^{-1})	Site			
	1	2	3	4
1650	77.9	65.5	77.2	74.4
1530	93.1	76.1	82.2	87.2
1450	111.0	96.0	98.5	102.0
1390	112.0	96.2	110.0	104.0
1250	114.0	98.8	102.0	105.5

SIMS. Figure 2 presents a "positive" SIMS spectra for
alloy A after saliva exposure. Comparison to the spectrum
from different alloys indicated the following. The peaks
related to alloying elements are Cu(AMU=63 & 65) for alloys
A, C, D, and E, Al(AMU=27), Mn(AMU=56), and Fe(AMU=54,56,
57,58) for alloy E, and B(AMU=11), Si(AMU=28,29,30), Cr
(AMU=50,52,53,54) and Nb(AMU=93) for alloy F. Nickel(AMU
=58,60,61,62,64) occurs with alloys E and F. Other strong
peaks include Na (AMU=23), K(AMU=39,41), and Ca(AMU=40,42,
43,44). Peaks also occur at AMU of 12, 13, 14(N), 15(NH),
16, 27(CNH , C_2H_3, or Al), and 43(C_2H_3O). Peaks also
occur for CrO(AMU=68), NiO(AMU=74), and CuO(AMU=80).
Tables IV and V present comparisons between SIMS peak

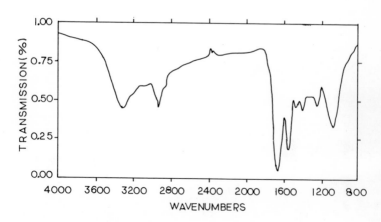

Figure 1.   FT-IR spectra from surface of alloy A which was retrieved from in vivo usage.

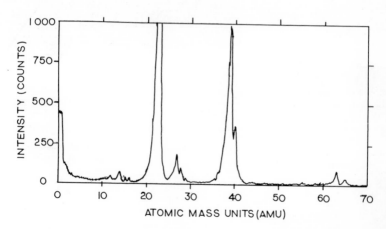

Figure 2.   Positive SIMS spectra from surface of alloy A which was exposed to the supernatant of human saliva for several hours.

Table IV. Positive SIMS Peak minus Background Intensities (counts) for Saliva Exposed Alloy Surfaces

AMU	A	C	D	E	F
			Alloy		
11					209
12	10	57	76	133	19
13		19	57	104	
14	48	48	104	209	133
15			20	228	86
16		28	10	48	95
23	1714	1666	171	152	1332
27	123	57	28	476	190
28	58		28	266	352
39	666	514	304	1856	323
40	571	209	542	1808	238
43				76	
52					1475
55				48	
56				48	
58				19	875
63	57	95	104	1428	
68					67
74					38
93					57
107-10	37,28	84,75	28		
116-22			56		

Table V. Positive SIMS Peak minus Background Intensities (counts) for As-Polished Alloy Surfaces

AMU	A	C	D	E	F
			Alloy		
11					283
12	113	85	28	85	
13	113		28	56	
14	283	198	226	141	113
15	368	311	226	198	141
16	57		56	28	28
23	2518	5377	4245	4811	6226
27	452	622	1301	6226	1641
28	113	141	198		1198
39	509	453	877	962	850
40	226	283	538	113	1301
43		57	226	141	95
52					5377
55	198	198	396	1641	
56	141	170	368	1584	
58				452	3962
63	3537	651	1068	5094	
68					849
74				85	283
93				85	198
107-10	680,481	3962,3900	339,311		
116-22		4000	1075		

intensities for saliva exposed and as-polished surfaces,
respectively. Peaks with atomic mass units higher than
those shown in Figure 2 include Ag(AMU= 107,109), Pd(AMU=
104,5,6,8,10), Sn(AMU=116-20,22,24), and SnO(AMU=132-40).
All five of the as-polished surfaces exhibited higher Na
intensities and higher intensities for AMU= 27(C_2H_3, CNH,
or Al) and 43(C_2H_3O). The alloying elements contained
in the substrate are also higher in intensity for the as-
polished surfaces. This is clearly evident in Tables IV
for Cr, Mn, Fe, Ni, Cu, Nb, Ag, Pd, and Sn, as well as for
CrO, NiO, CuO, and SnO. Peaks also occurred with four out
of the five alloys at AMU = 55 and 56. For alloy E these
are from Mn and Fe, while for the others, CaO is involved.
    Figure 3 presents the "negative" SIMS spectra for
alloy A in the saliva exposed condition. Peak intensities
for C(AMU=12), CH(AMU=13), O(AMU=16), OH(AMU=17), F(AMU
=19), and CN(AMU=26 or C_2H_2) are higher for the
exposed condition. However, the Cl(AMU=35,37) peak was
higher in the as-polished condition. Similar trends
took place with alloys C and F, while alloys D and E had
higher O and Cl intensities with the saliva exposed state.
Tables VI and VII present comparisons of the various peak
intensities for both saliva exposed and as-polished states.

Table VI. Negative SIMS Peak minus Background Intensities
(counts) for Saliva Exposed Alloys

(AMU)	Alloy				
	A	C	D	E	F
12	3966	5476	1019	1952	1322
13	5477	7365	1301	1840	2172
16	10953	11993	4160	2009	12842
17	3399	3871	1075	707	7460
19	3021	3305	3679	2066	566
24	661	1133	283	509	−
25	472	755	141	−	−
26	2455	3116	509	1211	94
32	566	944	378	424	378
35	2266	3772	907	509	236

Table VII. Negative SIMS Peak minus Background Intensities
(counts) for As-Polished Alloys

(AMU)	Alloy				
	A	C	D	E	F
12	906	708	1726	1132	962
13	1528	1443	2377	2037	1726
16	3679	3226	5264	5716	5150
17	906	934			
19	1415	3113	3339	4358	2575
24	452	340	339	2830	85
25	340	283	226	2264	113
26	1358	1075	311	2264	85
32	339	452	481	396	339
35	3906	4528	5086	4245	2094

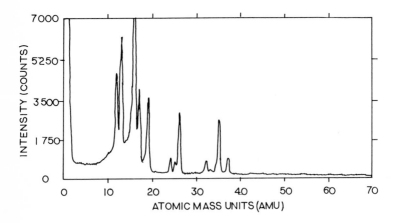

Figure 3.   Negative SIMS spectra from surface of alloy A which was exposed to the supernatant of human saliva for several hours.

Table VIII presents the change in peak intensities minus background for alloy A exposed to saliva after a 4 hr argon ion beam etch.

Table VIII. Percent Change in SIMS Spectrum for Saliva Exposed Alloy A after Argon Ion Beam Etch

Positive SIMS(AMU)

14	16	17	18	23	27	28	39	40
-40	-50	-99	-99	+30	+667	-87	+17	+113

Negative SIMS(AMU)

12	13	16	17	19	24	26	32	35
-38	-18	-13	-49	+17	-99	-14	-50	+50

IEF. Figure 4 presents a photogrraph of polyacrylamide gel plate stained with coomassie blue for saliva supernatants that were in contact with the eleven powders in the adsorption experiments. The twelfth pattern is for an unexposed saliva control sample. All patterns appeared similar and as many as 25 protein bands were discernable by inspection. However, only a limited number were easily detectable by observation. Proteins were detected within the pH range of 4.4 to 8.9. The proteins occurring between pH = 5.2 - 5.7 and at 6.6, 6.8, and 7.2 were most intense and discernable. A schmetatic representation of the patterns and the pH gradient corresponding to the plate dimensions are shown in Figure 5.

The IEF patterns for the H_3PO_4 and EDTA extracts from the eleven powders were very weak in intensity. The patterns from the enamel and hydroxapatite extracts, even though just barely discernable, were still the most intense from all of the extracts. These patterns indicated several protein lines at the same positions as occurring with the the control and supernatants. These lines corresponded to the pH = 5-6 range and close to 7. All remaining extract solutions only gave very weak indications for the existence of several bands and within the pH = 5-6 range. Schematics for these patterns are also shown in Figure 5.

Discussion

Some of the data from this project supports a selective adsorption process, while some other data tends to support the opposite viewpoint. The data from the SIMS work shows variability in the surface compositions among the various substrate materials, although some trends were also seen. The data from the IEF analyses indicated non specific adsorption, since the same proteins appeared to be bound to all of the different substrate compositions used. The FT-IR results showed that differences in adsorbed film composition can occur even on the same substrate material, and suggested the possibilities for other factors besides substrate composition to be important in adsorption. Some of these other factors may include surface smoothness and preparation.

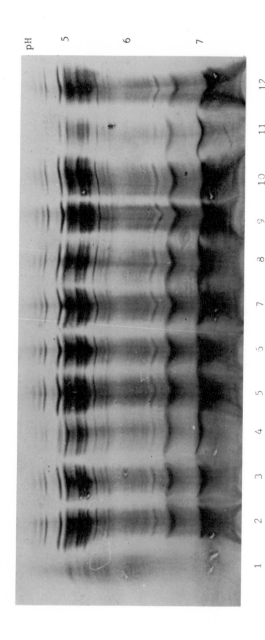

Figure 4. Photograph of polyacrylamide gel plate showing IEF patterns developed for saliva samples after contacting various powders (1–11). Pattern 12 is for unexposed saliva control.

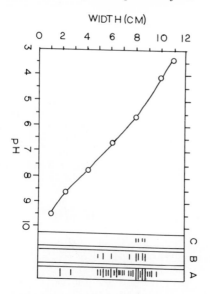

Figure 5.   Gradient in pH developed across plate
shown in Figure 4. Schematics A, B, and C refer to
patterns for saliva control, phosphate extract
from enamel or hydroxyapatite, and EDTA extracts
from all powders.

The effects from short time saliva exposures upon
surface characteristics of the alloys are evident from the
SIMS spectrum. This is most clearly seen by the lowering
in the alloying element characteristic intensities. This
most definitely establishes an adsorbed layer or film
of some kind. The characteristics of the adsorbed films
included variabilities in organics as well as inorganic
contents. The variations in organic contents are seen
in the positive SIMS data by AMU = 27(CNH or C_2H_3) and
43(C_2H_3O) and in the negative SIMS data by AMU = 12(C),
13(CH), 16(O), 17(OH), 24(C_2), and 26(CN). Certainly
the variations in some of these species with substrate
material supports selective adsorption. For the inorganic
contents, the high levels of Na, K, and Ca not only in the
outermost top surface layers, but also their increased
contents with film depth(argon etching time) may provide
electrical conductivity pathways through the flim to the
substrate. The release of substrate ions via corrosion
may then affect the adsorption processes of film onto
substrate and hence make adsoption a selective process.
These effects may be accentuated or minimized by the
variabilities in the concentrations of Na, K, and Ca
with substrate alloy. The variations in Cl with substrate
and its increased concentration with film thickness are
also likely to affect the adosorption processes.
The high levels of ions remaining on the surfaces

after preparations (as-polished condition) is also very
likely to affect the adsorption processes. For all five
alloys, the Cl concentration on the as-polished surfaces
were very much higher than on the surfaces after saliva
exposures. This reduction in outermost Cl may be partly
due to release of Cl into solution, or diffusion and
penetration further into the film thickness. Also, the
higher levels of organics retained on the as-polished
surfaces suggest that the state of the as-polished surface
may have more of an influence than the actual substrate
on the adsorption processes. Since the top surface layers
interact with the proteins and ions in saliva, it may be
that these top layers of ions may be controlling processes
related to adsorption. When viewed on the atomistic scale
the adsorption of salivary proteins may just be a remodi-
fication of the outermost top surface layers, already very
complex even before protein contact. One then should then
be concerned with surface alloy preparations instead of
the actual substrate material.
    The inability to unequivocally establish distinct and
intense IEF patterns for the extracts from the various
powders remained after intensified measures for better
definitions. A number of available methods were used that
are known to influence IEF patterns. Some additional
efforts undertaken included extract dialysis and concentration prior to IEF. Besides, in applying the extracts to
the gel plates, not one but four and more application
strips were applied to the gels, thus permitting at least
four times the amount of extract to be applied. The use
of agarose instead of polyacrylamide gels permitted the
very highly sensitive silver staining technique(25) to be
used. However, even here the resulting patterns for the
extracts lacked definition and clarity which is required
for photography exhibition. The similarities in features
from all the IEF extract patterns suggested the same
salivary proteins, with isoelectric points around pH =5,
to be involved with the adsorption processes. The enamel
and hydroxyapatite extracts showed better IEF definitions
and are considered to adsorb proteins better. Some concern
can always be raised regarding possible carry over from
supernatant used in adsorption, through rinsing, and into
extraction solutions. In adsorption experiments designed
with the agarose gels, more thorough rinsings of the
powders were performed prior to extraction. The rinsings
consisted of three separate 5 ml H_2O rinses and with each
vigorously shaken in tube with mixer. The IEF patterns
with the agarose plates were not that much different than
obtained with polyacrylamide. These results are in line
with a study (26) indicating the IEF patterns of the
extracted proteins from nine hydroxyapatites with various
surface properties were all similar and within the acidic
pH range observed here.
    The high carbohydrate contents analyzed by FT-IR on
all in-vivo exposed surfaces are likely due to products
of bacteria that colonize the surfaces after several days

of in-vivo usage. Hence, for these surfaces information obtained is likely to be different than for alloys exposed for short times. The variations in protein to carbohydrate ratios for surface scrapings from different sites on alloy surface exposed to saliva for only 1 hour are likely from adsorption of different amounts of the same species, or adsorption of different types of proteins, carbohydrates, or glycoproteins. Hence, this data reflects directly the nature of the adsorbing species.

The importance of the released corroded ions from alloys in affecting adsorption must be stressed. It was shown,(27) that for a variety of metallic salts added to whole saliva that Cu^{2+} and Zn^2 generated precipitates with proteins which were most similar to whole saliva. The precipitates with Ca^{2+} were missing some of the protein bands and the precipitates with Sr^{2+}, Ba^{2+}, Fe^{2+}, Fe^{3+}, and Al^{3+} contained only a strong acidic protein. Some cations like Cu^{2+} and Sn^{2+} are also considered (3) to inhibit adsorption through their binding to charged groups on pellicle or bacteria.

Comparison of the IEF patterns obtained in this study to the results obtained (25) with human palatine saliva indicates general agreement. About the same number of protein bands occurred with both studies and with acidic proteins being the most intense in the protein patterns.

Acknowledgments

The FT-IR spectrum of the retrieved in-vivo samples were courtesy of the National Center for Biomedical Infrared Spectroscopy, Battelle Laboratories, Columbus, Ohio.

This investigation was supported in part by USPHS Research Grant DE 05761 from the National Institute of Dental Research, Bethesda MD, 20205.

Literature Cited

1. Ericson, T.; Pruitt, K. M.; Arwin, H.; Lundstrom, I.
    Acta Odontol. Scand. 1982, 40, 197-201.
2. Baier, R. E.; Glantz, P. -O. Acta Odontol. Scand.
    1978, 36, 289-301.
3. Skjorland, K. Acta Odontol. Scand. 1982, 40, 129-34.
4. Sonju, T.; Christensen, T. B.; Knrnstad, L.; Rolla, G.
    Caries Res. 1974, 8, 113-22.
5. Bennick, A.; Cannon, M.; Madapallimattam, G.
    Caries Res. 1981, 15, 9-20.
6. Vogel, J. C.; Belcourt, A.; Gillmeyh, S. Caries Res.
    1981, 15, 243-49.
7. Rolla, G. Swed. Dent. J. 1977, 1, 241-51.
8. Ingersoll, C. E. J. Dent. Res. 1976, 55, Pt121(B).
9. Tabak, L. A.; Levine, M. J.; Mandel, I. D.; Ellison,
    S. A. J. Oral Path. 1982, 11, 1-17.
10. Hay, D. I. Arch. Oral Biol. 1967, 12, 937-46.
11. Meckel, A. H. Arch Oral Biol. 1965, 10, 585-97.
12. Pruitt, K. M. Swed. Dent. J. 1977, 1, 225-40.
13. Tabak, L. A.; Levine, M. J.; Jain, N. K.; Bryan,
    A. R.; Cohen, R. E.; Monte, L. D.; Zawacki, S.;

Nancollas, G. H.; Slomiany, A.; Slomiany, B. L. Arch. Oral Biol. 1985, 30, 423-7.

14. Mayhall, C. W. Arch. Oral Biol. 1970, 15, 1327-41.
15. Hay, D. I. Arch. Oral Biol. 1973, 18, 1517-29.
16. Leach, S. A.; Lyon, R.; Appleton, J. In Surface and Colloid Phenomena in the Oral Cavity: Methodological Aspects; IRL Press Ltd.: London, 1982; pp 63-78.
17. Hannesson Eggen, K; Rolla, G. Scand. J. Dent. Res. 1982, 90, 182-88.
18. Bennick, A.; Chau, G.; Goodlin, R.; Abrams, S.; Tustian, D.; Mandapallimattam, G. Arch. Oral Biol, 1983, 28. 19-27.
19. Stiefel, D. J. J. Dent. Res. 1976, 55, 66-73.
20. Juriaanse, A. C.; Booij, M.; Arends, J.; Ten Bosch, J. J. Arch. Oral Biol, 1981, 26, 91-6.
21. Arends, J.; Jongebloed, W. L. Swed. Dent. J. 1977, 1, 215-24.
22. Ellison, S. A. In Saliva and Dental Caries; Kleinberg, I.; Ellison, S. A.; Mandel, I. D., Ed.; Sp. Supp. Microbiology Abstracts, 1979; pp 13-29.
23. Sonju, T.; Glantz, P. -O. Arch. Oral Biol. 1975, 20, 687-91.
24. Mueller, H. J.; Barrie, R. M. J. Dent. Res. 1985, 64, Pt 1753.
25. Shiba, A.; Sano, K.; Nakao, M.; Kobayashi, K.; Igarashi, Y. Arch. Oral Biol. 1983, 28, 363-4.
26. Wilkes, P. D.; Leach, S. A. J. Dent. 1979, 7, 213-20.
27. Voegel, J. C.; Belcourt, A. Arch. Oral Biol. 1980, 25, 137-9.

RECEIVED January 3, 1987

Chapter 28

Reversible-Irreversible Protein Adsorption and Polymer Surface Characterization

Adam Baszkin, Michel Deyme, Eric Perez, and Jacques Emile Proust

Physico-Chimie des Surfaces et Innovation en Pharmacotechnie, UA Centre National de la Recherche Scientifique 1218, Université Paris—Sud, 5 rue J. B. Clément, 92296 Châtenay-Malabry, France

The present review paper describes the work on adsorption of proteins performed in the authors'laboratory. Behavior of two proteins of different type : collagen (rigid rod-like molecule) and mucin (flexible molecule) was investigated at the solid-solution interfaces using surface force and in situ adsorption/desorption measurements. The in situ adsorption/desorption technique, based on the use of ^{14}C labeled proteins, allows to distinguish between irreversibly and reversibly adsorbed protein. Results obtained from these experiments provide direct data on the structure and orientation of adsorbed collagen or mucin molecules at the studied surfaces. It is shown that the degree of reversibility of the adsorbed protein depends on the type of protein and surface. Since the nature, distribution and orientation of polymer functional groups have a direct influence upon the mechanism of protein adsorption, emphasis is given to the techniques of characterization of polymer surfaces developed in the laboratory.

Characterization of Polymers

Quantification of functional sites on polymer surfaces. The principle of the method developed in our laboratory for these purposes is based upon the use of radioactive isotopes emitting soft-β radiation (^{14}C, ^{45}Ca). When a polymer film bearing functional sites capable of adsorbing $S^{14}CN^-$ or $^{45}Ca^{2+}$ ions is placed in contact with a solution containing one of these ions, the measured radioactivity above the solution-polymer interface would come from the molecules adsorbed in excess at the interface plus that of a thin layer of solution. As the mean free path of these radiations in aqueous solution is respectively equal to 0.16 mm and 0.59 mm, all radiation originating from the solution below this depth is attenuated. To allow for the radioactivity from the adjacent thin solution layer a separate experiment is performed in which instead of a surface treated polymer film an untreated polymer film of the same thickness which does not adsorb SCN^- or Ca^{2+} ions is used.

Figure 1 illustrates the apparatus used for these measurements. Detailed description of a β-radiotracer adsorption method is given in the paragraph dealing with mucin and collagen adsorption at interfaces.

The calcium/thiocyanate method was extensively used in the laboratory for quantification of functional groups on different surface modified polymers. Figure 2 exemplifies a series of such typical polymeric surfaces. While the calcium adsorption isotherms on poly(maleic acid) grafted polyethylene surfaces yielded the amounts of dissociated COOH groups, thiocyanate adsorption isotherms were used to determine the amounts of quaternized polyamine groups on surfaces. Depending on the grafting conditions, pH and ionic strength of the aqueous adjacent phase, the surface density of functional groups on these polymers varied in the range of 10^{15} - 10^{17} sites/cm^2.

The calcium/thiocyanate isotherms, combined with contact angle measurements, reveal, on a molecular level, any rearrangement or reorientation of surface functional groups produced by the variation or alternation of the polymer adjacent phase. The main references to these studies are given in (1-5).

Thin wetting films. The Scheludko's technique of thin wetting films was adapted to study the behavior of proteins at various interfaces and in particular on contact lenses.

The principle of the method, shown in Figure 3, is the follo-wing : the sample is placed at the bottom of the cell made of optical glass which is filled with a protein solution of a given concentration. An air bubble is formed in the solution by means of a capillary tube. The pressure of the air bubble is maintained constant using a mercury pump adjusted with a manometer. The dis-tance, h, between the capillary tube and the sample may be adjusted so as to obtain a thin film of the solution between the air and the sample. The film thickness, h, varies between 20-150 nm, while its diameter is about 300 m. The cell is fixed on the table of a metallographic microscope and the kinetics of the failure or of the formation of the liquid film is observed directly or photographed by means of a movie camera.

Two main parameters are studied : (1) the break-up time which is defined as the time which elapses from the moment when the radius of the thin film formed becomes constant (about 100 μm) and the appearance of the first hole : and (2) the kinetics of dry spot

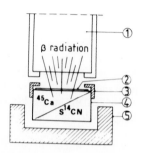

<u>Figure 1.</u> Schematic representation of adsorption measuring appa-
ratus. (1) gas flow counter ; (2) floating polymer film ; (3)
teflon window ; (4) circular glass container ; (5) support.

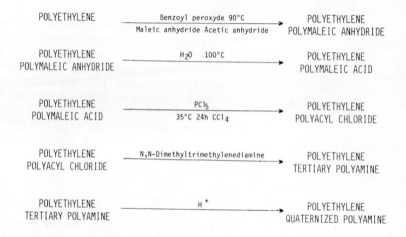

<u>Figure 2.</u> Outline of the surface reactions on poly(maleic anhy-
dride) grafted polyethylene.

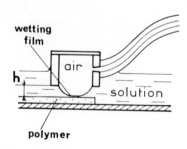

<u>Figure 3.</u> Schematic drawing illustrating formation of a thin
film on a solid substrate.

formation. This latter parameter is characterized as the variation of the mean diameter of a hole formed in the liquid film with time.

An example of the dry spot formation of mucin films on a silicone contact lens is illustrated in Figure 4.

The specificity of the method relies on its dynamic character. Spectacular results may be obtained with its use in many situations were the static wettability measurements are not sensitive enough. In particular, the method allows the detection of even minor surface modifications or changes in solution parameters. These alterations are evidenced by the stability and thickness change of the wetting films.

Various contact lenses were characterized by means of the wetting thin film technique and the results of these investigations are described in (6, 7).

Mucin and Collagen at Interfaces

When biomaterials come into contact with various biological fluids (blood, saliva, tears) protein adsorption at the solid-liquid interface is the first phenomenon which occurs. This primary adsorption process then exerts a profound influence over subsequent events and may give rise to such well recognized and undesired processes as thrombus formation, formation of dental plaque or dry spot formation in the case of contact lenses.

Although the phenomenon of protein adsorption at the liquid/air and solid/liquid interfaces has been the subject of a large number of investigations during the past several years, the answer to the key questions : what is the behavior of proteins at interfaces and why do proteins behave as they do at interfaces, remains unclear and further research effort has to be directed toward understanding of the mechanism of protein adsorption. In particular there is still a lack of direct experimental evidence on the organisation of various adsorbed protein layers and on their composition when protein adsorption takes place from multicomponent mixtures.

In our laboratory, two techniques have been extensively used for studying protein behavior at various interfaces. The first technique consists of _in situ_ measurement of protein adsorption with ^{14}C labeled proteins ; the second technique based on multiple-beam interferometry measures surface forces between two mica sheets with adsorbed proteins (Tabor-Israelachvili technique). While the in situ measurements enable quantitation of protein adsorption, force-distance measurements provide direct experimental data on the extension of adsorbed protein layers towards the solution and on their conformation.

Reported below are the adsorption and surface force experiments with two proteins of different type (mucin and collagen). Each of these proteins was isolated in the laboratory from animal organs.

Figure 4. Example of the kinetics of mucin film rupture on a silicone contact lens. Rate of camera motion 4 frames/s.

Mucin was extracted from bovine submaxillary glands and collagen was isolated from rat tail tendons. The experimental protocols describing extraction, isolation and the ^{14}C radiolabeling procedures of these proteins were reported in Refs.(8, 11).

Mucin. Bovine submaxillary mucin (BSM) belongs to the class of glycoproteins. It is a large flexible macromolecule believed to exist in a bottle-brush form. It has a molecular weight of about 4 x 10^6 and consists of a long polypeptide core with numerous disaccharide and oligosaccharide side chains. The oligosaccharides, mainly sialyl-N-acetylglucosamine, are linked to the peptide through glycosidic bonds between N-acetyl-glucosamine and the hydroxyl group of serine or threonine. The length of the molecule is about 800 nm and its radius of gyration measured by light scattering is 140 nm.

The main function of mucin from secretions of submaxillary glands, along with similar mucoproteins found in the respiratory, gastrointestinal, reproductive tracts and also in the tear liquid, is to lubricate epithelial cells and protect them from the external environment. The role of the mucous glycoproteins as a macromolecular surfactant is therefore of great importance in the science and technology of biomaterials. Such different biosurfaces as dentures, contact lenses or intrauterine contraceptive devices, in spite of different functions, have one common feature, namely that all are placed on a mucosal surface.

Collagen. Different types of collagens have one common feature : their molecule is composed of three continuous helical polypeptide chains wound together over most of their length. The triple helix of this glycoprotein is stabilized by intermolecular hydrogen bonds. The soluble collagen used in our experiments is Type I collagen. Its molecular weight is 300,000 and its molecule can be regarded as a rigid rod, about 300 nm long and 0.15 nm in diameter.

Collagen molecules undergo self-assembly by lateral associations into fibrils and fibers and are able, therefore, along with other biological functions, to ensure the mechanical support of the connective tissue. Collagen also plays an important role in many bioadhesion processes. Collagen molecules bound to implant materials enhance adhesion of epidermal cells to the surfaces of biomaterials and prevent implant failure.

In Situ Adsorption/Desorption Measurements. The techniques to study adsorption/desorption of proteins at interfaces are similar to those initially developed for quantification of functional groups at polymer surfaces and described above.

To measure adsorption of proteins at the solution/air interface the apparatus shown in Figure 5A is used. A circular glass container is filled with a ^{14}C protein and covered with a teflon window. The gas flow counter measures the radioactivity and continuously displays it on a recorder as a function of time. To allow for the radioactivity originating from the solution (A_b) close to the

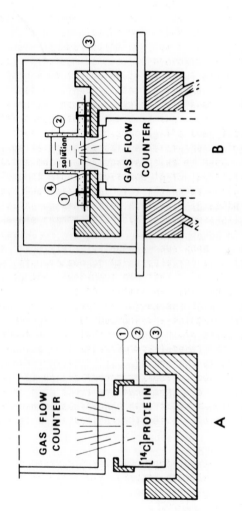

Figure 5. Adsorption measuring devices. (A) For adsorption at solution–air interface; (B) for adsorption on polymer or mica surfaces. (1), (3) Supports ensuring reproducibility of geometrical conditions and tightness; (2) glass container; (4) polymer or mica film.

solution/air interface, a separately run experiment is performed. Instead of a ^{14}C protein the glass container is filled with a solution of a non-adsorbing substance, containing the same radioactive element $K^{14}CNS$, and its radioactivity is measured in the same geometrical conditions as those in the experiments with proteins. The radioactivity A_b can then be calculated from

$$A_b = A_b' \frac{cp}{c'p'}$$

where c and c' are the concentrations of the protein solution and of the non-adsorbing solution respectively, p and p'are their respective specific activities, and A_b' is the radioactivity of the non adsorbing solution.

Subtraction of A_b from the total measured radioactivity (A_t) gives the radioactivity of protein molecules adsorbed in excess at the interface (A_{ad}) for each of the protein concentrations studied (Figure 6). At low protein solution concentrations (< 0.005 mg/ml), A_b is very small and A_t represents almost entirely the adsorbed quantity (A_{ad}) ; for the adsorption at higher protein concentration (> 0.5 mg/ml), A_b represents about 50% of the adsorbed value.

To measure protein adsorption on polymers two techniques are used. The first technique with the polymer films floating on the protein solution surface is illustrated in Figure 1. The second technique, which can be used with either a polymer or a mica thin film, is shown in Figure 5b. Both techniques give identical results indicating that with the use of the measuring device as illustrated in Figure 5b and under the conditions in which our adsorption experiments were performed, neither mucin nor collagen precipitation was observed. To measure the A_b value, the same procedure as above, with $^{14}CNS^-$ ions, is used. In addition, the A_t value has to allow for the absorption of radiation by the sample. The magnitude of the correction for each sample is determined with the help of a ^{14}C methyl methacrylate solid source placed above the polymer or mica window and in the same geometrical conditions as for the adsorption measurements. The necessary checks have been done to ascertain that protein labeling with ^{14}C did not cause any change of their surface activities and different protein adsorbability on the surfaces studied (8, 11, 13). For both adsorption techniques, the radioactivity measured, in counts/minute, is converted to the amount in milligrams of protein per square meter. This is dcne by depositing, drying on mica surfaces known amounts of labeled protein which, when counted, yielded a calibration factor per unit amount of protein. Knowing the conversion factor, and the area of the sample exposed to protein solution, the amount of protein adsorbed in mg/m^2 is obtained.

The in situ desorption experiments on polymer or mica surfaces are performed using the apparatus shown in Figure 5b. The protein solution is pumped out of the cell and simultaneously replaced by water or a buffer solution. Multiple replacement operations lead to a negligible protein concentration in the cell. The A_b value being zero, the measured radioactivity after allowing for the absorption of radiation by a solid sample is directly converted into the surface concentration of the irreversibly adsorbed protein. The loosely bound protein (reversibly adsorbed protein) as a fraction of the total adsorbed layer, is thus obtained by subtraction of the irreversible adsorption from the total adsorption value.

Mucin and collagen adsorption/desorption data on different interfaces have been reported previously. They include adsorption studies at the solution/air interface and adsorption/desorption data on hydrophobic and surface modified hydrophobic polymers (8-12). Adsorption of mucin was also studied on silicone and poly(vinyl pyrrolidone)-grafted contact lenses (10) and on mica surfaces (13, 14). It was shown that increasing solution concentration of these proteins tended to increase the initial rate of adsorption as well as the amount adsorbed at later times. It was also shown that modification of hydrophobic polymers by different treatments capable of generating functional groups at their surfaces (oxidation, superficial grafting of polar monomers) enhances adsorption of mucin and of collagen. The adsorbed amounts of these proteins increase with increasing surface density of functional sites on such polymers (8).

However, in spite of these similarities, the adsorbed amounts and the structure of the adsorbed mucin and collagen layers on the surfaces studied are entirely different. The behavior of these proteins is analyzed here on the hydrophobic polyethylene surface (water contact angle $\theta_{H2O} = 95°$), on the surface modified polyethylenes : oxidized polyethylene ($\theta_{H_2O} = 74°$) and poly(maleic acid) grafted polyethylene ($\theta_{H_2O} = 74°$) and on the hydrophilic mica surface ($\theta_{H_2 O} = 0°$). Acidic pH = 2.75 (for collagen) and slightly alkaline pH = 7.2 (for mucin) were chosen in order to minimize the association of these proteins in solution and to make possible the analysis of their adsorbabilities in comparable conditions.

Figure 7 shows typical adsorption kinetics of mucin and collagen on polyethylene and mica surfaces followed by displacement of the protein solution by a buffer solution. It may be noted that collagen adsorption on polyethylene is five times higher than that of mucin and that the initial adsorption rates are protein diffusivity dependent. The diffusion coefficients for mucin and collagen were reported by us previously. At a protein solution concentration of 0.05 mg/ml they are respectively equal to 1.5×10^{-13} $m^2.sec^{-1}$ and 1.0×10^{-12} $m^2.sec^{-1}$ (11, 13).

Desorption experiments, given also in Figure 7, clearly indicate that the adsorbed mucin and collagen layers differ substantially in their nature. While on polyethylene mucin is entirely

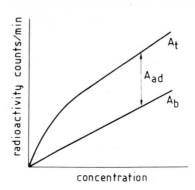

Figure 6. Schematic representation of adsorption measuring technique. A_t = total radioactivity measured ; A_b = radioactivity measured at the solution/air or solid/solution interface with a non-adsorbing substance ; A_{ad} = radioactivity corresponding to the amount adsorbed at the interface.

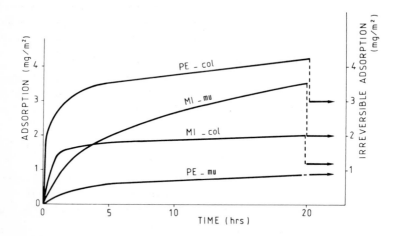

Figure 7. Adsorption kinetics for mucin (Mu) and collagen (Col) on polyethylene (PE) and mica (Mi). Protein solution concentration 0.05 mg/ml. Temp. 20°C. Collagen adsorption from 0.2 M NaCl - 0.1 M CH$_3$COOH buffer at pH = 2.75. Mucin adsorption from 10^{-3} M phosphate buffer with 0.15 M NaCl at pH = 7.2. Dotted lines and arrows indicate desorption and the amount after desorption.

irreversibly adsorbed, one fourth of the adsorbed collagen can be eluded from this surface. A significantly higher adsorption of mucin in comparison to collagen is observed on mica surfaces.

The surface density/solution concentration isotherms, not shown in this paper, reflect also the differences in the behavior of mucin and collagen upon their adsorption at solid interfaces. While the collagen isotherms on polyethylene and surface-grafted polyethylene show a plateau of adsorption at solution concentrations higher than 0.05 mg/ml, no plateau values for mucin adsorption are observed on polyethylene and surface oxidized polyethylene.

The desorption-adsorption relationship for mucin and collagen on mica is represented in Figure 8. This relationship for mucin is linear and clearly indicates that half of the adsorbed quantities can be desorbed. On the contrary, for collagen, this relationship shows a threshold value (1.1 mg/m^2) corresponding to the maximum irreversibly adsorbed value. Above this value all adsorbed collagen can be desorbed (the slope of the straight line is one).

Finally, Figure 9 presents the desorption-adsorption relationship for mucin and collagen on polyethylene and surface-modified polyethylene. The adsorption of mucin on untreated polyethylene is typically irreversible and the maximum adsorbed quantity is equal to 2.2 mg/m^2. In contrast, low but continuous desorption of mucin with increasing adsorbed concentrations is observed on surface oxidized polyethylene.

Collagen desorption-adsorption linear functions on polyethylenes exhibit clearly defined transition points. All collagen which adsorbs in addition to the irreversibly adsorbed layers (represented for polyethylene and grafted polyethylene by their abscissa values 3.0 and 3.7 mg/m^2) can be entirely desorbed. Above the transition points the desorption-adsorption slopes are equal to one.

The contrasting features of mucin and collagen adsorption are summarized in Table I.

To explain the mechanism of collagen adsorption on polyethylene surface, we have formulated a simple hypothesis according to which the reversible adsorption of collagen molecules is realized by their attachment to the irreversibly adsorbed layer. The intramolecular collagen-collagen bonds, much weaker than protein-polymer bonds, break during the washing out procedure causing the release of a fraction of the adsorbed protein. From a comparison of the dimensions of collagen molecules and the irreversibly adsorbed quantities, it is most unlikely that these molecules are adsorbed in their flattened "side-on" orientation. The most realistic picture of the irreversible adsorption would imply binding of rigid rod-like collagen molecules in a tilted "end-on" position. We believe that they are bound to polyethylene essentially via hydrophobic interactions. The angle which they form with the surface may vary between 0° and 90° and would to a large extent depend upon the type and number of protein-surface bonds. Thus for the poly(maleic acid)

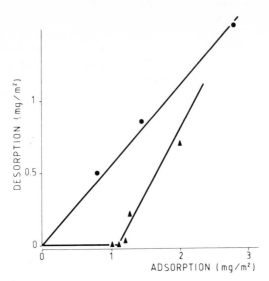

Figure 8.   Desorption-adsorption relationships on mica surfaces.
(•) mucin ;   (▲) collagen. Adsorption time 20 hrs ;   temp. 20°C.
Adsorption conditions as indicated in Figure 7.

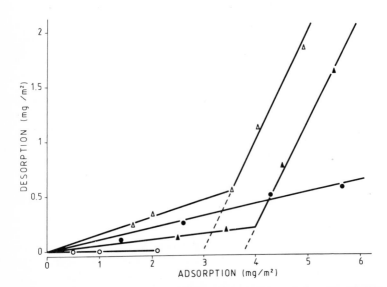

Figure 9.   Desorption-adsorption relationships (Δ) polyethylene-
collagen ;   (▲) poly(maleic acid) grafted polyethylene-collagen ;
(o) polyethylene-mucin ; (•) surface oxidized polyethylene-mucin.
Adsorption time 20 hrs ; temp. 20°C. Adsorption conditions as
indicated in Figure 7.

Table I.   Features of mucin and collagen adsorption on
polyethylene, surface treated polyethylene and mica surfaces

Surface	Protein	
	Mucin	Collagen
Polyethylene	low entirely irreversible adsorption	high irreversible adsorption followed by complete reversible adsorption
Surface treated polyethylene	continuous, very high irreversible adsorption (90% of total adsorption)	high irreversible adsorption followed by complete reversible adsorption
Mica	continuous equivalent irreversible and reversible adsorption	low irreversible adsorption followed by complete reversible adsorption

grafted polyethylene which bears functional sites on its surface, in addition to hydrophobic interactions via direct hydrogen bond formation may occur. The presence of these interactions would increase the amounts of irreversibly adsorbed collagen molecules and "tighten" the interactions with the polymer. This seems to be the case, since the irreversible adsorption is higher (3.7 mg/m^2 instead of 3.0 mg/m^2 for the polyethylene) and less collagen desorbs (the slope of the lower part of the collagen curve in Figure 9 is 0.05 while that of polyethylene is 0.2). The increase of irreversible adsorption by 0.7 mg/m^2 between the surface grafted and untreated polyethylene may be attributed to the appearance of additional bonds, most probably hydrogen bonds between the polymer surface and adsorbed collagen.

The amount of collagen adsorbed irreversibly to mica is about one third of that measured on polyethylenes (1.1 mg/m^2). Since mica is an entirely hydrophilic surface it may be considered that hydrogen bonds between the hydrated ions present on its surface and collagen chains would account for the major part of the binding mechanism. Otherwise this quantity (1.1 mg/m^2) is comparable with the adsorption increase (0.7 mg/m^2) due to the introduction of hydrophilic sites on a polyethylene surface.

Mucin may be regarded as a flexible macromolecule. In contrast to collagen it can assume a much greater number of different surface induced conformations. Its affinity for a particular surface with respect to others, would involve formation of a greater number of surface bonds or a greater mean energy per bond or both. The degree of reversibility of its adsorption would depend upon the type of configuration which a particular surface may induce. The untreated, hydrophobic polyethylene would be a representative example of a surface inducing mucin configurations leading to a strong and irreversible adsorption. Increasing the number of polar anchoring sites on polyethylene yields high irreversible levels of adsorption. The occurrence of configuration different from that on polyethylene would most probably involve formation of multiple hydrophobic and short range hydrogen bonding interactions between mucin and the surface. When the short range hydrogen bond interactions increase, as in the case of mica, the reversibility of mucin adsorption increases.

Surface Force Measurements. This technique enables the measurement of the force (10 mN accuracy) versus distance (0.1-0.2 nm accuracy) between two curved mica surfaces. The forces between two solid surfaces across an aqueous solution are highly sensitive to the structure of the solid/liquid interfaces. When such surfaces are covered with adsorbed protein layers, then, the analysis of the force/distance profiles may reveal the formation of protein bridges between the two surfaces, the occurrence of steric interactions, or any possible protein conformation change.

Figure 10 shows force/distance profiles between mica surfaces bearing collagen or mucin adsorbed layers. The experiments were performed by first measuring forces between bare mica surfaces across a pure electrolyte solution and then injecting protein solution into the measuring cell. After allowing 3 hours for the protein to adsorb, the forces were measured against distance. The experimental conditions were chosen in order to ensure the same protein (mucin or collagen) surface concentration at the mica/aqueous solution interface (2 mg/m^2) and to allow therefore the comparison of protein effects on the surface forces.

The collagen force/distance profiles clearly indicate that on approaching the mica surfaces, no forces are present down to the 320 nm separation distance. On decreasing the distance, repulsive forces increase smoothly.

An entirely different type of behavior is exhibited by adsorbed mucin layers. Forces are observed beginning at 450 nm (attractive interactions, inset in Figure 10). These are followed by a steep repulsion beginning at 100 nm. The attractive interactions with mucin have been explained by bridging mechanisms (14, 15). The repulsive regime can be interpreted in terms of steric interactions due to the confinement of adsorbed mucin molecules between two walls (reduced number of configurations). Other experiments, not reported in this paper, show that the solution ionic strength may profoundly

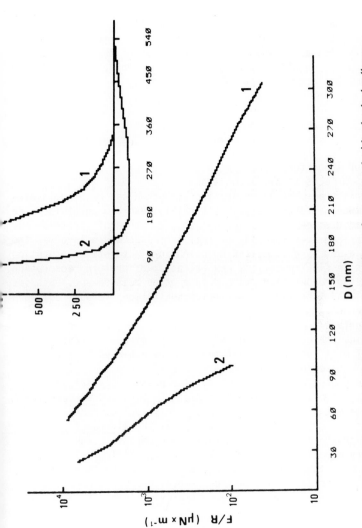

Figure 10. Forces (logarithmic scale) between mica surfaces covered with adsorbed collagen (1) and mucin (2). Inset: linear representation of forces. The forces are plotted as F/R where R is the radius of curvature of surfaces. Surface concentration of both proteins after adsorption in the force-measuring cell for 3 h was 2.0 mg/m². Mucin solution concentration: 0.05 mg/ml + 0.15 M NaCl; pH = 5.5. Collagen solution concentration: 0.05 mg/ml in 0.1 M NaCl + 0.05 M CH₃COOH; pH = 2.7.

influence mucin configurations. Also, bridging interactions are highly dependent on mucin surface density (15).

For collagen, repulsive forces start at a distance which is smaller than twice the length of the collagen stiff rod. The force/distance profiles for collagen confirm therefore the interpretation of the adsorption-desorption data. Collagen molecules are attached to the mica surface in a tilted end-on orientation. This orientation varies as a function of compression of the surfaces. The repulsive forces result from the head-to-head contacts of collagen rods or from the interpenetration of the layers.

Conclusions

Various surface chemistry techniques described in this paper and used in our laboratory can be used to characterize protein adsorbability at surfaces of different types.

The in situ adsorption/desorption technique can be used to distinguish between irreversibly and reversibly adsorbed protein. The desorption/adsorption ratio depends on the nature of protein and surface. Further development of this technique would involve the design of a new type of measuring cell for adsorption/desorption measurements in flow conditions and extension of protein adsorption experiments to competitive protein adsorption measurement. Studies of albumin and fibrinogen competitive adsorption versus collagen (12) have recently been initiated.

The surface force measurements between two adsorbed protein layers on mica are now being investigated on polymer surfaces. The coverage of mica surfaces by polymers may be achieved by successive dipping of these supports through a polymer monolayer (Langmuir-Blodgett technique) or by a direct polymer plasma polymerization on mica sheets. Also studies of interactions between two surfaces each bearing a different adsorbed protein are anticipated.

Studies on polymer monolayers spread at the air-water interface are now in progress in our laboratory. Biocompatible and biodegradable polymers used as nanoparticles carrying biologically active substances are characterized using the surface balance, surface potential and protein adsorption/desorption measurements. The combined data of all these measurements provide information on drug and protein penetration/delivery with these polymers.

Literature Cited

1. Baszkin, A.; Ter-Minassian-Saraga, L. J. Polymer Sci. 1971, 34, 243-252.
2. Baszkin, A.; Deyme, M.; Nishino, M.; Ter-Minassian-Saraga, L. Prog. Colloid & Polymer Sci. 1976, 61, 97-108.

3.   Baszkin, A.; Nishino, M.; Ter-Minassian-Saraga, L. J. Colloid Interface Sci. 1976, 54, 317-328.
4.   Leclercq, B.; Sotton, M.; Baszkin, A.; Ter-Minassian-Saraga, L. Polymer 1977, 18, 675-680.
5.   Eriksson, J.C.; Gölander, C.G.; Baszkin, A.; Ter-Minassian-Saraga, L. J. Colloid Interface Sci. 1984, 100, 381-392.
6.   Baszkin, A.; Boissonnade, M.M.; Proust, J.E.; Tchaliovska, S.D.; Ter-Minassian-Saraga, L.; Wajs, G. J. Bioengineering 1978, 2, 527-537.
7.   Proust, J.E. Proc. International Tear Film Symposium, Lubbock Texas, 1984, in press.
8.   Proust, J.E.; Baszkin, A.; Boissonnade, M.M. J. Colloid Interface Sci. 1983, 94, 421-429.
9.   Baszkin, A.; Proust, J.E.; Boissonnade, M.M. In Biomaterials & Biomechanics 1983; Ducheyne, P.; van der Perre, G.; Aubert, A.E., Eds.; Elsevier: Amsterdam, 1984; p 379.
10. Baszkin, A.; Proust, J.E.; Boissonnade, M.M. Biomaterials 1984, 5, 175-179.
11. Deyme, M.; Baszkin, A.; Proust, J.E.; Perez, E.; Boissonnade, M.M. J. Biomed. Mater. Res. 1986, 20, 951-962.
12. Deyme, M.; Baszkin, A.; Proust, J.E.; Perez, E.; Albrecht, G.; Boissonnade, M.M. J. Biomed. Mater. Res. in press.
13. Perez, E.; Proust, J.E.; Baszkin, A.; Boissonnade, M.M. Colloids & Surfaces 1984, 9, 297-306.
14. Proust, J.E.; Baszkin, A.; Perez, E.; Boissonnade, M.M. Colloids & Surfaces 1984, 10, 43-52.
15. Perez, E.; Proust, J.E. J. Colloid Interface Sci. in press.

RECEIVED January 28, 1987

Chapter 29

Protein Adsorption on Solid Surfaces: Physical Studies and Biological Model Reactions

Hans Elwing[1], Agneta Askenda[1], Bengt Ivarsson[1,3], Ulf Nilsson[2], Stefan Welin[1], and Ingemar Lundström[1]

[1]Laboratory of Applied Physics, Linköping University, S-581 83 Linköping, Sweden
[2]The Blood Centre, University Hospital, S-751 85 Uppsala, Sweden

The paper describes some of our studies of protein adsorption on solid surfaces. An emphasis is made on newly developed experimental techniques and on recent biological model experiments. We therefore discuss the use of a wettability gradient along a solid surface to investigate, in a convenient way, the influence of surface energy on the adsorption of protein molecules. The behavior of the complement system at solid surfaces is also discussed, with special attention to surface induced conformational changes of human complement factor 3. The protein adsorption studies described were all at the liquid – solid or solid – air interface. Lateral scanning ellipsometry was made to evaluate the surfaces with a wettability gradient. Experiments were most often made on (modified) silicon surfaces. The experimental results are also discussed in relation to the proposed theoretical models for protein adsorption.

During a number of years we have applied surface orientated analytical methods to the study of protein adsorption on solid surfaces. These investigations include in situ studies with ellipsometry, surface potential and capacitance measurements (1,2) We have applied spectroscopic techniques like infra-red reflection absorption spectroscopy (IRAS, 3-5) and ESCA (5-7) to investigate details in the interaction between organic molecules and surfaces. Spectroscopic techniques have also been used to

[3]Current address: Pharmacia, S-751 82 Uppsala, Sweden

0097-6156/87/0343-0468$06.25/0
© 1987 American Chemical Society

study metal surfaces which had been implanted in humans for different length of time (8,9). The input from the physical measurements has been used to develop some simple dynamic models for protein adsorption (10-12). Furthermore we have used mainly ellipsometry to study biological model reactions related to antigen-antibody reactions on solid surfaces (13) and the behavior of the immune complement system at solid surfaces (14-16). Several new analytical methods have been developed during the course of this work, notably the use of a gradient in the surface energy along a solid substrate to study the influence of surface energy on protein adsorption. (17). The influence of surface induced conformational changes on subsequent biological processes are under study. It should also be mentioned that our interest for protein adsorption on solid surfaces has led to the development of simple methods and instrumentation for medical diagnostic purposes (18-20).

The main purpose of our present studies is to investigate the details of protein-surface interactions with relevance to questions regarding biocompatibility, fouling and the possible development of (implantable) biosensors.

The solid surfaces used are evaporated metal films or polished silicon wafers, which are well suited for the optical studies. Several types of proteins have been used like human fibrinogen, immunoglobulins, complement factor 3 and others, including lysozyme, egg albumin and beta lactoglobulin. Detailed studies of the interaction between amino acids, like glycine, histidine and phenylalanine, and metal surfaces have been made with IRAS and ESCA. In some experiments deposition of organic material from whole blood, serum or plasma has been studied. In a short review it is not possible to cover all of the present and past research activities and to give a detailed account of the experimental methods used. Some of our work on protein adsorption on metal surfaces was recently summarized in ref 21, where also a number of methods used to study protein adsorption on metal (oxide) surfaces were described. We have therefore chosen to describe two of our more more recent developments, namely the so called "wettability gradient method" for the study of protein adsorption, desorption and exchange on solid surfaces, and the study of surface induced activation of the immune complement system, mainly the conformational change of complement factor 3 (C3) upon adsorption. Our experimental results are discussed in relation to the assumed models for protein adsorption on solid surfaces. The virtues and shortcomings of a simple dynamic model for protein adsorption are described in this context.

The wettability gradient method.

General It has long been known that solid surface wettability or energy plays a critical role in protein adsorption on solid surfaces. Most methods for the investigation of adsorption and desorption of proteins at solid surfaces involve the use of solid surfaces with a constant given chemical composition. The action of a specific surface constituent is usually investigated with the use of several preparations of the surface. This procedure is time consuming, uncertain and expensive. We have used another approach in which the specific surface constituent is attached to a flat solid surface in a gradient. The quantification of protein adsorption on the gradient surface is made with the use of lateral scanning ellipsometry. Ellipsometry is described in (21) and (22). "The gradient method" has been used for the investigation of the dependence of solid surface wettability on protein adsorption. The wettability gradient in these experiments is made by diffusion of methylsilane on silicon surfaces with a spontaneously grown layer of silicon dioxide. The surfaces so formed are hydrophilic at one end and hydrophobic at the other and in between there is a gradient in surface energy, 10-15 mm long. The distribution of the wettability along the gradient could be determined with the use of a capillary rise method on similarly treated glass plates or indirectly by means of ellipsometric determinations of adsorbed fibrinogen (in preparation). Both this methods give an estimate of the advancing contact angle with water. Experiments on protein adsorption and desorption reactions are performed by incubating the gradient plates in protein solution, followed by incubation in various detergents or other test solutions. All plates are finally rinsed in distilled water and dried with N_2. The amount of adsorbed protein along the gradient is determined by means of an ellipsometer (Rudolph Research, Auto Ell 2), equipped with a device for lateral scanning along the silicon surface. The adsorbed amount, Γ, was estimated from the mearured thickness of the organic layer using a density of 1.37 g/cm^3 and a refractive index of 1.6 of the dried protein layer (17).

<u>Protein adsorption and desorption on the gradient surfaces.</u>
Typical results of the dependence of the adsorbed amounts of protein on the wettability of the silicon surface with water are shown in Figure 1. Of the protein used, fibrinogen showed the largest quantitative difference between the hydrophobic and hydrophilic parts of the surface. Lysozyme and γ-globulin showed smaller differences. The reproducibility with regard to the appearance of the adsorption profile was usually very good between gradient plates from the same diffusion batch. There were .however. differences between plates from different diffusion batches as illustrated in Figure 1. This difference is most probably due to differences in the wettability distribution in plates from different batches.

The gradient method has been applied to the investigation of protein desorption effects of detergents and other agents (<u>17</u>). An experiment with desorption of γ-globulin induced by detergents is illustrated in Figure 2. It was observed that Tween 20, a non-ionic detergent caused a maximum desorption effect at a contact angle of about 40^0. At the hydrophilic side of the gradient there was very little desorption induced by Tween and at the hydrophobic side of the gradient the desorption decreased with decreased surface wettability. Desorption of adsorbed γ-globulin ,induced by an ionic detergent, SDS on gradient plates has also been studied . In contrast to Tween 20, SDS had a general high desorption activity on both the hydrophilic and hydrophobic side of the gradient (data not shown). Tween or SDS themselves caused no measurable adsorption on the gradient plates since they were probably removed during the rinsing procedure.

<u>Protein exchange reactions on the gradient surfaces.</u>The gradient method has been used for the investigation of exchange reactions on solid surfaces.(<u>23</u>). In these experiments the gradient plates were first incubated in fibrinogen (1 g/L, dissolved in phosphate buffered saline solution, PBS at pH 7.3) for one hour followed by incubation in γ-globulin for 4 hours under gentle stirring. Some of the plates were then incubated in antiserum against γ-globulin, diluted 1/25, for 30 min. The plates

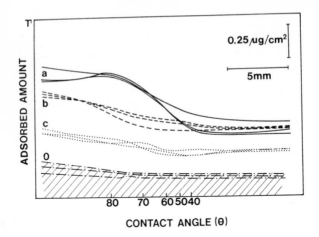

Figure 1: Ellipsometrically registered amount of adsorbed protein on silicon surfaces at the air/solid interface. a:fibrinogen b: γ-globulin and c:lysozyme. The proteins, dissolved in phosphate buffered saline solution (PBS) at pH 7.3 was adsorbed on the surfaces for 30 min at a concentration of 1g/l. The amounts was determined on dried surfaces in air. The lateral distance is given as a bar in the drawing. The corresponding contact angle values as estimated with the use of a capillary rise method are given on the X-axis. The lower dashed dotted "0"-lines indicate the ellipsometrically registered silicon dioxide layer including the methyl gradient. The adsorbed amount of protein is given by the deviation from this line in the figure. Each line in the drawing represents the average amount of adsorbed proteins measured on triplicate plates. Three different triplicate experiments were performed for each protein, involving a new preparation of gradients plates at each experiment.

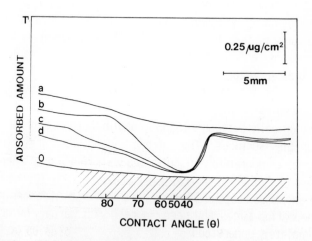

Figure 2:   Ellipsometrically registered   Tween 20 induced desorption of human γ-globulin adsorbed on gradient plates. a: buffer only, b: 0.005% Tween 20, c: 0.05% and d: 0.5%. For further explanation, see Figure 1 and text.

were finally rinsed and dried and the amount of organic material along the gradient was determined with ellipsometry. The experiment was also reversed in that the first incubation of the gradient plates was made with γ-globulin and the second incubation with fibrinogen and adsorbed fibrinogen was detected by using of antiserum against fibrinogen, diluted 1/25) The results of representative experiments are given in Figures 3 and 4. Adsorption of γ-globulin on the fibrinogen coated surface occured only at the hydrophilic end of the gradient as indicated by the deposited antibodies. In the reverse experiment it was noted that adsorption of fibrinogen on γ-globulin coated surfaces was extended into the hydrophobic side of the gradient. These results indicate that fibrinogen is more readily immunologically detecable than γ-globulin both on hydrophilic and hydrophobic surfaces. The strong adsorption preference of fibrinogen compared to γ-globulin is a well known phenomenon and has been described in several publications, e.g 34-28 The details of the exchange reactions depend, however,also on concentrations and incubation times and need more investigations before they are completely understood.

We are aware of the possibility that the adsorbed proteins may partially be removed by serum proteins during incubation with antiserum, especially at the hydrophilic side of the gradient. We do not believe ,however, that this has any major influence on possible qualitative conclusions.

The two examples above of the application of the gradient method illustrate its analytical properties for the investigation of mechanisms of macromolecular adsorption and desorption reactions. Some of the described phenomena could probably not have been observed without the use of the gradient method. Ellipsometric analysis of protein interaction on gradient surfaces may in principle also be performed at the liquid /solid interface, a procedure which perhaps further will increase the prescision and accuracy of the determinations. The air/solid measurement method makes however the gradient method very rational since the ellipsometer only is used a short time at the end of each experiments. At present, about 30 gradients including 1800 automatically performed ellipsometrical determinations, can quite easily be analyzed by one person in one day.

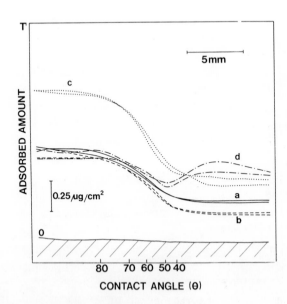

Figure 3: Exchange reactions of human γ-globulin (HGG) on gradient surfaces precoated with human fibrinogen (HFG). The adsorbed amount of protein, determined ellipsometrically is shown versus the wettability along the gradient. The curves (duplicate experiments) represent a:adsorption with HFG, b: adsorption of HFG followed by incubation of HGG, c: as in b) with an additional incubation in anti-HFG, d) as in b) with an additional incubation in anti- HGG. The adsorbed amount of organic material is given by the deviation from the "0" -line. For further explanations, see Figure 1.

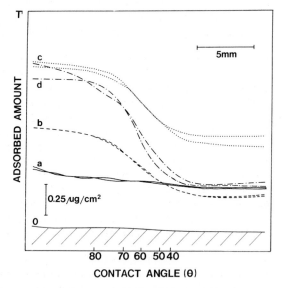

Figure 4: Exchange reactions of HFG on gradient surfaces precoated with HGG. The curves (duplicate experiments) represent a: adsorption of HGG, b: adsorbtion of HGG followed by incubation in HFG, c: as in b with an additional incubation in anti-HFG, d: as in b with an additional incubation in anti- HGG. For further explanation see Figure 3.

Surface activation of the immune complement system.

Identification of serum complement activation with the use of ellipsometry.

Activation of the immune complement system in blood, has been described in connection with haemodialysis and leukapheresis   It results in various "down stream symptoms" such as transient leukopenia (29-32). Activation of complement on a surface implies sequential activation of several factors. The activation process shows similarity with the clotting system of blood in that restricted proteolysis of some of the factors is one of the main regulation mechanisms and that activation is of the cascade type. C3 is the quantitatively dominating complement protein and has a central position in the complement activation cascade.

Activation of the complement system has been studied in vitro on various model surfaces. Most methods used are based on the analysis of various complement degrading factors. Activation of the complement system means however also that some of the factors, especially C3, are deposited on the solid surface. Therefore, we have used ellipsometry   to analyse complement activation on solid surfaces. The solid surfaces were either hydrophilic silicon surfaces with a spontaneously grown layer of silicon dioxide or silicon surfaces made hydrophobic by means of the attachement of methyl groups. Ellipsometric analysis was performed at the   liquid-solid interface. Some experimental results   are shown in Figure 5. It was noted that   incubation of hydrophobic surfaces in serum resulted in a slow deposition of material from the serum on the surface. Further incubation of antibodies against C3 resulted in deposition of C3 antibodies indicating that C3 had been adsorbed from serum. Incubation with serum containing EDTA did not cause any   accumulation of organic material from serum. (The complement system is inactivated by EDTA due to chelating of Mg^{2+} and Ca^{2+} ions which are needed for the normal function of the complement system). Instead the adsorption rapidly reached a plateu. In addition, the anti C3 response was insignificant. Precoating   the hydrophobic surface with γ-globulin before addition of serum resulted furthermore in a pronounced increase of the deposited organic material compared to the experiment without precoating of the surface. The same sets of experiments

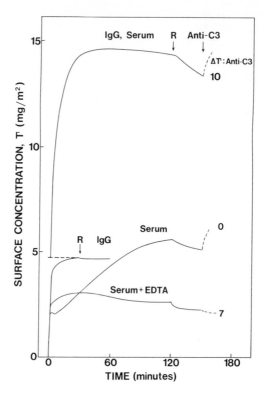

Figure 5: Kinetic measurements of the deposition of organic material from 10% human serum on hydrophobic silicon. "Anti-C3" denotes further incubation with anti-C3, diluted 1/50 at the end of the experiment. The kinetics of anti-C3 deposition is not shown but the additional adsorbed amount, Γ, after 120 min adsorption is given at the end of each curve. The upper curve represents deposition of organic material from serum on surfaces preadsorbed with γ-globulin. The dashed line marked "IgG" indicates the amount of preadsorbed γ-globulin in the γ-globulin ,serum experiment. The middle curve represents deposition from serum only and the lower curve from serum + 10 mM EDTA. The reproducibility in Γ was within ±15%.

were repeated on hydrophilic surfaces. Both the deposition reaction from serum as well as the subsequent reaction with C3 were smaller on hydrophilic surfaces compared to hydrophobic surfaces. The reproducibility of the results was usually good provided that the same serum preparation was used. There were however differences between serum from different individuals as well as serum from one individual   collected at different occasions. The characteristic slow growth of organic material on the surface was, however, always accompanied by subsequent anti-C3 deposition. Presence of EDTA in the serum   resulted always in the absence of subsequent anti-C3 deposition. These experiments show that ellipsometry is useful for the investigation of complement activation on solid surfaces. The method constitutes a useful   complement to other methods which are based on the analysis of soluble complement factors. As illustrated, it was in addition possible to perform real time measurements of the complement deposition reaction which is an uniqe feature of <u>in situ</u> ellipsometry.

<u>Surface induced conformational changes of complement factor 3.</u>
It is likely that the interaction of C3 with a surface is one of the recognition mechanisms that leads to surface induced complement activation. It has been demonstrated that C3 undergoes major antigenic alteration if C3 is denatured with SDS (sodium dodecyl sulphate) (<u>33</u>) Similar alteration occurs when C3 is activated and deposited on   biological target surfaces like erytrocyte membranes (<u>34</u>). It has been shown that there are three antigenic subtypes of C3 (<u>33</u>) .C3(N) which is present on the native but not on the SDS denatured or biologically activated form of C3, C3(D) which is present only on the SDS denature, biologically activated and surface deposited form, and C3(S) which is present on both native and denatured molecules. One of the most used ,commercially avaliable, antiserum preparation is ,anti-C3c, and contains a mixture of antibodies against C3(S) and C3(N) subtypes. The use of the described subsets of antibodies constitutes an useful analytical probe for investigating of the changed antigenicity of C3 on solid surfaces.

Purified C3 was adsorbed on hydrophilic and hydrophobic silicon surfaces followed by incubation with the various subtypes of antibodies. The amount of deposited organic material was measured with ellipsometry on the dried surface. It was observed in these experiments that C3 exposed S-determinants and possible also N-determinants on both hydrophilic and hydrophobic surfaces, since anti-C3c was adsorbed both on the hydrophilic and hydrophobic surfaces. . D-antigens could be detected only on C3 adsorbed on hydrophobic surfaces. Thus, it is obvious that C3 undergoes conformational changes on the hydrophobic surface similar to SDS denaturation or biological activation (to be published).

Experiments including specific binding of the different anti-C3 antibodies were also performed on the gradient surfaces. C3 (10 ug/ml in PBS) was adsorbed on the gradient plates for 30 min and some of the plates were also incubated for 30 min in the different antiserum preparations diluted 1/25. A representative experiment is illustrated in Figure 6. C3 was adsorbed on both the hydrophilic and the hydrophobic side of the gradient but to a slightly lesser extent on the hydrophilic side. Deposition of anti-C3c (anti-C3(N) and anti-C3(S)) , occured on both the hydrophilic and hydrophobic side of the gradient The rabbit anti-C3(D) antibodies used in the experiments were directed either to the α or β polypeptide chain. Ellipsometric visualization of deposited antibodies was here enhanced by subsequent incubation for 30 min in anti- rabbit immunoglobulin diluted in phosphate buffer. As illustrated in Figure 6, anti-α and anti-β antibodies reacted apparently little with C3 deposited at the hydrophilic side of the surface but there was an increased binding capacity of both anti-α and -β at the hydrophobic side of the gradient.

The difference in anti-C3(D) binding between the hydrophobic and the hydrophilic end of the gradient is not only due to the different amounts of C3 adsorbed , since a slightly larger amount of anti-C3c was bound to C3 on the hydrophilic side of the gradient . There are thus more C3(N) determinants on the hydrophilic side of the gradient since the C3(S) determinants per adsorbed molecule should be the same on both sides of the gradient. The C3(N) determinants are gradually lost against the hydrophobic side of the gradient. The presence of C3(D) determinants on C3 adsorbed on the hydrophobic silicon surface does not prove that C3 is biologically active on this surface e.g. bind its inhibitors. This possibility is presently under investigation.

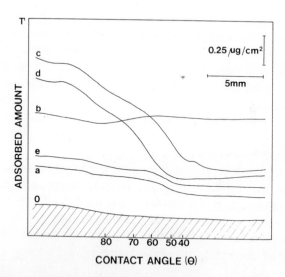

Figure 6: Antibody binding to C3 adsorbed on gradient surfaces. The plates were first incubated in 10 µg/mL of C3 and thereafter in different antisera. The adsorbed amount of protein, determined ellipsometrically is shown versus the wettability along the gradient. Each line in the drawing represent the average of three gradient plates. a: plates incubated with C3 for one hour, b: plates incubated with C3 followed by incubation with anti-C3c for 30 min. c: and d: plate incubated with C3 followed by incubation in anti-α (c: in the drawing) or anti-β (d: in the drawing) for 30 min followed by incubation with anti rabbit immunoglobulin for 30 min. e: plates incubated with C3 followed by incubation with anti rabbit immunoglobulin for 30 min (control).

Some notes on theoretical models for protein adsorption.

The behavior of protein molecules at solid surfaces is very complex. The interaction between the surface and the protein is determined both by the nature of the protein, the surface and the medium outside the surface. The situation is further complicated by the fact that exchange reactions between protein molecules of the same or different kinds take place on the surface. Except for these exchange reactions most protein molecules appear to be irreversibly adsorbed. Although the details of the interaction between protein molecules and surfaces are not known it is assumed that general properties of the surface and the protein such as hydrophobicity, charge density, ion binding, hydration etc. are involved. For reviews, see e.g (21,35-37).

Protein adsorption is a dynamic phenomenon, where protein molecules diffuse to the solid surface, adsorb and eventually change conformation. Furthermore the number of contact points and hence the strength of the protein surface interaction may increase with time. This in turn means that the probability for spontaneous desorption decreases with time. Exchange reactions are other dynamic phenomena where, in a dynamic equilibrium "irreversibly" adsorbed molecules may be exchanged by protein molecules in the solution. In this case the total energy is not necessarily changed (it may actually decrease) and protein exchange may therefore be thermodynamically much more probable than spontaneous desorption. The details of an exchange reaction are probably very complicated.

A possible mechanism is that the bonds initially formed between adsorbed molecules and the surface are broken sequentially and replaced by bonds between a new molecule from the solution and the surface. An adsorbed protein molecule may have several different conformations, which are determined by the surface properties. It is also likely that the "average" conformation is determined by the total amount of protein on the surface. Furthermore, the packing and conformation of the protein molecules are not necessarily homogenous along the surface. Surface aggregation and domain formation may occur. It is therefore very difficult to formulate a theoretical model for protein adsorption. The adsorption is not governed only by thermodynamic factors but also by kinetic and dynamic processes on the surface.

It thus matters how a protein is added to a solution during an adsorption experiment (1,38) Protein adsorption isotherms are also often non-Langmurian in shape, which may have several physical origins. We have tested a very simple dynamic model which gives rise to a new type of adsorption isotherm with a shape very similar to the experimentally observed ones (11,12) Even if the model can not explain in detail all the observed phenomena, for example the exchange reactions, it is a starting point for further developments. In the model it is assumed that a protein molecule adsorbs in one conformation and then undergoes a conformational change to another (more extended) conformation. The molecule is assumed to cover an area of a_1 and a_2 respectively in the two conformations (Figure 7) By allowing mathematically for a small desorption the following expressions for the coverage of molecules of form 1 and 2 respectively were obtained, with the notations in Figure 7 (11),

$$\theta_1 = \left[\left(\frac{k_1 C_1 + S_1 + r_1}{2(a r_1/r_2 - 1)S_1}\right)^2 + \frac{k_1 C_1}{(a r_1/r_2 - 1)S_1}\right]^{1/2}$$
$$- \frac{k_1 C_1 + S_1 + r_1}{2(a r_1/r_2 - 1)S_1} \quad and \quad \theta_2 = \frac{S_1 r_1 \theta_1^2}{r_2(k_1 C_1 - S_1 \theta_1)}$$

where $a \equiv a_2/a_1$ and θ_1 and θ_2 are the coverages of molecules in conformation 1 and 2 repectively. The maximum surface density of adsorbed molecules is $N_0 = 1/a_1$ (i.e. all molecules in conformation 1). The adsorbed amount of organic material e.g. the surface concentration ($\mu g/cm^2$) is proportional to $\theta_1 + \theta_2$. If the rate constant r_1 and r_2 are small we have in practise irreversible adsorption. If the r's are mathematically zero θ_1 and θ_2 are determined only by the kinetics of the adsorption process and we get as a first approximation by neglecting possible diffusion limitations (10):

$$\theta_1 + \theta_2 = \frac{k_1 C_1}{S_1} \ln\left[\frac{k_1 C_1}{S_1}/\left(\frac{k_1 C_1}{S_1} - \theta_1\right)\right]$$
$$and \quad 1 - \theta_1 - a\theta_2 = 1$$

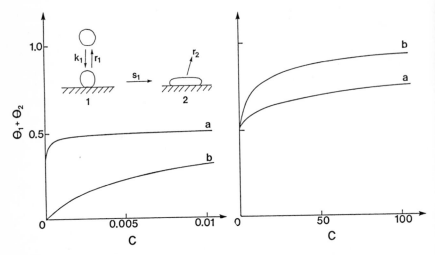

Figure 7 : Example of adsorption isotherms calculated from the dynamic model (illustrated in the insert) assuming (small) spontaneous desorption. The isotherms are drawn in two different scales and for two different relative magnitudes of the desorption terms.

a: $r_1/s_1 = 0.01$; $r_2/s_1 = 0.0001$

b: $r_1/s_1 = 0.1$; $r_2/s_1 = 0.01$

In both cases a $\equiv a_2/a_1 = 2$

C is the normalized concentration of proteins in solution (with conformation 1), $C \equiv K_1, C_1/s_1$. The rate constant s_1 implicitly N_0, the maximum density of adsorbed molecules; it is in a more detailed theory s_1 replaced by "s_1, N_0" (12)).

One interesting feature of the dynamic model is that it generates adsorption isotherms which are similar to experimentally observed isotherms. The shape of the isotherm is determined by the size of the "surface denaturation" time constant s_1 and a_2/a_1. The isotherms may have shapes normally taken as a sign of (lateral) interaction between adsorbed molecules or a heterogenous composition of the protein solution, see Figure 7 (12).

It has been suggested that protein exchange reactions could be described by a protein concentration dependent desorption of already adsorbed molecules i.e. by replacing r_1 by $r_1 = (r_{1d}) + r_{1e}C_1$ and r_2 by $r_2 = (r_{2d}) + r_{1e}C_1$ and r_2 by $r_2 = (r_{2d}) + r_{2e}C_1$ (12). This together with the assumption that with time a fraction of the protein molecules on the surface becomes truly irreversibly bound ("non exchangable") could explain the difference between direct and subsequent addition of the protein to the solution during an adsorption experiment. The difference between a hydrophobic and a hydrophilic surface was then to be found in the fraction of irreversible bound protein molecules (12). The use of a protein dependent desorption is, however, naive in the sense that this actually means that the surface coverage of proteins is first decreased and then the surface is covered with new molecules. The protein exchange mechanism is more likely a gradual phenomenon as described earlier in this section. Furthermore, the simple model with concentration dependent desorption is an equilibrium model which does not consider surface aggregation phenomena, phase transitions and domain formation on the surface. Thus, it is not possible, with the present model to explain all the details of protein adsorption and exchange on the surface.

Conclusions

Our results on protein adsorption on solid surfaces support the view that protein molecules, in general, change conformation more on hydrophobic surfaces and are more strongly bound on these surfaces than on hydrophilic surfaces. The use of a wettability gradient along a given substrate surface enables us to study, in a convenient way, the interaction between proteins and surfaces with different surface energies. The gradient method is most probably not limited to a gradient in wettability but can be extended to other types of gradients as well. It is however necessary to develope techniques to evaluate the microscopic details of the gradient surface itself to be able to make use of the full potential of the method.

The immune complement system appears to be an interesting model system for the study of surface induced biological phenomena closely related to conformational changes of surface deposited proteins. The study of C3 adsorption and conformational changes on surfaces has already yielded detailed information about surface-protein interaction. The fact that the deposition kinetics of organic material from human serum depends on the nature of the surface and that this deposition contains a large amount of C3 makes further studies of the complement system worthwhile, not the least in connection with biocompatibility.

Acknowledgments

Our research on protein adsorption on solid surfaces has been performed mainly through grants from the National Swedish Board for Technical Developement.

Literature Cited

1. Jönsson, U.; Ivarsson, B.; Lundström, I.; Berghem, L. J. Colloid Interface Sci. 1982 ,90, 148-63.
2. Ivarsson, B.; Hegg, P.-O.; Lundström, K.I.; Jönsson, U. Colloids and Surf.1985 ,13, 169-92.
3. Liedberg, B.; Ivarsson, B.; Lundström, I. J. Biochem. Biophys. Meth. 1984, 9, 233-43.
4. Liedberg, B.; Ivarsson, B.; Lundström, I.; Salaneck, W. Prog. Colloid Polym. Sci. 1985, 70, 67-75.
5. Liedberg, B.; Lundström, I.; Wu, C.R.; Salanec, W.R. J. Colloid Interface Sci. 1985, 108,123-32.
6. Lundström, I.; Salaneck, W.R. J. Colloid Interface Sci. 1985, 108, 288-91.
7. Sundgren, J.-E.; Bodö, P.; Ivarsson, B.; Lundström, I. J. Colloid Interf. Sci. 1986, 113, 530-43.
8. Sundgren, J.-E.; Bodö, P.; Lundström, I.; Berggren, A.; Hellem, S. J. Biomed. Mat. Res. 1985,19, 663-71.

9.   Sundgren, J.-E.; Bodö, P.; Lundström, I. J. Colloid
      Interface Sci. 1986, 110, 9-20.

10.   Lundström, I. Physica Scripta 1983, T4, 5-13, 1983.

11.   Lundström, I. Prog. Colloid Polym. Sci. 1985, 70,
      76-82.

12.   Lundström, I.; Ivarsson, B.; Jönsson, U.; Elwing, H. In
      Polymer Surfaces and Interfaces; W.J. Feast and H.S.
      Munro, Eds.; John Wiley and Sons, Chichester 1986,
      (in press)

13.   Lundström, I.; Elwing, H. J. Theoret. Biol. 1984,110,
      195-204.

14.   Elwing, H.; Dahlgren, C.; Harrison, R.; Lundström, I. J.
      Immunol. Methods 1984, 71, 185-191.

15.   Elwing, H.; Ivarsson, B.; Lundström I. Eur. J. Biochem.
      1986, 156, 359-65.

16.   Elwing, H., Ivarsson, B. and Lundström I. J. Biomed.
      Mat Res. (in press)

17.   Elwing, H., Welin, S., Askendahl, A., Nilsson U. and I.
      Lundström.   accepted in J. Colloid Interface Sci.

18.   Welin, S.; Elwing, H.; Wikström, M.; Arwin, H.;
      Lundström, I. Anal. Chim. Acta 1984,163, 263-67.

19.   Arwin, H.; Lundström, I. Anal. Biochem. 1985,145,
      106-112.

20.   Liedberg, B.; Nylander, C.; Lundström, I. Sensors and
      Actuators 1983,4, 299-309.

21.   Ivarsson, B.; Lundström, I. CRC Critical reviews in
      Biocompatibility 1986, 2,1-96.

22.   Azzam, R.M.A.; Bashara, N.M. Ellipsometry and
      Polarized Light, Elsevier/North-Holland,
      Amsterdam,1977.

23.   Elwing, H., Welin, S., Askendahl, A. and Lundström, I.
      (submitted)

24.   Vroman, L.; Adams, A.L. Surf. Sci. 1969, 16, 438-46.

25   Lee, R.G.; Adamson, C.; Kim, S.W. Thromb. Res. 1974, 4,
      485-90.

26   Brash, J.L.; Davidson, V.J. Thromb. Res. 1976, 9,
      249-59.

27   Brash, J.L.: Uniyal, S. J. Polymer Sci. 1979, 66,
      377-89.

28   Vroman, L.; Adams, A.L. J. Colloid. Interface Sci.
      1986, 111, 391-402

29.   Craddock, P.R.; Fehr, J.; Dalmasso, A.P.; Brigham, K.L.;
      Jacob, H.S. J. Clin. Invest. 1977, 59, 879-88.

30.   Arnaout, M.A.; Hakin, R.M.; Todd, R.F.; Dana N.; Colten,
      H.R. New. Eng. J. Med. 1985, 312, 457-62.

31.   Hammerschmidt, D.E.; Craddock, P.R.; McCulluogh, J.;
      Kronenberg, R.S.; Dalmasso, A.P.; Jacob H.S. Blood,
      1978, 51, 721-30.

32.   Herzlinger G. A. In Biocompatible polymeres, metals
      and composites; M. Szycher Eds.; Technomic
      Publishing Co. Inc., 851 New Holland Ave, Box 3535,
      Lancaster, Penn. USA. 1982 . p 89.

33.   Nilsson, U.R.; Nilsson, B. J. Immunol. 1982,129,
      2594-97.

34.   Nilsson, B; Nilsson U.R. Scand. J. Immunol. 1985, 22,
      703-10.

35.   Mac Ritchie, F. In Advances in Protein Chemistry;
      Anfinsen, C.B.; Edsall, J.T.; Richards F. M., Eds.; vol.
      32, Academic Press, New York, 1978, p 283.

36.   Hoffman, A.S. Blood-biomaterial interactions - an
      overview, Advances in Chemistry Series No199,
      American Chemical Society; Washington, DC. 1982;
      p.3.

37.   Biomedical Polymers: Protein Adsorption; J. D.
      Andrade, Ed.; Plenum Press, New York, 1985

38.   Soderquist, M.E.; Walton, A.G. J. Colloid Interface
      Sci. 1980, 75, 386-97.

RECEIVED April 29, 1987

ROLE OF PROTEIN ADSORPTION
IN BLOOD-MATERIAL INTERACTIONS

Chapter 30

Protein Adsorption at the Solid–Solution Interface in Relation to Blood–Material Interactions

John L. Brash

Departments of Chemical Engineering and Pathology, McMaster University, Hamilton, Ontario L8S 4L7, Canada

The author's work on the adsorption of plasma proteins using radiolabeling, hydrodynamic thickness measurements and elution/identification methods is reviewed. This work was motivated by the need to understand the role of protein adsorption in thrombus formation following blood-foreign surface contact. Studies of single protein systems showed that adsorption occurs in monolayers, that adsorption is inherently, but only slowly reversible and that some alterations of protein structure can occur upon adsorption. Work with protein mixtures showed the strong preferential adsorption of fibrinogen relative to albumin and I_gG. From plasma there is a relative lack of adsorption of the abundant proteins. Adsorption of fibrinogen from plasma was shown to be transient; it is displaced by contact phase clotting proteins. Part of the fibrinogen adsorbed from plasma was found to be degraded by surface generated plasmin implying that surfaces have thrombolytic as well as thrombogenic properties.

Work by the present author on protein adsorption has been motivated by the desire to understand the sequence of events leading to clot and thrombus formation on solid surfaces in contact with blood. The present review of these studies is organized on the basis of the medium in which the protein is dissolved, beginning with single proteins in buffer and progressing to plasma. By and large this is also the chronological order in which the work was carried out since it was believed that the more complex would only become comprehensible through an understanding of the simple. This belief has been only partially corroborated by our findings. There is no question that single protein data have enormously aided the interpretation of blood studies, for example by defining the "capacities" of surfaces for proteins and by providing data on protein denaturation at surfaces. At the same time events such as the transient adsorption of fibrinogen which occur in blood were not predicted from studies with simpler media.

We begin with some comments on the various experimental methods used in our studies. Investigations of single proteins in buffer are then discussed, including kinetics and isotherms, reversibility, denaturation, and the structural status of adsorbed proteins. Results of competitive adsorption studies using mixtures of proteins in buffer are then described. We next discuss our more recent studies of protein adsorption from blood plasma using both radiolabeled proteins and elution techniques. Finally, data on the effect of red blood cells on protein adsorption are summarized.

0097-6156/87/0343-0490$06.00/0
© 1987 American Chemical Society

Experimental Methods

In our experimental studies to measure adsorbed amounts we have mostly used methods based on radiolabeling with isotopes of iodine. We have favored the iodine monochloride (1) and lactoperoxidase (2) methods of labeling mainly because of various reports stating that proteins are only minimally, if at all, altered by these reagents (3). A certain amount of controversy has surrounded the use of radiolabeled proteins for adsorption studies (4-6). In this regard we have used a simple test to validate the method for any system under study. Adsorption is measured for a series of solutions of constant total concentration but varying in the ratio of labeled to unlabeled protein. If the surface concentration is invariant, then we conclude that radiolabeling does not influence adsorption. In the overwhelming majority of systems we have studied, this has proved to be the case, and thus we conclude that radiolabeling is a valid and widely applicable technique.

Most of our experiments have utilized surfaces in the form of tubing segments of about 0.25 cm diameter (7). The tubing is initially filled with buffer which is subsequently displaced by the protein solution or plasma in the absence of any air-solution interface. At the end of the adsorption period the protein solution is displaced by buffer, the tube is rinsed in a standard manner with buffer, and then counted for bound radioactivity. The necessity for rinsing of course introduces uncertainty since loosely bound or reversibly bound protein may be desorbed. Various attempts have been made to eliminate this uncertainty and in particular the use of unlabeled protein solutions of the same concentration for rinsing should be mentioned. We have found that very often on surfaces where desorption may be a problem, the adsorbed amount after rinsing with unlabeled protein is less than after rinsing with buffer showing that in such systems exchange of bound and dissolved protein may be more rapid than desorption.

In early work (8) we used infrared spectroscopy coupled with attenuated total reflection optics. This work was done before the availability of infrared equipment based on Fourier transform methods. Due to their relative speed these methods now permit in situ, real time measurements with a resolution of 1 sec or less (9), and continue to yield valuable data, particularly in the hands of the Battelle group in a series of studies dating from 1979 (10). In our early infrared work we had to be content to rinse and dry the surface before obtaining the infrared reflection spectrum. Nevertheless the values of surface concentration were remarkably close to those determined more recently. Infrared studies of proteins suffer generally from the fact that the main features of protein spectra are similar for all proteins and therefore it is difficult to distinguish one from another.

In studies with colleagues in Strasbourg (11) we used the hydrodynamic method based on capillary flow measurements to obtain estimates of adsorbed layer thickness. This method has been most widely used to measure the thickness of adsorbed synthetic polymers (12). It depends on the difference in flow through capillaries due to the reduction in diameter caused by adsorption. Capillary diameters of about 1 μm are suitable for detecting protein monolayers. Capillaries of this size are subject to blocking by dust particles and we found it necessary to work with networks of many capillaries in parallel as opposed to single capillaries since blockage of a few capillaries made little difference to the total flow. The difficulty then lies in defining the diameter of the capillary before adsorption since a distribution of diameters usually exists in the network. For this reason we concluded that this technique is better adapted to measuring changes in thickness due to variables such as temperature, pH etc. rather than absolute thickness.

Adsorption from Solutions of Single Proteins in Buffer

Early work using infrared spectroscopy. In this work (8) we measured isotherms at 37°C for fibrinogen, albumin and I$_g$G on various polymeric materials. As in many other investigations before and since, we noted the existence of plateaux in the isotherms and by suitable calibration we concluded that the plateaux correspond to monomolecular layers for each of the three proteins. The precision of the method was insufficient to permit detailed analysis of the isotherms although we were bold enough to claim that, at least on average, the molecules in the layer could not be drastically denatured since their average dimensions were not substantially changed compared to those in solution. In retrospect this work may have had its main value in drawing attention to protein adsorption as an important part of blood-material interactions (we also showed the development of protein layers on the surface after whole blood contact) and to the potential value of quantitative measurements of these phenomena.

<u>Kinetics and isotherms using radioiodine labeling</u>. In considering the kinetics of any interfacial phenomenon the question of transport versus reaction control must always be kept in mind. Our studies have not contributed much in this regard due to the fact that adsorption is relatively rapid. Our method requires separation of the surface from the solution before measuring and is consequently not capable of the 1 s or so time resolution needed to study initial adsorption kinetics. Using total internal reflection fluorescence methods, Robertson et al. (13) have shown that adsorption is in general transport controlled at low coverage but at higher coverage it becomes reaction limited due to "crowding" effects.

We have found generally that adsorption is as much as 75% complete in a few minutes (7,14,17) and reaches "equilibrium" (see below for a discussion of protein adsorption "equilibrium") in about one hour. Data on the glass-fibrinogen system were obtained under laminar flow conditions in tubes (14). It was found that the amount of fibrinogen adsorbed at a given time from 5 to 30 min was independent of surface shear rate from 0 to 2100 s^{-1}. These data give a clear indication that transport is not the controlling step under these conditions. In support of this conclusion we found that surface coverage after 5 min was about 75% indicating an intrinsically rapid adsorption at the concentration studied (3 µM).

We have measured adsorption isotherms at room temperature for a range of systems (14-17). The main results from this work are that the isotherms generally show well-developed plateaux and although the amounts adsorbed at the plateau vary from surface to surface, they are all in the close-packed monolayer range. In this connection it must be mentioned that adsorbed amounts in close packed protein layers can vary because of varying surface-protein orientation and the possibility of conformational change. Apparent exceptions to the monolayer "rule" are hydrogel-like materials such as segmented polyurethanes based on polyethylene oxide soft segments. These materials appear to adsorb essentially no protein (18,19). However, it is difficult to decide whether this result is real or only apparent due to the possible loss of protein during rinsing since adsorption on such materials is likely to be rapidly reversible.

One of our studies (17) was done to investigate the influence of fixed electrical surface charge on protein adsorption, and data were obtained for the adsorption of fibrinogen on glass and various polyelectrolyte complex surfaces. Different plateaux were exhibited by the different surfaces, perhaps indicating different specific surface areas and/or site densities. However, an analysis of the initial or rising portion of the isotherms showed that the slopes were the same for all surfaces, and thus that the adsorption energies are independent of the fixed electrical charges. In this conclusion, perhaps running contrary to intuition which might suggest that charge repulsion would occur if surface and protein have the same charge sign, we are in agreement with several other studies (20,21). Of course attractive interactions between surfaces and proteins of the same charge sign can be explained by ion-bridging. Also Norde and Lyklema (22) have shown that extensive charge redistribution may occur upon adsorption, involving incorporation of small ions into the adsorbed layer.

Measurements of equilibrium layer thickness were also done on the fibrinogen-glass system (11). The thickness versus solution concentration data showed an overall increase corresponding to the adsorption versus concentration curve. However the thickness "isotherm" was "stepped" at thicknesses corresponding closely to side-on and end-on molecular orientations. The thickness data thus support the point of view that adsorption occurs at the monolayer level. Interestingly this study also showed a tremendous extension of the fibrinogen molecules away from the surface at high pH, although the solution conformation is known to be stable under these conditions.

Reversibility. This topic might be considered appropriately under the "isotherm" heading but due to the persistent controversy and discussion surrounding it, we give it separate treatment. Most investigators, ourselves included, have found that adsorption of proteins on solid surfaces is irreversible on a normal time scale in the sense that one cannot "descend" the isotherm (8,13,14,23). This is particularly true of hydrophobic solids. Some authors have contended that since proteins can be eluted in media other than those used for adsorption, e.g. detergents or buffers of high ionic strength, adsorption is therefore inherently reversible. We regard this as a questionable argument since the introduction of the eluent solution means that the process of protein leaving the surface is no longer simply the reverse of adsorption. Similarly it has been argued that irreversibility can be due to conformational change (24). Again, however, reversibility of adsorption per se cannot be judged if an irreversible conformational change accompanies adsorption. It has also been suggested (24) that due to the high values of the putative equilibrium constants one would not be able to measure the vanishingly small quantities of protein released into solution as a result of desorption. This argument ignores the fact that, if one is measuring the protein bound to the surface, a decrease should be easily detected but usually is not observed.

Our main contribution to this debate, besides our own observations of "irreversibility" (8,14), lie in our demonstration that while desorption is either non-existent or extremely slow, there is measurable exchange of protein between surface and solution while the two remain in contact (7,14,15,25). The use of labeled proteins is particularly well adapted for this kind of study since unlabeled molecules or molecules labeled with different iodine isotopes can be used initially in the different "compartments" of the system.

The major conclusions from these studies are: (1) that such exchange occurs in all systems thus far investigated, (2) that the extent of exchange varies with the surface, is usually greater for hydrophilic than for hydrophobic surfaces, and depends on factors such as protein concentration and flow, (3) that the rate of exchange varies similarly to the extent of exchange, with "half-lives" ranging from days (e.g. polyethylene-albumin) (7) to hours (e.g. glass-fibrinogen) (14). With polyelectrolyte complex surfaces and fibrinogen we found that slow-exchanging and fast-exchanging populations of molecules could be distinguished, the former corresponding to points low on the isotherm where less than monolayer coverage exists (25). This "multipopulation" behavior is illustrated in Figure 1.

Exchange phenomena have been noted by others with respect both to proteins and synthetic polymers (13,26-28). The occurrence of exchange in contrast to the lack of desorption into solvent is counter-intuitional since exchange perforce involves a desorption step. All of the evidence, particularly concentration dependence, indicates a mechanism involving interactions between protein in solution and adsorbed protein. A likely explanation due to Jennissen (29) is illustrated in Figure 2 and is based on the very plausible concept of "multivalent" binding between protein and surface. Exchange is envisaged as a cooperative effect whereby a single site of an adsorbed molecule "desorbs". Such single site exchange represents the initial step of whole molecule exchange. The concept of multivalent binding also provides an explanation for lack of desorption since the breaking of several bonds simultaneously is an unlikely occurrence. According to this explanation irreversibility is only apparent and has kinetic origins.

Our current view, taking all the above evidence into account, is that adsorption is inherently reversible. In addition to the self-exchange phenomenon, other evidence in

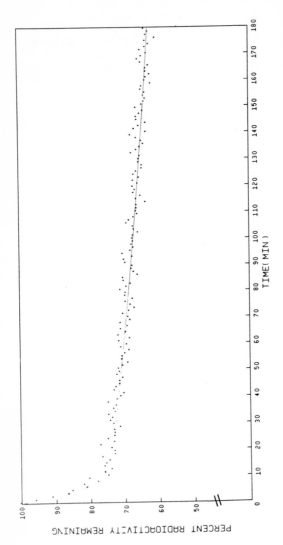

Figure 1. Exchange of dissolved and adsorbed fibrinogen on a positively charged polyelectrolyte complex at steady state (no change in adsorbed amount). Initially adsorbed radiolabeled fibrinogen is desorbing and unlabeled dissolved fibrinogen is adsorbing. The data clearly show a rapidly exchanging and a slowly exchanging population. (Reproduced with permission from Ref. 25. Copyright 1983, Academic Press.)

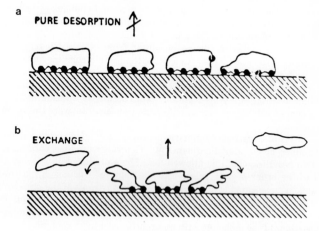

Figure 2. Schematic showing (a) low probability of desorption for a protein adsorbed via several binding sites and (b) concerted cooperative mechanism of exchange between dissolved and adsorbed protein. (Reproduced with permission from Ref. 66. Copyright 1985 Plenum.)

support of this view is that one can usually "climb" the isotherms in well delineated stages: if adsorption were truly irreversible, one would expect that the isotherm would rise instantly to the plateau or that in very dilute systems where there is insufficient protein to reach the plateau, complete depletion would be observed. Since this does not occur, reversibility is implied.

<u>Denaturation and Adsorption: Structural Status of Eluted Proteins</u>. Stemming to a large extent from the conventional wisdom that proteins adsorbed at the air-solution interface are denatured, for example in foaming phenomena, it has been widely speculated that denaturation accompanies adsorption on solids. Many investigations using both physical (<u>20,30</u>) and biological (<u>31,32</u>) criteria of denaturation have been conducted. Some systems have shown evidence of denaturation and others have not. For example Factor XII adsorbed on quartz was shown to undergo a conformational change (<u>33</u>) while thrombin adsorbed on cuprophan or polyvinylchloride retained most of its biological activity (<u>31</u>).

Our main contributions to the study of denaturation consist of experiments to evaluate the structural status of fibrinogen eluted from glass. Clearly, it would be preferable to evaluate structural alteration while the protein remains on the surface but such an approach presents formidable experimental difficulties, not the least of which is sensitivity. Therefore, we made the compromise of studying eluted protein and making the tacit assumption that the eluted protein is structurally similar to the adsorbed protein. The eluents used (high molarity buffers and surfactants) were shown not to affect the properties examined.

In one series of experiments, fibrinogen eluted from glass tubing or fritted glass filters was examined by circular dichroism (<u>34</u>). Controls were incorporated so that any effects of protein handling could be allowed for. The eluted protein showed a loss of α-helix content of the order of 50% relative to the "native" fibrinogen, so that surface interactions appear to be capable of disrupting α-helical regions of the protein. It should be noted that we could not distinguish in these experiments between indiscriminate desorption of whole protein with an overall reduction of α-helix content and preferential desorption of portions of the molecule with inherently low α-helix content.

A second series of experiments utilizing glass bead columns was designed so that adsorbed fibrinogen could be eluted in stages (<u>35</u>). After adsorption from 0.05 M Tris, pH 7.4, initial fractions were eluted using 1 M Tris, which, as had been shown with labeled protein, removes 80% of the adsorbed fibrinogen. The remaining 20% (as judged by labeled protein) was eluted with 2% SDS. The fractions were examined by SDS-PAGE under reducing conditions and it was found that the initially eluting fractions, presumably containing the fibrinogen molecules that are the least firmly bound, had undergone considerable chain degradation. The extent of degradation was less in later-eluting fractions, and for the SDS-eluted fractions there was very little difference from "native" fibrinogen. The fact that various fractions differ in extent of degradation suggests that several surface populations are present, in agreement with conclusions based on self-exchange and differential elutability as discussed above.

The gel band patterns of the degraded fractions bear a strong resemblance to those of early plasmin-induced degradation products of fibrinogen (FDP). Therefore, a possible explanation of our results is that traces of plasminogen present in the fibrinogen preparation are activated to plasmin by contact with glass. In this regard, we have found that fibrinogen purified on DEAE cellulose, which is reported to remove plasminogen (<u>36</u>) is less degraded after glass contact than is unpurified fibrinogen. These observations suggest rather strongly that plasminogen activation and subsequent fibrinogen degradation may be blood-material interactions of some significance that have not been recognized heretofore.

Mixtures of Proteins in Buffer: Competitive Adsorption

Most real protein-containing fluids of interest, such as blood, are multicomponent systems so that the influence of one protein on another may become important. With respect to adsorption, the main consideration is whether adsorbed amounts are in proportion to solution concentration or whether surfaces "select" one protein in preference to another. Presumably if proteins act independently of each other one should be able to predict adsorbed amounts in mixtures from relative affinities derived from single protein studies. Although there have been no systematic attempts to make such predictions it seems likely that they would fail. In general it has been found that preferential or selective adsorption occurs so that certain proteins may be enriched in the surface relative to the solution and vice versa. There have as yet been no attempts to determine the properties of protein-surface systems that govern the relative surface affinity of different proteins. More will be said on this topic when adsorption from plasma is discussed.

Mixture studies using two- or three-protein systems have been used as simple compositional models of complex biological fluids like plasma (<u>38</u>) or milk (<u>39</u>). In the case of plasma, which is relevant to our own work, mixtures of the three abundant proteins albumin, I_gG and fibrinogen in varying properties have often been used (<u>16,18,38,40,41</u>). We have carried out a number of investigations using mixtures of these proteins (<u>16,18,41</u>). Radiolabeling provides an excellent means of distinguishing one protein from another, and using the isotopes ^{125}I and ^{131}I two proteins may be measured simultaneously. For the binary system albumin-fibrinogen adsorbing on glass from Tris buffer (<u>16</u>), it was found that fibrinogen is preferentially adsorbed. In a series of mixtures of varying composition, the equilibrium ratio of surface mole fraction to solution mole fraction of fibrinogen ranged from 70 (solution mole fraction = 0.01) to 6 (solution mole fraction = 0.15). These data emphasize the very strong fractionation effect of the glass surface with these mixtures. Data for several other surfaces at a single solution composition (mole ratio fibrinogen: albumin = 0.005, close to the ratio in human plasma) showed surface enrichment of fibrinogen between 10- and 200-fold (<u>18</u>).

In ternary mixtures of fibrinogen, IgG and albumin (<u>41</u>), the preferential adsorption of fibrinogen was again observed. These experiments were conducted using proportions of the proteins the same as are found in blood, but at varying total concentrations. For each composition, three separate kinetic experiments were conducted in which each of the three pair combinations of the proteins were labeled, and the third protein was unlabeled. Over four hours, fibrinogen adsorption increased continuously, although the rate decreased, while IgG showed an adsorption maximum at about 5 minutes and then remained constant at a relatively low value. Albumin adsorption was effectively zero in these experiments. These data again demonstrate the overwhelming preference of glass for fibrinogen relative to the other abundant plasma proteins. Thus, it is tempting to suggest that from plasma or blood, fibrinogen would be preferentially adsorbed. This point should be kept in mind when results of adsorption from plasma are discussed below. It is also relevant to point out that the experiments with the ternary system were all conducted at total concentrations that are very dilute compared to plasma. As will be seen below, we have since found in plasma itself that adsorption of fibrinogen is greatly enhanced when the plasma is diluted.

Several other groups have also found preferential adsorption of fibrinogen from 2- and 3-protein mixtures (<u>13,38,40,42</u>). These studies have been done on a variety of surfaces with various mixture compositions and total concentrations and consistently confirm preferential adsorption of fibrinogen. Many of these measurements refer to equilibrium and involve relatively long adsorption times; information on the time dependence of relative quantities adsorbed in multi-protein systems is largely missing. One of the few such studies is that of Gendreau et al (<u>42</u>), using FTIR spectroscopic techniques. They found that for a 1:1 mixture of albumin and fibrinogen, albumin predominated in the first 7 minutes and then was gradually displaced by fibrinogen.

Adsorption of Proteins from Plasma

Clearly studies of single proteins and of mixtures in buffer solutions provide useful information as is evident from the above discussion. However, only studies using blood or plasma can give definitive answers regarding blood-material interactions. Accordingly much of our research over the past several years has been devoted to studies of protein adsorption from plasma or blood.

Our work with plasma has used two approaches. In the first, purified radiolabeled proteins are added in small amounts to the plasma as tracers and their adsorption measured as a function of time, plasma concentration and composition, and surface type. The surfaces are in the form of tubing segments, as for the single protein and mixture studies already discussed. This approach gives quantitative data on the adsorption of individual proteins. The second approach is more qualitative in nature and seeks to identify the multitude of proteins adsorbed out of plasma to different surfaces. In this work the surfaces are in particle form to give high surface-to-volume ratios, and after plasma contact they are eluted and the proteins are separated and identified by electrophoresis and other techniques. Our results with these two approaches are now described.

Studies using radiolabeled tracers. An initial series of experiments was done with albumin, fibrinogen and IgG, and a number of surfaces including glass, siliconized glass, polyethylene, polystyrene and several segmented polyurethanes (43). Fibrinogen was not detected on any of the hydrophilic surfaces. On polyethylene and siliconized glass fibrinogen adsorption was greatest at the shortest time (2 min) and then decreased to near zero. Only on polystyrene was fibrinogen adsorption substantial and constant with time. Albumin was also not detected on the hydrophilic surfaces but was adsorbed substantially on hydrophobic surfaces. IgG was detected on all surfaces though in relatively low surface concentrations. From these observations we concluded: (1) The plasma itself modifies adsorption. Therefore single protein or simple mixture studies cannot be used to predict plasma adsorption. (2) Fibrinogen is either not adsorbed or only transiently adsorbed on most surfaces. This is a surprising finding in view of the widely reported fact that fibrinogen is preferentially adsorbed from simple mixtures with albumin and IgG. (3) Since the major plasma proteins are only minimally adsorbed on surfaces that are known to adsorb compact monolayers of protein from solution, it appears that the less abundant proteins of plasma may be important components of adsorbed layers.

In a follow-on study using the trace-labeling approach (44), we have further investigated the adsorption of fibrinogen. As indicated above fibrinogen appeared to be transiently adsorbed to most surfaces with a "residence time" on the surface which varied with the material. On surfaces such as glass where we failed to detect any fibrinogen, we suspected this was due to very short residence times. Unfortunately our experimental technique is not suitable for observing residence times of less than a minute. However if fibrinogen adsorption is transient due to displacement by other proteins as suggested by Vroman et al (45), it seemed likely that by diluting the plasma sufficiently, the concentration of displacing species would decrease below effective levels so that fibrinogen residence times on the surface would be increased. Therefore we undertook a study of the effect of plasma dilution on fibrinogen adsorption.

The surfaces studied were glass, polyethylene and siliconized glass. On dilution of the plasma we found as predicted that fibrinogen adsorption gradually increased to a maximum at a plasma concentration about 1% of normal, the exact value depending on the surface and increasing with increasing surface hydrophobicity. Data for glass are shown in Figure 3. It was concluded that the increase of adsorption from 0 to 1% plasma is due to increasing availability of fibrinogen (an "isotherm" type of relationship) and that beyond the maximum the decrease of adsorption is due to displacement of fibrinogen by

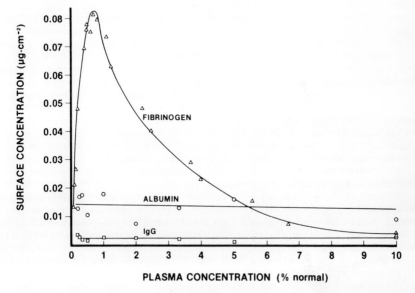

Figure 3: Adsorption of albumin, I_gG, and fibrinogen from plasma to glass as a function of plasma concentration. Plasma was diluted with isotonic Tris, pH 7.35 and adsorptions were for J min. Reproduced with permission from Ref. 44. Copyright 1984, F.K. Schattauer-Verlag.

other plasma components. Kinetics of adsorption at different plasma concentrations, as shown in Figure 4, also support this conclusion: at concentrations less than the adsorption maximum, kinetics is conventional with adsorption increasing onto a plateau, while at concentrations greater than the maximum, kinetics curves show maxima at times which increase as plasma concentration decreases (increasing fibrinogen residence times).

These results support the hypothesis of Vroman et al (45) regarding displacement of initially adsorbed fibrinogen and we now refer to this phenomenon as the "Vroman Effect". It may be noted that simultaneously with ourselves other groups have reported the same phenomena (46,47). The work of Bantjes et al (47) is notable in that it proposes high density lipoprotein as the fibrinogen displacing species. This is in contrast to Vroman's contention that components of the intrinsic coagulation system, e.g. high molecular weight kininogen (HMWK), are the displacing species (45, 48).

In recent studies (19,49) we have developed a more extensive data base on the Vroman Effect and have examined its phenomenology in detail. All surfaces studied, except for two hydrophilic polyurethanes, showed the Vroman Effect. However quantitative differences (peak height, peak position) were found among surfaces and there was some evidence that the height of the peak in the adsorption versus plasma concentration curves is correlated with gross thrombogenicity. This was especially true when data at longer times (24 h) were considered. The hydrophilic polyurethanes, which are generally regarded as relatively thromboresistant, showed no Vroman Effect from 0.07 to 10% plasma concentration. Any attempt to attach significance to this observation in terms of thromboresistance of surfaces is premature. For the moment we do not understand the significance of the Vroman Effect in relation to surface thrombogenesis and have postulated two possibilities: (1) the key surface property is prevention of coagulation so that if initially adsorbed fibrinogen is displaced by HMWK and other contact phase clotting factors (as suggested by Vroman et al. and by ourselves (see below)) then fibrinogen retention is desirable, (2) the key surface property is non-reactivity to platelets, so that since adsorbed fibrinogen is considered to be platelet-reactive, fibrinogen retention is undesirable.

As we and others continue this line of investigation it is becoming evident that the whole area of competitiveness of protein adsorption in the blood context, which at the moment seems hopelessly mired and anecdotal, may eventually reveal a rational, orderly nature. Vroman has postulated (50) that a rapid sequence of adsorption and displacement events occurs by which, over time, more abundant proteins are displaced by less abundant. The time frame of these events is such that albumin is adsorbed and replaced in a fraction of a second, thus accounting for the fact that it is not often observed on the surface after plasma contact. Clearly the time frame may be expected to vary with the surface so that in some cases the sequence will be relatively fast and in others relatively slow. Thus the apparent absence of a Vroman Effect for hydrophilic polyurethanes may reflect a very rapid sequence, such that a fibrinogen peak would be observed only at very short times or at low plasma concentrations (less than 0.05%).

The postulate of Vroman addresses only one variable, namely concentration, that may be expected to influence a protein's adsorption competitiveness. From known adsorption principles we may reasonably assume that protein affinity for the surface (free energy of adsorption), protein-protein interactions in the layer, and kinetic factors such as activation energy of adsorption will also influence competitive adsorption. The idea of a sequenced based at least partly on concentration seems reasonable, since initially diffusion would ensure relatively high concentrations of the abundant proteins, and only later would the true relative affinities of the proteins for the surface be expressed. This idea raises tantalizing questions such as whether all surfaces end up with the same protein layer or whether the sequence stops at different points on different surfaces.

In very recent work we have tried to identify the species that replace fibrinogen (51). The main approach in this work is to observe the adsorption of fibrinogen from plasmas that are deficient in various coagulation factors. If adsorbed fibrinogen is not

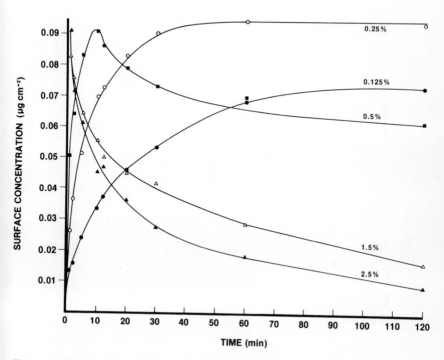

Figure 4: Kinetics of adsorption of fibrinogen from plasma to glass at various plasma concentrations (indicated on curves). Reproduced with permission from Ref. 44. Copyright 1984, F.K. Schattauer-Verlag.

replaced when plasma deficient in a particular factor is used, then that factor may be identified as a replacing species. For a glass surface the findings to date are: (a) HMWK-deficient plasma has a greatly diminished Vroman Effect thus confirming HMWK as a fibrinogen replacing species. (b) The Vroman Effect is not totally abolished in HMWK-deficient plasma suggesting there are other replacing components. (c) Plasminogen-deficient plasma gives a near normal Vroman Effect thus suggesting that fibrinogen is not substantially removed from the surface by plasmin attack. (This appeared to be a strong possibility given our data showing that adsorbed fibrinogen is degraded to some extent by plasmin.) (d) Factor XII-deficient plasma has a diminished Vroman Effect, although we believe this may be an indirect effect through kallikrein. (e) Activation of normal plasma with dextran sulfate (an activator of the contact system of coagulation) diminishes the Vroman Effect while the same treatment of Factor XI-deficient plasma has no effect. This suggests that HMWK is inactivated by Factor XIa.

Identification of proteins adsorbed from plasma.   As indicated above, our initial studies of adsorption of abundant plasma proteins led us to conclude that a major part of the adsorbed protein layer must consist of species as yet unidentified but other than IgG, fibrinogen or albumin. Since blood compatibility is believed to depend critically on the properties of the adsorbed protein layer it seemed to be a worthwhile goal to identify the proteins in this layer and particularly to try to pinpoint differences among various surfaces that might explain differences in compatibility. Our approach has been to expose the surface to plasma, then elute the adsorbed proteins and try to identify them. Since adsorption capacities are small, a large surface area is required to obtain useful quantities of protein for identification purposes. As a means of obtaining large areas we have used materials in particle form in packed columns. The experiment consists of incubation of the column with plasma, followed by sequential elution of adsorbed proteins with high molarity Tris buffer, then with sodium dodecyl sulfate (SDS). The eluted proteins are then concentrated and subjected to polyacrylamide gel electrophoresis (PAGE) to obtain molecular weight profiles. In addition, selected radiolabeled proteins were added to plasma in some experiments and the presence of these proteins (or fragments derived from them) in the eluates investigated by determining the location of radioactivity in the gels. Ouchterlony double diffusion immunoassays against a number of specific antibodies were also carried out. Results for various surfaces are as follows:

Glass (37): This typical hydrophilic surface shows complex PAGE patterns (Figure 5). Besides some albumin, IgG and fibrinogen, proteins found in the eluate were plasminogen and, most notably, plasmin-induced degradation products of fibrinogen (FDP) suggesting surface activation of plasminogen. In support of this conclusion it was found that FDP formation was less when either plasminogen-deficient plasma or Factor XII-deficient plasma was used. It should be noted in this connection that the contact factors of blood coagulation are known activators of the fibrinolytic system (52). Many other species are present in the glass eluates but are not yet identified, including a very strong band at 25,000 MW which appears to be a fragment of a larger protein.

Polystyrene: On this hydrophobic material fibrinogen has been identified as a major constituent of the layer in agreement with results from the trace-labeling studies. FDP are also observed although in lesser amounts than for glass. The major band at 25,000 seen for glass is absent. In general the eluates are considerably less complex than for glass, a result which may reflect the increased difficulty of eluting proteins from hydrophobic surfaces.

Modified (heparin-like) polystyrenes (53): This work represents a collaboration with the groups of M. and J. Jozefowicz of l'Université Paris-Nord, France who developed these materials (54,55). The materials in question are crosslinked polystyrenes to which

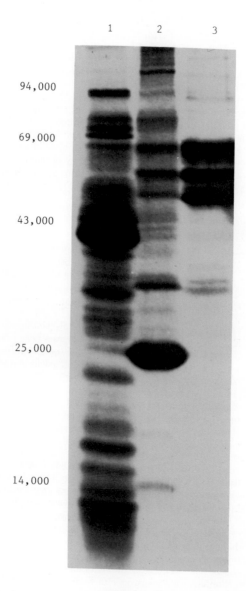

Figure 5: SDS-polyacrylamide gels of proteins eluted from glass after contact with normal human plasma. Lane 1: Pooled fractions eluted by 1M Tris, showing FDP. Lane 2: Pooled fractions eluted by SDS (note the very prominent band at 25,000 MW). Lane 3: Purified fibrinogen. Reproduced with permission from Ref. 37. Copyright 1986, John Wiley and Sons, Inc.

sulfonate, glutamic acid sulfamide or arginine sulfamide groups have been attached. They have been shown to possess anticoagulant properties and might therefore be expected to provide contrasting behavor to glass, a highly procoagulant surface, insofar as protein adsorption is related to activation of coagulation. Major findings with these materials are: (a) in contrast to glass there is no component at 25,000 MW. (b) There is considerable fibrinogen but also a large number of other components particularly at high MW. FDP and plasminogen are also found in the eluates.

General conclusions from this work to date are: (1) there are considerable differences in the proteins eluted from different surfaces, (2) the protein layers are complex and contain a large number of components, (3) a common feature for all materials so far examined is the formation of FDP implying activation of fibrinolysis at surfaces. The latter may well be a highly significant new finding in blood-material interactions, and suggests that a valid approach to thromboresistance may be to maximize fibrinolytic activity. Kusserow et al. (56) several years ago developed this approach by immobilizing urokinase on surfaces and more recently Senatore et al. (57) have tried to incorporate urokinase into small diameter vascular grafts.

The Influence of Red Blood Cells on Adsorption

Our original interest in this area was prompted by the fact that red cells are a major component of blood and as such might have some effect on protein adsorption. In addition we had shown (58) that red cells have a strong augmenting effect on platelet adhesion probably due to the increased mixing action deriving from red cell local motions in flowing blood. Similar effects of red cells on platelet diffusion have been noted by others (59).

In initial studies (60) we showed that, in contrast to platelet adhesion, adsorption of albumin and fibrinogen to polyethylene is inhibited in the presence of red cells (the red cell effect). In follow-on work we made a detailed study of the glass-fibrinogen system (61). In these experiments washed red blood cells were added to solutions of iodine-labeled fibrinogen and adsorption to the walls of glass tubing under flow conditions was measured. It was found that adsorption decreased as the red cell concentration increased, i.e. there is a "red cell effect" in this system also. It seemed possible that this inhibiting effect might be due to competitive adsorption by hemoglobin released from the red cells, but deliberately added hemolysate had no effect on adsorption. On the other hand hemoglobin-free ghost cells were as effective as whole red cells in inhibition of adsorption, suggesting that the red cell effect is membrane related, and is perhaps due to the deposition of membrane components which render the surface less adsorptive.

In subsequent studies (62) we pursued the possibility that membrane fragments may be deposited on the surface. Using both glass and siliconized glass bead columns, we have shown that materials elutable after red cell contact have UV-visible spectra and gel electrophoresis patterns that correspond very closely to those of red cell membranes, thus showing that membrane material is indeed deposited. The SDS-PAGE of proteins eluted from the column show that spectrin and actin are essentially missing, suggesting that the cytoskeleton remains intact as the cells interact with the surface. SEM of surfaces following cell contact shows the presence of filaments and particles whose dimensions correspond to the red cell "tethers" and microvesicles described in the literature (63). We have suggested that the deposition of membrane material as filaments and vesicles occurs through extrusion of part of the membrane, including the integral membrane proteins, on to the surface during contact. Whether the "red cell effect" found in our laboratory is important in whole blood contact situations has not been determined. In agreement with the red cell effect, Horbett has found reduced fibrinogen adsorption levels in ex vivo studies using baboons (64), whereas Chuang (65) has found higher levels of fibrinogen after clinical use of cuprophan dialysis membranes compared to in vitro plasma exposure.

Acknowledgments

The author wishes to thank his many colleagues and co-workers who have contributed to the work described in this article. Their names are clear from the reference list. Financial support from the Heart and Stroke Foundation of Ontario, and the Medical Research Council of Canada is also gratefully acknowledged.

Literature Cited

1.   McFarlane, A.S. J. Clin. Invest. 1963, 42, 346-61.
2.   Marchalonis, J.J. Biochem. J. 1969, 113, 299-305.
3.   Regoeczi, E. Iodine Labeled Plasma Proteins, Volume I, CRC Press: Boca Raton, Florida, 1984; Chapter 3.
4.   Crandall, R.E.; Janatova, J; Andrade, J.D. Prep. Biochem. 1981, 11, 111-38.
5.   Grant, W.H.; Smith, L.E.; Stromberg, R.R. J. Biomed. Mater. Res. Symp. 1977, 8, 33-38.
6.   Van Der Scheer, A.; Feijen, J.; Elhorst, J.K.; Krugers-Dagneault, P.G.; Smolders, C.A. J. Colloid Interface Sci. 1978, 66, 136-45.
7.   Brash, J.L.; Samak Q.M. J Colloid Interface Sci. 1978, 65, 189-201.
8.   Brash, J.L.; Lyman, D.J. J. Biomed. Mater. Res. 1969, 3, 175-89.
9.   Gendreau, R.M. Appl Spectrosc. 1982, 36, 47-49.
10. Gendreau, R.M.; Jakobsen, R.J. J. Biomed. Mater. Res. 1979, 13, 893-906.
11. de Baillou, N.; Dejardin, P.; Schmitt, A.; Brash, J.L. J. Colloid Interface Sci. 1984, 100, 167-174.
12. Rowland, F.W.; Eirich, F.R. J. Polym. Sci. Al. 1966, 4, 2401-21.
13. Lok, B.K.; Cheng, Y-L.; Robertson, C.R. J, Colloid Interface Sci. 1983, 91, 104-16.
14. Chan, B.M.C.; Brash, J.L. J. Colloid Interface Sci. 1981, 82, 217-25.
15. Brash, J.L.; Uniyal, S.; Samak, Q. Trans. Amer. Soc. Artif. Int. Organs 1974, 20, 69-76.
16. Brash, J.L.; Davidson, V.J. Thromb. Res. 1976, 9, 249-59.
17. Schmitt, A.; Varoqui, R.; Uniyal, S.; Brash, J.L.; Pusineri, C. J. Colloid Interface Sci. 1983, 92, 25-34.
18. Brash, J.L.; Uniyal, S. J. Polym. Sci. 1979, C66, 377-89.
19. Hudson, C.B.; Brash, J.L.; ten Hove, P. Trans. Soc. Biomaterials 1986, 9, 122.
20. Morrissey, B.W.; Stromberg, R.R. J. Colloid Interface Sci. 1974, 46, 152-64.
21. Norde, W.,; Lyklema, J. J. Colloid Interface Sci. 1978, 66, 257-65.
22. Norde, W.; Lyklema, J. J. Colloid Interface Sci. 1978, 66, 277-84.
23. MacRitchie, F. J. Colloid Interface Sci. 1972, 38, 484-88.
24. Norde, W.; MacRitchie, F.; Nowicka, G.; Lyklema, J. J. Colloid Interface Sci. 1986, 112, 447-56.
25. Brash, J.L.; Uniyal, S.; Pusineri, C.; Schmitt, A. J. Colloid Interface Sci. 1983, 95, 28-36.
26. Chuang, H.Y.K.; King, W.F.; Mason, R.G. J. Lab. Clin. Med. 1978, 92, 483-96.
27. Cheng, Y.L.; Darst, S.A.; Robertson, C.R. J. Colloid Interface Sci., to be published.
28. Pefferkorn, E.; Carroy, A.; Varoqui, R. J. Polym. Sci. Polymer Physics Ed. 1985, 23, 1997-2008.
29. Jennissen, H.P. Adv. Enzyme Reg. 1981, 19, 377-406.
30. Soderquist, M.E.; Walton, A.G. J. Colloid Interface Sci. 1980, 75, 386-97.
31. Chuang, H.Y.K.; Mohammad, S.F.; Sharma, N.C.; Mason, R.G. J. Biomed. Mater. Res. 1980, 14, 467-76.
32. Mizutani, T. J. Pharm. Sci. 1980, 69, 279-82.
33. McMillin, C.R.; Walton, A.G. J. Colloid Interface Sci. 1974, 48, 345-49.
34. Chan, B.M.C.; Brash, J.L. J. Colloid Interface Sci. 1981, 84, 263-65.

35.   Brash, J.L.; Chan, B.M.C.; Szota, P.; Thibodeau, J.A. J. Biomed. Mater. Res. 1985, 19, 1017-29.
36.   Lawrie, J.S.; Ross, J.; Kemp, G.D. Biochem. Soc. Trans. 1979, 7, 693-4.
37.   Brash, J.L.; Thibodeau, J.A. J. Biomed. Mater. Res. 1986, in press.
38.   Lee, R.G.; Adamson, C.; Kim. S.W. Thromb. Res. 1974, 4, 485-90.
39.   Arnebrant, T.; Nylander, T. J. Colloid Interface Sci. 1986, 111, 529-33.
40.   Horbett, T.A.; Hoffman, A.S. In Applied Chemistry at Protein Interfaces; Baier, R.E., Ed.; Advances in Chemistry Series No. 145; American Chemical Society: Washington, D.C., 1975; p. 230.
41.   Brash, J.L.; Uniyal S.; Chan, B.M.C.; Yu, A. In Polymeric Materials and Artificial Organs; Gebelein, C.G., Ed.; ACS Symposium Series No. 256; American Chemical Society: Washington DC, 1984; pp. 45-61.
42.   Gendreau, R.M.; Leininger, R.I.; Winters, S.; Jakobsen, R.J. In Biomaterials: Interfacial Phenomena and Applications; Cooper, S.L.; Peppas, N.A., Eds.; Advances in Chemisty Series No. 199, American Chemical Society; Washington, DC. 1982; p. 371.
43.   Uniyal, S.; Brash, J.L. Thromb. Haemostas. 1982, 47, 285-90.
44.   Brash, J.L.; ten Hove, P. Thromb. Haemostas. 1984, 51, 326-30.
45.   Vroman, L.; Adams, A.L.; Fischer, G.C.; Munoz, P.C. Blood 1980, 55, 156-59.
46.   Horbett, T.A. Thromb. Haemostas. 1984, 51, 174-81.
47.   Breemhaar, W.; Brinkman, E.; Ellens, D.J.; Beugeling, T.; Bantjes, A. Biomaterials 1984, 5, 269-74.
48.   Schmaier, A.H.; Silver, L.; Adams, A.L.; Fischer, G.C.; Munoz, P.C.; Vroman, L.; Colman, R.W. Thromb. Res. 1984, 33, 51-67.
49.   Wojciechowski, P.; ten Hove, P.; Brash, J.L. J. Colloid Interface Sci. 1986, 111, 455-65.
50.   Vroman, L.; Adams, A.L. J. Colloid Interface Sci. 1986, 111, 391-402.
51.   Brash, J.L.; Scott, C.F.; ten Hove, P.; Colman R.W. Trans. Soc. Biomaterials 1985, 8, 105.
52.   Mandle, R.J.; Kaplan, A.P. Blood 1979, 54, 850-62.
53.   Boisson, C.; Brash, J.L.; Jozefonvicz, J. Manuscript in preparation.
54.   Fougnot, C.; Jozefonvicz, J.; Samama, M.; Bara, L. Ann. Biomed. Eng. 1979, 7, 429-39.
55.   Boisson, C.; Gulino, D.; Jozefoonvicz, J.; Fischer, A.M.; Tapon-Bretaudiere, J. Thromb. Res. 1984, 34, 269-76.
56.   Kusserow, B.K.; Larrow, R.; Nichols, J. Trans. Am. Soc. Artif. Int. Organs 1971, 17, 1-5.
57.   Senatore, F.; Bernath, F.; Meisner, K. J. Biomed. Mater. Res. 1986, 20, 177-88.
58.   Brash, J.L.; Brophy, J.; Feuerstein, I.A. J. Biomed. Mater. Res. 1976, 10, 429-43.
59.   Turrito, V.T.; Benis, A.M.; Leonard, E.F. Ind. Eng. Chem. Fundam. 1972, 11, 216-23.
60.   Brash, J.L.; Uniyal, S. Trans. Am. Soc. Artif. Int. Organs 1976, 22, 253-59.
61.   Uniyal, S.; Brash, J.L.; Degterev, I.A. In Biomaterials: Interfacial Phenomena and Applications; Cooper, S.L.; Peppas, N.A., Eds.; Advances in Chemistry Series No. 199, American Chemical Society; Washington, DC, 1982; p. 277.
62.   Borenstein, N.; Brash, J.L. J. Biomed. Mater. Res. 1986, 20, 723-30.
63.   Hochmuth, R.M.; Mohandas, N.; Spaeth, E.E.; Williamson, J.R.; Blackshear, P.J., Jr.; Johnson, D.W. Trans. Am. Soc. Artif. Int. Organs 1972, 18, 325-32.
64.   Horbett, T.A. J. Biomed. Mater. Res. 1986, 20, 739-72.
65.   Chuang, H.Y.K. J. Biomed. Mater. Res. 1984, 18, 547-59.
66.   Andrade, J. D. Surface and Interfacial Aspects of Biomedical Polymers, Volume 2: Protein Adsorption; Plenum: New York, 1985.

RECEIVED March 27, 1987

Chapter 31

Platelet Activation by Polyalkyl Acrylates and Methacrylates: The Role of Surface-Bound Fibrinogen

Jack N. Lindon[1,2], Gerald McManama[1,5], Leslie Kushner[2,3], Marek Kloczewiak[1,2], Jacek Hawiger[1,2], Edward W. Merrill[4], and Edwin W. Salzman[1,2]

[1]Department of Surgery, Beth Israel Hospital, Boston, MA 02215
[2]Harvard Medical School, Boston, MA 02215
[3]New England Deaconess Hospital, Boston, MA 02215
[4]Department of Chemical Engineering, Massachusetts Institute of Technology, Cambridge, MA 02139

Platelet adhesion and activation by polyalkyl acrylates and methacrylates were found to increase with the length of the alkyl side chains and the resultant hydrophobicity of the polymers. For the methacrylates, platelet reactivity correlated with the affinity of anti-fibrinogen antibodies for fibrinogen adsorbed onto the surfaces in contact with fibrinogen solution, plasma (after short incubations only), or diluted plasma. Platelet reactivity did not correlate with total fibrinogen binding which did not vary among the methacrylates. Fibrinogen polymers, prepared by crosslinking fibrinogen with affinity purified antibodies against the E domain of fibrinogen, induced platelet aggregation which required intact metabolic activity and the availability of platelet membrane GP IIb/IIIa fibrinogen binding sites and did not require prior platelet activation. These results suggest that clusters of surface-bound fibrinogen molecules with minimal conformational alterations may provide a stimulus for platelet activation by surfaces.

Thrombotic complications are frequently encountered when blood is exposed to the surfaces of hemodialysis devices, heart-lung machines, arterial grafts, artificial heart components and other prosthetic devices. The blood platelets are particularly vulnerable to these adverse effects which may include a decrease in platelet count, shortened platelet survival and attendant higher platelet turnover, and altered platelet function. However the interaction of platelets with an artificial surface exposed to blood must be preceded by the interaction of the molecular components of plasma, particularly the plasma proteins, with the surface (1,2). This is due to the prepon-

[5]Current address: Washington University School of Medicine, St. Louis, MO 63130

derance of the plasma molecular components and their high diffusivity compared to the formed elements in blood. It is the nature of these protein-surface interactions, i.e., the composition of the adsorbed protein layer and the configuration of the proteins in it, which governs in large part the subsequent interactions of the various blood cells with the surface.

Numerous laboratories have shown that the plasma protein fibrinogen can potentiate the adhesion of platelets to surfaces (3,4) and that the ability of many surfaces to promote platelet adhesion tends to correlate with the amount of fibrinogen bound (5-7). When exposed to mixtures of purified proteins or to whole plasma, many surfaces adsorb fibrinogen preferentially (8-10), sometimes with subsequent desorption (10-13), which Vroman et al. (11) and Brash et al. (13) have reported to require high molecular weight kininogen. A popular theory to explain the mechanism of platelet activation by surfaces (14,15) suggests that the fibrinogen molecules adsorbed to a surface in contact with blood are conformationally altered, and that this partially denatured fibrinogen mediates platelet adhesion, activation and subsequent secretion.

Work in our laboratory has supported the idea that fibrinogen plays an important role in platelet-surface interaction. However, evidence is accumulating that fibrinogen molecules bound to the surface in an undistorted, "native" conformation are required for these interactions to occur. Much of our work on the mechanisms of platelet activation by surfaces has been facilitated by the finding (16) that two families of simple polymers - the polyalkyl acrylates and methacrylates - exhibit an extraordinary range of platelet reactivities, from the relatively non-reactive polymethyl acrylate and methacrylate to the highly reactive longer side-chain acrylates and methacrylates. These polymers (Fig. 1) consist of linear carbon backbones on which every second carbon possesses both a methyl group (on the methacrylates) or a hydrogen atom (on the acrylates) and a carboxylic acid group in ester linkage with an alkyl alcohol side chain. Many of the polymers' physical and chemical properties are determined by the lengths of these alkyl side chains.

We studied a series of highly purified homopolymers, prepared from alkyl acrylate and methacrylate monomers, which varied in the length of the alkyl side chains from 1 to 12 carbons. Platelet reactivity of these polymers was assessed using a bead column method (17) modeled after the Hellem glass bead test (18). Polyethylene columns (5 cm x 0.8 cm) were filled with small (0.35 - 0.50 mm diameter) glass beads and coated with polymer by deposition from a 1% solution in chloroform. The polymer-coated columns were exposed to flowing whole blood, maintained at 37°C and anticoagulated with sodium citrate. Platelet retention was determined by comparing the number of platelets entering and leaving the columns; platelet secretion was assayed by measuring extracellular serotonin (19), platelet factor 4 (20), and beta thromboglobulin (21) levels in the effluent blood plasma.

Columns packed with small polymer-coated glass beads, because of their large surface area to void volume ratio, provide a severe test of platelet-surface compatibility. A 2.5 ml column packed with 0.35 - 0.50 mm diameter beads provides ~200 cm^2 of surface area which, at a point in time, is exposed to ~0.6 ml of blood. This is equivalent to spreading a film of blood 30 um thick on a flat surface and, assuming a mean capillary diameter of 7 um, represents approximately one-twentieth the surface to volume ratio of a capillary bed.

Platelet retention in the polymer-coated bead columns was quantified by noting the percentage of platelets removed from aliquots 2 through 5 of 5 successive 1 ml effluent fractions. (Aliquot 1 was diluted with saline and was not counted.) Retention values obtained for the 14 polymers tested are shown in Table 1. These data are plotted in Fig. 2 as a function of both the alkyl side chain length and the glass transition temperatures (Tg's) of the polymers.

Many properties have been proposed to account for the consequences of blood-surface interaction. Some of these include surface topography, wettability, surface charge, surface free energy, the balance between hydrophobic and hydrophilic groups, chain mobility, crystallinity, the capacity for ionic and hydrogen bonding, and the presence of specific chemical groups at the surface. However, at present there is no general agreement as to which chemical or physical properties determine the behavior of artificial materials in contact with blood.

We found (16) a striking inverse correlation between platelet retention and the glass transition temperatures for each polymer family as well as a direct correlation between platelet retention and the lengths of the polymers' alkyl side chains (Fig. 2). Platelet reactivity, measured as secretion of serotonin, platelet factor 4 and beta thromboglobulin, paralleled that seen with the retention assay (16). The glass transition temperature is an expression of the temperature above which the polymer backbone loses stable van der Waals associations with nearby atomic groups and is a measure of chain mobility. The length of the alkyl side chain affects the polymer's capacity for hydrophobic interactions, presumably with adsorbed plasma protein. Although each polymer family showed inverse correlations between Tg and platelet retention, retention values for PMA, PEA and PPA were very different from those for PLMA, PDMA, PAMA and PBMA even though all 7 of these polymers had similar T_g's. On the other hand, platelet retention as a function of side chain length showed similar, and almost overlapping, correlations for both polymer families (Fig. 2). These results suggest that, for these groups of polymers, hydrophobicity is a more important determinant of platelet reactivity than is chain mobility. Earlier experiments with a large number of segmented polyether polyurethanes (22) also demonstrated that, as long as the polymer structure did not promote formation of hydrogen bonds or other polar interactions, increasing hydrophobicity correlated with increasing platelet-surface reactivity.

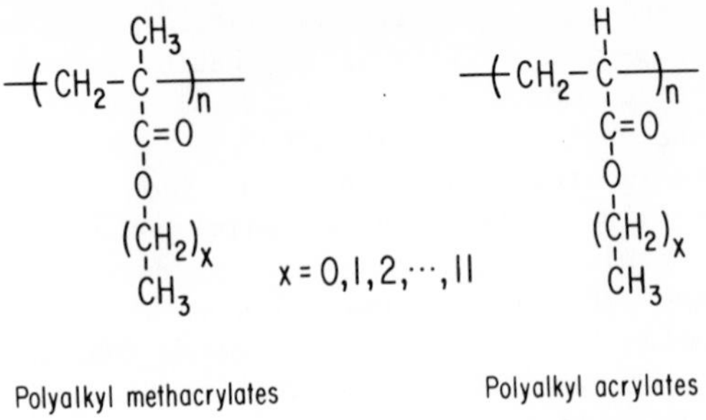

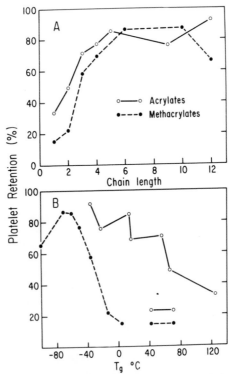

Polyalkyl methacrylates   Polyalkyl acrylates

Figure 1.   Structure of polyalkyl acrylates and methacrylates.
Polymers were synthesized by free radical polymerization initiated
by exposing alkyl acrylate and methacrylate monomers to high doses
(~10 Mev) of ionizing radiation using a Van de Graff accelerator.
Polymers were separated from unreacted monomer by precipitating in
methyl alcohol and redissolving in chloroform. Purified polymers
were dissolved in chloroform for use in coating glass beads
columns and glass cover slips.

Figure 2.   Platelet retention by polyalkyl acrylates and by polyalkyl methacrylates plotted vs length of the alkyl side chain (A)
or glass transition temperature, T_g (B). (Replotted from data in
Fig. 3 of Ref. 17.)

TABLE 1. Reactivity of Polyalkyl Acrylate and Methacrylate Polymers[a]

Platelet Retention (%)	Polymer[b]	PLMA	PDA	PHA	PAMA	PBA	PDMA	PPMA	PBMA	PLA	PPA	PEMA	PMMA	PEA	PMA
15	PMA	X	X	X	X	X	X	X	X	X	X	X	X		—
22	PEA	X	X	X	X	X	X	X	X	X	X	X	X	—	
33	PMMA	X	X	X	X	X	X	X	X	X	X	X	—	X	X
49	PEMA	X	X	X	X	X	X	X	X	X	X	—	X	X	X
58	PPA	X	X	X	X	X	X	X	X		—	X	X	X	X
66	PLA	X	X	X	X	X	X			—		X	X	X	X
69	PBMA	X	X	X	X				—		X	X	X	X	X
71	PPMA	X	X	X	X			—			X	X	X	X	X
76	PDMA	X	X	X	X		—			X	X	X	X	X	X
77	PBA	X	X	X		—				X	X	X	X	X	X
85	PAMA				—		X	X	X	X	X	X	X	X	X
86	PHA			—		X	X	X	X	X	X	X	X	X	X
87	PDA		—			X	X	X	X	X	X	X	X	X	X
92	PLMA	—				X	X	X	X	X	X	X	X	X	X

a) X = p < 0.05 (Duncan's multiple range test). For example, platelet retention by PBMA is significantly different from PAMA but not from PBA.

b) P means poly-; LMA, lauryl methacrylate; DA, n-decylacrylate; HA, n-hexylacrylate; AMA, n-amylmethacrylate; BA, n-butylacrylate; DMA, n-decylmethacrylate; PMA, n-propylmethacrylate; BMA, n-butylmethacrylate; LA, laurylacrylate; PA, n-propylacrylate; EMA, ethylmethacrylate; MMA methylmethacrylate; EA, ethylacrylate; MA, methylacrylate.

When each of the acrylate or methacrylate polymers was preincubated with whole plasma, the platelet reactivity of the surfaces upon
subsequent exposure to whole blood decreased significantly (Fig. 3)
On the other hand, with many other polymers this effect of plasma was
not seen. Of 20 varieties of segmented polyurethanes examined, none
showed this behavior (22), and platelet adhesion to polystyrene was
also unaffected by plasma pretreatment (23). The phenomenon of
plasma-induced passivation of methacrylate and acrylate polymers
presumably involves selective adsorption of specific plasma proteins
by the surfaces and/or a particular alteration of the adsorbed protein once bound.

Because of the wide range of platelet reactivities observed with
the acrylate and methacrylate polymer series, and the interesting
effects of plasma pretreatment, additional experiments were performed
with these polymers to investigate the role that fibrinogen might
play in platelet activation by surfaces. Fibrinogen adsorption was
carried out from single component solution or from whole or diluted
blood plasma. Two techniques were employed, one using radiolabeled
fibrinogen to measure total fibrinogen binding and a second using
radiolabeled anti-fibrinogen antibodies to determine the amount of
fibrinogen bound with minimal alteration in conformation (24).

Preliminary experiments with polymethyl, ethyl, propyl and butyl
methacrylate (PMMA, PEMA, PPMA, and PBMA) and pure solutions of
fibrinogen labeled with ^{125}I (25) showed identical binding isotherms
for these four polymers. Fibrinogen adsorption exhibited typical
"Langmuirian" binding at fibrinogen concentrations between zero and
50 ug/ml. However as the fibrinogen concentration was further increased to 2 mg/ml, small but measureable increases in binding were
observed and binding saturation was not obtained. Fibrinogen binding
was rapid, with most of the binding occurring within the first 30
seconds. At a concentration of 250 ug/ml, an incubation time of one
hour, and an incubation temperature of 22°C, fibrinogen binding (Fig.
4A) was approximately 0.3 ug/cm^2 and showed no statistically significant differences among the four polyalkyl methacrylates (p>0.1, n=9).
As described by Schmitt et al. (26), this level of binding probably
results from the formation of a monolayer of fibrinogen molecules on
the surface.

When the polyalkyl methacrylate surfaces were incubated with
whole plasma (Fig. 4B), fibrinogen binding occurred less rapidly than
with single component fibrinogen solutions. After exposure to plasma
to one hour, fibrinogen binding was approximately 0.25 ug/cm^2 for
each of the methacrylates, a level very close to that obtained with
the single component firinogen solutions. However, work in a number
of laboratories (8-13) has shown that the binding of fibrinogen to
surfaces exposed to whole plasma is a complex phenomenon which may
involve competition with other plasma proteins (11,13), particularly
high molecular weight kininogen (HMWK), over a very short time
period. Binding after long exposures may not, therefore, be physio-

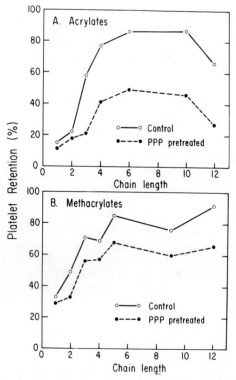

Figure 3. Passivation of polymers of the acrylate series (A) and the methacrylate series (B) by preliminary exposure to platelet-poor plasma (PPP) before whole blood: platelet retention vs alkyl side chain length. (Replotted from data in Fig. 5 of Ref. 17.)

logically relevant; while binding after very short exposures is
difficult to measure.

Brash and ten Hove (27) have shown that the early plasma protein
adsorption and desorption events occur more slowly and can be more
readily examined if the plasma is diluted. Using this approach, we
found that exposure of each of the methacrylate polymers to 1% plasma
(Fig. 4C) resulted in fibrinogen binding which showed maxima after 1
to 3 minutes of exposure followed by decreases in bound fibrinogen to
an apparent plateau at less than 0.1 ug/cm^2. These kinetics are
similar to those reported by Brash and ten Hove (27) for glass,
siliconized glass and polyethylene.

Thus, for the 4 polyalkyl methacrylates tested, the kinetics of
fibrinogen adsorption differed for fibrinogen solution, whole plasma
and diluted plasma. However for each one of these three media,
fibrinogen adsorption to each of the four methacrylates did not
differ significantly. The measured amounts of adsorbed fibrinogen
did not, therefore, correlate with the platelet reactivity of the
polymers, as previously assayed in polymer-coated bead columns.

Measurement of surface bound fibrinogen using polyclonal anti-
fibrinogen antibodies labeled with ^{125}I produced different results.
When the polyalkyl methacrylates were pre-exposed to a single component fibrinogen solution (250 ug/ml) for up to one hour and then to a
solution containing a low concentration (300 ng/ml) of radiolabeled
anti-fibrinogen antibody (Fig. 4D), significant differences were
noted among the four polymers. After the one hour pre-exposure,
antibody binding ranged from 1.26 ng/cm^2 for PMMA to 3.1 ng/cm^2 for
PBMA. At each pre-exposure time tested, antibody binding was significantly different for each of the four polymers (p<0.05, n=9). The
levels of antibody binding in these experiments were low because of
the low concentration of labeled antibody solution employed. At high
antibody concentrations (Table 2), the amount of bound antibody was
similar to that of the bound fibrinogen and the differences in antibody binding to the four polymers were no longer observed. These
results indicated that the antibody binding hierarchy for the four
methacrylate polymers was due to differences in binding affinities of
the adsorbed fibrinogen molecules for antibody rather than to differences in the number of fibrinogen molecules recognizable by the
antibodies (i.e., the number of antibody binding sites).

Antibody binding to polyalkyl methacrylates following exposure to
whole plasma for up to 60 minutes is shown in Fig. 4E. Short exposures to plasma for 15 seconds, 1 minute or 2.5 minutes resulted in
small but statistically significant differences in antibody binding
to each of the four methacrylates (p<0.05, n=14). The hierarchy of
antibody binding to the methacrylates exposed briefly to whole plasma
was identical to the hierarchy of antibody binding to polymers exposed to single component fibrinogen solutions. The differences in
antibody binding were not apparent when the polymers were pre-exposed
to plasma for longer times (i.e., for 8 or 60 minutes). Polymers

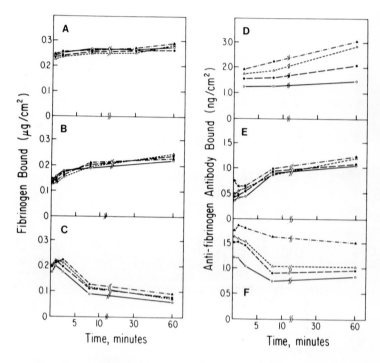

Figure 4. Fibrinogen binding and anti-fibrinogen antibody binding to polyalkyl methacrylates (○=PMMA, ●=PEMA, △=PPMA, ▲=PBMA). Polymer-coated glass cover slips were exposed to solutions spiked with ^{125}I-labeled fibrinogen (A-C) or were exposed first to unlabeled fibrinogen solutions and then to ^{125}I-labeled anti-fibrinogen antibody (D-F). (A) Purified fibrinogen solution (250 ug/ml). (B) Undiluted plasma. (C) 1% plasma. (D) Purified fibrinogen solution (250 ug/ml) followed by anti-fibrinogen antibody solution (300 ng/ml, ~800 cpm/ng). (E) Undiluted plasma followed by anti-fibrinogen antibody solution (100 ng/ml, ~8000 cpm/ng). (F) 1% plasma followed by anti-fibrinogen antibody solution (100 ng/ml, ~8000 cpm/ng). (Reproduced with permission from Ref. 24. Copyright 1986 Grune & Stratton.)

TABLE 2. Binding of Anti-fibrinogen Antibody (Fg-Ab) to Polyalkyl
Methacrylate Polymers Previously Exposed to Fibrinogen[a]

	Antibody Bound (ng/cm^2)			
[Fg-Ab][b]:	0.3 ug/ml	1 ug/ml	10 ug/ml	200ug/m
Polymer				
PMMA	1.26 ± .12[c]	5.7 ± .4	108 ± 5	250 ± 2
PEMA	2.09 ± .12	6.3 ± .3	102 ± 8	251 ± 1
PPMA	2.94 ± .05	8.1 ± .5	100 ± 7	234 ± 1
PBMA	3.07 ± .07	9.4 ± .8	105 ± 3	237 ± 1

a) Polymers were pre-exposed to a solution of purified fibrinogen
   (250 ug/ml) for 1 hour.
b) Values are the concentrations of anti-fibrinogen antibody
   solutions to which polymers were exposed following initial
   fibrinogen exposure.
c) Data represent means ± standard deviations: for the 0.3 ug/ml
   column, n = 9; for the 1, 10 and 200 ug/ml columns, n = 3.
(Reproduced with permission from Ref. 24. Copyright 1986 Grune
 & Stratton)

pre-exposed to plasma diluted to 1% (Fig. 4F) showed large, statistically significant differences (p<0.05, n=5) in their antibody binding capacities, and these differences persisted even after one hour of pre-exposure.

Differences in the binding affinities of anti-fibrinogen antibodies to fibrinogen molecules bound to different polymers might result from variations in the orientation of the fibrinogen molecules or from variations in the extent of the conformational alteration induced in the bound fibrinogen molecules by the energy of adsorption. If fibrinogen binds to the different methacrylates with no particular orientation but with the induction of various conformational changes, then a variety of antibodies should show the same hierarchy of affinity for the surface-bound fibrinogen molecules. If on the other hand, the different methacrylates adsorb fibrinogen with different orientations, then antibodies that recognize different parts of the fibrinogen molecule should show different hierarchies of antibody binding.

These two possibilities were tested by repeating the antibody binding experiments using five different antibody preparations, including Fab fragments of antibodies prepared against the D-domain (the terminal region) and the E-domain (the central region) of fibrinogen (Fig. 5). Although the level of antibody binding varied significantly from one antibody to another, each anti-fibrinogen antibody preparation exibited the same rank order of binding to the four polyalkyl methacrylates. A control anti-IgG antibody bound 10 to 100 fold less than the antifibrinogen antibodies and showed no polymer-to-polymer variation. The complexities of the multiple interactions of polyclonal antibodies with a large molecule like fibrinogen make these experiments difficult to interpret. However, these results are consistant with the hypothesis that the variations in antibody binding to fibrinogen adsorbed onto the different polymer surfaces are due to variations in conformation, not orientation, of the bound fibrinogen molecules, and that these variations in conformation are the result of differing degrees of structural change induced by the different polymers.

The rank order of binding of anti-fibrinogen antibodies, when the series of methacrylates was pre-exposed to fibrinogen solution, diluted plasma, or to whole plasma (for short times), showed significant positive correlations with the measurements of platelet reactivity induced by these polymers (Fig. 6). These correlations suggest that platelet activation induced by these polymers may depend upon the availability of conformationally unaltered, "native" fibrinogen on the surface, and may be less dependent on the total amount of bound fibrinogen.

One possible objection to the comparison of protein adsorption on flat surfaces with platelet adhesion and activation in polymer-coated bead columns is raised by the work of Vroman et al. (28) who showed that protein adsorption onto surfaces from plasma in narrow

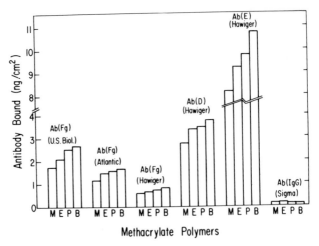

Figure 5. Binding of anti-fibrinogen antibodies to polyalkyl
methacrylates exposed to fibrinogen (M=PMMA, E=PEMA, P=PPMA, B=
PBMA). Two commercial anti-fibrinogen antibody IgG preparation
(US Biochemical and Atlantic Antibody), an anti-fibrinogen Fab
fragment [Ab(Fg)(Hawiger)], an anti-domain D Fab fragment [Ab(D)],
an anti-domain E Fab fragment [Ab(E)], and an anti-IgG antibody
[Ab(IgG)] were used. Polymer-coated cover slips were exposed first
to fibrinogen (250 ug/ml) and then to a ^{125}I-labeled antibody
preparation. Specific activities and concentrations of each of the
labeled antibody preparations were ∽1000 cpm/ng and 200 ng/ml.
(Reproduced with permission from Ref. 24. Copyright 1986 Grune &
Stratton.)

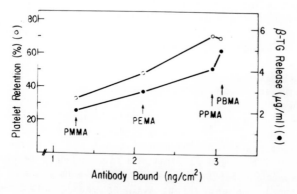

Figure 6. Correlation of anti-fibrinogen antibody binding with platelet activation by polyalkyl methacrylates. Polymer–coated glass cover slips were exposed first to fibrinogen solution (250 ug/ml) and then to ^{125}I–labeled antibody solution (300 ng/ml, ~800 cpm/ng). (Reproduced with permission from Ref. 24. Copyright 1986 Grune & Stratton.)

spaces, such as the spaces between small polymer-coated beads, may be
altered. They postulate that, in these narrow spaces, insufficient
HMWK may be present to complete the replacement of bound fibrinogen.
However, these effects were observed in a static system where protein
transport to the surface was limited by diffusion. In the bead
column system we have employed, the flow of blood and/or plasma
through the spaces between the beads is likely to mitigate these
diffusional limitations.

Thus it appears that fibrinogen's role in platelet adhesion to
surfaces may be similar to its essential role in platelet aggregation
(29-31). In whole blood or plasma, fibrinogen acts as a cofactor,
binding to GP IIb/IIIa receptors after activation of the platelet by
an agonist such as thrombin or ADP (32,33). The subsequent aggregation mechanism is unclear. Perhaps fibrinogen alone forms bridges
between GP IIb/IIIa receptors on contiguous platelets; more likely,
some other adhesive protein or proteins (e.g., von Willebrand factor
(34) or thrombospondin (35)) act in concert with fibrinogen to bind
platelets together.

Results presented above might even be compatible with the hypothesis that fibrinogen molecules properly arrayed on a surface may
themselves facilitate platelet adhesion and induce platelet activation. Adsorbed fibrinogen differs from fibrinogen in solution in two
important respects. First, on a surface it may be present in greatly
increased local concentration: e.g., an easily obtained surface
coverage of 0.5 ug/cm^2 is equivalent to a solution concentration of
110 mg/ml (26). Second, surface-bound fibrinogen molecules are likely to be closely arrayed and will be sterically constrained. Such
constrained molecules, if they possess the proper configurations,
would present multiple binding sites to their specific cell membrane
receptors. Individual fibrinogen molecules bind weakly or not at all
to unactivated platelets (31,33), but the binding of multivalent
ligands to multiple receptors has been shown to increase binding
affinity dramatically (36,37).

In order to test the effects of closely arrayed, sterically
constrained fibrinogen molecules on platelets, we sought to prepare
soluble fibrinogen polymers (38) as models of fibrinogen adsorbed
onto artificial surfaces. Although soluble fibrinogen polymers might
not maintain conformations identical to surface-bound fibrinogen, we
reasoned that the multivalent presentation of platelet binding loci
may mimic events which take place when circulating platelets interact
with fibrinogen-coated surfaces.

Fibrinogen polymers were prepared using divalent antibody fragments (Fab'$_2$) of affinity-purified antibodies directed against the E-
domain of the fibrinogen molecule (designated Fab'$_2$(E)). When mixed
with fibrinogen in approximately a 2:1 molar ratio, this antibody
fragment produced a polymer (designated Fg-Fab'$_2$(E)) which induced
dose-dependent, irreversible aggregation of gel-filtered platelets
(Fig. 7A). No cofactor or additional agonist was required to produce

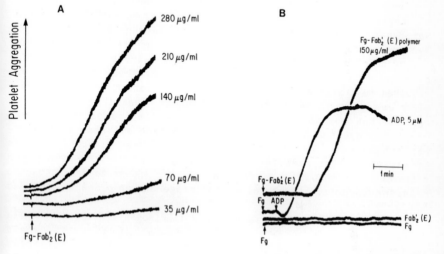

Figure 7.   (A) Dose response of Fg-Fab'₂(E) polymer-induced aggregation of gel-filtered platelets.   Values indicated are final
concentrations of the polymer.   (B) Aggregation of gel-filtered
platelets by Fg-Fab'₂(E) polymer and lack of aggregation by its
components.   Aggregation in response to Fg-Fab'₂(E) polymer (150
ug/ml) or fibrinogen (500 ug/ml) followed by ADP (5 uM) is contrasted to the lack of response to Fab'₂(E) antibody (300 ug/ml)
alone or fibrinogen (500 ug/ml) alone.   (Reproduced with permission from Ref. 38.   Copyright 1986 Grune & Stratton.)

this effect. Neither fibrinogen alone nor the antibody alone produced platelet aggregation (Fig. 7B), and ADP did not aggregate the gel-filtered platelet suspensions unless fibrinogen was also added. Platelet aggregation by Fg-Fab'$_2$(E) was blocked by antibodies specific for the D domain of fibrinogen and for the GP IIb/IIIa binding sites on the platelet (Fig. 8) suggesting that the polymer may function by binding its free D-domains to GP IIb/IIIa receptors on the platelets. Fg-Fab'$_2$(E)-induced aggregation was abolished by metabolic inhibitors (Fig. 8) - strong evidence that the aggregation response was not simply the result of agglutination.

Aggregation of gel-filtered platelets by Fg-Fab'$_2$(E) was attenuated but not eliminated (Fig. 8) by ADP scavenging systems and by 5'-fluorosulfonylbenzoyl adenosine (FSBA, which inhibits the binding of ADP to platelets (39)), suggesting that release of ADP from the platelet facilitates the polymer-induced aggregation, but is not required to initiate the response. The observation that no aggregation was seen in response to fibrinogen alone (Fig. 7B) argues against the possibility that these gel-filtered platelets were already subjected to ADP stimulation before exposure to the Fg-Fab'$_2$(E) polymer. Thus it appears that the polymer itself is capable of activating the platelets, presumably by binding to fibrinogen receptors, and that this activation does not appear to depend on the receptor-inducing effects of either exogenous or platelet-derived ADP.

It is not surprising that multiple fibrinogen binding sites held in close proximity, as provided by the Fg-Fab'$_2$(E) polymers, would produce a significantly greater effect than fibrinogen molecules acting independently. Dower et al. (37) have provided a detailed analysis of multivalency effects and showed that multivalent binding sites should always produce binding affinities many orders of magnitude higher than those of the monovalent ligands. Hornick and Karush (40) have provided experimental verification of this principle by showing that divalent anti-DNP (2,4-dinitrophenol) antibodies bind to DNP-coated phage particles with an affinity constant of 3.5×10^{11} $(mol/L)^{-1}$ compared with 6×10^6 $(mol/L)^{-1}$ for the monovalent Fab fragments.

Once bound to the platelet, the fibrinogen polymers may induce platelet activation via rearrangement of the GP IIb/IIIa receptors. According to the mobile receptor hypothesis of Jacobs and Cuatrecasas (41), receptor proteins move about freely in the plane of a resting membrane, but when they interact with their target ligands they may cluster into complexes which promote cell responses. Such receptor "patching" has been observed in platelets and several other cell systems (42,43). Polley (44) showed by electron microscopy that thrombin induces the clustering of GP IIb/IIIa receptors in platelet membranes. DeMarco et al. (45) suggested that GP IIb/IIIa clusters may result when platelets interact with von Willebrand factor; and Santoro and Cunningham (46) proposed similar mechanisms to explain

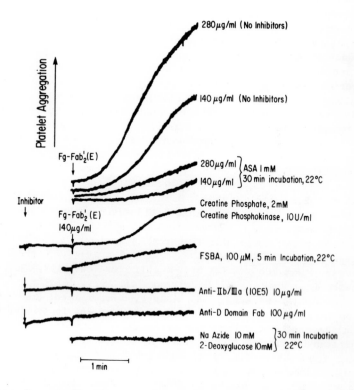

Figure 8. Effects of various inhibitors on Fg-Fab'$_2$(E) polymer-induced aggregation of gel-filtered platelets. (Reproduced with permission from Ref. 38. Copyright 1986 Grune & Stratton.)

collagen-induced platelet aggregation. These proposed mechanisms may be analogous to those responsible for the activation of platelets by the Fg-Fab'$_2$(E) polymers described here.

At this point little is known about how the formation of membrane receptor complexes might induce cellular responses in platelets. Perhaps the rearrangement of GP IIb/IIIa receptors alters interconnections between cytoskeletal and membrane components of the cell. Such changes in membrane and cytoskeletal structures might induce the activation of membrane phospholipases which, in turn, could generate elevated levels of cytoplasmic calcium (47) and/or diacylglycerol (48) and lead to the expression of additional GP IIb/IIIa receptors and to platelet secretion. This or a similar mechanism may prove to be the trigger by which platelets in contact with surface-bound fibrinogen are activated.

Literature Cited

1.   Baier, R.E.; Dutton, R.C.   J. Biomed. Mater. Res. 1969, 3, 191.
2.   Scarborough, D.E.; Mason, R.G.; Dalldorf, P.G.; Brinkhous, K.M. Lab. Invest. 1969, 20, 164.
3.   Packham, M.A.; Evans, G.; Glynn, M.F.; Mustard, J.F.   J. Lab. Clin. Med. 1969 73, 686.
4.   Zucker, M.B.; Vroman, L.   Soc. Exptl. Bio. Med. 1969, 131, 318.
5.   Brash, J.L.; Uniyal, S.   J. Polymer Sci. 1979, 66, 377.
6.   Young, B.R.; Lambrecht, L.K.; Mosher, D.F.; Cooper, S.L.   In Biomaterials: Interfacial Phenomena and Applications; Cooper, S.L.; Peppas, N.A., Eds.; Advances in Chemistry Series 199, American Chemical Society: Washington, DC, 1982, p 312.
7.   Nyilas, E.; Chiu, T.H.   Art. Organs 1978, 2(Suppl), 56.
8.   Lee, R.G.; Adamson, C.; Kim, S.W.   Thromb. Res. 1974, 4, 485.
9.   Horbett, T.A.; Hoffman, A.S.   In Applied Chemistry at Protein Interfaces; Baier, R.E., Ed.; American Chemical Society: Washington DC, 1975, p 230.
10. Brash, J.L.   In The Behavior of Blood and its Components at Interfaces; Vroman, L.; Leonard, E.F. Eds.; Ann. NY Acad. Sci. 1977, 283, 356.
11. Vroman, L.; Adams, A.L.; Fischer, G.C.; Munoz, P.C.   Blood 1980, 55, 156.
12. Horbett, T.A.   Thromb. Hemostas. 1984, 51, 174.
13. Brash, J.L.; Scott, C.F.; ten Hove, P.; Colman, R.W.   Trans. Soc. Biomaterials 1985, VIII, 105.
14. Chiu, T.H.; Nyilas, E.; Turcotte L.R.   Trans. Am. Soc. Artif. Int. Organs 1978, 24, 389.
15. Tomikawa, M.; Iwamoto, M.; Olsson, P.; Soderman, S.; Blomback, B.   Thromb. Res. 1980, 19, 869.
16. Brier-Russell, D.; Salzman, E.W.; Lindon, J.; Handin, R.; Merrill, E.W.; Dincer, A.K.; Wu, J.S.   J. Coll. Interface Sci. 1981, 81, 311.

17. Lindon, J.N.; Rodvien, R.; Brier, D.; Greenberg, R.; Merrill, E.; Salzman, E.W. <u>J. Lab. Clin. Med.</u> 1978, <u>92</u>, 904.
18. Hellem, A.J. <u>Scand. J. Clin. Lab. Invest.</u> 1960, <u>Suppl 51</u>, 1.
19. Spaet, T.H.; Zucker, M.B. <u>Am. J. Physiol.</u> 1964, <u>206</u>, 1267.
20. Ludlam, C.A.; Moore, S.; Bolton, A.E.; Pepper, D.S.; Cash, J.D. <u>Thromb. Res.</u> 1975, <u>6</u>, 543.
21. Handin, R.I.; McDonough, M.; Lesch, M. <u>J. Lab. Clin. Med.</u> 1978 <u>91</u>, 340.
22. Merrill, E.W.; Sa da Costa, V.; Salzman, E.W.; Brier-Russell, D.; Kushner, L.; Waugh, D.F.; Trudel, G.; Stopper, S.; Vitale, V. In <u>Biomaterials: Interfacial Phenomena and Applications</u>; Cooper, S.L.; Peppas, N.A., Eds.; Advances in Chemistry Series 199, American Chemical Society: Washington, DC, 1982, p 95.
23. Salzman, E.W.; Lindon, J.; Brier, D.; Merrill, E.W. In <u>The Behavior of Blood and its Components at Interfaces</u>; Vroman, L.; Leonard, E.F. Eds.; <u>Ann. NY Acad. Sci.</u> 1977, <u>283</u>, 114.
24. Lindon, J.N.; McMannama, G.; Kushner, L.; Merrill, E.W.; Salzman, E.W. <u>Blood</u> 1986, <u>68</u>, 355.
25. Knight, L.C.; Budzynski, A.Z.; Olexa, S.A. <u>Thromb. Haemostas.</u> 1981, <u>46</u>, 593.
26. Schmitt, A.; Varoqui, R.; Uniyal, S.; Brash, J.L.; Pusiner, C. <u>J. Coll. Interface Sci.</u> 1983, <u>92</u>, 25.
27. Brash, J.L.; ten Hove, P. <u>Thromb. Haemostas.</u> 1984, <u>51</u>, 326.
28. Vroman, L.; Adams, A.L.; Brakman, M. <u>Haemostasis</u> 1985, <u>15</u>, 300.
29. Cross, M.J. <u>Thromb. Diath. Haemorrh.</u> 1964, <u>12</u>, 524.
30. Niewiarowski, S.; Budzynski, A.Z.; Lipinski, B. <u>Blood</u> 1977, <u>49</u>, 635.
31. Marguerie, G.A.; Edgington, T.S.; Plow, E.F. <u>J. Biol. Chem.</u> 1980, <u>255</u>, 154.
32. Kloczewiak, M.; Timmons, S. Lukas, T.J.; Hawiger, J. <u>Biochem.</u> 1984, <u>23</u>, 1767.
33. Peerschke, E.I.; Zucker, M.B.; Grant, R.A.; Egan, J.J.; Johnson, M.M. <u>Blood</u> 1980, <u>55</u>, 841.
34. Plow, E.; Srouji, A.H.; Meyer, D.; Marguerie, G.; Ginsberg, M.H. <u>J. Biol. Chem.</u> 1984, <u>259</u>, 5388.
35. Asch, A.S.; Leung, L.L.; Polley, M.J.; Nachman, R.L. <u>Blood</u> 1985, <u>66</u>, 926.
36. Crothers, D.M.; Metzger, H. <u>Immunochem.</u> 1972, <u>9</u>, 341.
37. Dower, S.K.; Titus, J.A.; Segal, D.M. In <u>Cell Surface Dynamics: Concepts and Models</u>; Marcel Dekker: New York, 1984; p 277.
38. McManama, G.; Lindon, J.N.; Kloczewiak, M.; Smith, M.A.; Ware, J.A.; Hawiger, J.; Salzman, E.W. <u>Blood</u> 1986, <u>68</u>, 363.
39. Mills, D.C.B.; Figures, W.R.; Scearce, L.M.; Stewart, G.J.; Colman, R.F.; Colman, R.W. <u>J. Biol. Chem.</u> 1985, <u>260</u>, 8078.
40. Hornick, C.L.; Karush, F. <u>Israel J. Med. Sci.</u> 1969, <u>5</u>, 163.
41. Jacobs, S.; Cuatrecasas, P. <u>Trends Biochem. Sci.</u> September, 1977, p 289.

42. Taylor, R.B.; Duffis, P.H.; Raff, M.C.; de Petris, S. Nature
     (New Biol.) 1971, 233, 225.
43. Kahn, C.R.; Baird, K.C.; Jarrett, D.B.; Flier, J.S. Proc.
     Natl. Acad. Sci. USA 1978, 75, 4209.
44. Polley, M.J.; Leung, L.L.K.; Clark, F.Y.; Nachnam R.L. J. Exp.
     Med. 1981, 154, 1058.
45. DeMarco, L.; Zimmerman, T.S.; Ruggeri, Z.M. Clin. Res. 1985,
     33, 338A.
46. Santoro, S.A.; Cunningham, L.W. Thromb. Haemosta. 1980, 43,
     158.
47. Johnson, P.C.; Ware, J.A.; Cliveden, P.B.; Smith, M.; Dvorak,
     A.M.; Salzman, E.W. J. Biol. Chem. 1985, 260, 2069.
48. Burn, P.; Rotman, A.; Meyer, R.K.; Burger, M.M. Nature 1985,
     314, 469.

RECEIVED February 26, 1987

Chapter 32

Aspects of Platelet Adhesion to Protein-Coated Surfaces

Irwin A. Feuerstein

Departments of Chemical Engineering and Pathology, McMaster University, Hamilton, Ontario L8S 4L7, Canada

The work discussed here has shown that suspensions of platelets and red cells in a physiological medium can provide information for platelet surface interactions. Evidence is provided on the dynamic features of platelet-surface adhesion and detachment which indicates that more than one sequence of adhesion, detachment and re-adhesion can lead to the same net platelet adhesion. Surface generated substances, such as ADP and serotonin from platelets and thrombin from the coagulation pathway, may strongly influence the function of platelets approaching a surface. The supply of these substances depends on the presence of flow and continued arrival of platelets at a surface. The reactivity of surface-bound protein may be altered by platelet adhesion and detachment. This may occur as a result of deposition of cell membrane components, replacement of the original substrate with protein secreted from platelets or possibly by enzymatic digestion of surface bound protein.

The work to be discussed here deals with platelet adhesion to protein coated surfaces. The protein coating, the cells on the surface and the moving fluid adjacent to the surface may be viewed as a system of interacting components. Flow is an important feature of this system since it brings new protein and cells to the system, augments the transport of cells to the surface and can cause the detachment of adherent cells. Each component of the system may influence the conditions of the other components. The variation of the surface concentrations of proteins on a solid substrate continues to be studied and remains a key area of interest. However, the action of immobilized cells on the substrate needs to be examined more carefully as well as their contribution of secreted substances to the fluid phase adjacent to the substrate and to the substrate protein itself. Red cells, platelets and white cells may also adhere and detach from the substrate changing its make-up by yet another mechanism.

An Overall Picture of Platelet Adhesion to Protein-Coated Surfaces

Platelet adhesion to protein-coated substrates has been studied using whole blood (1), whole blood with anticoagulant (2) and with suspensions of washed platelets (3,4,5) and red cells (6) in a physiological medium. In order to maintain the surface concentration integrity of a precoated protein substrate and avoid the difficult to control action of thrombin on the substrate and platelets, washed cells in a solution containing Ca^{++} and Mg^{++} offer a good experimental alternative. The presence of Ca^{++} and Mg^{++} at physiological concentrations is an important condition for optimal physiological platelet

0097-6156/87/0343-0527$06.00/0

function (7,8). Anticoagulants which sequester divalent cations and heparin which often has unpredictable effects on platelets are other alternatives.

Using such suspensions of platelets and red cells and well controlled tube flow leads to an ordered pattern of platelet adhesion (6,9), see Figure 1. The net rate of platelet adhesion is at first constant at any position along the tube with this rate decreasing with distance from the tube's entrance. As time goes on the adhesion rates decrease and an invariant surface concentration of principally single platelets evolves with a surface concentration independent of position. It is not clear whether platelets are adhering and detaching, leaving their suface concentration constant, or not. Aggregates of platelets have not been observed under these conditions for glass coated with fibrinogen, fibronectin or albumin (6,10-12).

Secretion of Granule Contents from Adherent Platelets

If one views a segment of biomaterial while platelets are contacting and adhering and also examines the fluid adjacent to the surface, changes are continually occurring, see Figure 2. Blood cells carried by the moving suspension enter the region of view and leave downstream. Some cells are deposited on the surface and may remain adherent. Measurements of the granule content of adherent platelets before and after adhesion give information on the amount of secretion from these cells (13-15). The dense granules of platelets contain ADP and serotonin, known stimulatory agents for platelets, in addition to a number of other substances (16). The alpha granules of platelets contain proteins which include, amongst others: fibrinogen, fibronectin, thrombospondin, albumin, von Willebrand factor and several antiheparin proteins, β-thromboglobulin and platelet factor 4 (16-18). Low molecular weight substances such as ADP and serotonin will be transferred to the moving suspension and to the vicinity of moving suspended cells; proteins may be transferred to the adjacent substrate or to the suspension where they too may interact with cells.

A mathematical analysis coupled to platelet accumulation and secretion data has shown that a near-the-surface micro environment containing elevated levels of secreted substances can exist while accumulation and secretion occur. Levels of ADP and serotonin capable of activating platelets can exist adjacent to the substrate (16). The computations indicate that the region of higher concentration can range from 20-200 μm from the surface. The mathematical model describing this effect gives a concentration determined from contributions from all surface bound platelets. Concentrations one or two cell diameters into the flow may be higher than those calculated with this model. The computed results, however, represent those values of concentration which flow-borne platelets would encounter as they move in the vicinity of a biomaterial. The concentration values are relatively independent of flow rate (they depend on the wall shear rate to the 0.27 power). A possible picture is that of an unstimulated, resting, platelet moving into the vicinity of the active biomaterial where it is activated to a degree before surface contact with adherence and further activation. The formation of pseudopods, spiny extensions from the platelet's surface, occurs early as platelets are stimulated (19). These protrusions could aid with surface contact because: 1) platelets would not need to move as close to the substrate and 2) the hydrodynamic resistance of the smaller radius of curvature pseudopod, relative to the radius of curvature of the main body of the cell, would be less than the resistance of the main body of the cell.

Interaction of Platelets with the Coagulation Pathway

Since thrombin is formed upon the surface of stimulated platelets (20), and since Hageman Factor is activated at artificial surfaces (21), thrombin is likely to be preferentially formed at a surface during mural thrombogenesis with whole blood. It would then be open to the influence of convection and diffusion in a manner similar to

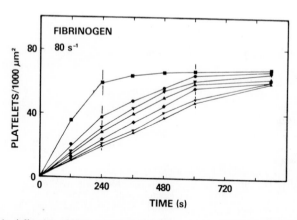

Figure 1: Adhesion of ^{51}Cr-labelled platelets on a fibrinogen-coated glass tube at a wall shear rate of 80 s[-1] Curves starting from the uppermost are for 0.5, 1.5, 2.5, 3.5, 5.0, 7.0 and 9.0 cm from the tube's inlet. (Reproduced with permission from Ref. 6. Copyright 1981 ASAIO Publications.)

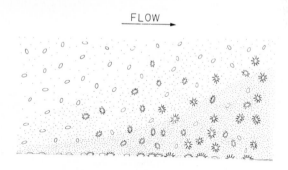

Figure 2: Schematic of platelet adhesion process.

that for substances released from surface-bound platelets. One may now ask, for flowing blood in contact with an artificial surface, what the effect of thrombin would be as a platelet stimulator. To deal with this question directly, while avoiding the use of whole blood which contains fibrinogen, which may form fibrin and trap platelets, we have chosen to prestimulate platelets with low concentrations of thrombin, in a physiological medium lacking fibrinogen, for a short time prior to flow. The thrombin may then be inactivated with substances such as hirudin (12) or Phe-Pro-Arg-CMK (11). This approach yields a uniform and known concentration of thrombin in the fluid surrounding the platelet.

Tests done at 1 and 3 minute flow times show no significant increase in platelet accumluation on fibrinogen- or albumin-coated surfaces as a result of thrombin addition. These results also indicate that for a period of flow up to one minute both control and thrombin prestimulated platelets adhere singly at the same average surface concentration. Thus, the early measurements of net platelet accumulation are not influenced by the presence of activated platelets while the later phase, aggregate formation is, see Figure 3, (11). The presence on the solid substrate of platelets which have been prestimulated is necessary for aggregates to form and grow. The ability of prestimulated platelets to form surface-bound aggregates while unstimulated platelets are unable to do this may be related to a difference of membrance receptor or adhesive protein availability. Our previous work has indicated that platelets adherent to surface-bound fibrinogen release up to 50% of their dense granule content (6). In that sense they may be viewed as stimulated. For adhesion to collagen-coated glass, the same study indicated a release of 80% or greater and the potential for aggregate formation. Others have shown that the availability of a membrance receptor and fibrinogen are suffucient for platelet aggregation in stirred suspensions (22). The membrane receptor has also been shown to become available for fibrinogen binding as a result of stimulation with a number of agonists such as ADP, collagen, arachidonic acid and thrombin (22,23). Fibrinogen which is stored in the alpha granules of platelets is also known to be released form stimulated platelets (16,17). These components then are likely not to be present in sufficient quantities to support cell-cell adhesion between flow-borne unstimulated platelets and platelets adherent to fibrinogen- and albumin-coated glass.

Over our 3 cm measurement region greater variation in cellular accumulation occurrred with thrombin present than without it, see Figure 4. Similar large variations may be observed on collagen coated-glass (6,14). This may be viewed as preferential aggregate formation followed by inhibition of accumulation causing a localizing phenomenon. On theoretical grounds a reduction in cellular accumulation can occur along a surface as a result of a depletion of platelets in the moving fluid near the surface due to upstream deposition (9). When aggregate formation occurs this effect may be augmented due to accumulation of platelets on to aggregates as well as on the solid substrate. The presence of aggregates may effectively increase the substrate surface area Preferential adhesion to aggregates may occur since they protrude into the fluid making them accesible to more platelets than the solid surface. Thus the intitial formation of small surface-bound aggregates may be the cause of the acceleration of platelet accumulation during the cell-cell contact phase.

Cell-Surface Dynamics

Previous work with flowing suspensions of red cells has shown that they can remove surface-bound molecules e.g. cholesterol (24), fibrinogen (25) and albumin (25). Adherent platelets have longer residence times on a substrate than do red cells and thus can provide different effects. Detached platelets are likely to be altered from their natural state and thus be predestined, in an artificial organ flow circuit, for removal from the circulation. Previous work using transmitted light microscopy which dealt with platelet adhesion and detachment was limited in that the flow rate range was low as was the concentration of red cells (26,27). Red cells are an important component of blood rheology and mass transfer for cells and solutes. Microscopic observations of cell-surface contacts in the presence of physiological hematocrit, 35-45%, is not feasible with transmitted light. To overcome this problem, a system utilizing video microscopy of fluorescently labelled platelets was developed. This required the inclusion of a video camera capable of low light level detection, see ref. 6 for a schematic

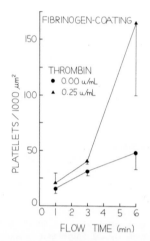

Figure 3: Platelet accumulation on the first 1 cm of fibrinogen-coated glass tubes versus time with and without thrombin prestimulation. (Reproduced with permission from Ref. 11. Copyright 1986 Pergamon Journals, Inc.)

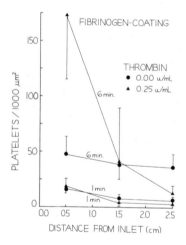

Figure 4: Platelet accumulation on fibrinogen-coated glass tubes versus distance from the tube's inlet with and without thrombin prestimulation. (Reproduced with permission from Ref. 11. Copyright 1986 Pergamon Journals, Inc.)

The measurements made with fluorescent video-microscopy result from a microscope objective which is focused on the inner surface of a glass tube having a wall thickness approximating that of a cover slip. Fluorescently labelled platelets may be viewed only when they are attached to the surface of the glass tube. The small depth of field of the objective, the lack of resolution of the video camera with respect to time and the ability of red cells to block incident and emitted fluorescent light are reasons for this. Two fluorescent dyes have been used; these are: 1) mepacrine which is actively taken up and stored in the dense granules of platelets and 2) carboxyfluoroescein diacetate which is lipophilic promoting its passage into the cytosol where it is enzymatically converted to a polar compound. Both dyes are compatible with the optical components used for fluorescein. Platelets are labelled in suspension after separation from plasma and other cells and washed after labelling. Aggregation tests with low concentrations of ADP and thrombin show these platelets to be as responsive as control suspensions (11,28). The flow system used is one in which a stagnant suspension of platelets and red cells is drawn through a vertically mounted tube by a syringe pump. The measurement region is located 0.5 cm downstream from the glass tube's (1.3 mm I.D.) entrance.

A video camera with a silicon intensifier target (SIT) tube was used to prepare video tapes which were analyzed one frame at a time (the time between frames is 33 ms). The location and arrival and detachment time of each cell were recorded. Notations were made of the movements of cells to new positions on the surface and a record was made of unoccupied grid spaces which had previously contained cells. The surfaces examined were albumin-, fibronectin- and collagen-precoated glass. Exposure times to flow were kept short, 1-4 min, in order to minimize the effects of protein desorption and exchange.

Studies on fibrinogen-coated glass were done at wall shear rates of $40\,s^{-1}$, $80\,s^{-1}$, and $944\,s^{-1}$(29). Data were reduced in order to give absolute rates of arrival, detachment and a net rate of adhesion as well as adhesion efficiencies which are normalized quantitites based on 100 platelets. Comparisons of the platelet arrival rates and net rates of adhesion among the three flow rates studied show a 3-4 fold increase over the shear rate range. This phenomenon is likely to be at least in part a result of the nonlinear increase in the transfer of platelets through the motions of rotating and translating red cells. The detachment rate of cells from the measurement region is greater at $944\,s^{-1}$ than at 40 or $80\,s^{-1}$. Although these rates are small relative to the convective input of platelets into the glass tube they may be important if they represent the rates of damaged cells returning to the circulation.

The efficiency for permanent adhesion on first contact is maximum at $80\,s^{-1}$ (comparison with $40\,s^{-1}$ and $944s^{-1}$) while the percents of arriving cells moving to new positions and departing from the measurement region are minimum at $80\,s^{-1}$. An overall balance, which includes movement of cells and adhesion at new sites, indicates a maximum in the efficiency for permanent adhesion at $80\,s^{-1}$ and a minimum in the percent of arriving platelets departing from the measurement region at that wall shear rate. The overall picture is thus one of 40-60 percent of platelets adhering on arrival with an additional 30 percent permanently adhering after several moves along the surface, see Table I. It is possible to think of these data in terms of forces promoting adhesion and forces promoting cell removal. The "sum" of these forces appears to change with increasing shear rate such that adhesion is favoured up to a point, while departure is favoured as the flow rate is increased.

Additional experiments were performed to compare fibrinogen- and albumin-coated glass with respect to the details of the platelet adhesion process (30). Using purified forms of fibrinogen and albumin it was demonstrated with ^{51}Cr-labelled platelets that the net rates of adhesion of platelets to these proteins are equal. The next task was to determine if the pathways leading to this were the same. With video-microscopy of mepacrine labelled platelets, we found a difference in pathways, see Table II. At a wall shear rate of $80\,s^{-1}$, the percent of platelets which adhere

Table I. Overall Accounting for Platelets Arriving on Fibrinogen-coated Glass

Flow rate (Shear Rate)	0.5 ml/min $(40 s^{-1})$		1.0 ml/min $(80 s^{-1})$		11.8 ml/min $(944 s^{-1})$
Exposure time to flow	0-2 min		0-2 min		0-1 min
No. of measurements	n = 15		n = 11		n = 10
First Contacts on Arrival					
Permanent adhesion	42.5(17.2)	$2P < 0.01$[a]	61.9(11.7)	$2P < 0.01$[a]	45.4(8.4)
Those moving to new sites in the measurement region[b]	43.0(15.4)	$2P < 0.05$	30.4(8.7)	N.S.	31.3(7.5)
Departure from the measurement region	14.5(10.2) 100.0	N.S.	7.7(7.8) 100.0	$2P < 0.001$	23.3(6.0) 100.0
Overall Balance					
Permanent adhesion	79.0(15.0)	$2P < 0.05$	91.2(8.3)	$2P < 0.001$	72.9(7.9)
Departure from the measurement region	21.0(15.0) 100.0	$2P < 0.05$	8.8(8.3) 100.0	$2P < 0.001$	27.1(7.9) 100.0

[a] The probabilities shown refer to t-tests for independent samples. Entries are Mean Percent (S.D.)

[b] The percent and S.D. of these platelets which adhere permanently is: (1) 85.5(12.2) for 0.5 ml/min, 0-2 min, (2) 96.2(5.7) for 1.0 ml/min, 0-2 min, (3) 85.3 (19.4) for 1.0 ml/min, 2-4 min. and (4) 88.9 (10.7) for 11.8 ml/min, 0-1 min.

TABLE II. Overall Accounting for Platelets Arriving on Albumin-coated and Fibrinogen-coated Glass at 1.0 ml/min (80 s^{-1}) - Basis 100 Platelets Contacting the Surface

Coating	Albumin	Fibrinogen	
Exposure time to flow	0-2 min	0-2 min	
Number of measurements	n = 12	n = 11.	
First Contacts on Arrival			
Permanent adhesion	45.2 (5.5)	62.2 (11.7)	2P<0.01
Those moving to new sites in the measurement region[a]	45.0 (6.7)	30.4 (8.7)	2P<0.01.
Departure from the measurement region	9.8 (6.0)	7.4 (7.8)	N.S.
	100.0	100.0	
Overall Balance			
Permanent adhesion	87.5 (8.7)	91.2 (8.3)	N.S.
Departure from the measurement regio	12.5 (8.7)	8.8 (8.3)	N.S.
	100.0	100.0	

Entries are Mean Percent (S.D.)

[a]   The Mean Percent (S.D.) of these platelets which adhere permanently: 94.0 (8.9) for albumin and 95.4 (5.7) for fibrinogen.

(Reproduced with permission from Ref. 30 Copyright 1986 F.K. Schattauer Verlag.)

permanently on first contact is less on albumin than on fibrinogen, while the percent of platelets which move to new positions is greater for albumin than for fibrinogen. The overall percent of cells permanently adhering is the same for both surface types. At a shear rate of $456 s^{-1}$, these differences are absent and the adhesion efficiency on initial contact and overall adhesion efficiency are less than corresponding values for each protein coating at $80 s^{-1}$. The mechanism of detachment and reattachment of surface-bound platelets is flow dependent and can lead to an equality of overall adhesion efficiency even when initial adhesion efficiencies are not equal.

Experiments with fluorescent video-microscopy have also been used to focus on the adhesion of platelets to unaltered fibrinogen and albumin and to regions on these protein substrates where platelets have detached, platelet-primed protein. On fibrinogen, no difference was found at low shear rates; at $944 s^{-1}$, it was shown that the adhesion efficiency was greater on the platelet-primed protein (29). On albumin, the platelet-primed protein was stickier at $80 s^{-1}$ and $456 s^{-1}$ (30).

Another question area which was probed was that of whether or not detached platelets were more sticky when they contacted a surface for the second time. This question was asked by making observations on unaltered fibrinogen and albumin. Here, it was demonstrated that second contacts are more efficient than first contacts (25,26). Possible explanations may be an alteration of the platelet's membrane resulting from detachment which makes it more adhesive or that the mechanical features of a cell moving along a surface to a nearby spot, 1-5 μm away, favours adhesion to a greater extent than for a cell contacting from the bulk suspension.

The alteration of surface-bound protein through adherence and detachment of platelets is but one way that a cell can alter a protein substrate. A number of cells which have been cultured on preadsorbed protein have been shown to alter this protein as well as protein provided by the cell to the surface (31, 32). Endocytosis and enzymatic digestion of surface-bound protein are possible mechanisms directly related to functions of adherent cells. Evidence exists to indicate that such processes may be occurring when platelets adhere to preadsorbed protein (33).

Acknowledgments

The author wishes to thank the Heart and Stroke Foundation of Ontario and the Natural Sciences and Engineering Research Council of Canada for continued support.

Literature Cited

1.   Park, K.; Mosher, D.F.; Cooper, S.L. J. Biomed. Mat. Res. 1986, 20, 580-612.
2.   Turitto, V.T.; Muggli, R.; Baumgartner, H.R. Asaio Journal 1979, 2, 28-34.
3.   Lahav, J.; Hynes, R.O. J. Supramolec. Struct. Cell. Biochem. 1981, 17, 299- 311.
4.   Mason, R.G.; Read, M.S.; Brinkhous, K.M. Proc. Soc. Exp. Biol. Med. 1971, 137, 680-682.
5.   Packham, M.A.; Evans, G.; Glynn, M.F.; Mustard, J.F. J. Lab. Clin. Med. 1969, 73, 686-697.
6.   Adams, G.A.; Feuerstein, I.A. Asaio Journal 1981, 4, 90-99.
7.   Mustard, J.F.; Perry, D.W.; Ardlie, N.G.; Packham, M.A. Br. J. Haematol. 1972, 22, 193-204.
8.   Kinlough-Rathbone, R.L.; Mustard J.F.; Packham, M.A.; Perry, D.W.; Reimers, H.-J.; Cazenave, J.-P. Thromb. Haemostas. 1977, 37, 291-308.
9.   Grabowski, E.F; Friedman, L.I.; Leonard, E.F. Ind. Eng. Chem. Fundam. 1972, 11, 224-232.
10. Adams, G.A.; Feuerstein, I.A. Trans. Amer. Soc. Artif. Intern Organs 1981, 27, 219-224.
11. Feuerstein, I.A.; Skupney-Garnham, L.E. Thromb. Res. 1986, 43, 497-505.
12. Feuerstein, I.A.; Dickson, V. Thromb. Haemostas. 1983, 50, 679-685.

13.   Whicher, S.J.; Brash, J.L. J. Biomed. Mat. Res. 1978, 12, 181-201.
14.   Adams, G.A.; Feuerstein, I.A. Am. J. Physiol. 1981, 240, H99-H108.
15.   Adams, G.A.; Feuerstein, I.A. Am. J. Physiol. 1983, 244, H109-H114.
16.   Holmsen, H., Weiss, H J. Annu. Rev. Med. 1979, 30, 119-134.
17.   Weiss, H.J.; Witte, L.D.; Kaplan, K.L.. Lages, B.A.; Chernoff, A.; Nossel, H.L.;
      Goodman, D.S.; Baumgartner, H.R. Blood 1979, 54, 1296-1319.
18.   Gartner, T.K.; Gerrard, J.M.; White, J.G.; Williams, D.C. Blood 1981, 58, 153-157.
19.   White, J.G. Blood 1968, 31, 604-622.
20.   Bevers, E.M.; Comfurius, P.; Van Rign, J.L.M.L.; Hemker, H.C.; Zwaal, F.A. Eur.
      J. Biochem. 1982, 122, 429-436.
21.   Griffin, J.H. In Interaction of Blood with Natural and Artificial Surfaces;
      Salzman, E.W., Ed.; Marcel Dekker: New York, 1977; pp. 139-170.
22.   Mustard, J.F.; Packham, M.A.; Kinlough-Rathbone, R.L.; Perry, D.W.; Regoeczi,
      E. Blood 1978, 52, 453-466.
23.   Di Minno, G.; Thiagarajan, P.; Perussia, B.; Martinez, J.; Shapiro, S.; Trinchieri,
      G.; Murphy, S. Blood 1983, 61, 140-148.
24.   Keller, K.H.; Yum, S.I. Trans. Amer. Soc. Artif. Intern Organs 1970, 16, 42- 47.
25.   Brash, J.L.; Uniyal, S. Trans Amer. Soc. Artif. Intern Organs 1976, 12, 253- 259.
26.   Richardson, P.D.; Bodziak, K. Trans. Amer. Soc. Artif. Intern Organs 1982, 28,
      426-429.
27.   Richardson, P.D.; Mohammed, S.F.; Mason, R.G.; Steiner, M.; Kane, R. Trans.
      Amer. Soc. Artif. Intern Organs 1979, 25, 147-151.
28.   Feuerstein, I.A.; Kush, J. Trans. Amer. Soc. Artif. Intern Organs 1983, 29,
      430-434.
29.   Feuerstein, I.A.; Kush, J. Trans. ASME J. Biomech. Eng. 1986, 108, 49-53.
30.   Feuerstein, I.A.; Kush, Thromb. Haemostas. 1986, 55, 184-186.
31.   Avnur, Z.; Geiger, B. Cell 1981, 25, 121-132.
32.   Chen, W.-T.; Olden, K.; Bernard, B.A.; Chu, F.-F. J. Cell Biol. 1984, 98,
      1546-1555.
33.   Feuerstein, I.A.; Kush, J. Trans. Amer. Soc. Artif. Intern Organs 1985, 31,
      270-274.

RECEIVED February 18, 1987

Chapter 33

Capillary Perfusion System for Quantitative Evaluation of Protein Adsorption and Platelet Adhesion to Artificial Surfaces

Jean-Pierre Cazenave and Juliette Mulvihill

Biologie et Pharmacologie des Interactions du Sang avec les Vaisseaux et les Biomatériaux, Centre Régional de Transfusion Sanguine, Institut National de la Santé et de la Recherche Médicale U.311, 10 rue Spielmann, 67085 Strasbourg Cédex, France

A capillary perfusion system has been developed for in vitro quantitation of protein adsorption and platelet accumulation on artificial surfaces under controlled hydrodynamic conditions ($0-4,000 s^{-1}$). Using washed platelet suspensions or anticoagulated whole blood, platelet deposition and/or protein adsorption is followed by radioisotopic techniques, platelet deposition in whole blood being measured by surface phase radioimmunoassay with a monoclonal antibody against human platelet membrane glycoprotein IIb-IIIa. Applications include studies of the influence of preadsorbed proteins on platelet accumulation on artificial surfaces and screening of biomaterials for short term hemocompatibility.

Under normal conditions, the luminal surface of the endothelium lining of the cardiovascular system is non thrombogenic and does not allow platelet and leucocyte adhesion ([1,2]). In contrast, exposure of an artificial surface to blood leads to basic reactions which may culminate in thrombus formation ([3]). When flowing blood enters into contact with an artificial surface, a series of events is initiated : rapid adsorption of a layer of plasma proteins at the interface, platelet adhesion to the protein layer and conformational changes of adsorbed contact factors leading to activation of the coagulation system to form thrombin and fibrin. Hemodynamic and rheological conditions play an important role in the localisation, size, structure and turnover of the resulting thrombus. In high flow regions, it is composed mainly of platelets and fibrin.

The adsorption of a layer of plasma proteins is the first event which occurs when blood is exposed to an artificial surface ([4]). As a result, a platelet never sees or adheres to a bare surface. The nature of the adsorbed protein layer, which depends on the relative concentrations and mobilities of the proteins in plasma and on their affinity for the surface, will condition the subsequent platelet-surface interaction ([5]). Protein adsorption to foreign surfaces has been studied experimentally by measuring the adsorption isotherms of single purified proteins (6,7) known to be involved in thrombosis (fibrinogen, fibrin, contact factor XII, high molecular weight kininogen, von Willebrand factor, thrombin), inflammatory and immunological reactions (immunoglobulins, immune complexes, complement chemotactic fragments C5a and C3a) and cell adhesion (albumin, fibronectin, collagen). Adsorption of a protein at an interface may be followed by denaturation, conformational change or biological activation. Furthermore, adsorption from blood, which is a complex mixture of proteins, is complicated by competition between proteins, desorption and structural modifications of the protein layer with time and rheological factors (6,8).

The development of thromboresistant biomaterials is important in all situations where an artificial surface is exposed to blood, for example in cardiovascular implants (heart valves, shunts, vascular grafts and catheters), extracorporeal circulation systems, blood filters and blood storage containers. In order to predict the short term clinical performance of biomaterials, we have developed an in vitro technique for the determination of protein adsorption and platelet adhesion to artificial surfaces which satisfies the following requirements : absence of an air-solution interface, an essential condition to avoid activation of proteins or platelets by air contact; well characterised hydrodynamic conditions (Poiseuille laminar flow) with wall shear rates corresponding to those encountered in the cardiovascular system $(0-4,000\ s^{-1})$ (9); using radioisotopic methods, possibility of the simultaneous study of platelet adhesion and the adsorption of one or more plasma proteins; quantitation of platelet accumulation either with washed radiolabeled platelets or by surface phase radioimmunoassay using a monoclonal antibody directed specifically against the membrane glycoprotein complex IIb-IIIa, thus enabling experiments to be performed with whole blood.

Experimental Procedures

Capillary Perfusion System. The in vitro perfusion system (Figure 1) consists of two plastic syringes mounted in parallel on a syringe pump (Precidor 5003, Infors, Basel, Switzerland) and leading to a pair of glass capillary tubes (Microcaps, Drummond Scientific Co., USA). Prior to use, capillaries are cleaned by immersion for one hour in sulphochromic acid, followed by extensive rinsing in distilled water and oven drying at 100°C. The capillary internal diameter is constant to within 2%. Capillaries of internal diameter 0.56 mm are used for perfusion at wall shear rates from 800 to 4,000 s^{-1} and capillaries of internal diameter 0.80 mm for perfusion at shear rates less than 800 s^{-1}. A three way stop-cock at the joint between each syringe and its connecting capillary leads via a peristaltic pump (Minipuls 2, HP4, Gilson, France) to a reservoir containing rinsing buffer. The entire system is maintained at 37.0 ± 0.2°C under a thermostated hood (ITH I, Infors, Basel, Switzerland). Using radiolabeled proteins, platelets or anti-platelet antibody, protein adsorption or platelet accumulation is measured according to the radioactivity deposited on capillary segments of known internal surface area, determined by radioactive counting in a well-type gamma counter (1282 Compugamma, LKB, Sweden). Blood collection at the capillary exit enables quantitative determination of

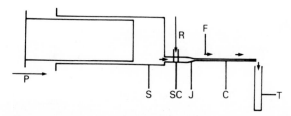

Figure 1: Schematic diagram of the in vitro capillary perfusion system. S : 50 ml plastic syringe containing protein or antibody solution, anticoagulated whole blood or washed platelet suspension; P : piston driven by syringe pump; SC : 3-way stop-cock; J : joint in silicone tubing; C : glass capillary (0.80 or 0.56 mm i.d.); F : direction of blood flow; T : plastic tube for blood collection; R : rinsing buffer from a reservoir at 37°C, flow controlled by a peristaltic pump. The entire apparatus is enclosed in a thermostated hood at 37°C.

factors indicating activation of platelets (platelet factor 4, β -
thromboglobulin) or the coagulation system (fibrinopeptide A).
Morphological examination of the capillary surface is also possible by
scanning electron microscopy (10).

Purification and Radiolabeling of Proteins. Human serum albumin is a
sterile solution (175 g/l) prepared by the Cohn method of ethanol
fractionation at the Centre Régional de Transfusion Sanguine de
Strasbourg. Fibrinogen is purified from human plasma by the technique of
Kekwick (11). Human fibronectin is purified by affinity chromatography
using gelatin immobilised on Sepharose 4B, according to Miekka (12). Von
Willebrand factor (vWF) is isolated by gel filtration of human plasma
cryoprecipitate and characterised according to the von Willebrand
antigen content, determined by immunoelectrophoresis, and the
ristocetin cofactor activity, measured in agglutination tests with
washed human platelets (13). The purity of plasma proteins, as verified
by electrophoresis in SDS-polyacrylamide gel (14), is always greater
than 98 %. Bovine collagen (type I, insoluble, Sigma Chemicals, St
Louis, USA) is a 0.25 % suspension in 0.522 M acetic acid, pH 2.8,
prepared by the method of Cazenave (13). Purified human plasma proteins
are radiolabeled with ^{125}Iodine (Commissariat à l'Energie Atomique,
Saclay, France) by the iodogen technique (15).

Production, Purification and Radiolabeling of Monoclonal Anti-human
Platelet Antibody 6C9. The monoclonal antibody 6C9 directed
specifically against the membrane glycoprotein complex IIb-IIIa of
human platelets is a development of the Central Laboratory of the
Netherlands Red Cross Blood Transfusion Service in Amsterdam. As
described in detail in previous publications (16), antibody is produced
in BALB/C mice, purified by ammonium sulphate precipitation and ion
exchange chromatography and characterised by serological and
immunochemical analysis. Radiolabeling with ^{125}Iodine is performed
using the iodogen technique.

Blood Collection and Preparation of Washed ^{111}Indium Labeled
Human Platelets. Blood is collected from healthy donors by
venipuncture using a large diameter (18/10) needle. Washed human
platelets are prepared by the method of Cazenave (13), a modification of
the technique of Mustard (17), and resuspended at 300,000/mm^3 in
Tyrode's buffer containing 2 mM Ca^{2+}, 1 mM Mg^{2+}, 0.35 % human serum
albumin, 0.1 % glucose and 20 µl/ml apyrase (13), pH 7.30. Radiolabeling
of platelets is carried out by incubation with ^{111}Indium-oxine (0.25
g/ml) for 15 min at 37°C, according to Eber (18). Under these conditions,
labeling yield is of the order of 90 % and platelet aggregation with ADP,
thrombin or collagen is not modified. Washed red blood cells from the
same donor are prepared as described by Cazenave (19). Immediately
before perfusion experiments, packed erythrocytes are added to the
suspension of washed ^{111}Indium labeled human platelets at a volume ratio
corresponding to a 40 % hematocrit. For perfusion studies with heparin
(10 U/ml, Roche, France) anticoagulated whole blood, the stoppered
tubes are kept at room temperature until required for perfusion, within
a maximum delay of two hours.

Protein Adsorption. Adsorption of human plasma proteins to glass capillary surfaces has been studied under static conditions at ambient temperature. Purified plasma proteins are diluted in rinsing buffer (Tyrode's buffer without calcium or magnesium, pH 7.30). The capillaries are first filled with rinsing buffer, the protein solution is introduced at low flow rate (150 s^{-1}) by displacement of at least four capillary volumes and the tubes are then stoppered and left for one hour, the time required to attain adsorption equilibrium. Using radiolabeled proteins for the measurement of adsorption isotherms, rinsing (10 min at 1,000 s^{-1}) is followed by radioactive counting. Precoating of glass capillary surfaces with plasma proteins or bovine collagen for subsequent platelet adhesion studies is performed in the same manner as the determination of adsorption isotherms, the bulk solution concentration of plasma protein being chosen to lie in the plateau region of the Langmuir-type adsorption curve. Bovine collagen is used as a 0.25 % suspension in 0.522 M acetic acid, pH 2.8. In the case of plasma proteins, the precoated tubes, rinsed and buffer filled, are left stoppered at ambient temperature until perfusion, within no longer than 18 hours. Collagen precoated capillaries are emptied to allow polymerisation of the collagen on contact with air (19) and may be stored at 4°C for up to 7 days. Care is taken at all times during protein adsorption or blood perfusion experiments to avoid the formation of air bubbles or an air-liquid interface, to prevent denaturation or activation of proteins or platelets on air contact.

Perfusion of Whole Blood or Washed Platelets Suspensions.
Capillaries are perfused with anticoagulated whole blood or with a suspension of washed [111]Indium labeled human platelets in Tyrode's-albumin buffer in the presence of a 40 % hematocrit. The blood or platelet suspension is introduced by displacement of rinsing buffer. Perfusion is continued for a fixed time interval, e.g. 2 or 5 min, at a steady flow rate corresponding to a predetermined wall shear rate, e.g. 50, 800, 2,000 or 4,000 s^{-1}. Rinsing with buffer (5 min at 1,000 s^{-1}) is followed by radioactive counting in the case of [111]Indium labeled platelets or by radioimmunoassay with [125]Iodine labeled anti-human platelet antibody in experiments with whole blood. The perfusate is collected in plastic tubes. Platelet count is determined before and after perfusion, while standard immunoenzymatic (ELISA) techniques with commercial kits (Asserachrom, Stago, France) are used to measure the production of fibrinopeptide A (FPA), platelet factor 4 (PF4) or β-thromboglobulin (βTG).

Quantitation of Platelet Deposition by Radioimmunoassay with Anti-human Platelet Antibody 6C9. In perfusion experiments with unlabeled platelets in anticoagulated whole blood, platelet deposition is quantitated by means of a surface phase radioimmunoassay using the monoclonal antibody 6C9, directed against the membrane glycoprotein complex IIb-IIIa of human platelets. An aliquot of purified [125]Iodine labeled antibody is diluted in rinsing buffer containing 0.1 % human serum albumin and introduced into the capillary tubes at low flow rate (150 s^{-1}) by displacement of at least four volumes of buffer. The tubes are stoppered and incubation is continued under static conditions at 37°C for 30 min, the time required for immunoadsorption to reach a plateau value. Background antibody adsorption is determined using capillaries perfused with platelet poor plasma. After rinsing (5 min at

$1,000\ s^{-1}$), radioactive counting yields the ^{125}Iodine radioactivity due to immunoadsorbed antibody for capillary segments of known surface area. Calibration of the radioimmunoassay is performed by means of double isotope experiments with ^{111}Indium labeled platelets and ^{125}Iodine labeled antibody. Capillaries are perfused with blood containing autologous washed ^{111}Indium labeled platelets and a linear correlation is obtained between adherent ^{111}Indium-platelets and immunoadsorbed ^{125}Iodine-antibody. For experiments using unlabeled platelets in whole blood, platelet adhesion is then calculated according to the relation:

$$\text{adherent platelets/mm}^2 = (^{125}I_A - {}^{125}I_B) \times \text{slope}_C/S \qquad (1)$$

where $^{125}I_A$ is the measured ^{125}Iodine radioactivity, $^{125}I_B$ the background antibody adsorption, S the surface area and slope$_C$ the slope of the linear correlation for radioimmunoassay (^{111}In-platelets versus ^{125}I-antibody), corrected for physical decay.

Results and Discussion

Flow Properties of Perfusion System. Under conditions of Poiseuille type laminar flow in capillary tubes, the Reynold's number (R_e) is given by the relation :

$$R_e = \frac{\rho <v> r}{\zeta} \qquad (2)$$

where ρ is the volumetric mass, r the tube radius, ζ the fluid viscosity and $<v>$ the average fluid velocity (flow rate/r^2). The onset of turbulent flow occurs at Reynold's numbers in excess of 2,500 (9). In our system, at maximum shear stress (4,000 s^{-1}), corresponding to a flow rate of 4.14 ml/min in capillaries of internal diameter 0.56 mm, the Reynold's number calculated from Equation 2 is 80. This figure lies well below the limit of turbulent flow. Furthermore, experiments with tracer dyes have shown no visible evidence of turbulence at wall shear rates in the range 50-4,000 s^{-1}, either within the capillaries or at entrance or exit points.

The length of the inlet region (L) at the entrance of the capillary is estimated as (9) :

$$L = 0.07\ R_e r \qquad (3)$$

which in our case is never greater than :

$$L = 0.07 \times 80 \times 2.8 \times 10^{-2}$$

$$= 0.16\ \text{cm}.$$

Since the minimum capillary length used in our system is 4 cm, the inlet region where the fluid flow is not of Poiseuille type represents 4 % or less of the total flow path studied. An entrance effect would be reflected in increased protein or platelet deposition under dynamic conditions. Using radiolabeled platelets, when capillaries are cut into 1.0 cm segments for radioactive counting, there is no increase in radioactivity of the first segment other than that which may be attributed to the known variation of platelet accumulation with tube length (see following paragraph). Sensitivity in these measurements is sufficient to ensure less than 1 % error in radioactive counting of capillary segments. The velocity profile in the perfusion system is therefore considered to be parabolic and wall shear rates (γ) are calculated according to the relation :

$$\gamma = \frac{4<v>}{r} \tag{4}$$

In platelet perfusion experiments, platelet accumulation on reactive adhesive surfaces such as collagen coated glass shows a steady decrease with distance from the capillary inlet, as previously observed in similar flow systems (20,21). This axial dependence does not, in our system, appear to follow a simple mathematical law as a function of distance from the capillary entrance (22). Platelet deposition is thus calculated as the mean deposition over the entire length ($\leqslant$ 10 cm) of a single capillary and in comparative studies capillaries of equal length and internal diameter are considered.

Static Adsorption of Plasma Proteins on Glass. Initial studies of the interaction of proteins with artificial surfaces concerned the highly simplified situation of static adsorption on glass from solutions of purified radiolabeled human plasma proteins. Albumin was chosen as a major plasma protein known for its non thrombogenic properties (5,6). Fibrinogen and fibronectin, on the contrary, are major proteins of plasma which enhance platelet and cellular adhesion (4,5,7,23-25).

Static adsorption isotherms on glass were determined at ambient temperature for purified [125]Iodine labeled human albumin, fibrinogen and fibronectin in Tyrode's buffer without calcium or magnesium, pH 7.30. Langmuir type adsorption was observed, the plateau region being attained for solution concentrations of 3.0 - 5.0 mg/ml (albumin), 0.5 - 1.0 mg/ml (fibrinogen) and 0.2 - 0.4 mg/ml (fibronectin). Plateau surface concentrations of protein (Table I) correspond approximately to monolayers of fibrinogen (26,27) and fibronectin (28) and to a bilayer of albumin (29). Affinity constants suggest an order of affinity for the glass surface : fibronectin > fibrinogen > albumin.

Washed Human Platelet Suspensions. The capillary perfusion system has proved to be of particular value for the study of the influence of a preadsorbed protein layer on subsequent platelet accumulation on artificial surfaces. Platelet deposition was measured for varying wall shear rates (150, 500 or 1,000 s^{-1}) and perfusion times (2, 5 or 10 min) on glass capillaries preadsorbed with purified human albumin, fibrinogen or fibronectin, using a suspension of washed [111]Indium labeled human platelets in the presence of a 40 % hematocrit. Bulk solution concentrations of plasma protein for preadsorption were chosen to lie in the plateau region of the previously determined adsorption isotherms. Results showed an increase in platelet accumulation with perfusion time on all three surfaces. On fibrinogen and fibronectin coated glass, platelet deposition increased with wall shear rate up to 1,000 s^{-1}, indicating that under these conditions adhesion was largely controlled by transport to the interface (30). Consistent with the

Table I. Static Adsorption Isotherms on Glass Capillary Tubes for
Purified Human Plasma Proteins

Protein	$\Gamma(\mu g/cm^2)$	$K \times 10^9$ (1/mole)
albumin	0,40	100
fibrinogen	0,40	174
fibronectin	0,25	313

Adsorption 1 hour, static conditions, 22°C.
Γ: Surface protein concentration in plateau region ; corresponding
solution concentrations : albumin 3.0 - 5.0 mg/ml, fibrinogen 0.5 -
1.0 mg/ml, fibronectin 0.2 - 0.4 mg/ml.
K: Affinity constant.

passive nature of albumin surfaces, accumulation of platelets on
albumin coated glass was lower than on the other surfaces and decreased
at high shear rate ($1,000 s^{-1}$), where it would appear that adhesion was
no longer governed by transport phenomena but rather by surface phase
biochemical reactions. At low wall shear rate ($150 s^{-1}$), platelet
deposition was greater on fibronectin than on fibrinogen. Conversely,
at high wall shear rate ($1,000 s^{-1}$), platelet accumulation was greatest
on fibrinogen. According to these studies, an albumin coated artificial
surface has low reactivity with respect to platelets, while a
fibronectin coated surface shows greater reactivity under hemodynamic
conditions corresponding to those of the venous circulation and a
fibrinogen coated surface enhanced reactivity under arterial
conditions (9).

In further experiments using washed human platelet suspensions,
platelet accumulation on albumin and fibrinogen coated glass was
compared with accumulation on glass capillaries coated with type I
bovine collagen for a perfusion time of 5 min and wall shear rates
ranging from 50 to $4,000 s^{-1}$. Surface reactivity followed the general
order : albumin < fibrinogen < collagen (Table II). On the highly
reactive collagen surface, platelet deposition increased continuously
with wall shear rate up to $4,000 s^{-1}$, indicating transport controlled
adhesion, whereas on the passive albumin surface, peak deposition was
observed at $800 s^{-1}$ and lowest values at $4,000 s^{-1}$, where the rate of
adhesion would be limited by biochemical reactions at the interface. The
fibrinogen surface showed intermediate behaviour, with a plateau level
of platelet accumulation from 800 to $4,000 s^{-1}$. Results are thus in
accord with the known high platelet reactivity of collagen, this protein
being an important constituent of subendothelium, where it is
considered to play the major role in platelet adhesion and subsequent
thrombus formation (31).

Platelet adhesion to subendothelium at high wall shear rate has
been shown to be mediated by vWF (32). Preliminary results in the
capillary perfusion system, using glass capillaries uncoated or coated
with purified human albumin or bovine collagen, showed an increase in
platelet deposition in the presence of vWF on the albumin surface at low
($50 s^{-1}$) and high ($2,000 s^{-1}$) wall shear rate, but on the glass and

Table II. Accumulation of Washed Human Platelets Labeled with
[111]Indium on Glass Capillary Surfaces Precoated
with Purified Human Albumin, Human Fibrinogen
or Bovine Collagen

Wall shear rate (s^{-1})	Adherent platelets/mm^2		
		Protein coating	
	albumin	fibrinogen	collagen
50	11,000	6,000	16,500
800	22,500	53,000	151,500
2,000	11,000	78,000	306,000
4,000	7,000	56,000	425,500

Preadsorption of protein coating 1 hour, static conditions, 22°C.
Perfusion 5 min at 37°C, platelets 180,000/mm^3 in Tyrode's-albumin
buffer, hematocrit 40 %.
Results are mean of four independent experiments.

collagen surfaces only at high shear rate (Table III). Experiments were
of two types : vWF was added to the platelet suspension (final
concentration 5 µg/ml) immediately before perfusion or the capillaries
were preadsorbed with vWF (5 µg/ml) before perfusion. Control values
were obtained by substituting buffer solution containing no vWF. The
enhancement of platelet adhesion by vWF was best seen on the more passive
albumin surface, where the concentration of adherent platelets after
perfusion for 2 min at 2,000 s^{-1} increased by a factor of 1.87 when vWF
was added to the platelet suspension and by a factor of 4 when vWF was
already present on the surface in the form of a preadsorbed layer.

<u>Anticoagulated Whole Blood</u>. In perfusion experiments using anticoa-
gulated whole blood, platelet accumulation on protein coated glass
capillary surfaces is determined by radioimmunoassay with a monoclonal
antibody directed specifically against the membrane glycoprotein
complex IIb-IIIa of human   platelets. The validity of this technique
has been discussed in detail elsewhere (<u>10</u>). Briefly, since the
glycoprotein complex IIb-IIIa exists only in platelets and endothelial
cells, when applied to the study of blood interaction in vitro with
artificial surfaces, the radioimmunoassay may be considered to give a
specific measurement of platelet deposition, regardless of the presence
of other blood cells. Calibration by the double isotope method results
in a linear correlation between adherent platelets ([111]Indium labeled)
and immunoadsorbed antibody ([125]Iodine labeled).   Linearity is
conserved up to at least 7.5×10^5 platelets/mm^2 and for a given
preparation of purified, radiolabeled antibody the slope of the
straight line relation, corrected for physical decay, remains constant
to within 10 % in separate calibration experiments. Thus antigen-
antibody specificity is not destroyed by platelet adhesion. Nor does
this specificity appear to be influenced by differing morphology of
adherent platelets, since the slope of the linear relation is not

affected by the nature of the adhesive surface, e.g. glass coated with albumin, fibrinogen or collagen. Background antibody adsorption in the absence of platelets is low, of the order of 10 % of values for surfaces bearing platelets. As applied to the quantitation of platelet accumulation in whole blood, this technique offers the advantage of avoiding the lengthy procedure of platelet separation, washing and radiolabeling, together with the possible selection of a particular population of platelets during this preparation.

Table III. Accumulation of Washed Human Platelets Labeled with [111]Indium in Presence and Absence of vWF on Glass Capillary Surfaces Uncoated or Precoated with Purified Human Albumin or Bovine Collagen

Wall shear rate (s^{-1})	Adherent platelets : ratio vWF/control buffer		
	Protein coating		
	uncoated	albumin	collagen
A. 50	0.98	1.18	0.94
2,000	1.33	1.87	1.49
B. 50	0.82	1.43	1.07
2,000	1.33	4.08	1.35

Preadsorption of protein coating 1 hour, static conditions, 22°C.
Perfusion 2 min at 37°C, platelets 180,000/mm^3 in Tyrode's-albumin buffer, hematocrit 40 %.
Results are mean of three independent experiments.
A. vWF added to washed platelet suspension (final concentration 5 µg/ml).
B. Preadsorption of vWF (5 µg/ml) 1 hour, static conditions, 22°C.

Typical results for a series of measurements of platelet accumulation by surface phase radioimmunoassay are presented in Table IV. Glass capillaries precoated with purified human albumin or fibrinogen or bovine collagen were perfused with heparin anticoagulated (10 U/ml) whole blood at varying wall shear rates (50-4,000 s^{-1}). As compared to equivalent experiments with washed platelet suspensions (Table II), platelet deposition was reduced on all surfaces. This diminution of surface reactivity, undoubtedly related to albumin passivation from plasma, was particularly important on albumin precoated glass, possibly due to favorable conditions for albumin adsorption by extension of the existing protein layer. In other respects, platelet deposition in whole blood followed trends similar to those observed in washed platelet suspensions. The order of surface reactivity was unchanged : albumin < fibrinogen < collagen. With rising wall shear rate, platelet accumulation on collagen showed a continuous increase up to 4,000 s^{-1}, as compared to a leveling out of values from 800 to 4,000 s^{-1} on the less reactive albumin and fibrinogen surfaces.

Table IV. Platelet Accumulation from Heparin Anticoagulated Whole
Blood on Glass Capillary Surfaces Precoated with
Purified Human Albumin, Human Fibrinogen or Bovine
Collagen. Quantitation of Platelet Deposition by
Radioimmunoassay with [125]Iodine Labeled
Monoclonal Antibody 6C9 against Human Platelet
Glycoprotein Complex IIb-IIIa

Wall shear rate (s^{-1})	Adherent platelets/mm^2		
		Protein coating	
	albumin	fibrinogen	collagen
50	90	4,500	16,000
800	150	27,000	39,000
2,000	170	25,500	43,500
4,000	100	31,000	79,000

Preadsorption of protein coating 1 hour, static conditions, 22°C.
Perfusion 5 min at 37°C, heparin anticoagulated (10 U/ml) whole blood,
platelets $173,000 \pm 15,000$/mm^3 ($\bar{x} \pm$ SEM, n=4).
Results are mean of four independent experiments.

Adaptation of Perfusion System to the Study of Biomaterials in
Catheter Form.   Although originally developed for use with glass
capillaries, the perfusion system is readily adapted to the study of
protein adsorption and platelet accumulation on polymeric biomaterials
in catheter form. As an example of this type of investigation, catheters
produced from mixtures of polyvinylchloride (PVC) with silicone based
copolymers (Hospal, Meyzieu, France) were compared with pure silicone
catheters. Using suspensions of washed [111]Indium labeled human
platelets in the presence of a 40 % hematocrit, platelet deposition
and release of βTG were determined for perfusion at flow rates
corresponding to a wall shear rate of 500 s^{-1} (Table V). Consistent with
the weak thrombogenicity of silicone polymers (3,33), platelet
accumulation and βTG secretion were lower on pure silicone than on the
mixed PVC/silicone surfaces.

A similar series of experiments concerned a blend of a terpolymer
of PVC with a cationic elastomer, containing or not ionically bound
heparin. Perfusion with washed platelet suspensions, under conditions
excluding thrombin generation from plasma, demonstrated the
activation of platelets by heparin (34), as reflected in a parallel
increase in platelet deposition and βTG secretion (Table V). The
perfusion system thus offers a simple means of in vitro screening of
biomaterials for short term interactions with platelets and plasma
proteins.

Table V. Platelet Interaction with Biomaterials in Catheter Form

Biomaterial	Adherent platelets/mm^2	βTG secretion (ng/ml)
A. Silicone	3,200	130
PVC/silicone (1)	4,000	170
PVC/silicone (2)	3,900	200
PVC/silicone (3)	3,600	180
B. PVC/EC	300	80
PVC/EC/H	500	130

Perfusion 5 min at 500 s^{-1}, 22°C, washed ^{111}Indium labeled human platelets 180,000/mm^3 in Tyrode's-albumin buffer, hematocrit 40 %. Results are mean of two independent experiments.
PVC/silicone (1), (2) and (3) : mixed PVC/silicone polymers containing respectively 15.9, 30.5 and 26.0 % silicone.
PVC/EC   : blend 50 % PVC, 50 % cationic elastomer.
PVC/EC/H : blend 45 % PVC, 45 % cationic elastomer, 10 % ionically bound heparin.

Clinical Application : Surface Passivation by Albumin Adsorption.
Passivation of artificial surfaces by preadsorption of human serum albumin is a technique frequently employed to improve the hemocompatibility of biomaterials used in clinical practice (3,35). The efficiency of this method is well illustrated by a study in our laboratory relating to therapeutic plasmapheresis (36). In the treatment of patients showing abnormal or excess immunoglobulins or immune complexes, plasmapheresis was hindered by blockage of the extracorporeal circulation system by platelet thrombi which formed in the filters and joints and in 44 % of cases led to premature termination of the plasmapheresis session. Perfusion tests in vitro indicated greatly reduced platelet accumulation on the PVC tubing of the extracorporeal system after preadsorption with purified human albumin. It was therefore decided to passivate the plasmapheresis circuits by preadsorption with 4 % human serum albumin, then to introduce the patient's blood by displacement of the albumin solution. Following adoption of this technique, blockage of the extracorporeal circulation system occured in only 5 % of cases treated.

Conclusions

In summary, the capillary perfusion system represents a simple in vitro technique for the quantitation of protein adsorption and platelet accumulation on artificial surfaces under well defined hydrodynamic conditions (0-4,000 s^{-1}). Radioisotopic methods enable simultaneous study of platelet deposition and adsorption of one or more plasma proteins. Experiments may be performed using washed platelet suspensions or anticoagulated whole blood, platelet deposition being measured in the latter case by surface phase radioimmunoassay with a monoclonal antibody against the human platelet glycoprotein complex IIb-IIIa. Activation of platelets or the coagulation system may be followed by determination of released PF4, βTG or FPA. The system is of particular value for the study of the influence of preadsorbed proteins on platelet accumulation on artificial surfaces, the screening of biomaterials for short term hemocompatibility and the evaluation of techniques such as albumin passivation designed to reduce the thrombogenicity of systems used in clinical practice.

Acknowledgments

We gratefully acknowledge the assistance of Mrs. A. Sutter-Bay, Mrs. J. Launay, Mrs. J. Eberhardt and Dr. S. Hemmendinger and the excellent secretarial assistance of Mrs. C. Helbourg and Mrs. M. Voyat. Monoclonal antibody 6C9 was a generous gift of Dr. H.G. Huisman (Central Laboratory of the Netherlands Red Cross Blood Transfusion Service, Amsterdam). Polymeric biomaterials were kindly supplied by Dr. C. Pusineri (Hospal, Meyzieu, France). This work was supported by a grant from CNRS GRECO n°48 "Polymères Hémocompatibles". J.-P. Cazenave is Directeur de Recherche, INSERM.

Literature Cited

1. Zetter, B.R. Diabetes 1981, 30 Suppl. 2, 24-28.
2. Cazenave, J.-P.; Klein-Soyer, C.; Beretz, A. Inter. Angio. 1984, 3, 27-32.
3. Salzman, E.W.; Merrill, E.W. In Hemostasis and Thrombosis: Basic Principles and Clinical Practice ; Colman, R.W.; Hirsh, J.; Marder, V.J.; Salzman, E.W., Eds.; J.B. Lippincott Co.: Philadelphia, Toronto, 1982; pp 931-943.
4. Vroman, L.; Adams, A.L.; Klings, M.; Fischer, G.C.; Munoz, P.C.; Solensky, R.P. Ann. N.Y. Acad. Sci. 1977, 283, 65-76.
5. Packham, M.A.; Evans, G.; Glynn, M.F.; Mustard, J.F. J. Lab. Clin. Med. 1969, 73, 686-697.
6. Brash, J.L. In Interaction of the Blood with Natural and Artificial Surfaces; Salzman, E.W., Ed.; Marcel Dekker Inc. : New York and Basel, 1981; pp 37-60.
7. Mosher, D.F. In Interaction of the Blood with Natural and Artificial Surfaces; Salzman, E.W., Ed.; Marcel Dekker Inc. : New York and Basel, 1981; pp 85-101.
8. Mulvihill, J.N.; Cazenave, J.-P.; Maennel, G.; Schmitt, A. ESAO Proceedings, 1982, pp 287-290.
9. Goldsmith, H.L.; Turitto, V.T. Thromb. Haemostas. 1986, 55, 415-435.
10. Mulvihill, J.N.; Huisman, H.G.; Cazenave, J.-P.; van Mourik, J.A.; van Aken, W.G. Thromb. Haemostas., submitted for publication.
11. Kekwick, R.A.; Mackay, M.E.; Nance, M.H.; Record, B.R. Biochem. J. 1955, 60, 671-683.
12. Miekka, S.I.; Ingham, K.C.; Menache, D. Thromb. Res. 1982, 27, 1-14.
13. Cazenave, J.-P.; Hemmendinger, S.; Beretz, A.; Sutter-Bay, A.; Launay J. Ann. Biol. Clin. 1983, 41, 167-179.
14. Laemmli, U.K. Nature 1970, 227, 680-685.

15. Regoeczi, E. Iodine-labeled Plasma Proteins; CRC Press : Boca
    Raton, Florida, 1984.
16. Modderman, P.; Huisman, H.G.; von dem Borne, A.E.G.Kr. Thromb.
    Hemostas. 1985, 54, 197.
17. Mustard, J.F.; Perry, D.W.; Ardlie, N.G.; Packham, M.A. Brit. J.
    Haematol. 1972, 22, 193-204.
18. Eber, M.; Cazenave; J.-P.; Grob, J.C.; Abecassis, J.; Methlin G. In
    Blood Cells in Nuclear Medicine, Part I. Cell Kinetics and
    Bio-distribution; Hardeman, M.R.; Najean, Y., Eds.; Martinus
    Nijhoff Publishers : Boston, 1984; pp 29-43.
19. Cazenave, J.-P.; Blondowska, D.; Richardson, M.; Kinlough-
    Rathbone, R.L.; Packham, M.A.; Mustard, J.F. J. Lab. Clin. Med.
    1979, 93, 60-70.
20. Sakariassen, K.S.; Baumgartner, H.S. Thromb. Haemostas. 1985,
    54, 109.
21. Adams, G.A.; Feuerstein, I.A. Thromb. Haemostas. 1984, 52, 45-49.
22. Mulvihill, J.N. Ph.D. Thesis, Université Louis Pasteur,
    Strasbourg, France, 1984.
23. Park, K.; Mosher, D.; Cooper, S.L. J. Biomed. Mater. Res. 1986, 20,
    589-612.
24. Grinnell, F.; Phan, T.V. Thromb. Res. 1985, 39, 165-171.
25. Grinnell, F.; Feld, M.; Minter, R. Cell 1980, 19, 517-525.
26. Fowler, W.E.; Erickson, H.P. J. Mol. Biol. 1979, 134, 241-249.
27. Estis, L.F.; Haschemeyer, R.H. Proc. Natl. Acad. Sci. USA 1980,
    77, 3139-3143.
28. Tooney, N.M.; Mosesson, M.W.; Amrani, D.L.; Hainfeld, J.F.; Wall,
    J.S. J. Cell. Biol. 1983, 97, 1686-1692.
29. De Baillou, N. Ph.D. Thesis, Université Louis Pasteur, Strasbourg,
    France, 1983; p 116.
30. Turitto, V.T.; Baumgartner, H.R. Methods in Hematology : Measure-
    ments of Platelet Function ; Harker, L.A.; Zimmermann, T.S., Eds.;
    Churchill Livingstone : Edinburgh, 1983; Vol. 8, pp 46-63.
31. Packham, M.A.; Mustard, J.F. In Progress in Hemostasis and
    Thrombosis; Spaet, T.H., Ed.; Grune and Stratton : Orlando,
    Florida, 1984; Vol. 7, pp 211-288.
32. Sakariassen, K.S.; Bolhuis, P.A.; Sixma, J.J. Nature 1979, 279,
    636-638.
33. Pitlick, F.A. In Interaction of the Blood with Natural and
    Artificial Surfaces; Salzman, E.W., Ed.; Marcel Dekker Inc. : New
    York and Basel, 1981; pp 185-199.
34. Salzman, E.W.; Rosenberg, R.D.; Smith, M.A.; Lindon, J.N.; Faureau,
    L. J. Clin. Invest. 1980, 65, 64-73.
35. Sigot-Luizard, M.F.; Domurado, D.; Sigot, M.; Guidoin, R.;
    Gosselin, C.; Marois, M.; Girard, J.F.; King, M.; Badour, B. J.
    Biomed. Mater. Res. 1984, 18, 895-909.
36. Mulvihill, J.N.; Cazenave, J.-P. Bull. Soc. Chim. France 1985, 4,
    551-554.

RECEIVED January 13, 1987

Chapter 34

Blood Protein–Material Interactions That Lead to Cellular Adhesion

C. A. Ward

Kinetics and Thermodynamics Laboratory, Department of Mechanical Engineering, University of Toronto, Toronto, Ontario M5S 1A4, Canada

A series of studies investigating the interaction between synthetic materials and blood indicates that the activation of the complement proteins is of major importance in the sequence of events leading to cellular adhesion, whereas fibrinogen plays only a minor role. This sequence of proteins is activated by the material itself and it is activated along the alternative pathway by the air nuclei that can be stabilized in the surface roughness of a synthetic material. If the air nuclei are removed during priming, the degree of complement activation by the biomaterial appears to be in proportion to the ("critical") surface tension of the material.

When a synthetic material is exposed to blood, a complicated sequence of events is initiated that can lead to thrombus formation. Two of the early events in the sequence are protein adsorption and subsequently cellular adhesion (1, 2); however the extent to which the cellular adhesion is controlled by the interaction between the adsorbed proteins and the biomaterial has not been clarified, nor has the role which the properties of the biomaterial play in the process.

The surface tension of the material could be reasonably expected to play a major role. Some of the early work (3) indicated that when a material with a small surface tension was exposed to blood there was less platelet adhesion than if the material exposed had a larger surface tension, but this was disputed in subsequent studies. The basic concept has been elaborated on (4); however it has remained controversial (5).

One of the difficulties with trying to determine if the amount of cellular adhesion to a material can be correlated with the surface tension is that the surface actually exposed to blood becomes a composite when materials with different surface tensions are used. For example, as materials with smaller surface tensions are exposed to blood, the contact angles between them and blood plasma become larger. If this contact angle is sufficiently large, tiny air nuclei (bubbles) can be trapped and stabilized in the surface roughness of the material (6, 7). Thus the blood is actually exposed to a composite

0097-6156/87/0343-0551$06.00/0
© 1987 American Chemical Society

surface: a portion is the blood-biomaterial interphase and the remainder is the blood-air interphase. The relative importance of these two interphases in promoting cellular adhesion is examined below.

A second difficulty in establishing this correlation is that of identifying the key plasma proteins involved in promoting cellular adhesion.   Plasma contains numerous proteins.   For example, twenty proteins are involved in the complement system alone (8).   The behavior of the different proteins varies widely once adsorbed. For example, the complement protein C3 can break into fragments, C3a and C3b when adsorbed on certain surfaces, i.e. surfaces which activate the complement system.   The fragment C3a returns to the fluid phase. These components can then have strong effects on the cellular aggregation.   The proteins which have received the most attention are the fibrinogen, albumin and γ-globulin; however, there is no evidence to indicate that they control the cellular adhesion.   Below the role of fibrinogen in promoting platelet adhesion is examined and compared to that of the complement system. The present evidence indicates that the complement system plays the more important role in promoting cellular adhesion.

Effect of Air Nuclei on Protein and Cellular Adhesion to Synthetic Materials

To determine if the air nuclei that can be trapped in the roughness of a synthetic material play a major role in cellular adhesion to the material, a series of experiments was conducted in which the air nuclei were removed before the material was exposed to a biological fluid. The biological fluids examined were blood, washed platelet suspensions and fibrinogen solutions.

As a model material, we chose silicone rubber. This material has certain properties that makes it of particular interest: 1) it is permeable to gases, such as oxygen and nitrogen;   2) it is used in artificial organs, such as the membrane oxygenator, and 3) it has a low surface tension that is characteristic of many synthetic biomaterials. Its surface tension is low enough so that one would expect the air nuclei to be trapped in its surface roughness during priming (6).

Priming to Remove Air Nuclei from Surface Roughness of Gas Permeable Materials

To remove the air microbubbles (nuclei) from the surface roughness of a gas permeable material, such as silicone rubber, a priming procedure was developed in which a priming buffer was brought in contact with one side of the material, while the other side of the material was exposed to a vacuum. The material was left in this configuration for a period of time to allow the air from microbubbles to be drawn through the material and removed by the vacuum pump, and to allow the priming buffer to enter and fill the space that had been occupied by the microbubbles. Afterwards, the vacuum was removed so that both sides of the material were exposed to atmospheric pressure before it was exposed to a biological fluid. A biological fluid could then be pumped in to displace the priming buffer; however the surface irregularities were then filled with the buffer, and the buffer in the surface irregularities could then mix with the solvent of the biological fluid, allowing the solvent to enter the surface irregularity. This brings about intimate contact between the synthetic material and the solvent of the biological fluid. Thereby the presence of an air-biological fluid interface would be eliminated. Thus if this method of priming, called denucleation priming, were adopted any adhesion to the material would be attributable to the material–biological fluid interface alone.

In the normal priming procedure in which a buffer is simply flushed past the material, and then it is displaced by the biological fluid, the air nuclei would be present in the surface irregularities.

Platelet Adhesion from Washed Platelet Suspensions

To assess the effect that any air nuclei have on cellular adhesion to silicone rubber, we first compared the platelet adhesion from blood and washed platelet suspensions when either of them were exposed to normally primed and to denucleation primed specimens of silicone rubber.

The washed platelet suspensions were prepared from swine blood (9, 10) and the platelet concentrations in them were set at the normal value (2.5×10^5 platelets/mm^3). As measured by the Lowry technique (11), the total protein concentration was reduced to an average of 0.018 mg/ml, which is a very small fraction of the normal value.

To expose the silicone rubber to washed platelet suspensions, two tubes of this material were primed, one was subjected to denucleation priming, and the other primed normally. The washed platelet suspensions were then pumped simultaneously through each at a Reynolds No. of 115 and a shearing rate of 78/s.

To assess the platelet adhesion to the differently primed samples of silicone rubber, they were prepared for viewing with the scanning electron microscope, and counts of the adhering platelets were made at different positions along the length of the tubes. It was found that the variation with position was small and the average for all positions was taken in each tube. The results for the differently primed tubes are shown by the second column in Table I. Before discussing these results, we consider the exposure of the differently primed samples of silicone rubber to blood.

Platelet Adhesion from Fresh Blood

After one tube of silicone rubber had been primed by the denucleation procedure and another primed normally, they were exposed to fresh swine blood. An animal was cannulated, the canula connected to the two tubes at a T-junction and fresh blood pumped simultaneously through the two tubes under the same flow conditions as used with the washed platelet suspensions. Also, the number of adhering platelets was assessed in the same manner. The results for the differently primed material samples are shown in the first column of Table I. Note that 1) the platelet adhesion on the denucleation primed specimen exposed to either a washed platelet suspension or fresh blood was approximately the same. Thus, if the air nuclei are removed from the surface roughness, it appears that the plasma proteins do not contribute significantly to the platelet adhesion. 2) By contrast, if the air nuclei are present, it appears the presence of the plasma proteins significantly enhances the platelet adhesion. This can be seen by comparing the platelet adhesion to normally primed specimens of silicone rubber when they were exposed to fresh blood or to washed platelet suspensions (see the second row of Table I). The maximum platelet adhesion occurs when both the plasma proteins and the air nuclei are present, i.e. when normally primed specimens were exposed to fresh blood.

These results support the conclusion that the consequences of the protein-air nuclei interaction lead to enchanced platelet adhesion to a material. This conclusion was supported further by studies of platelet adhesion and thrombus formation in membrane oxygenators (12-14). These latter studies were conducted with sheep using oxygenators that had membrane surface areas of 2.5 m^2. Some of the oxygenators were subjected to a denucleation priming procedure and others primed normally.

There were remarkable differences in the number of circulating platelets remaining in the blood after it had perfused the oxygenators for a period of time. Immediately after the blood perfused the normally primed oxygenators, there was a sharp reduction in the number of circulating platelets. A minimum of 27% of the control value was reached 32 mins. after the blood began to perfuse the normally primed oxygenators. The platelet count slowly recovered to 62% of the control value during the following four hours of perfusion (12). By contrast, the reduction in the number of circulating platelets when blood perfused the denucleation primed oxygenators was only slight. It fell to 87% of control value and remained relatively constant throughout the four hour perfusion period. Thus in the two cases where the number of circulating platelets reaches a minimum, there are approximately six times as many platelets adhering to the surface of the normally primed membrane as to the denucleation primed membrane.

Other factors were measured, such as the pressure drop across the oxygenator, the number of circulating leukocytes, the amount of thrombus on the membranes of the differently primed oxygenators. The measurement of each of these parameters consistently indicated that removing the gas nuclei from the surface roughness of the membrane significantly reduced the cellular adhesion to it (12-14).

Changes in the composition of the membrane of the oxygenator was also examined as a means of reducing the cellular adhesion to the membrane (14). However, the dominant effect was found to be the method by which the membrane of the oxygenator was primed. As with the results described above, the cellular adhesion was in each case significantly reduced by denucleation priming.

These results indicate that if the cellular adhesion to a material is to be related to the properties of that material, then it is important to remove the air nuclei from the surface roughness of the material. Otherwise, the air nuclei can dominate the interaction between blood and the material. Further, the immediate question that is raised when these results are considered is: how can gas nuclei bring about such a large effect on the cellular adhesion?   One possibility is that they affect the plasma proteins and, once altered, the plasma proteins promote platelet and other cellular adhesion.

Table I.   Platelet Adhesion from Whole Blood and Washed Platelet Suspensions on Normally Primed and Denucleation Primed Specimens of Silicone Rubber

Treatment of Silicone Rubber	Platelet Adhering/0.00176mm^2	
	Fresh Blood Mean ± SE	Washed Platelet Suspension Mean + SE
Denucleation Primed Specimen	3.2 ± 1	5.0 ± 0.3
Normally Primed Specimen	48.1 ± 2.8	11.2 ± 0.5

Adsorption of Fibrinogen and its Affect on Platelet Adhesion

Fibrinogen is one of the proteins to be considered as vulnerable to alteration when adsorbed at the interphase between blood and an air nucleus (15-16).   To examine this possibility, we have conducted two series of experiments.   In the first, we examined the amount of fibrinogen that is adsorbed on normally primed silicone rubber and the amount adsorbed on the same type of material after it has been primed so as to remove the gas nuclei (17, 18).   In the second series of experiments, the effect of preadsorbed fibrinogen on platelet adhesion was examined.

Fibrinogen was extracted from swine blood (19,20).   The purity was examined using ultracentrifugation and found to be greater than 94%. Its average clottability was approximately 97%.

The extracted fibrinogen was resuspended in a phosphate buffer. A portion of it was labelled with ^{125}I.   Two tubes of silicone rubber were prepared for exposure to the fibrinogen solution.   One was primed normally, and the other was subjected to the denucleation priming procedure.   The fibrinogen solution was pumped simultaneously through both tubes for a period of three minutes at a flow rate of 15 ml/min. The fibrinogen solution was then allowed to remain static in the tubes for a period of 12 min.   This was found to be sufficient to establish equilibrium (17,18).   The concentration of fibrinogen at each position was assessed.

The adsorption isotherm established by this method on the material subjected to denucleation priming is shown in Figure 1.   For comparison, the results reported by Morrissey and Stromberg (21) for bovine fibrinogen adsorbing on silica are also shown.   They used ellipsometry to measure the surface concentration of bovine fibrinogen. Note that the solution concentration which they considered did not extend as far as the ones we considered. However at the same solution concentrations, the results correspond well.   Since they used ellipsometry to measure the surface concentration and this is an optical technique, it would appear that the more invasive method of radioactive labelling nonetheless gives a valid assessment of the adsorption.

When the fibrinogen was dissolved in a phosphate buffer to the physiological concentration (2.8 mg/ml), the effect of the air nuclei increased the surface concentration of the fibrinogen by only approximately 10% (18). However, if one examines the fibrinogen isotherm, one finds an indication of a change in the interaction of the fibrinogen molecules when the surface concentration exceeds approximately 0.45 $\mu g/cm^2$, i.e. the second plateau (Figure 1). Thus if the fibrinogen adsorption at the site of an air nucleus exceeds 0.45 $\mu g/cm^2$, one possible explanation for the observed increase in platelet adhesion is that this new interaction changes the fibrinogen molecule and, as a result, it strongly promotes platelet adhesion.

To examine the effect of this interaction on platelet adhesion, porcine fibrinogen was first preadsorbed on the silicone rubber tubes, using the same experimental procedure as that used to establish the fibrinogen adsorption isotherm. After first preadsorbing the fibrinogen and without exposing it to air, the surface with the preadsorbed fibrinogen was then exposed to fresh porcine blood by cannulating an experimental animal at either the femoral or jugular vein. The blood was drawn directly into the tubes containing the preadsorbed fibrinogen at a flow rate of 7.5 ml/min. Afterwards, the adhering platelets were counted at positions 10, 20, and 30 cm from the entrance to the thermostated region.

The average platelet adhesion was plotted against the surface concentration of the preadsorbed fibrinogen and is shown in Figure 2. Note that the platelet adhesion increases almost linearly with increasing surface concentration of the preadsorbed fibrinogen, until the surface concentration reached the value at which the adsorption isotherm showed a second plateau (0.45 $\mu g/cm^2$). After this surface concentration of the preadsorbed fibrinogen is exceeded, further increases in the preadsorbed surface concentration of fibrinogen does not lead to any further increase in platelet adhesion. Thus rather than the fibrinogen in the second plateau giving rise to more platelet adhesion, it appears the additional fibrinogen is completely benign.

Activation of Complement System at the Plasma-Air Interphase

If the nine complement proteins of plasma are activated along either the classical pathway (C1, C2 ... C9) or along the alternate pathway (C3, C5 ... C9), they have the capacity for considerable amplification because the activation of one molecule in this sequentially acting, enzymatic cascade activates several of the next molecules in the sequence. Also, Polley and Nachman (22) have shown that the presence of C3a gives rise to platelet aggregation. This is the fluid phase component that results from the activation of the first step on the alternate pathway. The activation of the second step on this pathway produces the fluid phase fragment C5a and this component is known to strongly promote leukocyte aggregation (23,24). Thus it would appear that if the complement system is activated by the air nuceli in the surface roughness of a biomaterial, it is the type of system that could produce the increased cellular adhesion that is observed when the air nuclei are present in the surface roughness of a synthetic biomaterial exposed to blood.

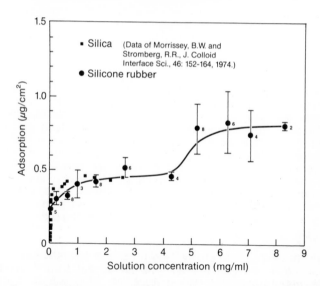

Figure 1.   Adsorption isotherm for fibrinogen adsorbing on denucleation primed silicone rubber (solid dots) and on silica (solid squares).   (Reproduced with permission from Ref. 17. Copyright 1986, Academic Press Inc.)

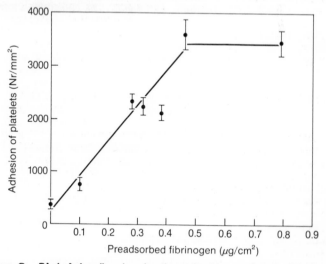

Figure 2. Platelet adhesion to denucleation primed silicone rubber after different surface concentrations of fibrinogen had been preadsorbed before the material was exposed to fresh blood. (Reproduced with permission from Ref. 17.   Copyright 1986, Academic Press Inc.)

As a first step toward investigating this possibility, a series of experiments was conducted in which air bubbles were incubated both with rabbit plasma (25) and with human plasma (Ward, C. A. et al. J. Appl. Physiol. in press). The same method of incubating the different types of plasma with air bubbles were used, but the degree of complement activation in human plasma was assessed by radioimmunoassays and that in rabbit plasma by the leukocyte aggregation test. Both experimental methods indicated similar results.

To perform the studies with human plasma, blood was collected from each of 34 male volunteers. The plasma was separated from the cells and divided into two portions. One portion, 1.5 ml, was placed in a 1.65 ml polypropylene tube. After capping, the remaining volume was filled with bubbles by vigorously shaking the capped tube. The polypropylene tube was then placed in a device that rotated the tube end-over-end at a rate of 22 RPM. This insured that the bubbles traversed the length of the tube twice during each rotation. Also, the tube was thumped during each rotation to insure the bubbles remained present. Another polypropylene tube was completely filled with the human plasma, capped carefully, and placed in the same device. This plasma sample served as a control. The device was then placed in a heat bath that maintained the temperature at 37°C. Both tubes were incubated while under rotation. There were no visible bubbles present in the control sample and this tube was not thumped.

To assess the complement activation in the human plasma, radioimmunoassays were used to measure C3a des Arg, C4a des Arg and C5a des Arg. During the complement activation process, the components C3a, C4a and C5a are first produced. Then each of these components is rapidly converted to the des Argine component. The radioimmunoasssay measures the des Argine component. We assume the concentration of the des Argine components is equal to the concentration of C3a, C4a and C5a.

Great variation was found in the sensitivity of the individuals to complement activation by air bubbles. For example, the concentration of C5a in the plasma samples from different individuals incubated with air bubbles was found to vary by over an order of magnitude (from 2.0 ng/ml to 66 ng/ml). A similar variation had been seen in the amount of complement activation occurring when rabbit plasma was incubated with air bubbles (25). Accordingly, the individuals who produced more than 25 ng/ml of C5a as a result of incubating their plasma with air bubbles were classified as Sensitive to complement activation and the others as Insensitive. Of the 34 individuals examined, 11 of them were found to be Sensitive.

In another study, this variation in sensitivity to complement activation was found to have important consequences. For example, the Sensitive individuals were found to be much more susceptible to decompression sickness than the Insensitive individuals (Ward et al. J. Appl. Physiol. in press).

In Figure 3, the concentration of C5a is shown for each group of individuals. One finds that the complement system underwent activation as a result of the presence of the air bubbles for both groups, although the degree of activation was significantly larger by the Sensitive group. A similar result was found for C3a; however the C4a concentration in the plasma samples incubated with air bubbles was found to be insignificantly greater than that in the control samples of plasma. Since C4 is only on the classical pathway, this would indicate that the complement activation by air bubbles is via the alternate pathway. Both C3 and C5 are on this pathway. Thus incubation of air bubbles with plasma produces the fluid phase metabolites of complement activation that are known to stimulate both platelet and leukocyte aggregation.

Complement Activation by Different Synthetic Biomaterials

In view of the above, one would expect that the air nuclei in the surface roughness of a synthetic material should also activate the complement system. If they do, then the complement system may be the series of proteins that is affected by the air nuclei with the consequence that cellular adhesion to the material is strongly promoted.

To examine this possibility, the degree of complement activation that results when rabbit plasma is incubated with polytetrafluoroethylene (PTFE) or silicone rubber or cellophane has been measured (26). Each of these materials was primed in two ways before it was exposed to the plasma. One method of priming removed the air nuclei from the surface roughness of the material and the other was simply the normal priming technique in which the material was immersed in the physiological saline before it was exposed to the plasma.

To measure the complement activation, the leukocyte aggregation test was used (23,24). In this procedure, a leukocyte suspension is formed from the rabbit's own polymorphonuclear (PMN) leukocytes, and the degree of complement activation is assessed from the degree of PMN leukocyte aggregation that occurs when a sample of plasma is injected into the suspension. In our experiments, one portion of the rabbit's plasma was incubated with zymosan, a substance known to strongly activate the complement system by the alternate pathway. Two other portions of the plasma were incubated with a specimen of the biomaterial being examined, one subjected to denucleation priming and the other to normal priming.

A portion of the PMN suspension (0.45 ml) was placed in an aggregometer, and then a sample of the plasma (0.05 ml) that had been incubated either with zymosan, or with a specimen of the differently primed biomaterials was injected into the suspension. The aggregation that followed was taken as a measure of the degree of complement activation. Since zymosan is a strong stimulant to activation, the measured amount of aggregation occurring when a sample of plasma was incubated with this substance was taken as the capacity of the complement system for activation. The degrees of aggregation measured in the other cases were then expressed as a percentage of the zymosan value. The results are shown in Table II.

For all three materials, the denucleation priming is seen to lead to a reduction in complement activation. It is statistically significant for both PTFE and silicone rubber but not for cellophane. Cellophane has a larger surface tension than either of the other materials; this may result in it having a smaller number of bubbles present in its surface roughness (6).

Once the bubble nuclei have been removed, the complement activation that results from incubating the material with the plasma proteins can be expected to be a function of the properties of the materials themselves because the plasma is only exposed to this surface. To examine this expectation, we have plotted the % complement activation given in Table II versus the "critical surface tension" for the material. This surface property is assumed to be an approximation to the true or thermodynamic surface tension and the value of the critical surface tension is available for these materials. The value of the thermodynamic surface tension for these materials remains controversial. This plot is shown in Figure 4.

These data show a clear trend of increasing complement activation with increasing surface tension of the material. Before one can be sure of this trend, more materials must be studied. However, it is interesting to note that the complement activation by cellophane is correlated with its critical surface tension. Cellophane is a polysaccharide and exposure of the polysaccharides to plasma had been thought to result in strong complement activation. Rather than there being anything special about the polysaccharides, it may be that the complement activating capacity of this molecular structure is only because they produce a surface with a relatively large surface tension.

Table II. Percentage Complement Activation Produced by Differently Primed Synthetic Materials

| MATERIAL | Priming Procedure | |
	Normal Mean % ± SE	Denucleation Mean % ± SE
PTFE (n = 5)	8.3 ± 1.6	3.1 ± 1.9
Silicone Rubber (n = 4)	34.5 ± 6.9	16.2 ± 2.8
Cellophane (n = 3)	46.0 ± 10.5	36.1 ± 7.1

Cellular Adhesion From Decomplemented Blood

The hypothesis can now be advanced that the air nuclei in the surface roughness of a synthetic material activates the complement system and as a result promotes cellular adhesion to the material. This hypothesis can be examined further by exposing the blood of normal animals and animals that have been decomplemented in-vivo to a specimen of the same synthetic materials. Fortunately, such experiments have already been performed by Herzlinger and Cumming (27).

Dogs were decomplemented in-vivo by injecting a factor derived from cobra-venom that is known to act as a strong convertase for C3. A rocket immunoelectrophoresis study of the plasma from animals that had been decomplemented with this technique showed that the C3

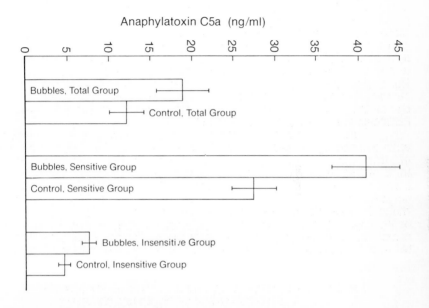

Figure 3. Activation of the complement system by air bubbles. C5a was produced by both sensitive and insensitive individuals when their plasma was incubated with air bubbles. (Reproduced with permission from Ref. 30. Copyright 1987 American Physiological Society.)

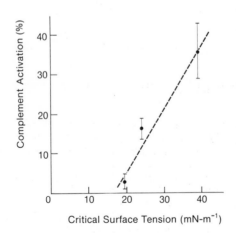

Figure 4. Activation of the complement system by different synthetic materials incubated with rabbit plasma. (Reproduced with permission from Ref. 26. Copyright 1984 Wiley.)

concentration in the plasma was reduced to only a small fraction of the normal value. Activation of C3 is necessary for activation of the complement system; thus removal of this plasma protein would prevent activation of the complement system. The decomplementation prodecure did not significantly change the concentration of fibrinogen in the plasma. Also, the platelets from decomplemented animals were found to aggregate almost normally in response to adenosine diphosphate and to undergo the release reactions.

To examine the cellular adhesion to synthetic materials, the "Stagnation Point Flow" cell was used (28). This device allows blood from the carotid artery of a dog to issue directly onto a specimen of synthetic material so that the flow is radially outward from the stagnation point. The shear stress increases in the radial direction in this flow configuration. The cellular adhesion is then characterized by the diameter of the circle within which there is cellular adhesion. The larger the diameter of this "white cell circle", the more cellular adhesion there is and the stronger the bond between the surface and the cells.

Blood from normal and decomplemented animals where exposed to specimens of both nylon 6, 6 and polymethyl methacrylate (PMMA). None of the specimens of the synthetic material were subjected to denucleation priming; however, if it is the complement proteins that are affected by the air nuclei, then the priming procedure should not matter for the decomplemented animals. The air nuclei would contribute to the cellular adhesion occurring when the blood of the normal animals was exposed to the biomaterial. However, if the hypothesis is valid, then there should be less cellular adhesion from the blood of the decomplemented animal.

The results reported by Herzlinger and Cumming are summarized in Table III. As may be seen, decomplementing the animals strongly reduces the cellular adhesion from their blood. We would emphasize that a strong reduction in cellular adhesion also resulted when the air nuclei were removed from the surface roughness of silicone rubber (e.g. see Table I). Also, that fibrinogen was present at the normal concentration in the blood of the decomplemented animals but, nonetheless, there was little cellular adhesion from the blood of these animals. These results strongly implicate the complement system as being responsible for the cellular adhesion that occurs when a biomaterial is exposed to blood.

The results listed above indicate that when a synthetic material – plasma interface is formed, the complement system is activated. The

Table III. Cellular Adhesion from Blood of Normal and
           Decomplemented Animals Exposed to Nylon
           and Polymethyl Methacrylate

Animal Preparation	Cellular Adhesion (White Cell Circle, μ)	
	Nylon	PMMA
Normal	>1600	>1500
Decomplemented	60	400

pathway by which the activation occurs is not yet clear. Herzlinger et al. (29) found some indication that in the case of PMMA the complement system is activated along the classical pathway. However, in the case of the air-plasma interface, it is clear from our study with radioimmunoassays that the activation is along the alternate pathway.

Conclusion

After reviewing the previously obtained data, the hypothesis is advanced that when a normally primed synthetic material is exposed to blood, the air nuclei trapped in the surface roughness of the material activates the complement proteins by the alternative pathway and that the products of this activation process (i.e. at least C3a, and C5a) promote cellular adhesion to the synthetic material. The evidence in favor of this hypothesis includes the following.

Studies have been conducted with both human plasma and with rabbit plasma in which a plasma sample was incubated with air bubbles and then assayed for complement activation. For those experiments conducted with human plasma, radioimmunoassays were performed for C3a, C4a, and C5a. It was found that the air bubbles did activate the complement system and that the activation was along the alternative pathway, i.e. C3a and C5a were produced but not C4a. For the studies with rabbit plasma, the assay technique was PMN leukocyte aggregation. The results with rabbit plasma were consistent with those found for human plasma.

In a series of studies, three synthetic materials were incubated separately with rabbit plasma after they had either been primed normally or subjected to denucleation priming which largely removes the air nuclei from the surface roughness. The plasma samples were then assayed for complement activation with the leukocyte aggregation test. It was found that removal of the air nuclei during priming indeed reduced the complement activation by each of the synthetic materials, as would be expected from the results listed above. Further, it was found that the amount of complement activation increased when the sample of synthetic material being incubated had a larger (critical) surface tension.

Denucleation priming then is one method of reducing the complement activation that results when a synthetic material is exposed to blood. Another method is by administering the animal a drug that removes the complement protein C3 from the animal's plasma *in-vivo*. Then when the animal's blood is exposed to a synthetic material, activation of the complement sequence is prevented. Both of these methods of reducing the complement activation have been found to reduce the cellular adhesion. In particular, when the air nuclei have been removed from the roughness of a synthetic material by subjecting it to denucleation priming and then exposing it to the blood of a normal animal, it has been found that the cellular adhesion is strongly reduced as compared to the cellular adhesion on the same type of synthetic material primed normally and then exposed to blood. An equally dramatic reduction in cellular adhesion to a synthetic material has been demonstrated to result when an animal is decomplemented *in-vivo* before its blood is exposed to a normally primed synthetic material, i.e. to a synthetic material with the air nuclei present in the surface roughness.

All of the above results are consistent with the hypothesis listed above and when taken together they lend it considerable support.

Although fibrinogen can not be ruled out as playing a role in promoting cellular adhesion as a result of the air nuclei, it appears unlikely to play a major role. The presence of the air nuclei in the surface roughness only gives rise to a 10% increase in the fibrinogen adsorption under physiological conditions. Also, the platelet adhesion appears to increase only linearly with increasing amounts of preadsorbed fibrinogen and then to become constant as the fibrinogen concentration is increased further. Thus the 10% increase in fibrinogen adsorption that results from the presence of the air nuclei seems unlikely to be responsible for the order of magnitude increase in cellular adhesion that results when both the plasma proteins and the air nuclei are allowed to interact (see Table I).

Acknowledgments

This work was supported by the Ontario Heart Foundation and the Defence and Civil Institute of Environmental Medicine, Downsview, Canada.

Literature Cited

1. Baier, R.E.; Dutton, R.C. J. Biomed. Mater. Res. 1969, 3, 191-206.
2. Vroman, L.; Adams, A.L. J. Biomed. Mater. Res. 1969, 3, 43-67.
3. Lyman, D.J.; Klein, K.G.; Brash, J.L.; Fritzinger, B.K. Thromb. Diath. Haemorrh 1970, 23, 120-128.
4. Baier, R.E. In Adhesion in Biological Systems; Manly, R.S., Ed.; Academic Press, New York, 1970, 15-48.
5. Ruckenstein, E.; Gourisankar, S.V. J. Coll. & Sci. Interface 1984, 101, 436-451.
6. Ward, C.A.; Forest, T.W. Ann. Biomed. Engin. 1976, 4, 184-207.
7. Ward, C.A.; Levart, E. J. Appl. Phys. 1984, 56, 491-500.
8. Kunkel, S.L.; Ward, P.A.; Caporale, L.H.; Vogel, C-W.; In Immunology: Basic Processes; Bellanti, J.A. Ed.; W. B. Saunders Co.; Toronto, 1985; Chapter 6, 106-116.
9. Ward, C.A.; Ruegsegger, B.; Stanga, D.; Zingg, W.; Herbert, M.A. Am. J. Physiol. 1977, 233, H100-H105.
10. Tangen, O.; Berman, H.J. Advan. Exptl. Biol. Med. 1973, 34, 235-243.
11. Lowry, O.H.; Rosebrough, N.J.; Farr, L.A.; Randall, R.J.; J. Biol. Chem. 1951, 193, 265-275.
12. Osada, H.; Ward, C.A.; Duffin, J.; Nelems, J.M.; Cooper, J.D. Am. J. Physiol. 1978, 234, H646-H652.
13. Osada, H.; Duffin, J.; Ward, C.A.; Nelems, J.M.; Cooper, J.D. Artif. Organs 1978, 2, 121-125.
14. Fountain, S.W.; Duffin, J.; Ward, C.A.; Osada, H.; Martin, B.A.; Cooper, J.D. Am. J. Physiol. 1979, 236, H371-H375.
15. Packham, M.A.; Evans, G.; Glynn, M.F.; Mustard, J.F. J. Lab. Clin. Med. 1969, 73, 686.-697.
16. Zucker, M.B.; Vroman, L. Proc. Soc. Exptl. Biol. Med. 1969, 131, 318.
17. Ward, C.A.; Stanga, D. J. Colloid Interface Sci. 1986, 114, 323-329.

18. Ward, C.A.; Stanga, D.; Zdasiuk, B.J.; Gates, F.L. Annals of Biomed. Engin 1979, 7, 451-469.
19. Cohn, E.J.; Strong, L.E.; Hughes, W.L.; Mulford, D.J.; Ashworth, J.N.; Melin, M.; Taylor, H.J. J. Am. Chem. Soc. 1946, 68, 459-475.
20. Blombäck, B.; Blombäck, M. Ark. for Kemi. 1957, 10, 415-443.
21. Morrissey, B.W.; Stromberg, R.R. J. Colloid Interface Sci. 1974, 46, 152-164.
22. Polley, M. J.; Nachmann, R.L. J. Exp. Med. 1983, 158, 603-615.
23. Craddock, P.R.; Fehr, J.; Dalmasso, A.P.; Brigham, K.L.; Jacob, H.S. J. Clin. Invest. 1977, 59, 879-888.
24. Craddock, P.R.; Hammerschmidt, D.; White, J.G.; Dalmasso, A.P.; Jacob, H.S. J. Clin. Invest. 1977, 60, 260-264.
25. Ward, C.A.; Koheil, A.; McCullough, D.; Johnson, W.R.; Fraser, W.D. J. Appl. Physiol. 1986, 60, 1651-1658.
26. Ward, C. A.; Koheil, A.; Johnson, W.R.; Madras, P.N. J. Biomed. Matr'l. Res. 1984, 18, 255-269.
27. Herzlinger, G.A.; Cumming, R.D. Trans. Am. Soc. Artif. Intern. Organs 1980, 26, 165-171.
28. Petschek, H.E.; Adamis, D.; Kantrowitz, A.P. Trans. Am. Soc. Artif. Intern. Organs, 1968, 14, 256-259.
29. Herzlinger, G.A.; Bing, D.H.; Stein, R.; Cumming, D. Blood 1981, 57, 764-770.
30. Ward, C.A.; McCullough, D.; Fraser, W.D. Relation Between Complement Activation and Susceptibility to Decompression Sickness; American Physiological Society: Bethesda, MD, 1987.

RECEIVED March 27, 1987

Chapter 35

Thrombin–Antithrombin III Interactions with a Heparin–Polyvinyl Alcohol Hydrogel

M. V. Sefton[1], G. Rollason[1], M. W. C. Hatton[2], M. F. A. Goosen[1], and B. A. H. Smith[1]

[1]Department of Chemical Engineering and Applied Chemistry, University of Toronto, Toronto, Ontario M5S 1A4, Canada
[2]Department of Pathology, McMaster University, Hamilton, Ontario L8S 4L7, Canada

A heparin-polyvinyl alcohol hydrogel (> 70% water) has been used to show that heparin, which is immobilized via the terminal serine group, can retain its biological activity in a number of clotting assays without being released into plasma (1-4). Reviewed here are studies of the interactions between thrombin, antithrombin III, and immobilized heparin, which suggested that the interaction between actively adsorbed thrombin and immobilized heparin was the first step in enzyme inactivation by antithrombin III, presumably by the formation of a surface bound heparin-thrombin-antithrombin III complex. Regeneration of surface heparin sites, which would be necessary for long term applications, was demonstrated by the slow removal of ^{125}I-thrombin (17-22% of bound radio-activity) as thrombin-antithrombin III complex (~ 70%) and thrombin-α-2-macroglobulin complex (~ 30%) with defibrinated or arvinized plasma. Thus, heparin-PVA has shown potential as a long term thromboresistant material.

Of the many methods reported in the literature for improving the thrombogenicity of biomaterials, heparinization has been one of the more intriguing and sometimes controversial. Ionically bound heparin was found to be released from the surface at biologically significant rates while early attempts at covalent immobilization led to inactivation of the heparin. This prompted early critics to conclude that heparinized materials could be effective only if enough heparin was released from the surface to create a microenvironment at the blood-material interface.

A more recently developed heparin-polyvinyl alcohol (heparin-PVA) hydrogel (> 70% water) has shown that heparin immobilized via the terminal serine group, which remains on many chains of commercially available heparin, can retain its biological activity in a

0097-6156/87/0343-0566$06.00/0

number of clotting tests without being released into the plasma (1-4). Subsequent ex vivo canine shunt experimentation (5) and mathematical model calculations (6) have confirmed this conclusion. In the absence of a heparin microenvironment we have presumed that a surface bound thrombin-antithrombin III complex is formed on the immobilized heparin with the heparin accelerating the rate of inactivation as happens in solution. Hence we have examined the interactions between thrombin, antithrombin III and the immobilized heparin in heparin-PVA with a view to understanding the mechanism of action of the immobilized heparin and the fate of inactive complex.

In addition to the question of heparin release, it has been thought that the heparinized surface may become saturated with the inactive complex which by tying up the active sites would prevent further thrombin from being inactivated. Alternatively, consumption of antithrombin III, prothrombin, or other clotting factors may result, leaving the blood systematically hypocoagulable. It is also conceivable that heparin-induced interruption of the intrinsic clotting system may prevent the normal failure mode of devices exposed to low blood flowrates (i.e., red thrombus formation) but only expose an underlying platelet aggregation or white thrombus formation problem. Only recently have these concerns been subject to critical analysis (7); until now their validity, although never verified, has been unquestioned. The work reviewed here focuses on the initial concern regarding the fate of the inactive complex. Other work dealing with platelets will not be discussed here.

Materials and Methods

Heparin-PVA. Heparin-polyvinyl alcohol (heparin-PVA) hydrogel was prepared as before (2) from an aqueous solution containing 10% PVA (20% acetylated PVA, Gelvatol 20-60, Monsanto Canada Ltd., Toronto, Ont.), 5% $MgCl_2 \cdot H_2O$, 0.5% glutaraldehyde, 3% formaldehyde, 4% glycerol and 1 or 2% sodium heparin (porcine mucosal, 176 U.S.P. U/mg, Canada Packers Ltd., Toronto, Ont.). Control gel was prepared without heparin. Films were cast onto petri dishes, air dried, and then cured at 70°C for 2 hours. The cured films were ground as previously described (8) within a nominal particle-diameter range of 105-250 μm. Unless stated otherwise, 1% heparin-PVA "beads" were used.

Heparin was released from 1% heparin-PVA gel "beads" at 1.7 x 10^{-2} μg/g wet gel/min. This rate was determined from the approximately linear portion of the heparin content versus PBS (phosphate buffered saline, pH 7.4) wash time curve (between 300 and 500 hours) by using the toluidine blue method (3). This release rate was 1/1000 of the minimum rate considered necessary for thromboresistance (9). After washing, the remaining heparin content was determined indirectly by mass balance to be approximately 7.1 mg/g wet gel for 1% heparin gels and 14 mg/g wet gel for 2% heparin gels.

Proteins. Thrombin and antithrombin III were obtained from various sources. Crude bovine thrombin was obtained from Parke-Davis (96 NIH U/mg, Detroit, MI) and Miles Laboratories (100 NIH U/mg, Elkart, IN). The crude bovine thrombin from Parke-Davis was purified on

SP-Sephadex as before (10) using a modification of the method of Lundblad et al. (11) to produce a product with a specific activity of 1850 NIH U/mg (12). Purified human α-thrombin (3000 - 3153 NIH U/mg) was the kind gift of Dr. J.W. Fenton II (N.Y. State Department of Health). Heat defibrinated citrated human plasma (54°C, 5 min) was used as a source of crude antithrombin III (0.2 mg/mL). Purified human antithrombin III (1000 U/mg; 1 mL plasma = 100 U) was the kind gift of Dr. M. Wickerhauser (American Red Cross, Bethesda, MD). A purified human antithrombin III of similiar activity to that of Wickerhauser (0.29 mg/mL) was also prepared from citrated human plasma (Red Cross Blood Bank, Toronto) using heparin-Sepharose (13).

Purified thrombin and antithrombin III were radiolabelled with ^{125}I using Enzymobeads (Bio-Rad Lab., Mississauga, Ont.) as described previously (14). Chromatography of radiolabelled thrombin on SP-Sephadex using gradient elution (10) demonstrated that more than 95% of the radioactivity coincided with plasma coagulant activity of the thrombin. The remaining 5% passed through the column and was considered to be free ^{125}I. The thrombin retained most of its biological activity. Although antithrombin III retained about 70% of its inhibitory activity after labelling, chromatography of radiolabelled antithrombin III on heparin-Sepharose showed that only 18-21% of the radioactivity bound to the column. Since most of the labelled antithrombin III did not have affinity for heparin-Sepharose, only a few experiments were made using radiolabelled antithrombin III.

Plasma Fractions. Fibrinogen-free plasma was prepared by treating human plasma, collected in acid citrate dextrose (ACD), with Arvin (also known as Ancrod; a gift from Dr. J.S. Burton of Berk Pharmaceuticals, Guilford, U.K.) as described before (15). Antithrombin III-free plasma was prepared by passing Arvin-treated plasma (100 mL) through a column of heparin-Sepharose (12 cm x 2.2 cm) at room temperature equilibrated with 0.02 M Tris·HCl, pH 7.6, containing 0.15 M NaCl. The pooled effluent, which possessed approximately 3% of the starting quantity of antithrombin III as shown by fluorogenic assay (16), was then passed a second time through a similar column of heparin-Sepharose. After two passages through heparin-Sepharose, the presence of antithrombin III was not detectable by either chromogenic assay or by immunodiffusion using antiserum to human antithrombin III raised in goats (Atlantic Antibodies, Westbrook, ME).

Human plasma free from α-2-macroglobulin was prepared by treating plasma with $(NH_4)_2$ SO_4 to 50% saturation. After 30 min at room temperature, the plasma was centrifuged and the precipitate discarded. The resulting supernatant, which was dialysed against 0.15 M NaCl at 4°C, contained good antithrombin III activity but was free from α-2-macroglobulin as shown by immunodiffusion using a specific antiserum raised in goats (Meloy, Springfield, VA).

Human α-2-macroglobulin was prepared from freshly drawn blood essentially as described by Harpel (17) except that, rather than a KBr gradient, a sucrose density gradient (minimum 0.2 M; maximum 1.0 M sucrose) was employed. After ultracentrifugation (Beckman L8-55; SW 28.1 rotor at 28,000 rpm for 16 hr), the α-2-macroglobulin activity was present largely in the hypophase. Immunodiffusion confirmed the presence of α-2-macroglobulin but antithrombin III was not detected.

Solutions of bovine serum albumin (Cohn fraction V, Miles Lab., Elkhart, IN) were made with PBS.

<u>Enzymatic Activity of Thrombin Bound to Heparin-PVA Beads.</u> Unless stated otherwise, heparin-PVA or PVA columns (1.4 cm diameter x 2.5 cm) were prepared with 3 g of unused "beads", equilibrated at room temperature with PBS and loaded with thrombin. The enzymatic activity of thrombin bound to the "beads" was quantified by loading 1 mL of chromogenic substrate (0.5 mg/mL, S2238 Ortho Diagnostics, Raritan, NJ or Chromozym TH, Boehringer Mannheim, Dorval, Que.) and determining the concentration of the reaction product (p-nitroaniline) by measuring the absorbance at 381 nm in the effluent not more than 10 minutes after elution from the column. The ratio of colour produced by enzymatic action to the total possible from the load of chromogen was termed "colour yield". Neither chromogen nor p-nitroaniline was observed binding to the columns.

<u>Thrombin Binding Affinity.</u> The affinity of thrombin for PVA or heparin-PVA was analyzed by loading crude bovine thrombin (Parke-Davis, 62 U, 18 nmoles) onto a PVA or heparin-PVA "bead" column, washing with 100 mL PBS and/or 15 mL 20% (w/v) bovine albumin, and measuring the residual thrombin activity (<u>8</u>). The thrombin activity after washing with 3 mg of crude antithrombin III (15 mL defibrinated plasma, 48 nmoles) instead of albumin was also measured. As well, the effect of precoating the gel columns with 20 mL of 10% (w/v) bovine albumin on the binding of purified human thrombin (1072 U, 9.4 nmoles) was determined.

Preliminary thrombin adsorption isotherms were prepared by incubating 100 mg of heparin-PVA (control PVA) beads in centrifuge tubes (15 mL) with 0.6 mL of thrombin solution (3-1200 U/mL). Unlabelled purified human thrombin was spiked with ^{125}I-labelled purified bovine thrombin. After adsorption for 15 minutes, the beads were washed twice with 5 mL of PBS and the supernatants counted. The amount of thrombin/gram of gel was determined.

<u>Loading Sequence.</u> Two similar columns of heparin-PVA beads were prepared (<u>8</u>). The first column was loaded with thrombin followed by antithrombin III; the sequence was reversed for the second column. The thrombin load was either crude bovine (62 U, 18 nmoles), pure bovine (23 U, 0.35 nmoles), or pure human (1072 U, 9.4 nmoles). The antithrombin III load was either crude human (3 mg; 48 nmoles, 15 mL of defibrinated plasma) or purified human (1.5 mg, 24 nmoles, 0.29 mg/mL). After loading each protein, 100 mL of PBS was passed through the column and the residual thrombin activity measured by the chromogenic substrate method.

<u>Displacement of Surface Bound Heparin-Thrombin-Antithrombin III Complex.</u> To determine whether the surface bound heparin became saturated with inactive thrombin-antithrombin III complex, the ability to displace bound inactive complex and thereby regenerate heparin was assessed.

A series of experiments were made with heparin-PVA bead columns. Thrombin (100 U crude bovine, 450 U purified human or 100 U purified bovine) was loaded onto the columns and the excess removed with PBS. Antithrombin III (2000-2500 U purified human) was then loaded, and the columns eluted with PBS, 5% (w/v) bovine serum albumin, or plasma fractions. In some cases heat defibrinated or arvinized plasma was used instead of purified human antithrombin III as a crude source of antithrombin III, as well as a displacing agent. The inactivation of thrombin by antithrombin III was shown in a separate experiment using the chromogenic substrate assay.

Further investigation of displacement was done by incubating 0.25 g of heparin-PVA "beads" in centrifuge tubes (15 mL, Corning Glass Works, Corning, N.Y.) with 8 U purified human thrombin spiked with purified bovine ^{125}I-thrombin. The displacing agents were added, vortexed, centrifuged, and decanted. Each set of beads was washed twice with 9 mL PBS followed by 5 mL heat defibrinated plasma. The beads were then washed with 10 mL of either PBS, 3 M NaCl in PBS, 1% PEG 7500 in PBS, or 6 M guanidine-HCl.

The radioactivity of the column eluents (collected in 1-5 mL fractions) and the supernatants from the centrifuge tubes were measured either by liquid scintillation (Beckman LS8000) or with a γ-counter (Beckman Gamma 3000).

Characterization of Displaced Protein. Two approaches were used simultaneously to characterize the ^{125}I-thrombin displaced from heparin-PVA columns: filtration on Sephadex G-200 and heparin-Sepharose affinity chromatography using the elution conditions described by Collen et al. (18) for the separation of antithrombin III from enzyme-antithrombin III complex. Radiolabelled antithrombin III displaced from the heparin-PVA column was characterized in a similar way by affinity chromatography on heparin-Sephrose. Detailed methods are presented elsewhere (14,7).

Results

Thrombin Binding Affinity. Bovine and human thrombins bound not only to heparin-PVA but also to PVA without heparin (8). Colour yields of 89% and 81% were obtained for the heparin-PVA and PVA gel respectively. Passing albumin through the same columns followed by chromogen lowered the colour yield for PVA to 54% while not significantly reducing the colour yield for heparin-PVA. Passing crude antithrombin III through a column of PVA previously loaded with crude bovine thrombin gave a colour yield of 50%, which was similar to that obtained when using albumin in place of crude antithrombin III. In contrast, crude bovine thrombin bound to heparin-PVA, while not being significantly inactivated (or desorbed) by bovine albumin, was significantly inactivated by crude antithrombin III as shown by the 24% colour yield. Precoating the gel columns with bovine albumin significantly reduced human thrombin binding to PVA, relative to heparin-PVA. Loading precoated columns with 1072 U purified human thrombin gave a colour yield of 78% with heparin-PVA and 38% with PVA alone.

Preliminary thrombin adsorption isotherms (purified human thrombin spiked with purified bovine ^{125}I-thrombin) have shown that thrombin is adsorbed proportionally to the mass of PVA or heparin-PVA in the range 3-1200 U/mL bulk concentration (Figure 1). Heparin-PVA beads adsorbed significantly more thrombin.

Thrombin Inactivation of Antithrombin III on Heparin-PVA Bead Columns.

Bovine and human thrombins loaded onto heparin-PVA columns, followed by chromogenic substrate, resulted in colour yields of 77-89%, which indicated a large amount of active thrombin was bound to the heparinized gel (8). After passing antithrombin III through each column, colour yields of 18-48% were obtained, indicating significant inactivation of thrombin by antithrombin III. However, using a second set of heparin - PVA columns with antithrombin III loaded first onto the heparinized gel followed by thrombin and chromogenic substrate gave colour yields of 78-91%, which was comparable to that of the loaded thrombin, indicating that essentially no thrombin had been inactivated.

Loading inhibitor before enzyme did not result in any significant decrease in thrombin activity. Only by passing antithrombin III through a heparin-PVA column that had been previously loaded with thrombin, could a reduction in thrombin activity be observed.

Displacement.

The displacement of ^{125}I-antithrombin III from a heparin-PVA column (2g, 14 mg heparin/g wet gel) previously exposed to thrombin (450 U purified human thrombin) by arvinized plasma is shown in Figure 2. Approximately 82% of the radioactivity and 43% of the antithrombin III activity associated with the 2000 U of pure human antithrombin III loaded into the heparin-PVA column that had been previously loaded with unlabelled thrombin did not bind, giving rise to the first peak in the chromatogram of Figure 2. On changing the PBS eluent to arvinized plasma there was a significant increase in radioactivity in the eluent, indicating the displacement from the column of radiolabelled antithrombin III previously bound to the thrombin or heparin in the heparin-PVA column. Approximately 8% of the antithrombin III left on the column after the PBS elution was displaced by arvinized plasma in the first hour. Similar displacement occurred with heat defibrinated plasma and a 5% (w/v) albumin solution.

Radiolabelled purified bovine thrombin, adsorbed to heparin-PVA, was exposed to a variety of eluents. Compared with PBS, all eluents were effective in displacing some of the bound thrombin (or thrombin-antithrombin III complex) although arvinized plasma was the most successful in this respect.

After washing with PBS approximately 48 ± 9 U (± SD) of thrombin (initial load = 100 U) was bound to the heparin-PVA column. Arvinized plasma (20 to 130 mL) was able to displace from 5.4 ± 1.3% (SD; n = 5) to 22% of the adsorbed thrombin in a progressive fashion (Figure 3). Bovine serum albumin, antithrombin III free plasma, antithrombin III deficient plasma, and α-2-macroglobulin-free plasma were only partially effective as displacing agents (2-5.6% removed with 20-50 mL of displacing agent). In contrast, purified antithrombin III and α-2-macroglobulin were ineffective, each inhibitor displacing less than 0.1% of the bound enzyme. However, as shown in a separate experiment, substantial inhibition by

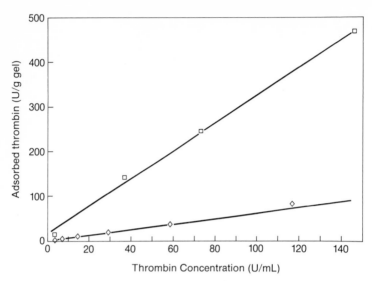

Figure 1.   Adsorption isotherms of ^{125}I-thrombin to heparin-PVA (□) and PVA (◊) "beads" (100 mg, 15 minutes, purified human thrombin spiked with purified bovine ^{125}I-thrombin).

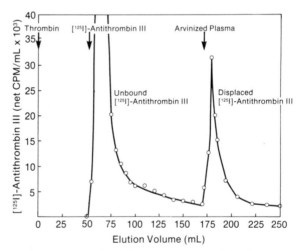

Figure 2. Displacement of ^{125}I-antithrombin III from heparin-PVA by arvinized plasma. Pure bovine thrombin (450 U) was loaded onto a heparin-PVA "bead" column (2 g, 14 mg heparin/g wet gel) followed by 2000 U of purified human ^{125}I-antithrombin III. After eluting with 120 mL PBS at 100 mL/hr, eluent was changed to arvinized plasma at 30 mL/hr. The first peak contained 43% of the biological activity and the second peak 8%. (Reproduced with permission from Ref. 7. Copyright 1982, American Society for Artificial Internal Organs.)

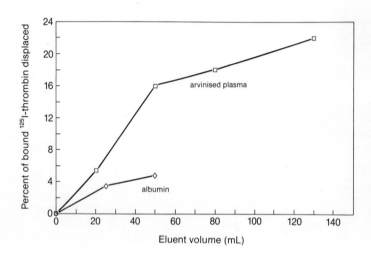

Figure 3. Displacement of purified bovine ^{125}I-thrombin from heparin-PVA "beads" by arvinized plasma. Each column was loaded with 100 U of thrombin, eluted with 100 mL of PBS and then further eluted with arvinized plasma or 5% (w/v) bovine albumin at a flowrate of 60 mL/hr. (Reproduced with permission from Ref. 28. Copyright 1984, Gordon and Breach, Science Publishers Inc.)

antithrombin III was observed. The passage of 40 nmol (2500 U, 1000 U/mg) purified human antithrombin III through a column, to which 1 nmol of purified human thrombin had been loaded, inactivated about 70% of the bound enzyme.

Further elution, after heat defibrinated plasma, with 1% PEG, 3 M NaCl or 6 M guanidine-HCl only removed another 7 - 16% of the bound radioactivity (Table 1). In all cases there was still a significant amount of ^{125}I- thrombin remaining on the heparin-PVA "beads".

Table I. Displacement of Purified Human Thrombin Spiked with Purified Bovine ^{125}I-thrombin from Heparin-PVA "Beads" in Centrifuge Tubes

Displacing Agent (10 mL)	% thrombin eluted[*]
PBS	3.8
3 M NaCl in PBS	7.2
1% PEG in PBS	9.3
6 M guanidine-HCl	15.8

[*] % ^{125}I-thrombin removed after heat defibrinated plasma had removed 16.6 ± 7% (SD; n = 6) of ^{125}I-thrombin not removed by PBS.

Characterization of Displaced Protein. With labelled antithrombin III, chromatography of the displaced radioactivity on heparin-Sepharose revealed that the bulk of the displaced radioactive material did not bind to heparin-Sepharose (Table II). With arvinized plasma as the displacing eluent, 65% of the ^{125}I-antithrombin III eluted in the void volume, compared with 49% of the control ^{125}I-antithrombin III (diluted in citrated plasma) that had not previously been used to inactivate thrombin; the latter unbound fraction was likely labelled impurities or inhibitor modified by radiolabelling to lose its heparin affinity. With 5% (w/v) albumin used as a displacing eluent, 78% of the ^{125}I-antithrombin III came out in the void volume. This increase in material that did not bind to heparin after displacement from heparin-PVA was attributed to post-complex antithrombin III, a modification of the original inhibitor resulting from the inactivation of thrombin. Neither thrombin-antithrombin III complex nor free antithrombin III were detected in the 5% (w/v) albumin displaced fractions while there was a barely detectable amount of complex (6%) and free antithrombin III (4%) in the material displaced by arvinized plasma. With the control ^{125}I-antithrombin III, 25% of the radioactivity was determined to be free antithrombin III and 2% as complex. The remainder (22-27%) was not recovered from the column.

Analyses of the effluent produced by passing arvinized plasma through a heparin-PVA "bead" column previously loaded with 100 U of purified bovine ^{125}I-thrombin showed that 46.8 ± 8.9% (SD, n = 5) of the recovered radioactivity (labelled thrombin) passed through

Table II. Heparin-Sepharose Chromatography of ^{125}I-antithrombin III Displaced from Heparin-PVA ($\underline{7}$)

Displacing Eluent	Fraction of Loaded Radioactivity			
	Void Volume	Intact Complex	Free Antithrombin III	Recovery from Column
Arvinized plasma	0.65	0.06	0.04	0.75
5% (w/v) albumin	0.78	--	--	0.78
Antithrombin III (Control)	0.49	0.02	0.25	0.76

heparin-Sepharose without affinity and 53.2 ± 8.5% was gradient-eluted as a low, flat peak, similar in profile to the standard thrombin-antithrombin III complex. Having corrected the unbound peak for thrombin-antithrombin III complex content, the results from heparin-Sepharose indicated that 36.2 ± 4.1% (SD, n = 5) of ^{125}I-thrombin in the heparin-PVA effluent was contained in ^{125}I-thrombin-α-2-macroglobulin complex and 63.8 ± 5.7% in ^{125}I-thrombin-antithrombin III complex. Free ^{125}I-thrombin was not observed. Analysis on Sephadex G-200 showed 27.1 ± 5.1% of the ^{125}I-thrombin was ^{125}I-thrombin-α-2-macroglobulin complex, and 72.9 ± 5.1% was ^{125}I-thrombin-antithrombin III complex.

Antithrombin III-free plasma (i.e. twice chromatographed through heparin-Sepharose) displaced ^{125}I-thrombin from heparin-PVA probably as thrombin-α-2-macroglobulin complex as indicated by the lack of affinity of ^{125}I-thrombin radioactivity for heparin-Sepharose and the preponderance of radioactivity in the 19S region after Sephadex G-200 chromatography. In contrast, antithrombin III deficient plasma (plasma once chromatographed through heparin-Sepharose and containing 3% of the normal plasma antithrombin III level) displaced both thrombin-antithrombin III and thrombin-α-2-macroglobulin complexes in a ratio which was not significantly different from those displaced by arvinized plasma. In this case the antithrombin III level, although low, was presumably still sufficient in quantity to interact with bound thrombin in a manner comparable to arvinized plasma. Alpha-2-macroglobulin-free plasma displaced radioactivity from heparin-PVA which, on Sephadex G-200, largely behaved as thrombin-antithrombin III complex; heparin- Sepharose analysis of the heparin-PVA effluent revealed a substantial proportion (23%) of unbound radioactivity which presumably was high molecular weight thrombin-antithrombin III complex without affinity for heparin-Sepharose. (More details can be found in ref. 14.)

Discussion

Mechanism of Inactivation. The results obtained with heparin-PVA suggest that the interaction between thrombin and immobilized heparin is the primary step in enzyme inactivation by antithrombin III.

Thrombin, loaded onto the heparin-PVA column, forms a complex mainly with immobilized heparin. Following the enzyme, antithrombin III could complex either with available heparin sites or with heparin/thrombin. The drastic reduction in colour yield indicated that antithrombin III had inactivated a substantial quantity of heparin/thrombin complex. Thus antithrombin III which had been shown to adsorb specifically to heparin (not significantly adsorbed to PVA) had higher affinity for thrombin sites than for heparin alone.

By using the reverse loading sequence thrombin could interact with heparin/antithrombin III complex or with heparin alone. The colour yield obtained was similar in this sequence to that obtained by only thrombin loading, indicating that none of the thrombin on heparin-PVA gel had been neutralized by the heparin/antithrombin III complex.

Thus the mechanism of inactivation appears to involve the binding of thrombin to heparin-PVA before it is inactivated by antithrombin III, which presumably binds to a neighbouring region in the heparin molecule. This is consistent with the mechanism proposed by Griffith et al. (19,20) and others that the interaction between heparin and thrombin brings about a conformational change in the enzyme that facilitates complex formation with antithrombin III. On the other hand it conflicts with the mechanism of Rosenberg and Damus (21,22), who suggested that the direct interaction between heparin and antithrombin III may be largely responsible for the kinetic effect of heparin. These differences may be due to the heterogeneity of heparin (difference between commercial heparin and antithrombin III affinity fractionated heparin) or a difference between immobilized heparin and heparin in solution. For example immobilization via glutaraldehyde to PVA may alter the ability of heparin to interact effectively with antithrombin III.

However, it is clear that thrombin adsorbed by PVA-heparin is biologically active and is inactivated by antithrombin III presumably through the formation of a surface-bound heparin-thrombin-antithrombin III complex. Furthermore, the biological activity of the heparinized gel and the mechanism of thrombin inactivation by antithrombin III on heparin-PVA have been verified using clotting assays (3).

Thrombin Affinity. Thrombin has been shown to adsorb to both PVA and heparin-PVA by using both chromogenic substrate and radiolabelled thrombin. As shown in the radiolabelled adsorption experiments, significantly more thrombin was adsorbed to heparin-PVA. Results of the chromogenic substrate experiments indicated that a significant portion (~ 30%) of the thrombin which was adsorbed to PVA was easily desorbed with albumin or defibrinated plasma (crude antithrombin III) or prevented from adsorption by pre-coating the PVA with albumin. These sites appear to have a lower affinity for thrombin, but a significant amount of thrombin was still bound to PVA with presumably higher affinity to other sites in the polymer. Presumably heparin-PVA has these PVA sites (low and high affinity) as well as the thrombin binding sites on heparin. It appears that specific binding to heparin predominates since thrombin was not removed significantly by albumin as it was for PVA. Formation of an inactive complex was shown when passing crude antithrombin III (defibrinated plasma) through a heparin-PVA column inactivated more than 70% of the bound thrombin.

Quantitative comparison of the amount of thrombin bound by chromogenic substrate and radiolabelled measurements differ significantly, with apparent biological activity being much less than the total labelled thrombin. Whether this reflects a limitation in using a chromogenic substrate to measure adsorbed protein activity or a real difference between biological activity and amount bound remains to be seen, but could be related to the multiplicity of binding sites apparent in heparin-PVA. This aspect of thrombin interactions with heparin-PVA is currently being followed up in detail with measurements of the thrombin adsorption isotherms and affinity to these materials.

Displacement. Although antithrombin III was able to inactivate the bound thrombin, it was not able to displace the complex. The most efficient displacement agent was arivinzed or heat defibrinated plasma. However, these displacing agents were only able to remove 17 - 22% of the ^{125}I-thrombin bound to heparin-PVA. The remaining radioactivity was presumably in part adsorbed thrombin which was not inactivated and thrombin-antithrombin III complex which was not removed from the heparin sites or at least from the gel. It appears that thrombin-antithrombin III complex was more easily removed than thrombin from heparin-PVA because of the lower recovery with antithrombin III deficient plasma and the fact that no free ^{125}I-thrombin was found in the displaced radioactivity; presumably the free ^{125}I-thrombin must have remained bound. Even stronger displacing agents, such as, 3 M NaCl and 6 M guandine-HCl were of limited use in displacing more ^{125}I-thrombin (only another 7 - 16% removed). The limited slow removal of ^{125}I-thrombin may be because of a diffusion limitation of a yet unknown displacing agent(s) in plasma into the "beads" or the thrombin- antithrombin III complex out; it should be noted that the hydrogel beads are permeable to thrombin and antithrombin III (23) and the heparin is immobilized throughout the beads.

Analysis of the radioactivity displaced by arvinized plasma indicated the presence of thrombin-antithrombin III complex (~ 70%) and what was presumed to be thrombin-α-2-macroglobulin complex (~ 30%). The bound thrombin is thought to react first with anti-thrombin III to produce a bound inactivated thrombin-antithrombin III complex, which is dislodged from heparin by a yet unknown plasma component(s), decomplexed by an unknown mechanism to react with α-2-macroglobulin. This mechanism is illustrated in Figure 4. After displacement, the increase in ^{125}I-antithrombin III which had lost its affinity for heparin-Sepharose was attributed to the production of a post complex antithrombin III on decomplexation of the inactive complex. This modified antithrombin III has been described by Lam et al. (24), Fish et al. (25) and Marciniak (26). Neither free ^{125}I-thrombin nor ^{125}I-antithrombin III were detected in the displaced eluent.

Although displacement of ^{125}I-thrombin was limited, some heparin sites were regenerated. Further experiments, not yet published (27), have demonstrated that the immobilized heparin appeared to retain its ability to accelerate the inactivation of thrombin at least over 10 cycles of exposure to thrombin and antithrombin III. This suggested that the regenerated heparin sites can retain their catalytic

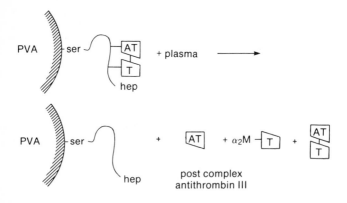

Figure 4.   Schematic illustration of fate of surface bound
complex.   Thrombin- antithrombin III complex is displaced and
dissociated by an unknown component(s) in plasma to produce
residual complex, post complex (inactive) antithrombin III and a
thrombin-α-2-macroglobulin complex. (Reproduced with permission
from Ref. 28.   Copyright 1984, Gordon and Breach, Science
Publishers Inc.)

activity and can be repeatedly regenerated. This may be sufficient to give heparin-PVA the capacity for long term nonthrombogenicity.

Conclusions

The results reported here, in conjunction with earlier results (1-6) indicate that immobilized heparin need not necessarily be lost from a surface in order to accelerate the inactivation of thrombin and reduce the thrombogenicity of a surface. For heparin-PVA, and perhaps for other covalent reactions that do not inactivate the heparin, the irreversibly bound heparin can accelerate the formation of a surface-bound inactive thrombin-antithrombin III complex. Furthermore, our results suggest that the inactive complex is not itself permanently bound to the surface, but rather can be displaced by a component or components in plasma.

While materials that lose heparin at a controlled rate can be clinically acceptable in short-term applications, the long-term use of heparinized materials requires stable immobilization techniques that do not compromise the biological function of the heparin. These investigations indicate that these requirements may not be mutually exclusive, and that bound heparin can potentially retain its biological activity over the long term. Heparinization can be an important means of preparing the materials needed for the development of improved cardiocirculatory assist devices and blood handling procedures.

Acknowledgment

The authors acknowledge the financial support of the National Heart, Lung and Blood Institute under grant No. HL25020.

Literature Cited

1. Goosen, M.F.A.; Sefton, M.V. Adv. Chem. 1982, 199, 147-160.
2. Goosen, M.F.A.; Sefton, M.V. J. Biomed. Mater. Res. 1979, 13, 347.
3. Goosen, M.F.A.; Sefton, M.V. J. Biomed. Mater. Res. 1983, 17, 359.
4. Cholakis, C.H.; Sefton, M.V. In Polymers as Biomaterials; Shalaby, S.W. et al., Ed.; Plenum Press, 1984.
5. Ip, W.F.; Zingg, W.; Sefton, M.V. J. Biomed. Mater. Res. 1985, 19, 161-178.
6. Basmadjian, D.; Sefton, M.V. J. Biomed. Mater. Res. 1983, 17, 509.
7. Goosen, M.F.A.; Sefton, M.V. Trans. A.S.A.I.O. 1982, 28, 451-455.
8. Goosen, M.F.A.; Sefton, M.V. Thromb. Res. 1980, 20, 543-554.
9. Idezuki, Y.; Watanabe, H.; Hagiwara, M.; Kanasugi, K.; Mori, Y.; Nagaoka, S.; Hagio, M.; Yamamoti, K.; Tanzawa, H.; Trans. A.S.A.I.O. 1975, 21, 436.
10. Hatton, M.W.C.; Berry, L.R.; Regoeczi, E. Thromb. Res. 1978, 13, 655-670.
11. Lundblad, R.L.; Uhteg, L.C.; Vogel, C.N.; Kingdon, H.S.; Mann, K.G. Biochem. Biophys. Res. Commun. 1975, 66, 482-489.

12. Hatton, M.W.C. Biochem. J., 1973, 131, 799-807.
13. Hatton, M.W.C.; Regoeczi, E. Thromb. Res. 1977, 10, 645-669.
14. Hatton, M.W.C.; Rollason, G.; Sefton, M.V. Thromb. Haemastas 1983, 50(4), 873-877.
15. Hatton, M.W.C.; Regoeczi, E. Biochem. Biophys. Acta 1974, 359, 55-65.
16. Mitchell, G.A.; Hudson, P.M.; Huseby, R.M.; Pochron, S.P.; Gargrilo, R.J. Thromb. Res. 1978, 12, 219-225.
17. Harpel, P.C. Meth. Enzymol. 1970, 45, 639-652.
18. Collen, D.; De Cock, F.; Verstraete, M. Eur. J. Clin. Invest. 1977, 7, 407-411.
19. Griffith, M.J. J. Biol. Chem. 1979, 254, 12044-12049.
20. Griffith, M.J. J. Biol. Chem. 1982, 257, 7360-7365.
21. Rosenbeg, R.D.; Damus, P.S. J. Biol. Chem. 1973, 248, 6490-6505.
22. Jordon, R.E.; Oosta, G.M.; Gardner, W.T.; Rosenberg, R.D. J. Biol. Chem. 1980, 255, 10081-10090.
23. Smith, B.A.H.; Sefton, M.V. J. Biomed. Mater. Res. (submitted for publication).
24. Lam, L.S.L.; Regoecozi, E.; Hatton, M.W.C. Br. J. Exp. Pathol. 1979, 60, 151.
25. Fish, W.W.; Orre, K.; Bjork, I. FEBS. Lett. 1979, 98, 103.
26. Marcinciak, E. Br. J. Haemotol 1981, 48, 325.
27. Rollason, G.; Sefton, M.V. (unpublished results).
28. Sefton, M.V.; Ip, W.F.; Rollason, G.; Hatton, M.W.C.; Zingg, W. Chem. Eng. Commun. 1986, 30, 141-154.

RECEIVED February 18, 1987

APPLICATIONS OF PROTEINS
AT INTERFACES

Chapter 36

Applications of Adsorbed Proteins at Solid and Liquid Substrates

Ivar Giaever and Charles R. Keese

Corporate Research and Development, General Electric Company,
Schenectady, NY 12301

The understanding of protein adsorption onto nonbiological substrates is an important problem in biotechnology that may lead to many practical advances. For example, solid-phase immunology tests such as enzyme-linked immunosorbant assays rely on a preadsorbed layer of either antigen or antibody on a plastic or glass surface. Protein-covered interfaces are also very important in tissue culture research, as the cells attach to a protein film adsorbed at the plastic surface of the tissue culture dish and not directly to the dish itself. In addition, because of the increasing use of artificial organs, understanding of the adsorption of protein from body fluids and the search for a nonthrombogenic surface has intensified. If successful, such surfaces will be of major importance in medicine. This paper presents a summary of work in our laboratory to understand the phenomenon of protein adsorption and to apply this understanding to biotechnological problems.

Adsorption of Protein on Solid Surfaces

It is generally agreed that proteins adsorb to most artificial surfaces in a monolayer; however, much confusion exists with regard to the desorption and replacement of protein. One reason for this is the variety of different buffers used in studying this phenomenon. In this laboratory we have confirmed earlier findings (1) that phosphate and borate-based buffers under certain conditions interfere with the adsorption process, and can also cause proteins to desorb. Thus to avoid this complication, these buffers have been generally avoided in our work, and Tris is used in most experiments requiring a buffer.

If protein is adsorbed from a saline solution onto a solid surface, we believe that the protein binds in a random orientation at the site of the molecule's first encounter with the surface. We simulated this process on a computer, approximating the proteins with a disk, and found that the final

0097-6156/87/0343-0582$06.25/0
© 1987 American Chemical Society

fractional coverage was only 0.547 compared to 0.907 for a close packed surface. In Figure 1, our computer calculation is contrasted with an experimental observation of the adsorption of ferritin onto a carbon surface. The agreement between fractional coverage measurement for the model and the experimental situation is excellent (2).

Detection of Adsorbed Protein Layers

Many ingenious methods have been introduced to study protein adsorption. If the kinetics of the adsorption process are important, the ellipsometric method introduced by Rothen (3) is probably the best. In this method protein adsorption can be studied *in situ* from a solution. The method has been used to study the kinetics of both the adsorption of protein in single layers and in double layers that can occur in the immune-reaction. When protein such as bovine serum albumin (BSA) was adsorbed from a dilute solution onto a surface, after a delay of a few seconds, steady-state diffusion controlled the adsorption process and, consequently, the amount bound to the surface increased linearly with time. However, as the surface became covered, adsorption slowed down, because it was now limited by the number of available sites on the surface. The final layer of BSA was roughly 2 nanometer thick.

Figure 2 is a good illustration of the power of the ellipsometric technique. Curve (a) on the figure shows the specific attachment of antibody to a preadsorbed BSA layer. Curve (b) is a repeat of the experiment, except roughly 90 sec. after the start of the run, additional BSA was added to the solution thus effectively neutralizing the specific antibodies.

For static measurement of protein films we have developed a method that relies on light scattering; the technique is referred to as the Indium Slide Method (4,5,6). When indium is evaporated onto a transparent surface such as glass or plastic in a vacuum, the indium atoms will condense upon the surface in small particles. The physical size of the indium particles depends mainly on the amount of indium evaporated, but also on the temperature of the substrate. The optimum size of the particles for this method is roughly equal to the wavelength of light, i.e. a few hundred nanometers in diameter. The test relies on the fact that visible light scattered by particles in this size range is markedly increased if the particles are covered with thin dielectric layers. Adsorbed protein acts as this dielectric layer and, in general, the more protein adsorbed the more the scattering increases. Thus it is possible to quantify the amount of protein adsorbed by measuring the amount of light transmitted through the slide with the help of a simple densitometer, or one can simply estimate the amount of protein by visual inspection.

Application of Adsorbed Protein Layers in Immunology

We have used the slides extensively for measuring various forms of the immune reaction, from screening for monoclonal antibodies

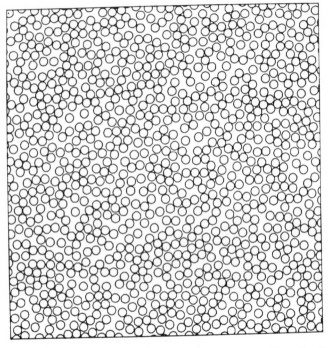

Figure 1a. Monte Carlo simulation of protein adsorption. At the jamming limit for disks, the final coverage is 0.547 of the available area.

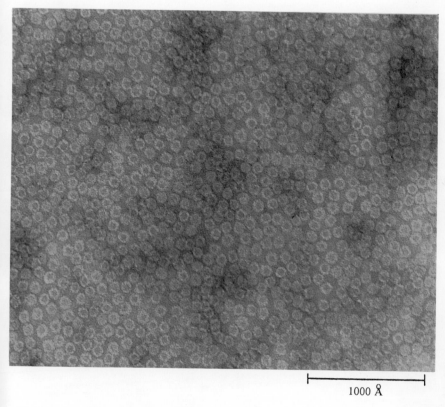

1000 Å

Figure 1b. Ferritin (horse spleen) adsorbed on carbon and stained with uranyl acetate.

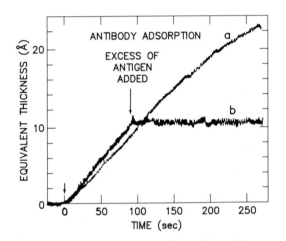

Figure 2 (a) Ellipsometric detection of the adsorption of rabbit antiserum to BSA onto a monolayer of BSA at a gold surface. (b) Same conditions as in curve (a), but at the indicated time, excess BSA was added to the antiserum dilution.

(R. Rej, I. Giaever and C.R. Keese, A Screening Technique for Monoclonal Antibody Production: Application of an Indium Slide Immunoassay, submitted for publication) to detecting hepatitis or schistosomiasis infections (4). In a typical test, antigen is first adsorbed onto the slide as a small spot from a solution of a relatively pure protein. The protein adsorbs in a monolayer, and if the slide is rinsed and dried, this layer can easily be detected by a change in the amount of transmitted light. Next, if desired, the slide is "masked" by adsorbing an inert protein over the remaining surface of the slide that has the same light scattering effect as the antigen. The slide is now exposed to a solution that may or may not contain antibodies to the adsorbed antigen. If antibody is absent, no additional protein will attach to any portion of the slide, and the light scattering will not change. On the other hand, if specific antibody is present, some of the antibody will bind to the preadsorbed antigen causing a distinct change in the transmitted light in that region. There are several variations of this procedure. For example, it is possible to enhance the effect by using a second antibody. If it is desired to detect antigen, it is necessary to do an inhibition test (5), as only a small fraction of the antibodies remain active if they are adsorbed on a surface.

Figure 3 shows a photograph of the indium slide applied to detection of rheumatoid factor (6). Figure 3a is a photograph of a naked indium slide. Figure 3b is a photograph of the slide following the adsorption of antigen spots, on the left is human IgG and on the right, rabbit IgG. Figure 3c shows the slide after it has been dipped into a solution of aldolase. The aldolase, which alters the light scattering with approximately the same intensity as the IgG molecules, adsorbs around the antigen "masking" them from view. Figure 3d is the appearance of a slide after it has been incubated in a serum that does not contain rheumatoid factor. Figure 3e, on the other hand, shows the slide after it has been incubated with a serum containing rheumatoid factors against both the human and rabbit antigen. The "antigen spots" are now visible because the monomolecular layer of adsorbed IgG is now covered with a layer of rheumatoid factor. Finally Figure 3f shows the result of incubation with a serum whose rheumatoid factor only reacted with the human IgG, a much more common occurrence.

Adsorbed Protein Layers and Cells in Tissue Culture

Since the 1950's it has been possible to grow mammalian cells isolated from a variety of different tissues and organisms in the laboratory. In tissue culture, cells divide and carry on a variety of biological activities while feeding on a rich nutrient medium that supplies all of the necessary molecules for their survival. Unlike bacterial cultures, most commonly cultured normal mammalian cells, such as fibroblasts, will not grow in suspension but require attachment to a rigid surface in order to undergo mitosis. Traditionally, this substrate has been glass or polystyrene that has been treated to render it

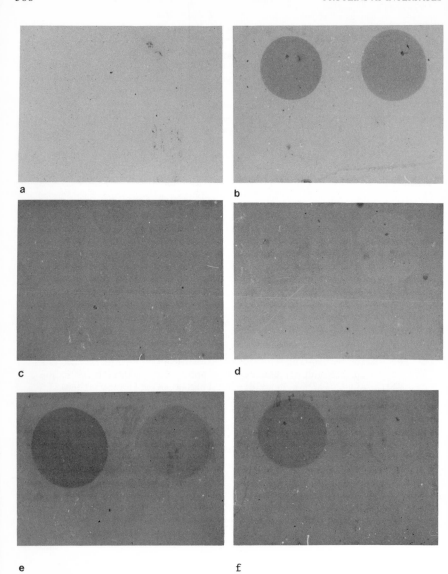

Figure 3 Detection of rheumatoid factor in human sera
using the indium slide immunoassay. (a) Indium slide
before spotting. (b) Antigen spots applied. (c) Spots
masked. (d) Slide exposed to control serum. (e,f) Slide
exposed to two different positive sera.

hydrophilic. In most tissue culture work, the medium contains large amounts (e.g. 10%) of plasma or serum in addition to defined components such as salts, sugars, vitamins, etc. Consequently, cells in tissue culture are grown in the presence of protein molecules, and the surfaces with which they interact are covered with a monolayer of protein that spontaneously adsorbs on the solid interface.

Our laboratory has been engaged in studies of the interactions of fibroblastic cells with these adsorbed protein layers. There are many facets to this study. When surfaces are initially inoculated, the cells are introduced as a monodisperse suspension with roughly a spherical morphology. Upon settling to the substrate, a complex series of events occur where the cells attach to the surface, spread into highly flattened, irregular shapes, and then crawl about (Figure 4a). Although the attachment and subsequent spreading and locomotion involve making and breaking contact and exerting forces upon the adsorbed proteins, as a general rule, we have found that the type of protein adsorbed at the surface has only subtle effects on cell behavior.

One exception to this generalization was observed when cells were grown on substrates covered with IgG molecules. In this case the ability of the cells to attach and spread upon the substrate was noticeably impaired. An even more pronounced effect was observed when the substrate was coated with a bimolecular protein layer consisting of IgG molecules specifically bound to an adsorbed antigen layer. In this situation, no cell attachment or spreading was detected for a wide variety of both normal and transformed cell lines (7).

Figure 4b demonstrates this effect and emphasizes the fact that cells in culture interact with interfacial protein layers and not directly with the solid substrate. To produce the effect shown, different protein layers were placed on a glass coverslip in defined regions using a UV lithographic technique (8). The substrate was then inoculated using standard tissue culture protocol with WI-38 human embryoic lung fibroblasts. Following overnight incubation the coverslip was gently rinsed with tissue culture medium to remove unattached cells. The remaining cells were fixed and stained to reveal their location. The background showing normal cell-substrate interaction is covered with a layer of BSA while the pattern (GE100) consists of a base layer of BSA covered with specifically attached IgG molecules and, consequently, is void of cells.

Monitoring Cell Attachment and Spreading Electrically

It is possible to detect small differences in cell-substrate interactions using weak electric fields and in this manner to quantitatively measure differences in the dynamics of cell attachment and spreading to defined protein monolayers. The details of the system have been previously described (9-10).

In brief, cells were cultured on gold electrodes under standard tissue culture conditions. To minimize the effect of

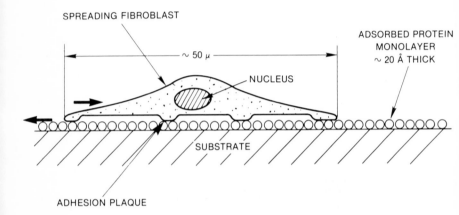

Figure 4a. The illustration depicts a spread fibroblast on a layer of adsorbed protein; the arrows represent forces generated by the microfilaments of the cell as an action–reaction pair. These forces are involved in the process of spreading as well as locomotion of the cells on the substrate.

Figure 4b. Haptotactic behavior of WI–38 cells on a specially prepared glass coverslip. The background, showing normal cell–substrate interaction, is covered with a layer of adsorbed BSA, while the pattern (GE100) is covered with IgG molecules specifically attached to a base layer of BSA.

solution resistance, the system was designed with one small electrode ($\simeq 10^{-4} cm^2$) and one large electrode; under these conditions the measurement is dominated by the electrolytic interface between the small electrode and the solution. An applied alternating electric field (normally 4000 Hz) produced a voltage drop of a few millivolt at the boundary of the solution and the electrode, and the current density was a few milliamperes/cm^2. Under these conditions, there were no detectable effects of the electric fields on cells as judged by cell morphology, length of generation time, etc. As fibroblasts attached and spread on this surface, however, the impedance of the electrode was observed to increase, reaching a maximum value about two hours after inoculation (using a sufficient number of cells for a confluent layer). At this point, the impedance decreased slightly, and after approximately another hour the average value stabilized. At this time, the impedance fluctuated about the mean as the fibroblasts crawled about, altering their contact with the electrode surface. Figure 5a illustrates these events with WI-38 cells in which the data is presented as the measured in- and out-of-phase potential across the small electrode as a function of time.

We have applied this new means of monitoring cell behavior to study the interaction of cells in culture medium with defined layers of adsorbed protein. Before the addition of the tissue culture medium containing serum, the small electrode was exposed for 15 min. to a 100 µg/ml solution of a selected protein. Following adsorption, the electrode was thoroughly rinsed free of unadsorbed protein and inoculated with a fibroblast suspension. Figure 5b presents data obtained when electrodes coated with adsorbed layers of plasma fibronectin, gelatin, BSA and fetuin were inoculated with WI-38/VA 13 cells, a transformed (cancerous) cell line derived from WI-38. As can be seen, there was a pronounced difference in the response of the cells to the different protein layers. Although the rate of change in the resistive component of the impedance was greatly reduced for BSA and fetuin, eventually the final change in impedance, and hence in cell-substrate interaction, appeared to be equivalent (data not shown in figure.) When different cell lines were compared, the ordering of the protein layers with regard to the rate of impedance increase varied, but in all cases examined, rapid initial change in impedance occurred when the protein coat was fibronectin; this protein has long been the leading candidate for the "glue" that connects cells to a surface.

In addition to studies involving the dynamics of cell attachment and spreading on protein-coated substrates, the system has also been employed to study cell locomotion as revealed by oscillations in the impedance observed following cell attachment and spreading. The belief that these are related to cell motion is supported by drug studies where compounds known to interfere with cell motion greatly reduced the amplitude of these fluctuations (9). An extensive search was undertaken to discover if dominant frequencies were present by digitally processing the signals. So far this search has been negative, but the power density spectrum of the

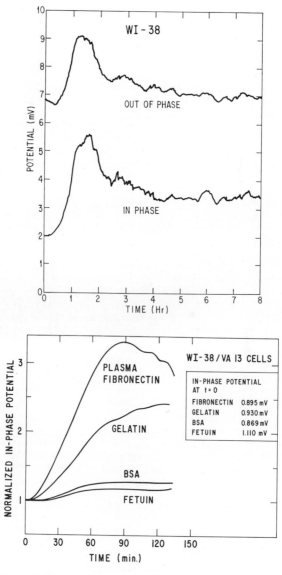

Figure 5   (a) Monitoring cell-substrate interactions in tissue culture using weak electric fields.   (b) The effect of different adsorbed protein layers on WI-38/VA13 cells monitored using weak electric fields.

cell-induced fluctuations has been obtained from this study. In general, the magnitude of the spectrum varies inversely with the square of the frequency of the fluctuation (Brownian noise) and is much larger for cancer cells than normal cells. Because there is a large amount of scatter in the data, we have not yet been able to relate the noise to the various proteins used as substrates.

Oil-water Interfaces as Substrates for Cell Growth

In 1964, Rosenberg introduced the use of a fluid substrate for the growth of both transformed and anchorage-dependent cells (11). In this method, a cell suspension was introduced over a hydrophobic liquid having a density greater than that of the aqueous medium, and then the cells, being of intermediate density, settled to the interface where they were observed. Using fluorocarbon fluids, Rosenberg found that such interfaces could serve as supports for attachment, spreading, and growth of a variety of cell lines. Again, as in the case of solid substrates, the interface was coated with a monolayer of adsorbed protein from the culture medium.

Since this initial observation, we have demonstrated that the adsorbed proteins on highly purified fluorocarbon fluids do not form adequate interfacial substrates unless the oil contains small amounts of specific surface active compound (12). The compound we have found to be most effective in this capacity is pentafluorobenzoyl chloride (PFBC). To produce an interfacial substrate that is adequate for the growth of most fibroblastic mammalian cells, this compound is added to the oil-phase to yield a final surface concentration of at least 0.25 μg per square centimeter (Figure 6a). The necessity for this (or similar compound) for cell growth has been thoroughly investigated in our laboratory because it affects the mechanical strength of the adsorbed protein. In order to achieve a spread morphology and to move about on a surface, cells in culture exert forces at their points of attachment. These forces are generated by an intracellular system of muscle-like fibers referred to as microfilaments and composed mainly of the muscle protein actin. If the adsorbed protein layer at the oil-water interface is unable to support such forces, it will yield causing the cells to retract to a rounded state. Hence, if one is to use fluorocarbon fluid-water interfaces as tissue culture supports, they must satisfy the minimal mechanical properties required to sustain the forces involved in cell spreading.

We have investigated the alteration in mechanical properties of the protein layer caused by the PFBC using a modified surface viscometer. The protein film was placed under a shearing stress by the application of a small torque to a teflon paddle wheel inserted into the interfacial boundary, and the angular deformation of the film was measured. From this data it was possible to obtain stress-strain curves and to determine the surface shear modulus and surface fracture point for the protein layer. In most studies protein was adsorbed to the interface of perfluorotributylamine from either a buffered

BSA solution or culture medium containing 10% serum.   Following a 24-hour period, to allow diffusion of PFBC to the interface, measurements were carried out.   The results are shown in Figure 6b where the conversion of an essentially fluid layer of adsorbed protein to an elastic film by the presence of increasing amounts of the acid chloride is shown.

We believe there are two possible mechanisms to account for this alteration.   Following adsorption of the   interfacial protein layer, the PFBC molecules diffusing to the interface will react with the acid chloride moiety either with the water or, more importantly, with functional groups on the surface of the adsorbed protein by a condensation reaction. In this manner what were formerly the most hydrophilic residue are chemically modified into groups that now are likely to be inserted into the oil-phase.   This alteration could result in severe denaturation of the protein molecules, such that adjacent molecules would tangle to form an elastic film.   It is also possible that the fluorine in the para position could undergo a nucleophilic replacement reaction.   In this manner the PFBC could be acting as a bifunctional crosslinking compound, joining adjacent proteins by covalent bonds.

We are now using interfacial protein layers to characterize some of the mechanical properties required of a cell substrate. To carry out these studies two different measurements are being made. First, the surface shear modulus and surface fracture point of 24-hour old adsorbed protein layers are measured as a function of the surface concentration of PFBC as described above.   Next, identical interfaces are inoculated with fibroblasts and incubated for 16 hours. Following this period, the cells at the interface are fixed and stained, and the projection area of the cells is measured with a Zeiss IBAS image analysis system.   From this data, the relative amount of cell spreading is calculated. By correlating the results of these two types of experiments, we expect to determine the minimal mechanical properties required of a substrate for different cell lines.   Conversely, these values should also allow us to infer the magnitude of the forces exerted by different cell lines upon the substrate.   In vivo these forces are thought to be associated with normal cell migration in development and wound healing. They have also been implicated in the process of metastasis whereby a cancer cell is able to leave its primary location and establish secondary tumors throughout the body.

Applications of Oil-water Interfacial Protein Layers

By utilizing fluorocarbon fluids containing pentafluorobenzoyl chloride, a liquid microcarrier system has been developed capable of use with a variety of cell types including normal human fibroblast. In this configuration, cells on the surface of a coarse oil dispersion ($\approx$ 150 μm diameter) exhibit exponential growth (Figure 7).   In addition, a microcarrier based on silicone oil has been formed and used to culture mouse fibroblasts (12-13).

These novel interfacial substrates may allow manipulation

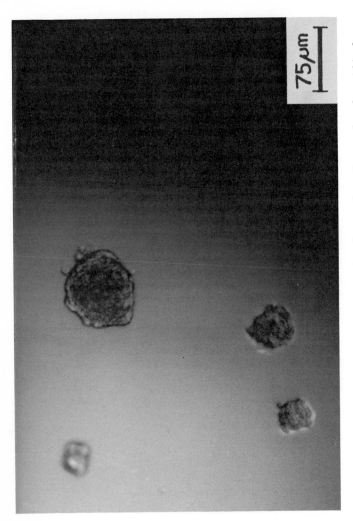

Figure 6a. 3T3-L1 mouse fibroblasts are shown at fluorocarbon oil–water interfaces 16 h after inoculation. In this photograph, where the cells have not spread but have clustered together in spheroids, the oil received no additive. *Continued on next page.*

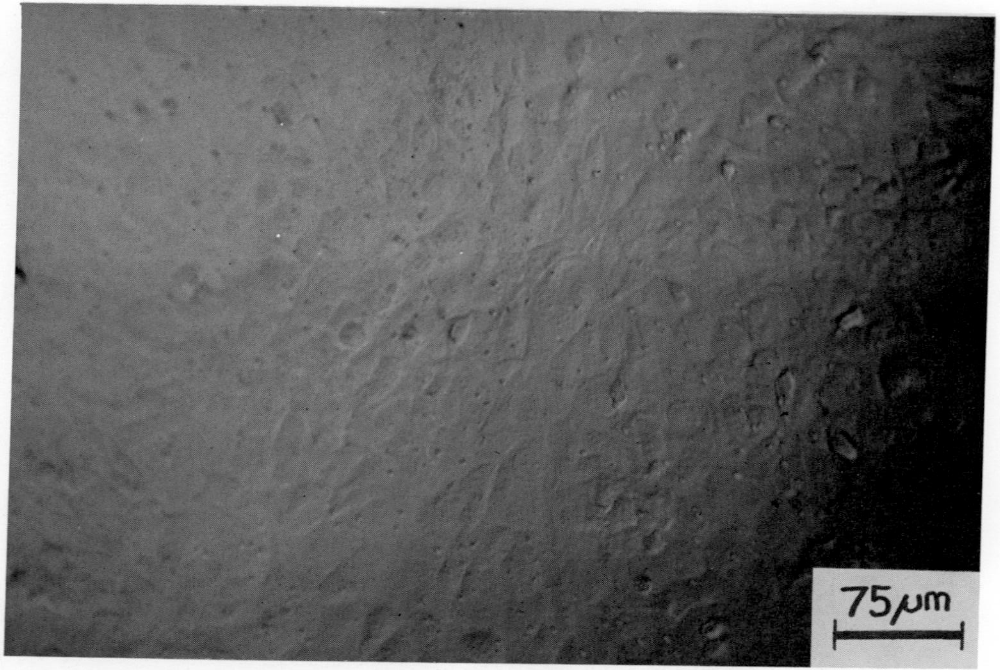

75 μm

Figure 6a.—*Continued.* 3T3-L1 mouse fibroblasts are shown at fluorocarbon oil–water interfaces 16 h after inoculation. The photograph shows a confluent layer of spread fibroblasts. PFBC was present in the oil phase at a surface concentration of 1 μg/cm².

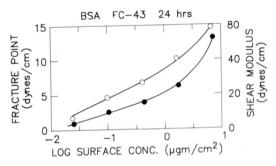

Figure 6b. Mechanical properties of adsorbed BSA layers at a fluorocarbon oil–water interface as a function of the log of the surface concentration of PFBC (○, surface fracture point; ●, surface shear modulus).

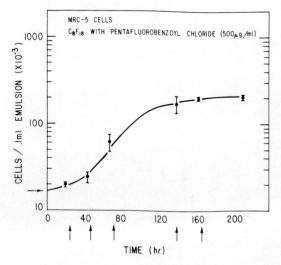

Figure 7a.   Growth curve for the human fibroblast MRC-5 at the interface of oil droplets (microcarriers).

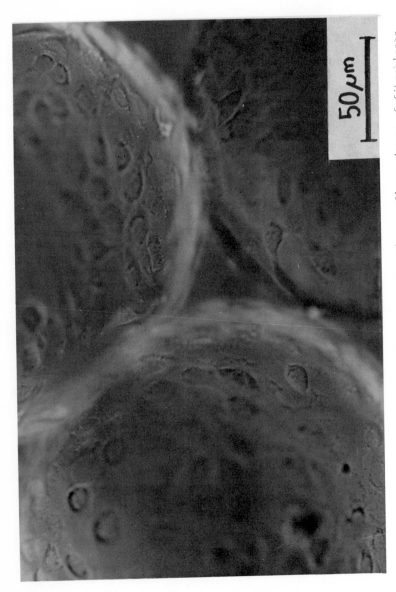

Figure 7b. Photograph of the droplets covered with a confluent layer of fibroblasts.

of cells in culture in ways previously not possible using solid-substrate culturing. The liquid microcarrier system permits mass culturing cells in arrangements commonly used with solid microcarriers but also allows one to mechanically harvest the cells by breaking the dispersion into its two component phases. Such an arrangement may prove to be of particular value in studies involving purification of large quantities of receptors and other surface molecules, where chemical methods of cell harvesting could damage the components of interest. Another interesting property of this arrangement is its capacity for delivery of water insoluble compounds to cells. By first dissolving such compounds in the oil-phase they could then continuously partition from the oil-phase into the cells membranes.

Adsorbed protein on fluorocarbon oil-water interfaces have also been used in our laboratory to develop a variation of the latex agglutination assay to detect immunological molecules. Fluorocarbon oil was emulsified by sonication in the presence of an antigenic protein that also serves as the emulsifying agent. Average particles had a diameter of the order of 1 μm and were highly stable without the addition of other emulsifiers or the additive, PFBC. When these droplets were combined with specific antiserum and allowed to slide by each other with a gentle rocking motion, agglutination could be observed and quantitated using a image analysis system. The system had a sensitivity of 1 μg/ml of antibody for a 15-min. reaction time. Interestingly, the most sensitive results were obtained when an impure antigen was used to stabilize the emulsion (14).

Acknowledgments

This work was carried out in part pursuant to a contract with the National Foundation for Cancer Research.

Literature Cited

1. Trurnit, H.J. Science III 1950, 1.
2. Feder, J.; Giaever, I. J. Colloid & Interface Science 1980, 78, 144.
3. Rothen, A.; Mathot, C. Hel. Chim. Acta 1971, 54, 1208.
4. Giaever, I.; Laffin, R.J. Proc. Natl. Acad. Sci., USA 1974, 71, 4533.
5. Rej, R.; Keese, C.R.; Giaever, I. Clinical Chemistry 1981, 27, 1597.
6. Giaever, I.; Keese, C.R.; Rynes, R.I. Clinical Chemistry 1984, 30, 880.
7. Giaever, I.; Ward, E. Proc. Natl. Acad. Sci., USA 1978, 75, 1366.
8. Panitz, J.A.; Giaever, I. Surface Science 1980, 97, 25.
9. Giaever, I.; Keese, C.R. Proc. Natl. Acad. Sci., USA 1984, 81, 3761.
10. Giaever, I.; Keese, C.R. IEEE Trans. Biomed. Engrg. 1986, 33, 242.

11. Rosenberg, M.D. Cellular Control Mechanisms and Cancer;
    Emmelot, P. and Muhlbock, O., Eds.; Elsevier: Amsterdam,
    1964, pp 146-164.
12. Keese, C.R.; Giaever, I. Proc. Natl. Acad. Sci., USA
    1983, 80, 5622.
13. Keese, C.R.; Giaever, I. Science 1983, 219, 1448.
14. Prather, T.L.; Grande, J.; Keese, C.R.; Giaever, I.
    J. Immunol. Methods 1986, 87, 211.

RECEIVED January 3, 1987

Chapter 37

Affinity Selection of Cells on Solid-Phase Matrices with Immobilized Proteins

Kazunori Kataoka[1], Yasuhisa Sakurai[1], and Teiji Tsuruta[2]

[1]Institute of Biomedical Engineering, Tokyo Women's Medical College, 8-1 Kawada-cho, Shinjuku-ku, Tokyo 162, Japan
[2]Department of Industrial Chemistry, Science University of Tokyo, 1-3 Kagurazaka, Shinjuku-ku, Tokyo 162, Japan

This paper reviews the present status of affinity separation of cells based on the biospecific interaction of cellular receptors with proteinaceous ligands immobilized on a solid-phase matrix. Special emphasis was placed on the development of new matrix materials for immuno-affinity chromatography of lymphocyte subpopulations. Our newly developed matrix of poly(2-hydroxyethyl methacrylate)/polyamine graft copolymer offered novel advantages in (1) elimination of non-specific adsorption of lymphocytes and (2) simple immobilization procedure of ligand protein through non-covalent adsorption. This matrix allowed a rapid separation of preparative quantities of pure and vital lymphocyte subpopulations (IgG-positive and -negative cells) in excellent yield.

Proteins non-covalently adsorbed or covalently linked on solid-phase matrices have many applications in the wide area of science and technology (1,2). Indeed, in the field of separations science and technology, affinity selection based on the biospecific interaction of biomolecules with proteinaceous ligands immobilized on a solid-phase matrix has been widely noted as the separation method with highest selectivity, and is commonly utilized for separation and purification of biopolymers including proteins and nucleic acids (3,4).

Recently, affinity selection methods have also been shown to allow efficient separation of viruses, bacteria, cellular organelles, and even whole cells in their vital form (4-7). These methods offer a special advantage in the separation of lymphocyte subpopulations based on the biospecific interaction of immobilized ligands with marker proteins specifically expressed on the plasma membrane surface of each subpopulation. As reviewed in many articles (6-13), the separation of lymphocyte subpopulations has become increasingly important in the diagnosis as well as in the therapy of immuno-diseases, in donor-recipient matching in transplantation, and in the

production of high-value bioactive compounds including interferon
and monoclonal antibodies. It is to be noted that the therapy of
auto-immune diseases, including systemic lupus erythematosus (SLE)
and rheumatoid arthritis, would be facilitated greatly if it were
feasible to deplete a particular subpopulation of lymphocytes by
extracorporeal hemoperfusion through a column containing
bioselective adsorbents.

     As will be described below, the first application of affinity
chromatography in cell separation, reported in 1969 (14,15), was
based on antigen-antibody reactions (immuno-affinity chromatography).
Cellular immuno-affinity chromatography has since been further
extended by a number of investigators, and in its present form,
offers a wider applicability because of an increased number and
variety of available antibodies.

     Other than antigen and antibody molecules, protein A, avidin,
and various types of lectins have often been utilized as specific
ligand proteins. Protein A is a protein of molecular weight 42,000,
and was orginally found on the cell-wall surface of *Staphylococcus
aureus*. It has a specific binding ability toward the Fc portion of
immunoglobulin, usually IgG (16). Avidin is a protein from egg
white, and shows an extraordinary affinity (association constant :
~10^{15}) with biotin. Cells reacted with biotinylated antibodies can
be selectively adsorbed on columns containing immobilized avidin (6).
Lectin is a general term for a group of sugar-binding proteins and
glycoproteins chiefly but not exclusively of plant origin. The word
"lectin" is derived from Latin "*legere*" whose meaning is "select
out". As to its meaning, each kind of lectin selectively binds
specific carbohydrate residues of cell-surface carbohydrates (11).

     In spite of the widespread utilization of affinity
chromatography in cell separation, there are still a considerable
number of problems to be solved. The most serious problem is that
there is always a substantial fraction of cells that are non-
specifically adsorbed on the matrix surface. The research on cell
affinity chromatography done in the last decade seems to be more
biased towards the improvement in operating conditions than to the
development of specially designed matrices for cell separation as
well as the characterization of immobilized proteins. Nevertheless,
there is no doubt that further advances in affinity chromatography
as an effective tool for cell separation virtually depend on the
detailed understanding of the features of matrix materials and
immobilized proteins, as well as their interacions at the interface.
In this respect, cell affinity selection based on the specific
interaction of cells with immobilized proteins on a solid-phase
matrix is now a major area of interest in the field of biomaterials
science.

     This paper briefly reviews the progress and present status of
cell affinity selection with special emphasis on immuno-affinity
chromatography, and then deals with our own studies on the
development of new matrix materials specially designed for cell
affinity chromatography.

Application of Immuno-affinity Chromatography in Cell Separation

In 1969, Wigzell et al (14) and Evans et al (15) first applied immuno-affinity chromatography in cell separation. This was only a year after the term "affinity chromatography" had been coined by Cuatrecasas et al (17). Wigzell et al used glass or poly(methyl methacrylate)(Degalan) beads as a solid-phase matrix, and immobilized protein molecules thereon by non-covalent physical adsorption. In their first report (14), cells forming antibodies to bovine serum albumin (BSA) were depleted from murine lymphnode lymphocyte populations by biospecific adsorption to a column packed with beads bearing adsorbed BSA. They further carried out the biospecific adsorption of lymphocytes expressing immunoglobulins (Ig) on their surfaces (so-called B cells) by using beads with adsorbed anti-Ig antibodies (18). Evans et al used polyesterurethane foam as a matrix materials instead of beaded-form matrices. Similar to Wigzell's method, ligand proteins were immbilized on the foam surface by physical adsorption.

The most serious problem with this earliest type of affinity column based on non-covalently adsorbed ligands, was non-specific adsorption of cells on the column. In order to decrease this non-specific adsorption, very high concentrations of ligand solution (>1mg/ml) were used for the treatment of the matrix. Subsequent to this physical adsorption of ligand proteins, the matrix was further incubated with isotonic saline containing at least 5% fetal calf serum (FCS). In spite of this rather complicated and time-consuming process of double-coating, there was still a considerable degree of non-specific adsorption (approximately 30 - 50% of the number of loaded cells). Furthermore, treatment of a ligand-immobilized matrix with FCS-containing medium might induce the desorption of the ligand from the matrix surface, because of the possibility of ligand exchange with serum proteins having higher affinity for the matrix surface. Also, recovery of bound cells from the column was not easily achieved unless the column bed was agitated mechanically. This mechanical agitation can cause a considerable decrease in cell viability.

To eliminate non-specific cellular adsorption, a search for matrix materials suitable for cell separation has been carried out by many investigators (5,7). These studies have revealed that hydrophilic polymer gels in bead form, which were originally developed as column matrices for affinity chromatography of biological macromolecules, have generally acceptable feature as matrices for use in cell affinity chromatography. These hydrophilic matrices include dextran (Sephadex) (19-21), agarose (Sepharose) (22-35), polyacrylamide (Bio-gel P) (6,36-39), poly(2-hydroxyethyl methacrylate) (Spheron) (40), and copolymers of acrylamide and N-acryloyl-2-amino-2-hydroxymethyl-1,3-propane diol (Trisacryl) (7,41). Among these, agarose and polyacrylamide are the most commonly used solid-phase matrices.

Although these hydrogel matrices showed far less non-specific adsorption than conventional matrices such as glass and poly(methyl methacrylate), they have not yet fulfilled all of the fundamental requirements. On these hydrogel matrices, ligands should be convalently immobilized to prevent ligand desorption. This process of covalent-linking of the ligand to the matrix is not only time-consuming but is also liable to form new sites for non-specific cellular adsorption due to the chemical modification of the matrix

surface. For example, it is well documented that CNBr-activated
agarose is commonly used for coupling of ligand proteins.
Nevertheless, excess imido-carbonate groups remaining after ligand
coupling could be derivatized, on reacting with water, to urethane
groups which may provoke a considerable degree of non-specific
cellular adsorption (32,42).

Affinity "Panning" Methods for Cell Separation

There is another type of affinity cell selection technique besides
column chromatography. In 1971, an affinity selection method using
nylon fibers was developed by Edelman et al for selecting murine
lymphocytes possessing receptors for concanavalin A (ConA) (43).
Nylon fibers were stringed under tension in a ring-shaped supporting
frame followed by incubation with a solution of ligand protein, in
this case ConA (0.025 - 1.0 mg/ml). After this adsorption procedure,
the fibers in the frame were rinsed with saline, and then incubated
with a cell suspension under gentle horizontal shaking. Cells bound
on the fibers could be recovered by plucking each fiber with a
needle. The density of ligand proteins on the fiber can be
controlled by incubating the fiber in a solution containing varying
proportions of a ligand protein and a spacer protein such as BSA
(44). Competitively adsorbed BSA molecules are interspaced with
ligand proteins on the fiber surfaces. This nylon-stringing method
was shown to be useful for the capture of a particular fraction of
receptor-positive cells out of a heterogeneous cell population (i.e.
positive selection) rather than their depletion from the cell
suspension (i.e. negative selection). The latter can be
accomplished more efficiently by the forementioned affinity column
chromatography.
      There is another method similar to this nylon-stringing method,
namely the affinity "panning" method, a term coined by Wysocki and
Sato (45), in which a plastic petri dish coated with ligand proteins
is used as adsorbent. Petri dishes made of polystyrene are most
frequently utilized in this method (10). Ligand proteins are non-
covalently adsorbed on the dish surface from their buffered solution.
As in the case of affinity chromatography, FCS treatment of the dish
subsequent to ligand adsorption is required to eliminate non-
specific cellular adsorption (45). Plasma-discharged polystyrene
dishes (tissue culture grade) were reported to provoke a more severe
degree of non-specific adsorption of lymphocytes compared with non-
discharge-treated polystyrene dishes (bacteriological grade) (45).
      Because the affinity panning method requires no special
instrument for separation, it has been widely carried out as a
laboratory technique for cell separation, especially for positive
selection of lymphocyte subpopulations. Nevertheless, there are a
number of disadvantages to this method as follows. First, the
quantity of cells applicable to the dish is limited, and thus the
method is not suitable for large scale separations. Secondly, a
relatively long time-period, ca. 60 min, is required to allow all of
the receptor-bearing cells to be stably anchored on the dish.
Thirdly, the operating temperature seriously affects the results of
the separation. Separations done at low temperature for example as
4°C usually give inadequate results because of the weak binding of

cells to the immobilized ligands on the dish. This may relate to the fluidity of membrane receptors. Receptors on cell surfaces seem to diffuse laterally to form a cluster or patch at the interface with immobilized ligands (46,47). This leads to multipoint interactions of cellular receptors with immobilized ligands, leading to higher affinity binding. Indeed, clustering of cellular receptors at the interface with the extracellular matrix was shown to occur for asialoglycoprotein-receptor mediated binding of hepatocytes to immobilized galactose residues on a polyacrylamide gel surface (48). This type of cluster formation may be one of the reasons for another disadvantage of panning, namely that release of "panned" cells from the dish surface usually requires vigorous pipetting and scratching. Lastly, prolonged incubation of cells with immobilized ligands at physiological temperature may provoke undesirable changes in cellular functions. This could be accerelated by clustering of cellular receptors at the interface (46,47).

Development of tert-Amine Derivatized Matrices for Cell Affinity Chromatography

Referring to the literature on cell affinity chromatography which has been published thus far, it could be concluded that there is at present no solid-phase matrix which fulfills both of the following fundamental requirements : (1) elimination of non-specific cellular adsorption, and (2) elimination of the need for any complicated and time-consuming process of covalent-binding of ligand proteins (in other words, stable immobilization of ligand proteins through simple adsorption should be possible). In this section, our approach to the development of new matrices fulfilling both of the above requirements will be described.

Through our systematic studies of cellular interactions with the surface of microdomain structured polymers using a microsphere column method, we found that a group of block and graft copolymers did not provoke the adsorption and subsequent contact-induced activation of platelets (49-54). Some of these copolymers showed excellent *in vivo* non-thrombogenicity (55, Yui, N.; Kataoka, K.; Sakurai, Y.; Aoki, T.; Sanui, K.; Ogata, N. J. Biomater. Sci., submitted). Of further interest is the fact that an albuminated surface of polystyrene/polyamine graft copolymers (SA copolymers) eliminated rat platelet adsorption (56). The structural formula of SA copolymer is shown in Figure 1. Coating of γ-globulin as well as albumin was effective for eliminating platelet adsorption on SA copolymer surface, although no such elimination was observed for a bare SA surface. Another interesting feature of platelet adsorption on SA copolymer surfaces is the elimination of shape change of adsorbed platelets (56,57).

These suppressive effects of SA copolymers on cellular adsorption and shape change were observed for rat lymphocytes as well as platelets (58,59). Figure 2 clearly demonstrates the elimination of adsorption of lymphocytes on albuminated surfaces of SA copolymers containing 9 wt% polyamine branches (SA9). A change in the back-bone structure of the graft copolymer from polystyrene to the more hydrophilic poly(2-hydroxyethyl methacrylate) (PHEMA)

decreased the non-specific binding of lymphocytes to a greater
extent. Also, this PHEMA/polyamine graft copolymer (HA copolymer),
whose structure is illustrated in Figure 3, could suppress the non-
specific adsorption of lymphocytes even in the absence of adsorbed
albumin.

Summarizing the above description, the surface of polyamine
graft copolymers with a definite amount of polyamine branches showed
an extremely small quantity of non-specifically adsorbed lymphocytes.
This advantageous characteristic of polyamine graft copolymers led
us to utilize them as solid-phase matrices for cell affinity
chromatography.

As the separation of lymphocyte subpopulations based on the
expression of immunoglobulin (Ig) molecules on their plasma membrane
surfaces is of considerable practical interest at the present time
(13), we have applied polyamine graft copolymers as solid-phase
matrices for the separation of IgG-positive (IgG$^+$) and -negative
(IgG$^-$) lymphocytes. The solid-phase matrix was prepared by coating
graft copolymers on glass beads of 48 - 60 mesh by solvent
evaporation techniques.

In our study, HA2, PHEMA/polyamine graft copolymer with 2 wt%
polyamine, was used as the main matrix material, because of its
superior feature of eliminating non-specific adsorption of
lymphocytes (Maruyama, A.; Tsuruta, T.; Kataoka, K.; Sakurai, Y. J.
Biomater. Sci., submitted). At a flow rate of 0.4 ml/min, a column
packed with 1g of HA2-coated glass beads (column length: ca. 10cm,
column inner diameter: 3mm) showed non-specific adsorption of
lymphocytes as low as 1% of the number of cells loaded. Selective
binding of IgG$^+$ cells on HA2 columns treated with goat anti-rat IgG
was successfully carried out as shown in Figure 4, demonstrating an
increase in the retention of IgG$^+$ cells with an increase in the
concentration of antibody in the PBS used for the treatment of the
column. This increased retention of IgG$^+$ cells with an increase in
the antibody concentration strongly suggests that selective binding
of IgG$^+$ cells on the HA2 column with adsorbed antibody is based on
the biospecific interaction of adsorbed antibodies (goat anti-rat
IgG) with IgG molecules expressed on the lymphocyte surfaces.

More than 80% of IgG$^+$ cells were retained on the column
pretreated with PBS containing 0.08 mg/ml of antibody, whereas non-
specific adsorption of IgG$^-$ cells was reduced to less than 5%.
Consequently, an IgG$^-$ cell population with more than 90% purity was
obtained as column effluent in approximately 95% yield by using the
HA2 column (3mm ID × 10cm length) pretreated with 0.08 mg/ml
solution of goat anti-rat IgG (Kataoka, K.; Sakurai, Y.; Hanai, T.;
Maruyama, A.; Tsuruta, T., in preparation). As summarized in Table
I, successful separation can be achieved even on an HA2 column
treated with a more dilute solution of antibody (0.04 mg/ml) by
controlling experimental conditions such as flow rate, column length,
and operating temperature as well as by adding bovine serum albumin
to the medium. As previously mentioned, conventional columns using
glass or poly(methyl methacrylate) beads usually have to be treated
with antibody solutions containing 1 mg/ml or more of antibody (18).
The range of antibody concentrations (0.04 - 0.08 mg/ml) in our
procedures is much lower than that generally used for the treatment
of conventional columns. This feature offers the practical

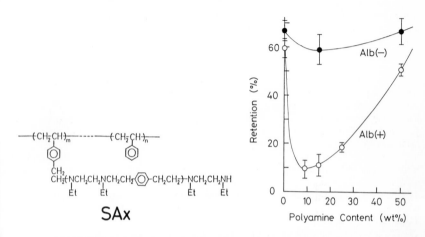

Figure 1 (left). Structural formula of polystyrene/polyamine graft copolymer (SA copolymer). x represents wt% of polyamine portion in the copolymer.

Figure 2 (right). Retention of rat mesenteric lymph node lymphocytes on SA copolymer column. O: Column with albumin coating. ●: column without albumin coating.

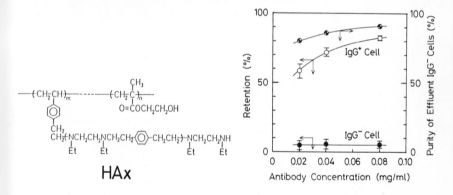

Figure 3 (left). Structural formula of poly(2-hydroxyethyl methacrylate)/polyamine copolymer (HA copolymer). x represents wt% of polyamine portion in the copolymer.

Figure 4 (right). Effect of antibody concentration of selective retention of IgG[+] lymphocytes on HA2 column. Column size: 0.3 cm ID x 10 cm length. Flow rate, 0.2 ml/min. Temperature: 23 °C. Loaded cell number: 1.0×10^{7}.

Table I. Selected Operational Conditions of HA2 Column

Antibody Conc. (mg/ml)	Temp. (°C)	Flow Rate (ml/min)	Column Length (cm)	Conc. of BSA (g/dl)	Effluent IgG⁻ Cell Purity (%)	IgG⁻ Cell Recovery (%)	Time of Separation (min)
0.08	23	0.2	10	0	90.4±1.0	94.6±2.8	7
0.04	23	0.3	20	0	90.1±1.1	94.4±2.6	11
0.04	23	0.2	20	0.1	92.4±1.6	91.7±4.1	14
0.04	4	0.1	20	0	93.1±0.9	95.0±1.1	28

advantage of reducing the quantities of expensive antibodies required. As HA2 itself showed negligible non-specific adsorption of lymphocytes, further treatment of the HA2 columns with FCS-containing medium is not required.

Rapid separation is another advantageous feature of our column system. As also summarized in Table I, separation can be done in times as short as 7 min since there is no need to incubate the lymphocyte suspension in the column. This time period of 7 min is less than one-third of that required for conventional affinity chromatography or for the panning method described in the preceding sections.

As well as IgG$^-$ cells, IgG$^+$ cells were readily recovered from the column by gentle pipetting of the column matrices. The eluted population contained approximately twice the fraction of IgG$^+$ cells as the initial population (Table II). It is to be noted that recovery of IgG$^+$ cells was more than 80%.

Table II. Recovery of IgG-positive Lymphocytes from Column

Expriment No.	Purity of Recovered IgG$^+$ Cell Population[a] (%)	Recovery of IgG$^+$ Cell[b] (%)
1[c]	56.0±2.6	79.4±3.5
2[d]	64.5±3.6	87.9±3.3

a) Initial fraction of IgG$^+$ cell: 30.7±2.4 %.
b) % of loaded number of IgG$^+$ cells into the column.
c) Flow rate: 0.2 ml/min, temperature: 23 °C.
d) Flow rate: 0.1 ml/min, temperature: 4 °C.

As briefly reviewed in this paper, affinity selection of cells using proteinaceous ligands immobilized on solid-phase matrices is a subject which requires detailed understanding of the nature of proteins at interface as well as the mode of their interaction with living cells. In this sense, this is surely a promising frontier which presents a strong challenge to researchers in the biomaterials field.

Acknowledgments

The authors express their thanks to Mr. A. Maruyama and Mr. T. Hanai for their intensive contribution to our work cited in this paper.

Literature Cited

1. Biomedical Applications of Immobilized Enzymes and Proteins; Chang, T.M.S., Ed.; Plenum: New York, 1977; Vols. 1&2.

2. Biomaterials Science; Tsuruta, T.; Sakurai, Y., Eds.; Nanko-do:
    Tokyo, 1982; Vols. 1&2.
3. Turková, J. Affinity Chromatography; Elsevier: Amsterdam, 1978.
4. Scouten, W.H. Affinity Chromatography, Bioselective Adsorption
    on Inert Matrices; John Wiley: New York, 1981.
5. Sharma, S.K.; Mahendroo, P.P. J. Chromatogr. 1980, 184, 471-
    499.
6. Basch, R.S.; Berman, J.W.; Lakow, E. J. Immunol. Methods 1983,
    56, 269-280.
7. Bonnafous, J.C.; Bornans, J.; Favero, J.; Mani, J.-C. In
    Affinity Chromatography, A Practical Approach; Dean, P.D.G.;
    Johnson, W.S.; Middle, F.A., Eds.; IRL Press: Oxford, 1985;
    p 191-206.
8. Böyum, A. Scan. J. Clin. Lab. Invest. 1968, 21, Suppl. 97,
    9-106.
9. Shortman, K. In Methods of Cell Separation; Catsimpoolas, N.
    Ed.; Plenum: New York, 1977; Vol. 1, P 229-249.
10. Fong, S. In Cell Separation: Methods and Selected Applications;
    Pretlow II, T.G.; Pretlow, T.P. Eds.; Academic: New York, 1983;
    Vol. 2, p 203-219.
11. Sharon, N. In Cell Separation: Methods and Selected
    Applications; Pretlow II, T.G.; Pretlow, T.P. Eds.; Academic:
    Orlando, 1984; Vol. 3, p 13-52.
12. Molday, R.S. ibid.; p 237-263.
13. Methods in Enzymology, Immunochemical Techniques Part G,
    Separation and Characterization of Lymphoid Cells; Di Sabato,
    G.; Langone, J.J.; Vunakis, H.V. Eds.; Academic: Orlando, 1984;
    Vol. 108.
14. Wigzell, H.; Anderson, B. J. Exp. Med. 1969, 129, 23-36.
15. Evans, W.H.; Mage, M.G.; Peterson, E.A. J. Immunol. 1969, 102,
    899-907.
16. Langone, J.J. J. Immunol. Methods 1982, 55, 277-296.
17. Cuatrecasas, P.; Wilchek, M.; Anfinsen, C.B. Proc. Natl. Acad.
    Sci., USA 1968, 61, 636-643.
18. Wigzell, H. Scand. J. Immunol., 1976, 5, Suppl. 5, 23-31.
19. Schlossman, S.F.; Hudson, L. J. Immunol. 1973, 110, 313-315.
20. Thomas, D.B.; Phillips, B. Eur. J. Immunol. 1973, 3, 740-742.
21. Chess, L.: MacDermott, R.P.; Schlossman, S.F. J. Immunol. 1974,
    113, 1113-1121.
22. Hellström, U.; Diller, M.-L.; Hammarström, S.; Perlmann, P.
    J. Exp. Med. 1976, 144, 1381-1385.
23. Irlé, C.; Pignet, P.-F.; Vassalli, P. ibid. 1978, 148, 32-45.
24. Nicola, N.A.; Burgess, A.W.; Metcalf, D.; Battye, F.L. Austral.
    J. Exp. Biol. Med. Sci. 1978, 56, 663-679.
25. Haller, O.; Gidlund, M.; Hellström, U.; Hammarström, S.;
    Wigzell, H. Eur. J. Immunol. 1978, 8, 765-771.
26. Warr, G.W.; Lee, J.C.; Marchalonis, J.J. J. Immunol. 1978, 121,
    1767-1772.
27. Ghetie, V.; Mota, G.; Sjöquist, J. J. Immunol. Methods 1978,
    21, 133-141.
28. Manderino, G.L.; Gooch, G.T.; Stavisky, A.B. Cell. Immunol.
    1978, 41, 264-275.
29. Carton, J.P.; Nurden, A.T. Nature 1979, 282, 621-623.

30. Marshak-Rothstein, A.; Fink, P.; Gridley, T.; Raulet, D.; Bevan, M.J.; Gefter, J.L. *J. Immunol.* 1979, 122, 2491-2497.

31. Schremph-Decker, G.E.; Baron, D.; Wernet, P. *J. Immunol. Methods* 1980, 32, 285-296.

32. Duffey, P.S.; Drouillard, D.L.; Barbe, C.P. *ibid.* 1981, 45, 137-151.

33. Ghetie, V.; Sjöquist, J. In *Methods in Enzymology, Immunochemical Techniques Part G, Separation and Characterization of Lymphoid Cells*; DiSabato, G.; Langone, J.J.; Vunakis, H.V. Eds.; Academic: Orlando, 1984; Vol. 108, p 132-138.

34. Hubbard, R.A.; Schluter, S.F.; Marchalonis, J.J. *ibid.*; p 139-148.

35. Hellström, U.; Hammarström, M.-L.; Hammarström, S.; Perlmann, P. *ibid.*; p 153-168.

36. Truffa-Bachi, P.; Wofsy, L. *Proc. Natl. Acad. Sci., USA* 1970, 66, 685-692.

37. Wofsy, L.; Kimura, J.; Truffa-Bachi, P. *J. Immunol.* 1971, 107, 725-729.

38. Baran, M.M.; Allen, D.M.; Russell, S.R.; Scheetz, II, M.E.; Monthony, J.F. *J. Immunol. Methods* 1982, 53, 321-334.

39. Braun, R.; Teute, H.; Kirchner, H.; Munk, K. *ibid.* 1982, 54, 251-258.

40 Tlaskalová-Hogenová, H.; Čoupek, J.; Pospíšil, M.; Tučková, L.; Kamínková, J.; Mančal, P. *J. Polym. Sci.; Polym. Sympo.* 1980, 68, 89-95.

41. Bonnafous, J.-C.; Dornand, J.; Favero, J.; Sizes, M.; Boschetti, E.; Mani, J.-C. *J. Immunol. Methods* 1983, 58, 93-107.

42. Heinzel, W.; Rahimi-Laridjani, I.; Grimminger, H. *ibid.* 1976, 9, 337-344.

43. Edelman, G.M.; Rutishauser, U.; Millette, C.F. *Proc. Natl. Acad. Sci., USA* 1971, 68, 2153-2157.

44. Rutishauser, U.S.; Edelman, G.M. In *Methods of Cell Separation*; Catsimpoolas, N. Ed.; Plenum: New York, 1977; Vol. 1, 193-228.

45. Wysocki, L.J.; Sato, V.L. *Proc. Natl. Acad. Sci., USA* 1978, 75, 2844-2848.

46. Kataoka, K. In *Biomaterials Science*; Tsuruta, T.; Sakurai, Y. Eds.; Nanko-do: Tokyo, 1982; Vol. 1, p 93-109.

47. Kataoka, K.; Sakurai, Y.; Tsuruta, T. *Makromol. Chem., Suppl.* 1985, 9, 53-67.

48. Weigel, P.H. *J. Cell Biol.* 1980, 87, 855-861.

49. Okano, T.; Nishiyama, S.; Shinohara, I.; Akaike, T.; Sakurai, Y.; Kataoka, K.; Tsuruta, T. *J. Biomed. Mater. Res.* 1981, 15, 393-402.

50. Okano, T.; Kataoka, K.; Abe, K.; Sakurai, Y.; Shimada, M.; Shinohara, I. In *Progress in Artificial Organs - 1983*; Atsumi, K.; Maekawa, M.; Ota, K. Eds.; ISAO Press: Cleveland, 1984; Vol. 2, p 863-866.

51. Yui, N.; Oomiyama, T.; Sanui, K.; Ogata, N.; Kataoka, K.; Okano, T.; Sakurai, Y. *Makromol. Chem., Rapid Commun.* 1984, 5, 805-809.

52. Yui, N.; Kataoka, K.; Sakurai, Y.; Sanui, K.; Ogata, N.; Takahara, A.; Kajiyama, T. *Makromol. Chem.* 1986, 187, 943-953.

53. Yui, N.; Kataoka, K.; Sakurai, Y. In *Artificial Heart*; Akutsu,

T.; Koyanagi, H.; Pennington, D.G.; Poirier, V.L.; Takatani, S.; Kataoka, K. Eds.; Springer: Tokyo, 1986; Vol. 1, p 23-30.

54. Akemi, H.; Aoyagi, T.; Shinohara, I.; Okano, T.; Kataoka, K.; Sakurai, Y. Makromol. Chem. 1986, 187, 1627-1638 (1986).

55. Okano, T.; Aoyagi, T.; Kataoka, K.; Abe,K.; Sakurai, Y.; Shimada, M.; Shinohara, I. J. Biomed. Mater. Res. 1986, 20, 919-927.

56. Kataoka, K.; Okano, T.; Sakurai, Y.; Nishimura, T.; Maeda, M.; Inoue, S.; Shimada, M.; Shinohara, I.; Akaike, T.; Tsuruta, T. Artificial Organs 1981, 5, Suppl., 532-539.

57. Kataoka, K.; Okano, T.; Sakurai, Y.; Nishimura, T.; Maeda, M.; Inoue, S.; Tsuruta, T. Biomaterials 1982, 3, 237-240.

58. Kataoka, K.; Okano, T.; Sakurai, Y.; Nihsimura, T.; Maeda, M.; Inoue, S.; Watanabe, T.; Tsuruta, T. Makromol. Chem., Rapid Commun. 1982, 3, 275-279.

59. Kataoka, K. In Polymers as Biomaterials; Shalaby, S.W.; Hoffman, A.S.; Ratner, B.D.; Horbett, T.A. Eds.; Plenum; New York, 1984; p 225-239.

RECEIVED February 18, 1987

Chapter 38

Fibronectin-Mediated Attachment of Mammalian Cells to Polymeric Substrata

Robert J. Klebe, Kevin L. Bentley, and Danelle P. Hanson

Department of Cellular and Structural Biology, University of Texas Health Science Center, San Antonio, TX 78284

Fibronectin and related cell adhesion proteins have been shown to mediate the attachment of many mammalian cells to natural and artificial substrata. Following a presentation of our current understanding of the molecular biology of cell adhesion, factors influencing the attachment and growth of cells on polymeric substrata are discussed.

In a study of 52 diverse materials, we found that 48 chemically dissimilar substrata bind fibronectin and, thereby, permit the attachment and growth of cells. Our results indicate that the ability of a polymer to bind fibronectin and other extracellular matrix proteins is an important determinant of the biological properties of the polymer. Poly(hydroxyethyl-methacrylate) was found to be unique in that it does not bind fibronectin and, thereby, does not support cell adhesion.

Various surface treatments of polymers are shown to alternatively inhibit or enhance the binding of fibronectin to polymeric substrata. We have also found that oxidation of the surface of "bacteriological" poly(stryrene) petri plates converts such surfaces into "tissue culture dishes".

By means of a new technique, termed cytoscribing, fibronectin can be deposited in precise patterns on substrata such that two dimensional tissues can be formed.

The objective of this paper is to review our knowledge of the biochemical events which lead to cell adhesion both _in vivo_ and to polymeric substrata _in vitro_. Prior to presenting studies involving cell adhesion, a brief review of our current knowledge of the molecules involved in fibronectin-mediated cell adhesion will be

presented. It should be noted that in addition to the now familiar
fibronectin system (1-3), progress over the last several years has
revealed a series of cell adhesion proteins which, while similar to
fibronectin in some respects, display greater cell type specificity;
i.e., laminin (4), vitronectin (5), chondronectin (6), epinectin
(7), uvomorulin (8), n-CAM (9), and l-CAM (9). As in the past, the
detailed knowledge which has been generated concerning fibronectin
should continue to facilitate studies of other cell adhesion
proteins.

The Molecular Basis of Cell Adhesion.

Over the last few years, important progress has been made in
understanding the structure and function of fibronectin and other
cell adhesion proteins (see ref. 1-3 for reviews). A model for the
molecular interactions which govern cell adhesion is presented in
Figure 1. This model will be discussed below.

Cell Surface Receptor Proteins.

Several groups have now been able to purify to homogeneity the cell
surface receptor proteins for laminin (10,11), fibronectin (12), and
vitronectin (13). The laminin, fibronectin, and vitronectin
receptors are integral membrane proteins with reduced subunit
molecular weights of 0.7 kD, 1.4 kD, and 1.25/1.15 kD, respectively.
While the fibronectin and vitronectin cell surface receptors both
bind a unique tetrapeptide (Arg-Gly-Asp-X) present in the cell
binding domains of fibronectin and vitronectin (12,13), both
receptors display specificity for their target proteins (13). It
has been shown that insertion of the vitronectin receptor into a
reconstituted lipid membrane results in an adhesive response to
vitronectin, but not fibronectin (13). The isolation of the
receptors for cell adhesion proteins should make it possible to
answer many basic questions about the biological function of cell
adhesion proteins. Biochemical approaches are now available to
determine the mechanism which results in the loss of fibronectin
following cell transformation (2) and the molecular basis of
fibronectin mediated transmembrane signalling which results in the
reorganization of actin stress fibers (14).

Role of Collagen, Gangliosides and Heparin in Cell Adhesion.

In addition to the recently isolated cell surface receptor proteins
for cell adhesion proteins (10-13), it has become clear that most
cell adhesion molecules also recognize other cell surface molecules.
Most cell adhesion proteins possess discrete binding sites for
collagen, heparin, and ganglioside (or other glycolipids). The
avidity of most cell adhesion proteins for one or more species of
collagen and the involvement of this interaction in cell adhesion
has been reviewed (1) and we will treat this area more thoroughly in
a subsequent section.

  Several lines of evidence indicate that gangliosides play an
important role in the interaction of fibronectin with cells.
Following the initial observation of Kleinman et al. (15) indicating
that the oligosaccharide moiety of gangliosides competitively

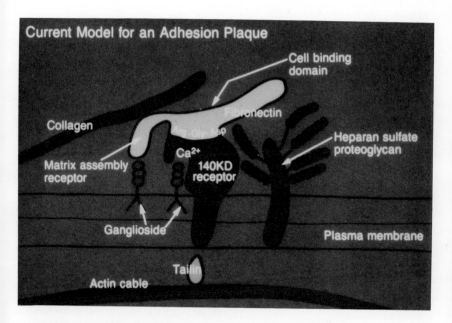

Figure 1.   Current model for fibronectin mediated cell
attachment. The molecules currently known to be involved in
fibronectin-mediated   cell   adhesion   are   presented.
Gangliosides are involved in two steps; namely, (a) the
initial interaction of fibronectin with the cell surface
involves a ganglioside binding site of the amino terminal
domain of fibronectin and (b) gangliosides and divalent
cations are required for the 140 kD cell surface receptor
protein to recognize the Arg-Gly-Asp sequence in the cell
binding domain of fibronectin. While interaction of the
fibronectin 140 kD receptor with the Arg-Gly-Asp sequence of
fibronectin results in the anchorage of cell to substratum,
cell spreading does not occur without the intercession of
either a heparin binding site of fibronectin or a heparin
binding protein. The heparin interaction leads to a
transmembrane signal that results in the organization of
actin cables and, thereby, cell spreading.

inhibits fibronectin mediated cell adhesion, several other studies have appeared which confirm and extend this finding in a variety of cell lines (16,17). Specificity in the interaction of gangliosides with fibronectin was demonstrated by the finding that the more highly sialated gangliosides (GT_{1b} and GD_{1a}) were more effective than GM_1 (18). The effects of gangliosides on cell adhesion could also be shown to extend to other fibronectin mediated phenomena; namely, cell spreading (16) and restoration of a normal morphology to transformed cells (16). Again, the more complex species of gangliosides proved to be the most effective in altering cell spreading and cell morphology (16). In addition to the effects of gangliosides on cell adhesion noted above, it has been demonstrated that gangliosides co-distribute with fibronectin on the cell surface. Development of monoclonal antibodies with recognize gangliosides GD_2 and GD_3 has permitted the localization of gangliosides to the adhesion plaques of cells (18). Via fluorescence microscopy, it has been shown that gangliosides integrate randomly into the cell membrane initially and then localize in regions rich in fibronectin (17).

Several more direct lines of evidence now clearly indicate a role of gangliosides in cell adhesion. First it has been demonstrated (a) that a mutant cell line which lacks gangliosides is also defective in its adhesive properties (17,19) and (b) that purified gangliosides can restore normal cell adhesion to the above mutant cell line (20). That purified gangliosides potentiate the interaction of fibronectin with a ganglioside deficient cell line provides clear evidence that gangliosides are important in fibronectin mediated functions. Second, it recently has been shown that the 140 kD fibronectin receptor co-purifies with gangliosides and requires gangliosides and divalent cations for binding to the Arg-Gly-Asp-X cell binding domain peptide (21). Third, we have recently developed a fluorescence polarization assay which clearly demonstrates that fibronectin itself binds gangliosides and, using this assay, we have localized the ganglioside binding site of fibronectin to the amino terminus of the molecule (22). Since the

binding site for the cell surface fibronectin receptor protein (12) is located near the middle of the molecule (23), the fibronectin receptors for gangliosides and the cell surface receptor protein are distinct entities. Recently, McKeown-Longo and Mosher (24) have demonstrated the existence of a "matrix assembly receptor" in the domain that we have shown contains the ganglioside binding site (22). Since the "matrix assembly receptor" is required for integration of fibronectin into the extracellular matrix (24), the ganglioside binding site of fibronectin may have a related, or possibly identical, function. Thus, several independent lines of evidence indicate that gangliosides are important in fibronectin biological activity (15-24).

Heparin Related Molecules as Fibronectin Receptors.

In addition to gangliosides, heparin related molecules have also been shown to be involved in fibronectin mediated cell adhesion. Heparin binding sites have been described on many cell adhesion proteins; e.g., fibronectin (25-26), laminin (27), and vitronectin

(28). The ubiquitous heparin binding sites of cell adhesion proteins have been found to bind to both heparin and heparan sulfate proteoglycan (29), the most abundant proteoglycan on the surface of many cells. It has been found that the heparin binding site of fibronectin is involved in cell adhesion (14,30,31), the related phenomena of cell migration (31), and the determination of cellular morphology (31,32). While the fibronectin cell surface binding domain (which lacks a heparin site) can attach cells to a substrate, it has recently been shown that under such conditions cells do not re-organize their actin stress fibers (14). In contrast, cells bound to a substratum containing the cell binding domain of fibronectin plus a heparin binding protein, partially reorganize their stress fibers (14). Thus, it has been suggested that the heparin binding site of fibronectin may be involved in the transmembrane signalling event which occurs during fibronectin mediated cell adhesion (14). We have recently found that the ability of heparin to bind to fibronectin depends on the presence of a specific sub-population of heparin molecules (25).

The discussion above indicates that at least four cell surface molecules are involved in fibronectin mediated cell adhesion; namely, (a) the recently isolated cell surface receptor protein (12), (b) heparan sulfate proteoglycan (14,31), (c) collagen(s) (1) and (d) gangliosides (15-24). This fact suggests that co-operative interactions between several receptor sites on the fibronectin molecule may be involved in fibronectin mediated cell adhesion (Figure 1). There are numerous other examples of co-operative interactions between fibronectin binding sites which have been established from biochemical studies. For example, heparin has been shown to be involved in both the rate and strength of the binding of fibronectin to collagen (33). Thus, the involvement of at least four of the 14 (or more) binding sites of fibronectin in the cell adhesive activity of the molecule is the rule rather than the exception.

Cell Biology of Fibronectin-Mediated Cell Adhesion

As indicated above, most, if not all, the molecules required for cell adhesion have now been identified; nevertheless, many important features of the echanism of cell adhesion remain at the phenomenological level. In the following section, a brief review of cell biological studies of fibronectin-mediated cell adhesion will be presented.

Due to the well-known observation that EDTA can dissociate cells and tissues, it has been recognized for some time that divalent cations are required for the adhesion of many cell types. During the adhesion of fibroblasts to fibronectin substrata, it has been demonstrated that (a) several divalent cations are active and (b) that fibronectin displays the following order of preference for divalent cations:

$$Mn^{2+}>Mg^{2+}>Co^{2+}>Ca^{2+}>Zn^{2+}>Ni^{2+}$$

with Ba^{2+} and Sr^{2+} being inactive (34). A mutant cell line with an altered adhesive response to divalent cations has been described (35). The recent observation that Ca^{2+} is required for the activation of the 140 kD fibronectin receptor (21) suggests a possible role for divalent cations in the biochemical mechanism of cell adhesion. Study of other ionic requirements for cell adhesion indicates that cell adhesion occurs optimally at neutral pH and under isotonic conditions (34). Inhibition of cell adhesion noted at high pH can be ascribed to the insolubility of both Ca^{2+} and Mg^{2+} at high pH.

It has been established for some time that cells require metabolic energy to both initiate an adhesive response to fibronectin as well as maintain their adhesion to fibronectin (36). Conditions which lower cellular ATP levels (metabolic inhibitors and low temperature) have been shown to both inhibit cell attachment and result in the rounding and detachment of cells from fibronectin coated substrata (36). It is well known that metabolically dead cells do not attach to fibronectin or other adhesive substrata; the above observation indicates that cell adhesion is not as simple as an antigen-antibody reaction and also provides additional support for the observation that cellular metabolic energy is required for cell adhesion (36).

The observation that interaction of fibronectin with cell surface molecules results in a transmembrane signal which leads to the organization of cytoplasmic actin cables (14) suggests a role for cellular metabolic energy in the mechanism of cell adhesion. Transmembrane signalling has been shown to require the cell binding domain of fibronectin plus one or more of the heparin binding sites of fibronectin (14). The mechanism by which fibronectin generates a transmembrane signal which affects actin organization is currently known.

Due to the established cell adhesive activity of fibronectin, it is not surprising that fibronectin also is involved in the control of cell motility and cell shape. Via the phagokinetic track method, it has been shown that the motility of cells increases with added fibronectin (31) and we have recently shown that heparin related molecules augment the action of fibronectin in this system (31). In addition to the stimulatory activity of fibronectin on the motility of several cell types _in vitro_, fibronectin has also been shown to be involved in the migration of neural crest cells _in ovo_ (37).

One of the earliest observations in the cell adhesion protein area was the loss of fibronectin from the surface of most malignant (but not necessarily transformed) cells (2). The involvement of fibronectin in the malignant spread of tumor cells has recently been demonstrated by the finding that Arg-Gly-Asp cell binding domain sequences can inhibit invasion of the lungs of mice injected with melanoma cells by over 90% (38).

Cell Adhesion to Natural and Artificial Substrata

Electron microscopy of cells cultured on plastic substrata reveals that the plasma membrane does not make direct adhesive contact with plastic but rather is separated from the plastic surface by a layer of proteinaceous material of about 100 angstrom thickness (39). Due to the presence in serum of fibronectin, vitronectin, laminin (1-3) and possibly other cell adhesion proteins, cells cultured in serum containing medium adhere to their substrata via one or more cell adhesion proteins.   Adhesion of tissues to implantable devises probably also depends on cell adhesion proteins.   In this section, we will briefly review our knowledge of the surface properties of natural and artificial substrata which control the activity of cell adhesion proteins.

While it has been shown that the cell surface receptor proteins for several cell adhesion proteins bind to an Arg-Gly-Asp cell binding domain sequence found in many adhesive proteins (12,13), it has been known for some time that appropriate surfaces greatly facilitate the interaction of fibronectin with the cell surface (1,40-43).   Thus, while fibronectin can bind directly to the cell surface (24), collagen increases the bindability of fibronectin to the cell surface by a factor of about six (44,45).   At present, the events which lead to binding of fibronectin are not entirely resolved; however, it is clear that more than recognition of an Arg-Gly-Asp sequence in fibronectin by the fibronectin receptor is required for binding of fibronectin to the cell surface.   For example, it has been shown by McKeown-Longo and Mosher (24) that an amino terminal fibronectin peptide derived from the amino terminal of the molecule can block fibronectin incorporation into the extracellular matrix of cells.   The "matrix assembly receptor" identified by McKeown-Longo and Mosher (24) does not contain the Arg-Gly-Asp cell binding domain of fibronectin and is currently thought to depend on a ganglioside binding site that we (22) and others (46) have demonstrated at the amino terminus of fibronectin. Binding of fibronectin to the cell surface involves, in addition to binding to the 140 kD fibronectin receptor, interaction with gangliosides (21,22,46), heparin related molecules (14), and collagen or a suitable substrate (40,44,45).   Below, we will briefly review our knowledge of the surface activation of fibronectin by natural and artificial substrates.

Collagen: The Natural Substratum for Fibronectin and Other Cell Adhesion Proteins.

While it is clear that fibronectin will bind to many man-made polymers, it is also clear that collagen is the natural substrate material for fibronectin and other cell adhesion proteins.   Collagen and fibronectin are the most abundant proteins in the extracellular matrix of cells; e.g., collagen represents 5-20% of the total protein in most tissues (47) while fibronectin makes up 30% of the protein on the cell surface (48).   The abundance and wide tissue distribution of collagen makes collagen an ideal substrate material for fibronectin.   The specificity of the interaction of fibronectin with collagen has been demonstrated by the finding that fibronectin possesses a receptor for collagen which recognizes a unique sequence in collagen I (49).   It is interesting that the collagen sequence recognized by fibronectin also contains the sole site on collagen cleaved by mammalian collagenase (49).   A sequence similar to that present in collagen I is also recognized in collagens II and III by fibronectin (49).   Laminin has been found to bind preferentially to collagen IV (50).

The specific interaction of fibronectin with collagen has functional significance.   While it has been demonstrated that fibronectin can bind to the cell surface without the addition of exogenous collagen, it has been found that collagen greatly enhances the amount of fibronectin bound to the cell surface (44,45). (Due to the fact that most cultured cells synthesize collagen (51), it is difficult to exclude the possibility that collagen is required for fibronectin binding to the cell surface). Collagen binding to fibronectin may result in the exposure of a cryptic Arg-Gly-Asp cell binding domain peptide of fibronectin or increase the binding constant of fibronectin for its cell surface receptors. With the isolation of the molecules involved in fibronectin-mediated cell adhesion, it should be possible to test the possibilities above in the near future.

While collagen is clearly the extracellular matrix material to which fibronectin binds in vivo, it has also been established that a wide variety of man-made polymers will promote fibronectin binding to the cell surface (41). The surface activation of fibronectin by artificial substrates may be similar to the well characterized surface activation of blood clotting factor XII, the Hageman factor (52,53). It has been shown that binding of the Hageman factor to glass results in a several hundred fold increase in the proteolytic conversion of the Hageman factor from a proenzyme into an active enzyme (52,53). The surface activation of fibronectin by many polymers may have a similar mechanism.

Binding of Fibronectin to Man-Made Polymers.

For many years, a wide variety of cells have been cultured on plastics; however, only recently have we begun to understand the molecular events that permit plastic substrates to promote the attachment and growth of cells. It has been known for many years that a suitable substratum is an absolute requirement for the growth of normal, but not transformed cells (the phenomenon of anchorage dependence (54)). The anchorage independence of transformed cells has become the basis of one of the most reliable assays for cell transformation; i.e., the agar cloning assay of Montager and McPherson (55).

The table presents materials which either bind or do not bind fibronectin as well as the ability of these materials to support cell attachment and growth. For a more complete treatment, see reference 41.

We have investigated the ability of 52 polymers, ceramics, metals, and waxes to support the attachment and growth of cultured cells (Table I and reference 41). Of the 52 chemically dissimilar materials studied, 48 proved to support the attachment and growth of cells. With the aid of a monoclonal antibody based ELISA procedure that detects fibronectin, we were also able to demonstrate that the ability of a surface to bind fibronectin distinguished the materials capable of supporting cell attachment and growth from inactive materials (41). Thus, the ability of a surface to bind fibronectin determines its anchorage and growth promoting properties.

Of the 52 materials we investigated, poly(hydroxyethyl-methacrylate) (poly (HEMA) proved to be unique (41). Poly(HEMA) was the only solid material that did not support fibronectin binding (and, thus, also did not permit cell attachment and growth). It is interesting that the addition of one part per million of an adhesive polymer to poly(HEMA) permits fibronectin binding to occur (41).

Table I.   Ability of Materials to Support Fibronectin
Binding, Cell Adhesion, and Cell Growth

Substrate Material	Cell Attachment	Fibronectin Binding	Cell Growth
A. Fibronectin Binding Materials			
Lux Culture Dish	100%	100%	100%
Collagen	90	174	108
Poly(styrene)	100	100	100
Poly(methylmethacrylate)	56	66	94
Poly(vinyl chloride)	100	36	91
Poly(vinyl acetate)	70	95	94
Teflon	60	-	74
Ethyl cellulose	60	55	88
Pyrex glass	79	100	92
Carnauba wax	83	71	109
Silicone grease	64	27	91
Aluminum metal	88	94	81
B. Materials Which Do Not Bind Fibronectin			
Bovine Serum Albumin	13	0	95
Agar	0	0	0
Poly(acrylamide)	20	0	-
Poly(hydroxyethylmethacrylate)	1	0	13

Thus, impurities on treated substrates can alter fibronectin binding
and cell adhesion to poly(HEMA) (41).   The surface distribution of
adhesive polymers in poly(HEMA) was not tested.
    While treatment of poly(styrene) and 48 diverse materials with
purified fibronectin renders such materials adhesive (41), treatment
of poly(styrene) with fibronectin in the presence of other serum
proteins is considerably less effective (42,43,56).   Again via ELISA
assay for fibronectin binding, it can be shown that less fibronectin
binds to plastic in the presence of other serum proteins (42).   This
observation is expected since it can be shown that bovine serum
albumin and other purified proteins compete with fibronectin for
non-specific binding sites on plastic.
    While both bacteriological petri plates and "tissue culture
dishes" are made of poly(stryrene)(57-59), the fibronectin binding
properties of these materials are quite different (42).   ELISA
assays for fibronectin binding indicate that "tissue culture dishes"
bind more fibronectin than bacteriological petri plates (42).   Thus,
"tissue culture dishes" bind more cells and promote better growth of
cells than bacteriological petri plates.   Since both types of
plastic petri plate are made of poly(styrene), we tested the
hypothesis that a proprietary surface treatment was involved.   By
dissolving "tissue culture dishes" in benzene and recasting the
polymer, we could show that the recast polymer now had the
fibronectin binding properties of bacteriological plastics (42).
While the initial attachment of cells to serum treated
bacteriological plastics is comparable to tissue culture plastics,
cell attachment decreases markedly at time points beyond 48 hours on
bacteriological plastics (Figure 2).   By treating bacteriological

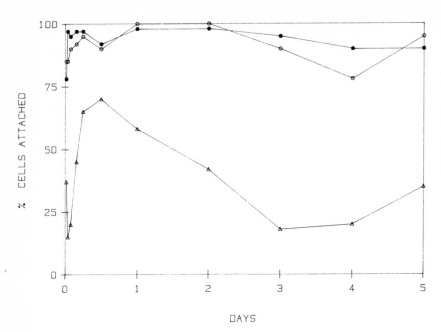

Figure 2. Cell attachment to bacteriological and tissue culture plastics. The attachment of CHO cells to bacteriological (opened triangles), tissue culture (opened circles), and H_2SO_4-oxidized substrata (closed circles) is presented over a 5 day period. Note that cell attachment decreases with time on bacteriological plates treated with serum. The detachment of cells with time observed on bacteriological plates explains the poor growth of cells observed on bacteriological plastics.

plastics with sulfuric acid, we could also show that bacteriological petri plates gained the fibronectin binding properties of "tissue culture dishes". Thus, it is possible to interconvert the properties of bacteriological petri plates and "tissue culture dishes" by different surface treatments (42). Curtis and co-workers have shown that surfuric acid and other oxidizing agents introduce several functional groups into poly(styrene) and that introduced hydroxyl groups are responsible for the alteration in biological properties noted (43).

While the binding of fibronectin appears to be a critical feature in the response of cells to many man-made materials, recent studies have shown that treatment of surfaces with other extracellular matrix components, such as laminin and type IV collagen (60), can greatly alter the biological properties of plastics. Treatment of plastic with a biomatrix derived from a basement membrane tumor can alter the morphology and differentiated state of several cell types (60). Hence, the biological properties of polymeric materials are dependent on the type of extracellular atrix molecule absorbed more than the chemical nature of the polymer itself.

Cytoscribing.

Recently, we have used the fibronectin binding properties of
materials and computer graphics to synthesize two-dimensional
tissues via a process we term cytoscribing (61). By adding
fibronectin to the cartridge of an ink jet printer or loading the
pen of a plotter with fibronectin, we can position fibronectin in
any desired pattern on a substratum (Figure 3). Following BSA
treatment which inactivates areas of the substratum untreated with
fibronectin, cells will attach only to areas which have been
cytoscribed with fibronectin (Figure 3)(61). Hence, by utilizing
differences in the fibronectin binding properties of materials, it
is possible to construct surfaces with different adhesive patterns
for different cell types. Thus, it is possible to construct two
dimensional tissues.

In addition to direct application of fibronectin to predefined
areas of a substratum, surfaces with differential adhesivity can be
constructed by photoengraving technology (Figure 3)(61). For
example, the photoengraving of a non-adhesive agar surface with a
photopolymerizable polymer results in the generation of regions of a
substratum which can bind fibronectin and, hence, can support cell
attachment (Figure 3). Due to the high precision of photoengraving
methods, it should be possible to utilize photoengraving as a
valuable tool in micropositioning cells on substrata (61).

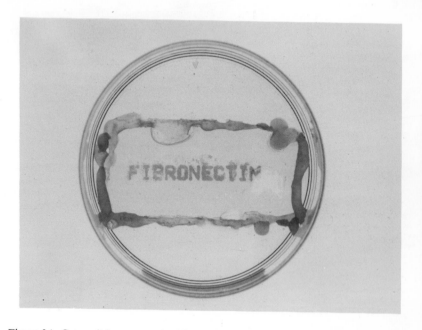

Figure 3A. Cytoscribing as a method for high-precision micropositioning of cell
populations. A plastic surface was cytoscribed with fibronectin by application of the
cell adhesion protein by an ink jet printer. Cells attach to those areas of the plastic to
which fibronectin had been applied.

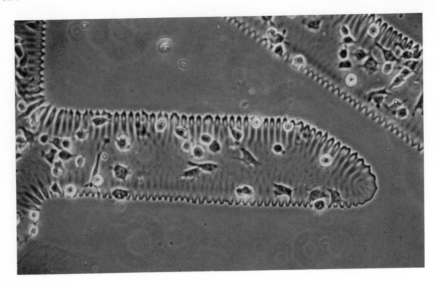

Figure 3B. Cytoscribing as a method for high-precision micropositioning of cell populations. Cytoscribing was carried out by photopolymerizing a polymer over a nonadhesive agar substratum. Cells attach selectively to the photoengraved region of the substratum.

Literature Cited

1.   Kleinman, H.; Klebe, R.; Martin, G. J. Cell Biol. 1981, 88, 473-485.
2.   Yamada, K.M. Ann. Rev. Biochem. 1983, 52, 761-779.
3.   Hynes, R. Ann. Rev. Cell Biol. 1985, 1, 67-90.
4.   Liotta, L.A.; P.H. Hand; C.N. Rao; G. Bryant; S.H. Barsky; and J. Schlom. Exp. Cell Res. 1984, 140, 315-322.
5.   Suzuki, S.; Oldberg, A.; Hayman, E.G.; Pierschbacher, M.D.; and Ruoslahti, E. EMBO Journal 1986, 4, 2519-2524.
6.   Hewitt, A.T.; Varner, H.H.; Silver, M.H.; Dessau, W.; Wilkes, C.M.; Martin, G.R. J. Biol. Chem. 1982, 257, 2330-2334.
7.   Enenstein, J.; Furcht, L.T. J. Cell Biol. 1984, 99, 464-470.
8.   Boller, K.; Vestweber, D.; Kemler, R. J. Cell Biol. 100, 327-332.
9.   Edelman, G.M. Proc. Natl. Acad. Sci. USA 1984, 81, 1460-1464.
10. Rao, N.C.; Barsky, S.H.; Terranova, V.P.; Liotta, L.A. Biochem. Biophys. Res. Comm. 1983, 111, 804-808.
11. Brown, S.S.; Malinoff, H.L.; Wicha, M.S. Proc. Natl. Acad. Sci. USA 1983, 80, 5927-5930.
12. Pytela, R.; Pierschbacher, M.D.; Ruoslahti, E. Cell 1985, 40, 191-198.
13. Pytela, R.; Pierschbacher, M.D.; Ruoslahti, E. Proc. Natl. Acad. Sci. USA 1985, 82, 5766-5770.
14. Lark, M.W.; Laterra, J.; Culp, L.A. Fed. Proc. 1985, 44, 394-403.
15. Kleinman, H.K.; Martin, G.R.; Fishman, P.H. Proc. Natl. Acad. Sci. USA 76, 3367-3371.

16.   Yamada, K.M.; Kennedy, D.W.; Grotendorst, G.R.; Momoi, T.   J. Cell. Physiol. 1981, 109, 343-351.

17.   Spiegel, S.; Yamada, K.M.; Hom, B.E.; Moss, J.; Fishman, P.H. J. Cell Biol. 1985, 100, 721-726.

18.   Cheresh, D.A.; Harper, J.R.; Schulz, G.; Reisfeld, R.A.   Proc. Natl. Acad. Sci. USA 1984, 81, 5767-5771.

19.   Spiegel, S.; Yamada, K.M; Hom, B.E.; Moss, J.; Fishman, P.H. J. Cell Biol. 1986, 102, 1898-1906.

20.   Yamada, K.M.; Critchley, D.R.; Fishman, P.H.; Moss, J.   Expt. Cell Res. 1983, 143, 295-302.

21.   Ruoshlati, E. (personal communication)

22.   Thompson, L.K.; Horowitz, P.M.; Bentley, K.L.; Thomas, D.D.; Aldrete, J.F.; Klebe, R.J.   J. Biol. Chem. 1986, 261, 5209-5214.

23.   Pierschbacher M.D.; and Ruoslahti, E. Nature 1984, 309. 30-33.

24.   McKeown-Longo, P.J.; Mosher, D.F.   J. Cell Biol. 1985, 100, 364-374.

25.   Bentley, K.; Klebe, R.; Hurst, R.; Horowitz, P.   J. Biol. Chem. 1985, 260, 7250-7256.

26.   Stathakis, N.; Mosesson, M.   J. Clin. Invest. 1977, 60, 855-865.

27.   Edgar, D.; Timpl, R.; Thoenen, H.   EMBO Journal 1984, 3, 1463-1468.

28.   Hayman, E.G.; Pierschbacher, M.D.; Ohgren, Y.; Ruoslahti, E. Proc. Natl. Acad. Sci. USA 1983, 80, 4003-4007.

29.   Sekiguchi, K.; Hakomori, S.I.; Funahashi, M.; Matsumoto, I.; Seno, N.   J. Biol. Chem. 1983, 258, 14359-14365.

30.   Klebe, R.; Mock, P.   J. Cell Physiol. 1982, 112, 5-9.

31.   Klebe, R.J.; Escobedo, L.V.; Bentley, K.L.; Thompson, L.K. Cell Motility and Cytoskeleton 1986, 6, 273-281.

32.   Schreiber, A.B.; Kenney, J.; Kowalski, W.J.; Friesel, R.; Mehlman, T.; and Maciag, T.   Proc. Natl. Acad. Sci., USA 1985, 82, 6138-6142.

33.   Hormann, H.   Klin. Wochenschr. 1982, 60, 1265-1277.

34.   Klebe, R. J.; Hall, J.; Rosenberger, R.; Dickey, W.D.   Exptl. Cell Res. 1977, 110, 419-425.

35.   Klebe, R. J.; Rosenberger, P.; Naylor, S.L.; Burns, R.L.; Novak, R.; Kleinman, H.   Expt. Cell Res. 1977, 104, 119-125.

36.   Klebe, R. J.   J. Cellular Physiol. 1975, 86, 231-236.

37.   Thiery, J.P.   Cell Differ. 1984, 15, 1-15.

38.   Humphries, M.J.; Olden, K.; Yamada, K.M.   Science 1986, 233, 467-470.

39.   Grinnell, F.   Int. Rev. Cytol. 53, 65-144.

40.   Klebe, R. J.   Nature 1974, 250, 248-251.

41.   Klebe, R. J.; Bentley, K.L.; Schoen, R.C.   J. Cellular Physiology 1981, 109, 481-488.

42.   Bentley, K.L.; Klebe, R.J.   J. Biomed. Mater. Res. 1985, 19, 757-769.

43.   Curtis, A.S.G.; Forrester, J.V.; McInnes, C.; Laurie, F. J. Cell Biol. 1983, 97, 1500-1506.

44.   Pearlstein, E.   Int. J. Cancer 1977, 22, 32-35.

45.   Rennard, S.I.; Wind, M.L.; Hewitt, A.T.; Kleinman, H.K.   Arch. Biochem. Biophys. 1981, 206, 205-212.

46. Matyas, G.R.; Evers, D.C.; Radinsky, R.; Morre, D.J. Expt. Cell Res. 1986, 162, 296-318.
47. Harkness, R.D. Biol. Rev. 1961, 36, 399-
48. Baum, B.; McDonald, J.A.; Crystal, R.G. Biochem. Biophys. Res. Comm. 1977, 79, 15.
49. Kleinman, H. K.; McGoodwin, E.B.; Martin, G.R.; Klebe, R.J.; Fietzek, P.; Woolley, D.E. J. Biol. Chem. 1978, 253, 5462-5646.
50. Kleinman, H.K.; McGarvey, M.L.; Hassell, J.R.; Martin, G.R. Biochemistry 1983, 22, 4969-4974.
51. Bornstein, P.; Sage, H. Ann. Rev. Biochem. 1980, 49, 957-1003.
52. Griffin, J.H. Proc. Natl. Acad. USA 1978, 75, 1998-2002.
53. Dunn, J.T.; Silverberg, M.; Kaplan, A.P. J. Biol. Chem. 1982, 257, 1779-1784.
54. Stoker, M.G.P.; O'Neill, C.; Berryman, S.; Waxman, V. Int. J. Cancer 1968, 3, 683-
55. Macpherson, I.; Montagnier, L. Virology 1964, 23, 291-297.
56. Grinnel, F.; Feld, M.K. J. Biol. Chem. 1982, 257, 4888-4893,
57. Matsuda T.; Litt, M. Modification and characterization of polystyrene surface used for cell culture; in Biomedical Applications of Polymers, H.P. Gregor, Ed., Plenum Press, New York, 1975, pp. 135-146.
58. Falcon Labware Catalog, Becton Dickinson and Co., 1978, p. 15.
59. Klemperer, H.G.; Knox, P. Attachment and growth of BHK and liver cells on polystyrene: effect of surface groups introduced by treatment with chronic acid; Lab. Pract., 1977, 26, 179-180.
60. Hadley, M.A.; Byers, S.W.; Suarez-Quian, C.A.; Kleinman, H.K.; Dym, M. J. Cell Biol. 1984, 101, 1511-1522
61. Klebe, R.J. J. Cell Biol. 1987, (In press).

RECEIVED January 29, 1987

Chapter 39

Film, Foaming, and Emulsifying Properties of Food Proteins: Effects of Modification

John E. Kinsella and Dana M. Whitehead

Institute of Food Science, Cornell University, Ithaca, NY 14853

The spectrum of surface active behavior displayed
by food proteins directly reflects differences in
structural and physicochemical properties among the
proteins originating from various sources i.e.
meat, milk, legumes. Chemical or enzymatic
modification of model food proteins has indicated
that alteration of specific structural features
e.g. net charge, disulfide bonding, size, does
influence film formation, foaming and emulsifying
properties.

The specific sequence of amino acids in a protein determines its
structure, conformation, and physicochemical properties. The
structure of protein is categorized as primary, secondary,
tertiary, or quaternary, depending on the progressive state of
spatial arrangement of polypeptide chains of the protein ([1,2]).
Although the primary structures of almost all major food proteins
are known, the exact conformation of only a few native proteins
e.g. β-lactoglobulin has been elucidated ([3-5]).
    In an aqueous environment, the component polypeptides of a
protein tend to fold in a characteristic fashion to form local-
ized secondary structures i.e. α-helix, β-pleated sheet, β-turns,
or random coil ([6]). The integrity and stabilization of secon-
dary, tertiary, and quaternary structures of a given protein are
dependent on different forces. An understanding of the various
forces responsible for the native structure of proteins is
fundamental in comprehending how they affect the conformation and
functional properties of proteins ([7]).
    The non-covalent forces involved in stabilizing the second-
ary and tertiary structure and influencing the functional
behavior of proteins include: hydrogen bonding, van der Waal's
forces, electrostatic interactions, and hydrophobic associations. Covalent disulfide bonds are also important in maintaining structural integrity of some food proteins via intramolecular
and intermolecular bonds e.g. glycinin, β-lactoglobulin ([8]).

The hydrogen bond (H-bond) is ionic in nature and stabilizes secondary structures. Hydrogen bonding is involved in protein--protein associations in films.

Electrostatic interactions are the major forces after hydrogen bonding. The involvement of electrostatic interactions in the functional behavior of food proteins is indicated by the effect of pH on several properties i.e. solubility, emulsion stabilization, film formation, and foaming. Modification of cationic groups and introduction of anionic groups significantly alters the physical properties of food proteins and improves certain functional properties (9).

Van der Waal's forces are general, nonspecific, short-range forces which are operative between closely apposed groups in adjacent polypeptides. The involvement of these forces in film formation has not been established.

Hydrophobic interactions which are enforced (entropy driven) by the nature of water are the principle forces behind protein folding (6). They facilitate the establishment of other stabilizing interactions (7,10). Hydrophobic interactions, being of fundamental importance to protein structure, are very relevant to the functional properties of many food proteins, especially caseins. These forces affect solubility, gelation, coagulation, micelle formation, film formation, surfactant properties and flavor binding (7,10).

Protein Behavior at Interfaces

Film Formation

The surface active properties of proteins are related to their ability to lower the interfacial tension between air/water or oil/water interfaces. Surface activity is a function of the ease with which proteins can diffuse to, adsorb at, unfold, and rearrange at an interface (11,12). Thus, size, native structure and solubility in the aqueous phase are closely correlated with the surface activity of proteins in model systems (13-16).

In model systems, interfacial film formation is enhanced by exposed hydrophobic regions on the protein. Thus proteins with molecular 'flexibility' show superior surface activity, as displayed by the caseins.

Protein conformation at an oil/water interface is not fully understood. Model structures have been proposed based on the polarity of amino acid residue side chains, which depict the polypeptide chain in three segments: 'trains' of amino acid residues in contact with the interface and 'loops' (and 'tails') of residues protruding into either bulk phase, depending on their polarity (13,15,17).

A study with the hydrophobic signal peptide of E. coli lambda phage in phospholipid monolayers, showed a preference for α-helical conformation when the peptide was inserted into the lipid phase (18). However, interaction with the lipid surface without insertion induced the peptide to adopt the β-structure (18). These observations, obtained with circular dichroism and Fourier transform-infrared (FT-IR) data, provide the first direct evidence for interconversions between various conformational states, under specific environmental conditions, when the peptide approaches a lipid surface (18).

During the initial stages of film formation there is rapid diffusion (nanosecond) of proteins from solution to the interface. This is thermodynamically favorable because some of the conformational and hydration energy of the protein is lost at the interface (12). Initially, at low protein concentrations, there is no barrier to adsorption and for protein molecules that are readily adsorbed at the interface, the rate of adsorption is diffusion controlled. But after some time, especially at high surface protein concentrations, there is an activation energy barrier to adsorption (14,19), which may involve electrostatic, steric and osmotic effects close to the interfacial or surface layers. Under the latter conditions, the ability of the protein molecules to interpenetrate and create space in the existing film and to rearrange at the surface is rate-determining. The capacity of proteins to unfold at an interface depends very much on the conformational stability of flexible segments of the protein molecule (13). Where there is extensive intramolecular associations and disulfide bonding, unfolding at the interface tends to be limited, and formation of an interfacial membrane takes relatively longer e.g. soy proteins compared to caseins (10,15,20).

Food Protein Films in Model Systems

In order to elucidate relationships between surface active and film forming properties of food proteins, it is useful to examine the surface active properties of proteins whose physical and molecular properties are well characterized e.g. β-casein, bovine serum albumin (BSA), lysozyme (17), and β-lactoglobulin (b-Lg) (21). These represent a range of tertiary structures for soluble proteins. Lysozyme is a rigid and roughly ellipsoidal molecule, whereas the hydrophobic β-casein molecule is mostly a random coil structure. The b-Lg molecule consists almost entirely of anti-parallel β-sheet strands organized into a flattened cone (5).

The kinetics of protein adsorption at an interface can be measured by monitoring surface concentration and surface pressure i.e. depression of surface tension (γ) as a function of time (17). β-casein is more surface active than serum albumin or b-Lg and much more so than lysozyme. This reflects not only the rate of diffusion of the native protein to the interface, but also its molecular 'flexibility' and amphipathic nature (15,17,22).

Comparisons of the surface adsorption behavior of β-casein, lysozyme and BSA have been well-documented (13-15,22). Rates of adsorption at any given surface pressure (2 to 20 mN/M) and protein concentration (0.01 mg/dl to 1 mg/dl) reflect differences in unfolding of the protein molecules as well as differences in their isoelectric points. β-casein has a pI of approximately 5.3 (average of the genetic variants) and forms a dilute monolayer of tightly packed molecules (>7.7 A^2/residue) until the protein concentration is 1 mg/dl. Further adsorption is averted because of charge repulsion and steric factors i.e. net negative charge on the protein at pH 7 (13,23,24). Lysozyme retains extensive native structure at the interface as it forms a concentrated film and multilayers of the protein accumulate above 1 mg/dl. Lysozyme multilayer films display greater viscosity, resistance to shear and lower compressibility (larger dilatational modulus) than β-casein films, reflecting a greater degree of crosslinking (disulfide bonds) and intermolecular associations (23). The pI of lysozyme is close to neutral pH, thus electrostatic repulsive forces and steric factors are negligible at this pH.

Benjamins et al (24) studied the effects of aging on the elasticity of β-casein and K-casein films. The dilatational modulus of K-casein was larger than that of β-casein and increased by a factor of three with film age, whereas the dilatational modulus of β-casein films changed little with time (24). K-casein unfolds less at the air/water interface since it has less random structure than β-casein. This can also be interpreted in terms of K-casein having less direct contact with the film surface at any given protein concentration (25). Significant protein-protein interactions i.e. steric/electrostatic repulsion, are believed to occur between segments of polypeptide chains which extend both above and below the plane of the air/water interface in surface protein films (16,26)

The surface viscosity or resistance to shear stress of the surface film is an index of its mechanical strength and is an important parameter related to the stability of films and foams (21). Surface yield stress of BSA films were determined using the above parameters (23). Maximum values were obtained in the pH range of 5-6, near the isoelectric point of BSA, and decreased rapidly above pH 6.0 [Table 1]. These observations again reflect the enhanced intermolecular interactions between protein components in the film as the isoelectric point is reached and the enhanced electrostatic repulsion between neighboring molecules as the pH is raised above the isoelectric point, i.e. net charge on the protein surface was increased.

The general validity of these relationships has also been demonstrated with β-lactoglobulin (21), ribulose 1,5-biphosphate carboxylase (25), and soy glycinin (27).

The adsorption of soy protein at an interface is relatively slow compared to casein, and the rate is affected by ionic strength, being higher at 0.2 M than at zero NaCl where the subunits may be dissociated. Conceivably the reduction of the zeta potential and electrostatic repulsion (from 0 to 0.2 M salt) facilitates penetration and subsequent surface packing (28). The rate of penetration of additional molecules into the film indicated that the soy proteins initially adsorbed and spread easily at the surface (29). However, this seems inconsistent with the highly stable disulfide linked tertiary structure of soy glycinin (30) and it is perhaps the conglycinin component that forms the initial interfacial film (31).

The behavior of proteins at interfaces influences the formation of foams and emulsions (32). Stabilization of foams and emulsions depends, to a great extent, on the formation, rheological, and mechanical properties of the interfacial film (22). Factors which ensure optimum film properties in simple systems may retard film formation or cause destabilization in foams or emulsions (33); for example, many rheological properties of films are maximum in the isoelectric pH range of specific proteins, yet most proteins have minimum solubility in this pH range (34). Thus, environmental and processing factors which

TABLE 1

Effect of pH on Some Film and Foaming Properties
of Bovine Serum Albumin

		Film		Foam
pH	Surface Pressure	Surface Yield Stress (dyne/cm)	Film Elasticity	Drainage half-life (min)
4.0	2.8	3.0	2.2	5.0
5.0	15.8	3.8	5.0	8.0
5.5	19.0	4.0	5.2	9.6
6.0	14.0	4.3	5.4	8.5
7.0	10.0	3.0	2.3	6.3
8.0	2.0	2.2	1.8	6.0

From Kim and Kinsella, 1985 (23)

alter the conformation and stability of proteins greatly affects film formation and properties (34).

Because protein-based foams depend upon the intrinsic molecular properties (extent and nature of protein-protein interactions) of the protein, foaming properties (formation and stabilization) can vary immensely between different proteins. The intrinsic properties of the protein together with extrinsic factors (temperature, pH, salts, and viscosity of the continuous phase) determine the physical stability of the film. Films with enhanced mechanical strength (greater protein-protein interactions), and better rheological and viscoelastic properties (flexible residual tertiary structure) are more stable (12,15), and this is reflected in more stable foams/emulsions (14,33). Such films have better viscoelastic properties (dilatational modulus) (35) and can adapt to physical perturbations without rupture. This is illustrated by β-lactoglobulin which forms strong viscous films while casein films show limited viscosity due to diminished protein-protein (electrostatic) interactions and lack of bulky structure (steric effects) which apparently improves interactions at the interface (7,13,19).

In the case of the major cytoplasmic protein of leaves, ribulose 1,5-biphosphate carboxylase (RUBISCO), the surface rheological properties and foam stability were maximum at pH 5.5, close to the isoelectric point (pH 4.8) and all parameters measured were greater than any other protein studied (25). This may be related to the large molecular size of RUBISCO, i.e. 560 000 daltons, and its disulfide stabilized globular structure.

Hydrophobic interactions between proteins and fats are very critical in emulsion formation. The emulsifying activities of various proteins e.g. BSA, trypsin, ovalbumin, and lysozyme have been reported to be correlated with their average net hydrophobicities (36). It has been suggested that the conformational properties i.e. 'flexibility' of whey proteins are important in adsorption and possibly affects their emulsifying ability (37). Subsequent studies revealed that the conformational stability of β-lactoglobulin varied depending on pH i.e. its conformation was more rigid and resistant to denaturation at pH 3 than at pH 7 (38). The low emulsifying and surface activity of b-Lg at acidic pH was assumed to be due to low denaturability ('flexibility') of the molecule (38) and might be an important factor in governing surface active properties (14).

It has been suggested that protein 'flexibility' is an important structural factor governing emulsifying and foaming properties (39). Chemically induced cross-linking of BSA and lysozyme greatly reduced the foaming power and foam stability of both proteins and there was a similar though less marked change in emulsifying activity and emulsion stability (40). The monomeric cross-linked proteins were resistant to proteolysis and heat-induced conformational changes, suggesting that molecular flexibility may play a role in foaming and emulsion properties (40).

Effect of Modification on Film, Foam, and Emulsion Properties

Intentional modification of protein structure through alterations of protein net charge, hydrophobicity, hydrogen bonding, and disulfide bonding, provide approaches for studying the importance of specific structural features on film structure and stability, foaming capacity, and emulsion stabilization (20).

Modification of Net Charge

Succinylation: The net charge on protein molecules affects the solubility and the extent of protein-protein interactions. Both are important in achieving optimal film and surface properties (8,9). The solubility of a protein results from an equilibrium between protein-solvent and protein-protein interactions (41). Conditions which favor protein-solvent interactions generally increases the solubility of a protein. Chemical derivatization of the ϵ-amino groups of lysine residues in proteins with succinic anhydride improves the functionality of food proteins (9).

Progressive succinylation of glycinin or β-lactoglobulin increased the amount of unordered structure, electronegativity (42), specific viscosity (43), hydration and solubility (44), and enhanced the foaming and emulsifying properties (44) of the succinylated proteins.

Film strength as reflected in surface yield stress of glycinin was increased by moderate succinylation (<50%) but decreased following extensive succinylation. The surface yield stress also decreased with increasing pH [Table 2]. Both observations are consistent with the thesis that with an increase in net negative charge, excessive repulsion reduces the formation of a continuous cohesive viscoelastic film. Dynamic film elasticity as measured by tensiolaminometry (20) revealed that limited succinylation enhanced elasticity, but at high levels of succinylation, elasticity decreased. These data reflect the diminished cohesiveness of the film as net charge repulsion between proteins in the film was increased.

Foaming behavior was improved by limited succinylation of glycinin (44). The foam stability, expressed as half lifetime of liquid in the foam, i.e. drainage, significantly increased upon succinylation, particularly at low (up to 50%) levels of succinylation. However, at 100% succinylation, foam stability was greater than that observed for the native glycinin. This may reflect the enhanced stability of the bubbles caused by charge repulsion which impeded approach and coalescence of contiguous bubbles. The increase in net negative charge on the protein may have resulted in greater water binding and retention in the lamella as reflected by the increased viscosity of modified glycinin [Table 2].

pH: The effects of pH on the surface active properties of BSA and β-lactoglobulin reflect the importance of net charge. BSA is surface active and its ability to depress surface tension is affected by net charge i.e. pH. The initial surface pressure development of BSA was markedly pH dependent showing a sharp maximum around pH 5.5 close to the isoelectric point of BSA (23) (Fig.1). The rate of adsorption and surface pressure development of b-Lg is significantly greater in the pH range 4.5-6.0 (21) (Fig.2).

Thus the optimum adsorption and film formation of both proteins occurs near their isoelectric points where the surface charge tends toward neutral. The rate of adsorption of protein

TABLE 2

Effects of Succinylation on Some Molecular and
Surface Active Properties of Succinylated Soy Glycinin

| | (%) Extent of Succinylation of Glycinin | | | |
	0	25	50	100
Net charge at pH 7.5	-250	-290	-330	-400
Specific viscosity ($\times 10^{-2}$)	0.5	0.8	1.5	8.0
Surface Hydrophobicity	200	330	360	290
UV Absorbance (270 nm)	0.9	0.8	0.75	0.67
Surface Pressure (5 min dyne/cm)	3.0	16.0	14.5	5.5
Surface Yield Stress (dyne/cm)				
pH 5.0	-	5.3	5.0	4.5
pH 6.0	3.5	4.6	4.0	0.2
pH 8.0	3.2	4.1	3.5	0.2
Film Elasticity (dyne/cm)	3.9	4.6	4.4	1.9
Foam Stability (half-life, min)	5.0	14.5	11.0	8.0

From Kim and Kinsella, 1986 (27)

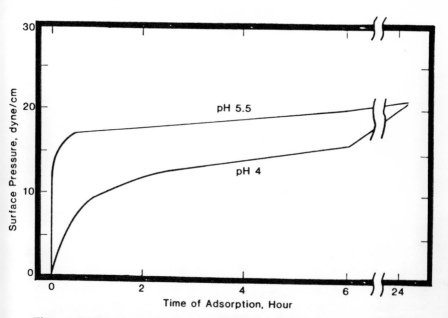

Figure 1. Surface pressure development of bovine serum albumin at two pH values. Protein 5×10^{-3}% in citrate buffer (10mM). (Reproduced with permission from Ref. 23. Copyright 1985 Institute of Food Technologists.)

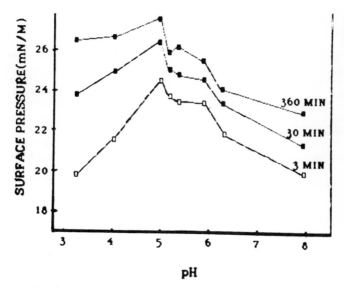

Figure 2. Relationship between pH and surface pressure of β-lactoglobulin films at different time intervals. (Reproduced from Ref. 21. Copyright 1985 American Chemical Society.)

at the interface is increased near the pI of a protein when the protein remains soluble (12, 45), because the proteins have decreased electrostatic repulsion at the interface. The compact protein molecules can pack more easily into the interfacial film, facilitating favorable protein-protein interactions.

Phosphorylation: Chemical phosphorylation of proteins should alter the net charge and amphipathic nature of the proteins and this may modify the surface active properties, e.g. emulsifying power of the protein. Although little work has been done regarding the film forming and foaming properties of phosphorylated food proteins, the effects of chemically phosphorylated β-lactoglobulin on the native structure, solubility, viscosity, and emulsification of the protein have been reported (46, 47). Bovine b-Lg was phosphorylated with phosphorus oxychloride (POCl$_3$) at pH 8.5 to give a product that contained up to 14 mol of phosphorus/mol protein (46). Phosphorylation of b-Lg resulted in a decrease (approx. 5%) of the α-helical content compared to the native protein molecule (46). The structural changes brought about by phosphorylation were not reversed by removal of the phosphate groups, as indicated by circular dichroism measurements (46).

Emulsions prepared with the phosphorylated derivative of b-Lg were up to 30% more stable at pH 7 and pH 5 compared with corresponding emulsions prepared with native protein (47). The viscosity of an emulsion prepared with phosphorylated b-Lg at pH 5 was about double that prepared with native protein (47).

Glycosylation: Little is known about the effects of non-charged hydrophilic substituents on the surface active behavior of proteins. Cumper (45) proposed that hydrophilic interactions between proteins and the aqueous phase were important in surface properties and foam stability. Thus, the glycoproteins of egg white may account for its superior foaming properties, particularly foam stability.

The effects of glycosylation on β-lactoglobulin were assessed in order to observe changes in the surface active behavior and structural parameters resulting from modification (48-50). The free amino groups of b-Lg were modified to varying degrees with maltosyl residues using the cyclic carbonate method and the carboxylic groups were modified with glucosaminyl residues using the carbodiimide method (48). Data in Table 3 shows that glycosylation increased the molecular weight of b-Lg with a concomitant reduction in the number of charged groups (48).

Changes in the relative viscosity for both derivatives of b-Lg i.e. maltosyl-b-Lg (M-b-Lg) and glucosaminyl-b-Lg (G-b-Lg) were observed and are shown in Table 3. The deviations seen in M-b-Lg and G-b-Lg are directly correlated with the number of residues modified and the size of the added carbohydrate substituents; at >10 residues modified with maltose, the viscosity increases to a greater extent compared to G-b-Lg. This reflects some conformational changes in the protein (48, 50).

Circular dichroism measurements revealed significant reduction in the α-helical content and an increase in the unordered structure of M-b-Lg, reflecting significant perturbation of secondary structure in the derivatized proteins (50).

TABLE 3

Effects of Glycosylation on Film, Foaming, and Emulsifying Properties of β-Lactoglobulin

Property	Relative Changes in Properties				
	b-Lg	M-b-Lg		G-b-Lg	
Residues modified	0	7(NH$_2$)	11(NH$_2$)	6(COOH)	16(COOH)
Molecular weight of added carbohydrates	-	4,352	11,220	1,014	2,544
Circular dichroism (%) -helix	10	5	2	8	4
-structure	40	42	41	44	40
remainder	50	53	57	48	56
Surface Pressure (mN/M) 0.1% 10 min					
pH 3.5	22.0	23.0	23.0	24.0	19.0
pH 5.3	24.5	23.5	23.0	21.7	23.0
pH 7.0	22.0	21.5	21.5	22.7	22.5
Rate of Adsorption (min^{-1})					
pH 3.5	.28	.29	.31	.40	.31
pH 5.3	.46	.32	.36	.36	.39
pH 7.0	.44	.38	.37	.43	.37
Foaming: Initial volume of Liquid in Foam (ml)					
pH 3.5	10.0	12.4	13.1	9.8	9.6
pH 5.3	5.4	12.9	13.7	9.3	10.1
pH 7.0	8.4	14.3	14.1	7.2	8.5
Drainage of Foam after 10 min (%)					
pH 3.5	64	56	50	65	58
pH 5.3	74	49	56	64	50
pH 7.0	68	63	65	53	56
Foam Strength after 10 min (sec/ml)					
pH 3.5	1	74	67	20	14
pH 5.3	0	78	63	32	74
pH 7.0	6	57	40	34	49

From Waniska, 1981. (<u>49</u>)

Apparently, hydrophilic interactions between the modified proteins in water increased with a concomitant reduction in ionic and hydrophobic interactions resulting in destabilization of the native structure of the protein (50).

Glycosylation affected the surface active properties of b-Lg; surface pressure progressively increased with time reaching equilibrium after 360 minutes. The rates of adsorption, penetration and rearrangement of the protein molecules in the interface were estimated from surface pressure data. The area cleared during adsorption tended to increase with pH and with extent of modification of b-Lg, suggesting a more expanded molecular structure with increasing pH and bulkiness of the added modifying group. The glycosylated derivatives of b-Lg had inferior surface properties in the isoelectric pH range, whereas the surface films appeared to be more condensed at lower pH's and more expanded at higher pH values. The carbohydrate moieties enhanced hydrogen bonding between the proteins and solvent and altered the number of charges on the protein which caused changes in the nature and magnitude of the forces acting between molecules in the interfacial film.

The presence of glycosyl moieties reduced the sensitivity of protein to pH effects (41). Changes in the hydrodynamic volume of proteins would be expected to reduce the rate of diffusion of the modified protein e.g. G-b-Lg to the interface thereby slowing the rate of surface adsorption and, finally, loss of conformational energy during modification i.e. less secondary structure may result in a decreased gain in free energy of these proteins upon adsorption at the interface.

The properties of foams generated from modified b-Lg were studied. In contrast to native b-Lg, the maltosyl derivatives showed little pH sensitivity and the amount of liquid in these foams was much greater than in native b-Lg indicating enhanced hydration of the glycosylated derivatives. Significantly less of the liquid drained from these foams during the initial 10 minutes reflecting higher foam stability in the modified proteins, particularly in the isoelectric pH range .

In general, foams made from b-Lg were quite unstable which is in contrast to the superior surface pressure and film surface viscosity of b-Lg compared to the glycosylated derivatives (49). Apparently glycosylation of b-Lg improved foam stability possibly by enhancing water holding ability by the glycosyl residues and, in addition, the disordered structure of the derivatized protein may have facilitated greater entanglement and protein-protein interactions in the surface film.

Disulfide Bond Reduction:
Disulfide bonds stabilize the tertiary structure of proteins and impede the ability of protein to rearrange and interact at the interface by conferring structural constraints e.g. lysozyme, glycinin. The limited film forming and foaming properties of soy glycinin has been suggested to be due to its compact disulfide linked globular structure (20,32). The effects of reductive modification of disulfide bonds and the effects on conformation, film forming and foaming properties of glycinin have been studied (20,27).

Glycinin (which has 6 inter- and 12-14 intramolecular disulfide bonds) was treated with 5 and 10 mM dithiothreitol (DTT) which reduced seven and >95% of the disulfide bonds, respectively (27). Gel electrophoresis indicated that 5 mM DTT reduced mostly the intermolecular linkages but that some of the intramolecular disulfide bonds were also reduced (27). The specific viscosity was significantly increased following reduction with 5 mM DTT whereas it was only slightly altered following treatment with 10 mM DTT (27). Analysis of the UV spectra indicated a blue shift from 277 to 275 nm with a concomitant decrease in absorbance, evidence that with reduction the apolar, internal chromophores (tryptophan) were initially exposed to a more polar environment. Intrinsic fluorescence data supported the contention that reduced glycinin is structurally less compact with changes occurring in the net hydrophobicity depending on the extent of reduction i.e. 5 or 10 mM DTT (27). [Table 4]

The improved surface active properties of the modified glycinin enhanced stability of foams formed from these proteins (20). Using the column aeration technique (50), foams were formed of either conglycinin, glycinin, an equal mixture of both, or reduced glycinin (10 mM DTT or metabisulfite). The most highly reduced glycinin foamed most rapidly. Foam stability, estimated from the rate of drainage of liquid from the foam, was significantly increased with the extent of reduction of glycinin at all pH values studied, particularly at pH 6 and 7 (20). Increasing the pH, which also enhanced the foaming properties of native glycinin, progressively diminished the stability of foams made from reduced glycinin. Destabilization is caused by drainage of lamellar fluid, coalescence and rupture of the lamella (14-16).

The above data indicate positive relationships between protein conformation, net hydrophobicity, surface pressure, surface yield stress, film elasticity, and foam stability.

Enzymatic Modification

The functional properties of food proteins may be improved by the use of specific enzymes to partially hydrolyze the proteins or to add specific functional groups to the proteins (51). Modification reactions employing enzymes e.g. trypsin, papain, are especially attractive because they may be carried out under mild conditions, are not likely to lead to toxic products, and the necessity for their removal after completion of the reaction may be bypassed (52). A very limited amount of research has been conducted on enzymatic modification as it affects the structural and surface properties of food proteins.

Improvement of protein solubility can be achieved through limited digestion of food proteins e.g. whey proteins, soy with various food-grade proteases (53). The complete solubilization of heat-denatured cheese whey protein (α-lactalbumin) by trypsin, within certain pH limits (pH 5-7), and partial solubilization by treatment with papain or neutral protease (54) has been reported.

Partial proteolysis of soy protein isolate with neutral protease from _Aspergillus oryzae_ altered certain functional properties (55). Solubility was increased in the enzyme-treated soy isolate at both neutral pH and at the isoelectric point (pH

TABLE 4

Summary of Relative Changes in Molecular and Surfactant
Properties of Glycinin Following Reduction of Disulfide Bonds

Properties	Glycinin		
	Native 20 SS Bonds	Reduced 13 SS Bonds	Fully Reduced 0 SS Bonds
Relative Hydrophobicity	220	900	870
Specific Viscosity ($\times 10^{-3}$)	4.2	8.0	7.3
UV Absorbance (272 nm)	0.90	0.65	0.79
Fluorescence Intensity (336 nm)	46	52	57
Surface Pressure (dyne/cm 5 min)			
pH 6	2	16	13
pH 7	5	14	15
pH 8	4	13	15
Surface Film Yield Stress (dyne/cm)			
pH 6	3.5	4.6	5.0
pH 7	3.7	3.9	4.7
pH 8	3.2	3.5	4.0
Surface Film Elasticity (dyne/cm)			
pH 6	3.7	5.8	6.3
pH 8	4.3	4.5	5.3
Foam Stability (half-life, min)			
pH 6	3	16	60
pH 7	6	13	25
pH 8	7	10	12

From Kim and Kinsella, 1986. (27)

4.5). Limited enzyme treatment significantly reduced the viscosity of concentrated protein solutions and emulsification capacities were increased (55). The enzyme-treated proteins had slightly increased water absorption and foaming properties, but foam and emulsion stabilities were decreased (54).

Hydrolysis of whey protein concentrate (WPC) with pepsin, pronase or prolase decreased emulsifying capacity; specific foam volume was increased by very limited hydrolysis (56,57), but decreased by more extensive hydrolysis, while foam stability was greatly decreased by limited hydrolysis (57) (Fig.3).

In general, foaming agents originating from proteins are prepared by partial hydrolysis which results in increased foaming power (56); however, foam stability is generally decreased by this treatment (16) because of the change of molecular structure and change in molecular size during hydrolysis (58).

A correlation between content of hydrophobic amino acids and surface activity of five different food proteins partially hydrolyzed with 0.1% pepsin has been reported (58), but exceptions were noted. Protein hydrolysates exhibiting large surface absorption were correlated with large foam stability and a large external hydrophobic region. It was concluded that protein hydrolysates with large surface hydrophobic regions adsorbed more readily at interfaces and rates of surface desorption were lower. However, secondary structures, as measured by optical rotatory dispersion and infrared spectra, and the content of the total hydrophobic amino acids in the protein hydrolysates showed no correlation with their foam stabilities (58).

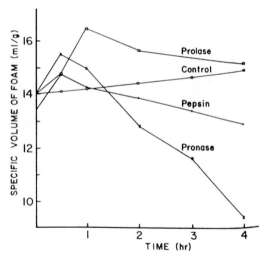

Figure 3. Effect of enzymatic hydrolysis on specific volume of foam obtained by whipping a heated whey protein sol (4% w/w, 85 C, 6 min. whipping. (Reproduced with permission from Ref. 57. Copyright 1979 Institute of Food Technologists.)

Conclusions

Because of the complexity of actual food systems it is difficult to predict the performance of a protein based on its behavior in a simple model system. Thus, pH, salts, or level of surfactant, e.g. monoglyceride, in a food system could adversely affect the apparent surface activity of protein.

The need for continuing research to further elucidate basic physicochemical properties to clarify structure-function relationships is evident. Information drawn from studies of chemical and enzymatic modification of food proteins and improvement of methodology for obtaining accurate data will further the objective of better utilization of plant and animal proteins as important food ingredients.

Acknowledgments

This paper was supported in part by NSF grant CBT-850623, the National Dairy Board, and a special award given to John E. Kinsella from the General Foods Foundation.

Literature Cited

1. Anfinsen, C.B. Science. 1973, 181, 223-230.
2. Anfinsen, C.B. and Sheraga, H.A. Adv. Prot. Chem. 1975, 29, 205-300.
3. Swaisgood, H.E. In Developments in Dairy Chemistry; Fox, P., Ed.; Applied: London, 1982; Vol. 1, pp. 1-59.
4. Creamer,L.K., Parry, D.A.D., Malcolm, G.N. Arch. Biochem. Biophys. 1983, 227, 98-105.
5. Sawyer, L., Papiz, M.Z., North, A.C.T. Biochem. Soc. Trans. 1985, 13, 265-266.
6. Tanford, C. Adv. Prot. Chem. 1968, 23, 121-282.
7. Kinsella, J.E. CRC Crit. Rev. Food Sci. Nutr. 1984, 21, 197-262.
8. Kinsella, J. E., Damodaran, S., German, B. In New Food Proteins; Altschul, A., Wilkie, H., Eds.; Academic:New York, 1985; Vol. 5, p. 116-120.
9. Kinsella, J.E. and Shetty, J.K. In Functionality and Protein Structure; Pour-El, A., Ed., ACS Symposium Series No. 92; American Chemical Society:Washington, D.C., 1979; pp 37-56.
10. Kinsella, J.E. CRC Crit. Rev. Food Sci. Nutr. 1976, 7, 219-279.
11. Bull, B.B. J. Colloid Interfac. Sci. 1972, 41, 305-310.
12. MacRitchie, F. Adv. Prot. Chem. 1978, 32, 283-315
13. Phillips, M.C. Chem. Ind. (London). 1977, 5, 170-176.
14. Graham, D.E. and Phillips, M.C. In Foams; Akers, R.J., Ed.; Academic:New York; 1976; pp 237-255.
15. Phillips, M.C. Food Technol. 1981, 35, 50-57.
16. Halling, P.J. CRC Crit. Rev. Food Sci. Nutr. 1981, 13, 155-215.
17. Graham, D.E. and Phillips, M.C. J. Colloid Interfac. Sci. 1979, 75, 427-439.
18. Briggs, M.S., Cornell, D.G., Dluhy, R.A., Gierasch, L.M. Science. 1986, 233, 206-208.

19. Graham, D.E. and Phillips, M.C. J. Colloid Interfac. Sci. 1979, 75, 403-414.
20. German, J.B., O'Neill, T.E., Kinsella, J.E. JAOCS. 1985, 62, 1358-1366.
21. Waniska, R.d. and Kinsella, J.E. J. Agric. Food Chem. 1985, 33, 1143-1148.
22. Kinsella, J.E. Food Chem. 1981, 7, 272-288.
23. Kim, S.H. and Kinsella, J.E. J. Food Sci. 1985, 50, 1526-1530.
24. Benjamins, J., De Feijter, J.A., Evans, M.T.A., Graham, D.E., Phillips, M.C. Discuss. Faraday Soc. 1975, 59, 218-229.
25. Barbeau, W.E. and Kinsella, J.E. Colloids and Surfaces. 1986, 17, 167-183.
26. Graham, D.E. and Phillips, M.C. J. Colloid Interfac. Sci. 1980, 76, 240-249.
27. Kim, S.H. and Kinsella, J.E. J. Agric. Food Chem. 1986, 34, 623-627.
28. Tornberg, E. In Functionality and Protein Structure; Pour-El, A., Ed.; ACS Symposium Series No. 92; American Chemical Society:Washington, DC, 1979; pp 105-137.
29. Tornberg, E. J. Sci. Food Agric. 1978, 29, 762-770.
30. Draper, M. and Catsimpoolas, N. Cereal Chem. 1978, 55, 16-22.
31. Kella, N.K.D., Barbeau, W.E., Kinsella, J.E. J. Agric. Food Chem. 1986, 34, 251-256.
32. Pomeranz, Y. Functional Properties of Food Components; Academic: New York; 1986; p. 167.
33. Friberg, S. Food Emulsions; Marcel Dekker: New York, 1976; pp. 35-50.
34. Waniska, R.D. and Kinsella, J.E. J. Agric. Food Chem. 1981, 29, 826-831.
35. Graham, D.E. and Phillips, M.C. J. Colloid Interfac. Sci. 1980, 76, 227-239.
36. Kato, A. and Nakai, S. BBA. 1980, 624, 13-20.
37. Shimizu, M., Kamiya, T., Yamauchi, K. Agric. Biol. Chem. 1981, 45, 2491-2496.
38. Shimizu, M., Saito, M., Yamauchi, K. Agric. Biol. Chem. 1985, 49, 189-194.
39. Kato, A., Komatsu, K., Fujimoto, K., Kobajashi, K. J. Agric. Food Chem. 1985, 33, 931-934.
40. Kato, A., Yamaoka, H., Matsudomi, N., Kobayashi, K. J. Agric. Food Chem. 1986, 34, 370-372.
41. Kuntz, I.D. and Kauzmann, W. Adv. Prot. Chem. 1974, 28, 239-282.
42. Shetty, K.J. and Rao, M.S.N. Int. J. Peptide Prot. Res. 1978, 11, 305-313.
43. Rao, H.G. and Rao, N.S. Int. J. Prptide Prot. Res. 1979, 14, 307-315.
44. Franzen, K.L. and Kinsella, J.E. J. Agric. Food Chem. 1976, 24, 788-795.
45. Cumper, C.W. Trans. Faraday Soc. 1953, 49, 1360-1370.
46. Woo, S.L., Creamer, L.K., Richardson, T. J. Agric. Food Chem. 1982, 30, 65-70.

47. Woo, S.L. and Richardson, T. J. Dairy Sci. 1983., 66, 984-987.
48. Waniska, R.D. and Kinsella, J.E. Int. J. Peptide Prot. Res. 1984, 23, 467-476.
49. Waniska, R.D. Ph.D. Thesis, Cornell University, Ithaca, New York, 1981.
50. Waniska, R.D. and Kinsella, J.E. J. Food Sci. 1979, 44, 1398-1402.
51. Whitaker, J.R. In Food Proteins: Improvement Through Chemical and Enzymatic Modification; Gould, R.F., Ed.; American Chemical Society: Washington, DC, 1977; p. 95.
52. Phillips, R.D. and Beuchat, L.R. In Protein Function in Foods; Cherry, J.P., Ed.; ACS Symposium Series No. 147; American Chemical Society:Washington, DC, 1981; p 275.
53. Adler-Nissen, J. J. Agric. Food Chem. 1976, 24, 1090-1093.
54. Monti, J.C. and Jost, R. J. Dairy Sci. 1978, 61, 1233-1237.
55. Puski, G. Cereal Chem. 1975, 52, 655-664.
56. Adler-Nissen, J. Proc. Biochem. 1977, 12, 18-23.
57. Kuehler, C.A. and Stine, C.M. J. Food Sci. 1979, 39, 379-382.
58. Horiuchi, T., Fukushima, D., Sugimoto, H., Hattori, T. Food Chem. 1978, 3, 35-42.

RECEIVED January 29, 1987

Chapter 40

Interfacial Behavior of Food Proteins Studied by the Drop Volume Method

E. Tornberg

Swedish Meat Research Institute, P.O. Box 504, S-244 00 Kävlinge, Sweden

Many food items contain emulsions and foams, which are often
stabilised by proteins forming a protective membrane at the
interface. By preparing the food, adsorption of the available
proteins, - by virtue of their surface activity -, is performed at
the liquid/air (foams) and/or at the liquid/liquid (emulsions)
interface. One way to study the interfacial behaviour of food
proteins at those interfaces is to follow the interfacial tension
decay accomplished by the adsorption of the proteins.

So far a considerable amount of work has been devoted to the
study of spread protein films at these types of interface.
However, the study of the adsorption of proteins at the interface
from a subphase of known concentration is more similar to the
conditions prevailing during formation of emulsions and foams.

Methods

The interfacial tension decay of food proteins adsorbing from a
subphase has, in this study, been monitored with an apparatus
based on the drop volume technique (1, 2, 3). The following
procedure was used (for details cf. ref. 2).

A drop of a certain volume, corresponding to a certain
interfacial tension (γ) value, is expelled rapidly, and the time
necessary for the interfacial tension to fall to such a value that
the drop becomes detached is measured. This procedure is repeated
for differing drop sizes, i.e. for different values of the
interfacial tension. A plot of the interfacial tension as a
function of time (t) can then be made, as seen in Figure 1. This

0097-6156/87/0343-0647$06.00/0
© 1987 American Chemical Society

has been done for the adsorption at the air/water interface of an ultrafiltrated and spray-dried whey protein concentrate (WPC) dispersed in 0.2 M NaCl solution at pH 7 for different initial subphase concentrations. By use of this procedure the advantages of the drop volume method, as opposed to the Wilhelmy plate method can, be exploited; i.e. no problems with the contact angle ($\underline{1}$), which makes it especially suited for studies at the liquid/liquid interfaces, the possibility of measurement at elevated temperatures (t > 25°C) ($\underline{1}$), and the formation of a clean interface in short time-periods with an initially uniform concentration of proteins ($\underline{3}$).

These types of measurement have been used to follow the interfacial tension decay or the rise in surface pressure, $\pi_t = \gamma_0 - \gamma_t$ (γ_0 = initial interfacial tension of the clean interface) with time for a variety of food protein preparations. They have also been studied at different interfaces [air/water (A/W) and soybean oil/water (O/W) interfaces] and when the charge density of the proteins varies (pH, ionic strength). All the measurements have been carried out at a temperature of 25°C.

The kinetics of the interfacial tension decay

In Figure 2 three representative γ-t-curves are demonstrated. The slowest decay is exerted at the A/W-interface by a WPC dispersed in 0.2 M NaCl at pH 7, denoted (0.2-7), at a protein concentration of 10^{-3} wt%. As can be seen from this curve there is an induction period before the interfacial tension starts to fall, which can be even more pronounced at lower concentrations and for other proteins ($\underline{4}$). J.A. de Feijter ($\underline{5}$) has recently suggested a mechanism for this behaviour. He found, when measuring simultaneously the surface pressure and the surface concentration (Γ by ellipsometry), that the Γ-t-measurements did not give rise to an induction period, whereas the π-t-curve could. Moreover, the surface concentration at the end of the induction period was about 1-1.5 mg/m^2 for the proteins studied (Lysozyme, BSA and Ovalbumin), i.e. about monolayer coverage. This means there will be no appreciable increase in surface pressure until almost monolayer coverage, or expressed differently at the beginning of the condensed phase. Rearrangements of the proteins within the adsorbed film will then increase the surface coverage and thereby the interfacial pressure ($\underline{6}$). However, these rearrangements will with time be restricted by the increased incompressibility of the film. Therefore, the rise in surface pressure will fall off, as seen from the curves in Figure 2.

By plotting the rate of the surface pressure increase, $\log\frac{d\pi}{dt}$, as a function of π the change in compressibility of the protein film can more easily be followed. This is illustrated in Figure 3 for the three curves in Figure 2. Graham & Phillips ($\underline{7}$) have shown that slow conformational changes occur in a film, when it is highly incompressible. Therefore, the 'kinks' observed in the $\log\frac{d\pi}{dt}$-π-curves are suggested to arise from a relatively abrupt increase in the incompressibility of the film.

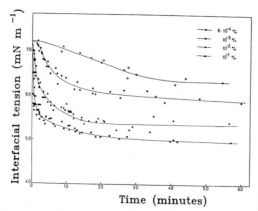

Figure 1. Time-dependence of interfacial tensions at the air-water (A/W) interface for WPC at different subphase concentrations. The WPC is dispersed in 0.2 M NaCl solution at pH 7 (0.2-7).(Reproduced with permission from Ref. 4. Copyright 1978 Blackwell Scientific Publications.)

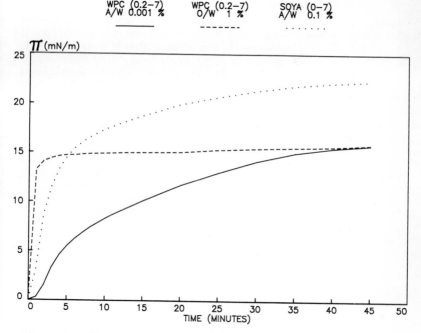

Figure 2. Time-dependence of the interfacial pressure, π 40, for three curves. They represent WPC dispersed in (0.2-7) adsorbing at the A/W and the soya bean oil/water (O/W) interfaces at a subphase concentration of 10^{-3} and 10^{0} wt%, respectively, and a soya protein isolate adsorbing at the A/W-interface at a subphase concentration of 10^{-1} wt%.

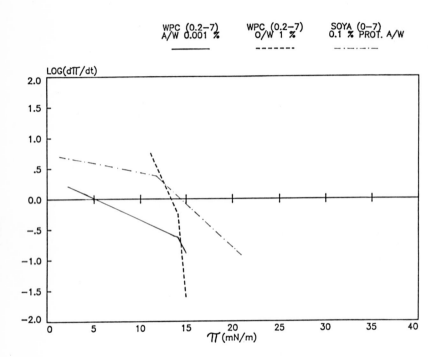

Figure 3. Log$\frac{d\pi}{dt}$ as a function of π for those curves given in Figure 2.

Consequently, this will be followed by a decrease in the rate of the surface pressure build-up.

Another interesting γ-t-curve, observed for extracted meat proteins at certain concentrations is a peculiar, stepwise behaviour not earlier shown for proteins (8). This can be seen in the middle diagram in Figure 4. To be able to elucidate this behaviour we have to find out what the extracted juice of meat proteins consists of.

The beef muscle is heavily comminuted (Moulinex), centrifuged (25,000 x g) and the supernatant is collected as a meat juice of about 10% of the original weight and with a protein content ranging from 11 to 15%. Through TCA-precipitation it was found that the non-protein nitrogen content of this meat juice was about 25%. A beef muscle consists on average of 75% water, 18% protein, 3% fat and 4% other substances, which to 45% of their content are composed of non-protein nitrogen (creatin, amino acids and dipeptides). The proteins in the meat juice were identified by electrophoretic separation as consisting of sarcoplasmic proteins to 96% and high molecular weight proteins to 4%, mainly titin (MW ≈ 1000 kdalton). The sarcoplasmic proteins, which are the soluble proteins of the sarcoplasm (mostly the enzymes of the glycolytic pathway), constitute about 30 to 35% of the total muscle protein. Evidently, the molecular weight distribution of the components within the meat juice covers such a wide range as 1000 kdalton down to 100-200 dalton.

The upper diagram in Figure 4 gives the lowering of the interfacial tension at the A/W-interface of the meat juice in (0.2-7) at a concentration of 10^{-2} wt%. The interfacial tension decay is relatively rapid and high. By lowering the protein concentration of the meat juice by one decade to 10^{-3} wt% the stepwise character of the γ-t-curve emerges. A very quick lowering of the interfacial tension is followed by an induction period and later on another, slower decrease in surface tension is obtained, more like the usual behaviour of proteins at these low concentrations. Therefore, the first quick decay in surface tension was suspected to originate from the non-protein nitrogen fraction. This was confirmed by TCA-precipitation of the meat juice and thereafter registration of the γ-t-curve of the supernatant at the concentration of 10^{-3} wt%. The result can be seen in the lower diagram in Figure 4. Furthermore, measurements of the interfacial tension decay of pure amino acids (for example α-alanine) at the same concentration give similar results (8).

Therefore, it is suggested that the stepwise character of the γ-t-curve originates from competitive adsorption between the amino acids and dipeptides in the non-protein nitrogen fraction and the high molecular weight proteins in the protein fraction. It is not so much that the probability of adsorption is much higher for the proteins than for the smaller peptides, but rather that the rate of desorption decreases markedly with increasing molecular weight (9). The proteins will then remain for longer time-periods at the interface compared to the smaller peptides, resulting in exclusion of the latter and a larger interfacial tension decay.

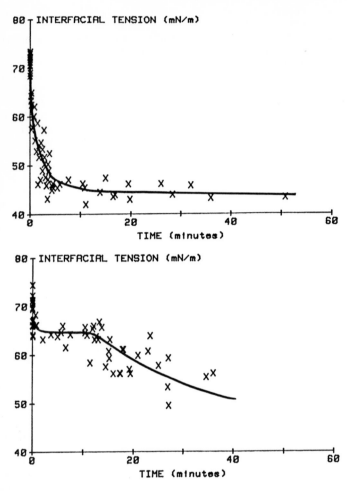

Figure 4. The lowering of the interfacial tension at the air–water interface of meat juice in (0.2–7) at a concentration of 10^{-2} wt% (upper) and 10^{-3} (lower). *Continued on next page.*

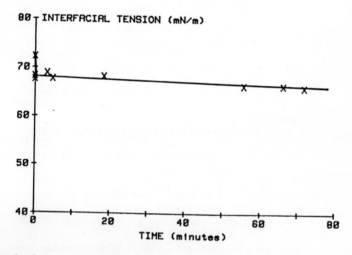

Figure 4.—*Continued*. The lowering of the interfacial tension at the air–water interface of meat juice in (0.2–7) at a concentration of 10^{-3} wt% after TCA precipitation.

The concentration dependence of the interfacial tension decay

The concentration dependence of the surface pressure attained after 40 minutes (π_{40}) can vary substantially between proteins, which Figure 5 illustrates (4). The measurements have been made at the A/W-interface. A mildly produced soy protein isolate (kindly provided by Central Soya), a commercially available sodium caseinate (DMV, Holland) and an ultrafiltrated and spray-dried WPC were used. They were studied when dispersed in distilled water and in 0.2 M NaCl solution at pH 7 denoted as (0-7) and (0.2-7), respectively. Analysis of the proteins is given in (4).

At high concentrations (10^0-10^{-1} wt% of initial subphase concentration), the surface activity of all the proteins is high and almost equal, whereas at lower concentrations the differences in surface behaviour of the proteins become evident. The caseinate (0.2-7) system is most effective as a surface active agent and is more or less independent of the concentration in the concentration range of 10^{-1}-10^{-3} wt%. The contrary is observed for the soy proteins, which gradually lose their surface activity with decreasing subphase concentration. WPC (0-7) and caseinate (0-7) have a rather similar concentration dependence in this range, and the curves are in between those of the caseinate (0.2-7) and the soy proteins. The addition of 0.2 M NaCl to the WPC dispersions does not raise the surface activity of the WPC far beyond that of the caseinate (0.2-7). The increase in lowering of interfacial tension due to the addition of salt was also observed for the other two proteins.

Log$\frac{d}{dt}\pi$-π-curves also change with protein concentration, as illustrated in Figure 6. The curves represent the rate of surface tension decay performed by WPC (0.2-7) adsorbing at the soya bean oil/water interface at different subphase concentrations (10). The π_{40}-C_p-dependence for WPC under these conditions is a more continuous increase of π_{40} as a function of protein concentration (C_p) than the more abrupt behaviour of caseinate (0.2-7), as shown in the previous Figure. As can be seen from Figure 6 a decrease in protein concentration firstly lowers the rate of surface tension decay and secondly the first barrier of lowered compressibility of the film turns up at a lower value of π. At lower concentrations the adsorbed molecules have more time to expand or unfold at the interface than at higher concentrations. The more unfolded the macromolecule the higher the number of attachments and the higher the kinetic barrier for conformational changes at the interface. Moreover, native, globular proteins form a more condensed and more tightly packed interfacial film than unfolded proteins (11), i.e. the former type of protein film should have a higher surface coverage and therefore cause a greater surface pressure than the latter. This means that with decreasing concentration the incompressibility of the film will start to rise at a lower π, which leads to a slower and lower interfacial tension decay.

What does the different π-C_p-dependence of the proteins, as is obvious in Figure 5, tell us about their interfacial behaviour? Firstly, we compare the two extremes, i.e. the interfacial behaviour of the caseinate (0.2-7) and the soy protein (0-7).

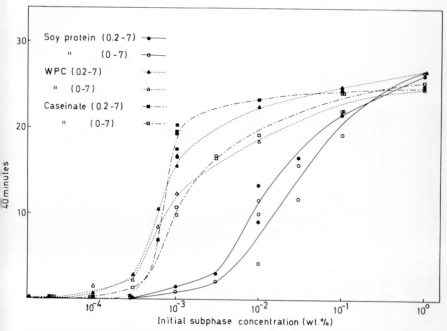

Figure 5. The surface pressure attained after 40 min., $\pi_{40\ min.}$, as a function of the initial subphase concentration for the soy protein isolate, the WPC and the sodium caseinate dispersed in (0-7) and (0.2-7), respectively. (Reproduced with permission from Ref. 4. Copyright 1978 Blackwell Scientific Publications.)

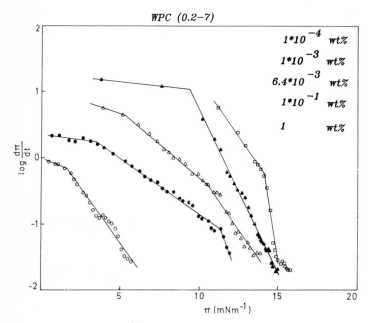

Figure 6.   Log$\frac{d\pi}{dt}$ as a function of π for WPC (0.2-7) adsorbing at the soya bean oil/water interface at different subphase concentrations (10).

Their $\log\frac{d\pi}{dt}-\pi$-curves at the concentrations of 10^{-1} and
10^{-2}wt%, respectively, can be seen in Figure 7. From the Figure
it can be deduced that the kinetics of the interfacial tension
decay of the caseinate (0.2-7) at the two concentrations are
similar. Moreover, the lowering of the interfacial tension by
these proteins under these conditions is relatively independent
over a wide concentration range (see Figure 5). This suggests that
those conformational changes that occur on adsorption of caseinate
(0.2-7) are so quick that an increase of the protein concentration
from about 3×10^{-3} to 1 wt% will give rise to almost the same
type of kinetics and therefore similar configuration of the
adsorbed proteins at the interface.

However, when lowering the subphase concentration below
10^{-3} wt% the surface activity of the caseinates is more
drastically lost than for the WPC, as revealed by Figure 5.
Evidently, the caseinates do not seem to be able to spread
sufficiently at the interface to form a monolayer at these low
concentrations. This seems though to be the case for the whey
proteins, which are more surface active than the caseinates in
this concentration region.

In Figure 7 it can be seen that, for the soy proteins
dispersed in (0-7), the $\log\frac{d\pi}{dt}-\pi$-curves differ substantially
with regard to the concentrations investigated. According to
Figure 5, $\pi_{40\text{ min}}$ for the soy protein (0-7) is also very
concentration-dependent. The low surface pressure attained for
these proteins at the concentration of 10^{-3} wt% suggests such
slow conformational changes occurring on adsorption that almost no
monolayer coverage is obtained after 40 minutes. By increasing the
concentration, films containing a mixture of unfolded and
essentially native molecules are formed, where the latter
configuration of the proteins predominates the higher the
concentration.

Comparison of proteins adsorbing at the A/W- and the O/W-interface

An overview of the interfacial pressure attained after 40 minutes
as a function of the protein concentration can be seen in
Figure 8, for the three proteins soy protein, WPC and sodium
caseinate in (0-7) and (0.2-7) and soya bean
oil water (O/W) interface (10).

An interesting feature to be observed in Figure 8 is that at
the high concentration range π_{40} can be as much as 10 mN m^{-1}
higher at the A/W-interface than at the O/W-interface, whereas in
the low concentration range the interfacial pressure is higher at
the O/W- than at the A/W-interface. This is the case for all the
proteins studied. However, the subphase concentration range in
which the surface activity of the proteins at the O/W-interface
exceeds that of the A/W-interface differs between the proteins.
For the soya proteins this happens at a concentration of about
10^{-2} wt% and for the other two proteins at $\approx 10^{-3}$ wt%.

Such great differences in the surface pressure obtained
between the O/W- and A/W-interfaces are not consistent with the
findings of Graham and Phillips (13). Moreover, their recorded

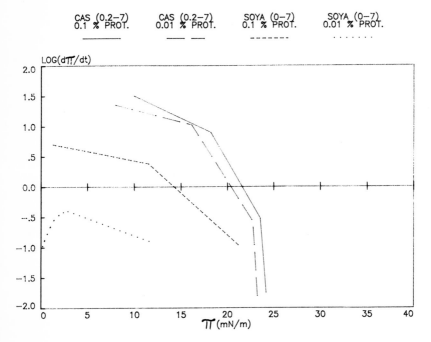

Figure 7. Log$\frac{d\pi}{dt}$ as a function of π for caseinate (0.2-7) and soy protein (0-7) adsorbing at the A/W-interface at the concentrations of 10^{-1} and 10^{-2} wt%, respectively.

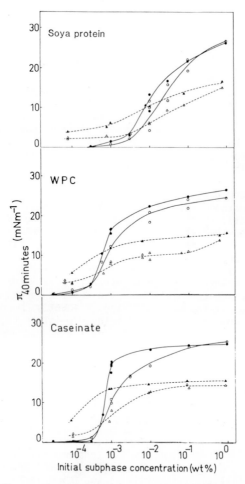

Figure 8. The surface pressure attained after 40 min., π_{40}, as a function of the initial subphase concentration for the proteins soy protein, WPC, and caseinate adsorbing at the ($\bullet$, $\circ$), A/W- and ($\blacktriangle$, $\triangle$) O/W- interfaces; closed symbols (0.2-7); and open symbols (0-7). (Reproduced with permission from Ref. 10. Copyright 1982 Blackwell Scientific Publications.)

π-C_p-curves, although rather similar in pattern to ours, are generally displaced two decades lower in concentration. Although Graham and Phillips have worked with other proteins (Lysozyme, BSA and β-casein) we would like to suggest that these differences mainly originate from variation in methodology. They have recorded the equilibrium surface pressure using the Wilhelmy plate method.

In order to have an initial clean interface for adsorption studies with this technique, one usually injects the protein solution into the subphase. In the drop volume method the cleaning of the interface is performed by forming a drop rapidly in between those to be measured, in such a way that the volume detached corresponds to γ_0, i.e. the clean interface. The main difference between the drop volume and the Wilhelmy plate method, when using the injection technique, is that the fresh interface is exposed to an initially uniform concentration of protein in the former method, whereas this is not the case in the latter. This difference promotes the existence of more unfolded proteins in the adsorbed films formed with the Wilhelmy plate method as opposed to the drop volume method. This could be the explanation for the higher surface activity shown by the proteins at the very low concentrations and the less marked difference in behaviour at the A/W- and O/W-interfaces as obtained by measurements with the Wilhelmy plate method.

When looking at the kinetics of the interfacial tension decay at the two studied interfaces some interesting features emerge. Firstly, no induction period is found in the γ-t-curves for the proteins adsorbing at the O/W-interface. Evidently, monolayer coverage is performed more quickly at the O/W-interface, that it is undetectable within the time limits of the method used. As the initial subphase concentration is the same at the two interfaces this behaviour suggests that the proteins are more unfolded at the O/W-interface.

Secondly, the $\log\frac{d\pi}{dt}$-π-curves look different at both interfaces when studied at the same protein concentration. This is illustrated in Figure 9 for three proteins at different concentrations. When comparing interfaces it can be seen from Figure 9 that at the very beginning of the process (i.e. at low π) the lowering of the interfacial tension is faster at the O/W- than at the A/W-interface. But as the process proceeds the lowered compressibility of the film at the O/W-interface reduces the increase in interfacial pressure more than at the A/W-interface, which in the end results in a higher interfacial pressure being obtained at the A/W-interface. This behaviour is consistent with more expanded proteins at the O/W-interface as opposed to the A/W-interface according to the same reasoning as before.

The different π-C_p-dependence for the two interfaces as observed in Figure 8 might also be attributed to more unfolded proteins existing at the O/W-interface. At the high concentrations the higher degree of unfolding at the O/W-interface will lead to a less packed interface and consequently a lower π_{40}. However, at the lower concentrations only those proteins unfolded to such a degree that they are able to form a monolayer can give any appreciable rise in surface pressure. Probably, therefore, higher

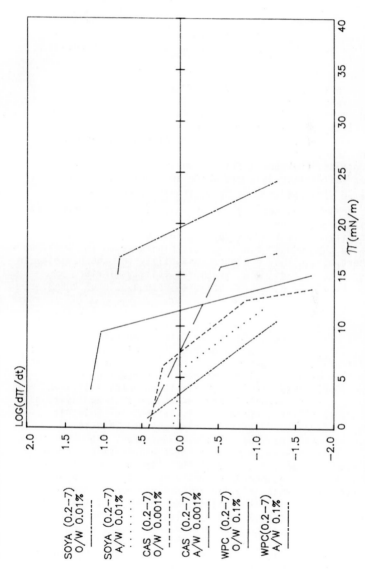

Figure 9. Log $\frac{d\pi}{dt}$ as a function of π for the soy protein (0.2–7), and WPC (0.2–7) adsorbing at the A/W- and O/W- interfaces. (Reproduced with permission from Ref. 10. Copyright 1982 Blackwell Scientific Publications.)

surface activity is observed for proteins adsorbed at the
O/W-interface as opposed to the A/W-interface in this low
concentration region.

Comparison of proteins of varying charge density (pH, ionic strength)

The charge density of the proteins can be varied by change in pH
and ionic strength. As judged from Figure 8, the addition of salt
to 0.2 M NaCl in the protein solution raises, in general, the
surface activity of the proteins studied at every concentration
both for the A/W- and the O/W-interface. By elucidating the
kinetics of the interfacial tension decay ($\underline{4}$, $\underline{10}$), it was found
for the proteins, that the initial lowering of the interfacial
tension was faster when salt was added and the first 'kink' in the
$\log\frac{d\pi}{dt}-\pi$-curve appeared at a higher π. The first observation
suggests a somewhat higher degreee of unfolding at the interface
in (0.2-7), but that does not seem to create a lowered
compressibility of the film as opposed to the protein film at
(0-7). Due to the reduced electrostatic hindrance on the addition
of salt there is probably a higher possibility for the proteins to
come closer without causing an incompressibility barrier.
Therefore, the addition of salt promotes both a quicker and a
higher interfacial tension decay.
    We have also made some preliminary measurements ($\underline{14}$) in
lowering the pH nearer the isoelectric point (IEP) of the
proteins, as results appearing in the literature ($\underline{15}$) show optimum
surface activity in the neighbourhood of the IEP. The results of
our measurements can be seen in Figure 10, where the $\pi-C_p$-curves
are plotted for WPC and sodium caseinate dispersed in (0-7),
(0,2-7) and (0-6), respectively. The latter denotes distilled
water at pH 6. It can be noted from the Figure that by lowering
the pH from 7 to 6, the surface activity of both proteins is
raised, especially at the lowest concentrations. This is
consistent with the behaviour on salt being added, i.e. an
increased unravelling of the proteins at the interface without
causing a lowered compressibility of the film. However, the degree
of unfolding seems to be more pronounced in (0-6) than in (0.2-7),
as the π_{40} attained is clearly larger for the former than for
the latter at very low concentrations. Moreover, in comparison
with the film formed in (0.2-7) the proteins adsorbing in (0-6)
give rise to a lower surface pressure in the concentration region
of 10^{-2} and 10^{-3} wt%. This suggests that increased
incompressibility appears at a lower π for protein films formed in
(0-6), which is also consistent with more spread proteins in the
films formed in (0-6) than in (0.2-7).

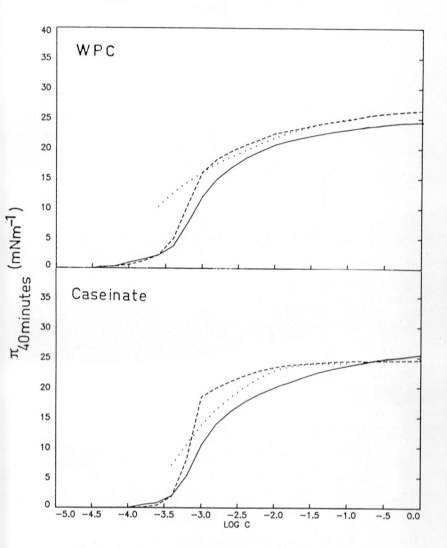

Figure 10. The surface pressure attained at the A/W–
interface after 40 min, π_{40}, as a function of the
initial subphase concentration for WPC and sodium
caseinate dispersed in (0.2-7) (– – –), (0-7) (——) and
(0-6) (·····), respectively.

Literature Cited

1. Tornberg, E. J. Coll. Interface Sci. 1977, 60, 50.

2. Tornberg, E. J. Coll. Interface Sci. 1978, 64, 391.

3. Tornberg, E.; Lundh, G. J. Coll. Interface Sci. 1981, 79, 76.

4. Tornberg, E. J. Sci. Fd. Agric. 1978, 29, 762.

5. de Feijter, J.A. Proceedings from an International Symposium on 'Food Emulsions and Foams', 24-26 March, Leeds, 1986.

6. de Feijter, J.A. J. Coll. Interface Sci. 1982, 90, 289.

7. Graham, D.E.; Phillips, M.C. J. Coll. Interface Sci. 1979, 70, 403.

8. Tornberg, E.; Olsson, A. To be published.

9. MacRitchie, F. J. Coll. Interface Sci. 1985, 105, 119.

10. Tornberg, E.; Granfeldt, Y.; Håkansson, C. J. Sci. Fd. Agric. 1982, 33, 904.

11. MacRitchie, F. Adv. Protein. Chem. 1978, 32, 283.

12. Olson, D.G.; Parrish, F.C.; Stromer, M.H. J. Food Sci. 1976, 41, 1036.

13. Graham, D.E.; Phillips, M.C. J. Coll. Interface Sci. 1979, 75, 415.

14. Tornberg, E.; Ediriweera, N. To be published.

15. Waniska, R.D.; Kinsella, J.E. J. Agr. Food Chem. 1985, 33, 1143.

RECEIVED January 28, 1987

Chapter 41

Caseins and Casein Micelles at Interfaces

Douglas G. Dalgleish

The Hannah Research Institute, Ayr KA6 5HL, Scotland, United Kingdom

The stabilizing properties of individual caseins,
sodium caseinate and casein micelles are described with
respect to the formation and behavior of emulsions. In
particular, attempts are made to relate the properties
of the emulsions (or rather the interfacial proteins)
to the properties of the proteins and protein complexes
when they are in their solution or suspended state. In
this, the stabilizing action of κ-casein in the
different emulsions is described an important factor
being its susceptibility to atack by rennet, which may
serve as an indicator of its conformation on the
interface.

Two forms of the casein proteins are used to stabilize oil/water
emulsions. Most simply, a mixture of the proteins (as in sodium
caseinate), or the individual proteins themselves, allows the
formation thin layers, if not monolayers, of protein at fat/water
interfaces (1). By contrast, in preparations such as homogenized
milk, the entities which bind to and stabilize the fat/water
interface are much larger and more complex, and may be considered as
intact or semi-intact casein micelles (2). These particles are
highly aggregated complexes of the four caseins (α_{s1}-, α_{s2}-, β- and
κ-casein), linked together by inorganic calcium phosphate (3). They
have molecular weights in the region 10^7 - $>10^9$, and diameters of
20-600 nm (4), compared with the monomeric caseins which have

0097-6156/87/0343-0665$06.00/0

molecular weights in the region of 20 000. It is possible to
consider the casein micelle as an interfacial system, since there
are differences between the proteins comprising the interior core of
the particle (the α_s - and β-caseins, which have overall a
hydrophobic character), and the protein which defines the surface,
the amphiphilic κ-casein, which appears to bind to the core and
provide steric stabilization of the particles (5). κ-casein in this
state has different properties from its state on oil/water
interfaces, but the stabilizing properties of κ-casein in casein
micelles are relevant not only to the micelles themselves, but to
the emulsions formed using the pure protein, sodium caseinate, or
casein micelles. It has therefore been essential to understand the
properties of "interfacial" κ-casein in casein micelles.

The caseins (or more generally, caseinate) have been long
recognized as excellent surfactants, and have been studied from that
point of view (6). Less effort has been devoted to the comparison
between the properties of the individual caseins when they are in
free solution and when they are bound to fat/water interfaces. It
is these relationships which we find of particular interest, since
the behavior of the caseins in solution has been extensively
studied. Although our studies of caseins at interfaces are
relatively recent, they were established within the framework of
earlier research on the caseins and casein micelles generally.
Especially, emulsions stabilized by individual caseins have not been
widely studied, probably because small-scale homogenization
equipment is not widely available, although such instruments have
been described (7). The work described here was enabled by the
development of a small-volume homogenizer by the group at the
Procter Department of Food Science at the University of Leeds, and
we acknowledge the help and collaboration of E. Dickinson and G.
Stainsby in the studies of the emulsions involving individual
caseins.

This brief review comprises three subject areas: (i) the
structure and properties of the κ-casein surface layer in casein
micelles; (ii) the properties of the protein fraction in homogenized
milks (i.e. basically intact casein micelles adsorbed at fat-water
interfaces); (iii) the properties of caseinate and individual
caseins adsorbed at the interfaces. In this, we are at present less

involved in determining how the proteins bind to the interface than
in the properties of the emulsions and their proteins relative to
their native properties in solution or suspension.

The Surface of the Casein Micelle – κ-Casein at an Interface

Casein micelles, as seen by electron microscopy, are approximately
spherical, although surface detail is difficult to distinguish. From
analysis of the distribution of κ-casein between micelles of
different sizes (7), it appears that the κ-casein is to be found on
the surface, with the micellar "core" (8) being formed of the other
(α_{s1}-, α_{s2}- and β-caseins). The behavior of the particles cannot be
described by models involving charge interactions (10),and so a
model involving steric stabilization was postulated in 1975 (5):
this was later reinforced by studies of the voluminosity (i.e. the
weight of water incorporated in the structure per unit weight of
protein) of micelles (11). The presence of a "hairy" stabilizing
layer of κ-casein was confirmed by the observation that the the
removal of the polypeptide "hairs" by the proteolytic action of
chymosin decreased the diameter of micelles (12). We have made a
number of studies of the properties of this hairy layer.

Calculations based on measurements of the changes in diameter
and electrophoretic mobility of casein micelles during controlled
renneting, showed that the hairy layer has a true thickness of 10–15
nm, but because it is partially draining, the hydrodynamic thickness
is about 5 nm (13). Studies of the charge density show that the
density of hairs is such that each hair occupies some 480 nm^2 of the
micellar surface. Only about 10–15% of the κ-casein in the micelles
is involved directly in the formation of the hairy layer, although,
since κ-casein is probably oligomeric in the micelle (14), this may
not invalidate the theory that all of this protein is near the
micellar surface. The large area per hair may be partly explained
by the presence of oligomeric κ-casein or there may be other casein
molecules between the molecules of κ-casein. The hairy layer
collapses when micelles are introduced into buffers containing
concentrations of ethanol of about 15%, after which no further
change in diameter is caused by removal of the hairs with rennet
(15). This, and similar studies allow the stabilizing layer to be

considered as a weak gel, whose structural strength and overall
thickness is dependent on the interactions between the hairs,
mediated by such factors as the ionic strength, and the calcium
concentration (16). These two descriptions are by no means
contradictory, insofar as the individual hairs on the surface will
interact and structurally correlate with one another.

Thus, κ-casein in its native stabilizing role exists, probably
as small disulphide linked polymers, bound to the micellar surface
(the ill-defined boundary between the hydrophobic interior of the
micelle and the aqueous phase). C-terminal polypeptides (61
residues) of the protein project from the surface into the solution.
In this position, the macropeptide moiety of the protein is
conformationally free (17), constrained only by its interactions
with its neighbours (16), and the bond 105-106 of the protein is
held in a particularly advantageous position for attack by enzymes
such as chymosin (18). The importance of this will be apparent when
emulsions stabilized by κ-casein are being discussed. The enzymic
action has a relatively small but detectable effect on the
hydrodynamic diameters of the particles, and a large effect on their
electrophoretic mobilities, which decrease by between one-third and
one-half, depending on the solution conditions (19).

Homogenized Milks - Casein Micelles at Interfaces

Of the caseins, only κ-casein is capable of exerting a stabilizing
effect in systems which contain calcium ions in appreciable
quantities. This is the case in milk, where the other caseins are
effectively rendered insoluble by their binding to calcium
phosphate, and where there are also appreciable pools of calcium
ions. When the calcium is removed, all of the caseins act as
surfactants, and can stabilize emulsions. Even in milk, the casein
micelles can bind to and stabilize unprotected fat surfaces, as in
homogenized milk.

From the studies of Walstra and co-workers (2,20), it appears
that the fat globules in homogenized milk are stabilized by
apparently largely intact casein micelles although some smaller
caseinate complexes can also be involved. However, homogenized
milks differ in a number of respects from whole or skim milks,

especially their behaviour during renneting and their response to heating. Thus, although the adsorbed casein micelles are virtually intact, the stabilized fat globules do not share the properties of the stabilizing agents (homogenization of micelles does not introduce differences in behaviour of the micelles (21)).

The particles in homogenized milk can be fractionated by centrifugation: this allows either a coarse fractionation into particles which float or sink (i.e. are more or less dense than water), or a more detailed fractionation, when differential centrifugation is used (22). Generally, although depending on the homogenization pressures, approximately half of the fat fraction floats. The particles in this fraction have a lower than average protein load, and the dividing point between the two fractions appears to be when the fat surfaces are saturated with casein micelles, so that the less dense fraction will have gaps between the stabilizing micelles (23). Nevertheless, all of the particles had larger protein loads than could be explained by monolayer coverage by monomeric caseins. All of the fractions isolated had hydrodynamic radii which were greater by about micellar dimensions than the globules of fat, as could be demonstrated by measuring the diameters of the particles before and after dissociation of the micelles by treatment with EDTA (23). This confirms that the micelles bound to the fat globules were largely intact. In contrast to the observations of Walstra (20), we found little evidence that the serum proteins of the milk were bound to the fat surface.

The micelles on the fat surfaces cannot be completely intact, because the original hydrophilic κ-casein surface of the micelle is unlikely to bind to the fat surface. Homogenization must cause partial disruption of the micelles (21), allowing hydrophobic points of contact with the freshly exposed fat surfaces. There is no evidence that casein micelles interact with polystyrene latices to form a model system, for example (Dalgleish, unpublished results). Thus, although the micelles which bind to the fat in homogenized milk appear to be intact, their surfaces must have suffered some distortion, particularly of the sterically stabilizing κ-casein molecules which are near to the point of interaction of the micelle and the fat surface.

The particles in homogenized milk showed similar electrophoretic properties to casein micelles (19), despite the slightly higher mobilities of the latter (23). The small differences may result from more complex hydrodynamic properties of the micelle/fat complexes, or may reflect small differences in the conformation of the surfaces of the bound micelles. It was impossible to estimate how intact were the hairy layers of the stabilizing micelles, because the experimental error in determining the diameters of the composite particles is larger than the expected changes in diameter during renneting (about 10 nm). The kinetics of the changes in electrophoretic mobility during renneting were similar in casein micelles and homogenized milk particles (23), suggesting that the κ-casein was still close to the surface of shear in the composite particles, and also that its susceptibility to rennet attack, and therefore its conformation, had not been altered.

The destabilization of the fat particles by rennet did, however, depend on their composition. Particles which had a full covering of micelles aggregated only late in the enzymic reaction, when most of their κ-casein had been destroyed, as do casein micelles (24). Conversely, the more sparsely covered fat particles began to aggregate at an early stage in the enzymic reaction, indicating that they could be destabilized by the breakdown of only a small fraction of their κ-casein (25). The reason for this is unclear. Although larger particles have an incomplete covering of casein micelles, they are unlikely to have portions of their surfaces completely bare of protein: it is probable that much smaller casein complexes cover the spaces between bound micelles. Since the casein micelles are the overall stabilizing agents, these small complexes need not be κ-casein. They might therefore stabilize the surface only to a small extent. Such a model explains the observation that the changes in electrophoretic mobility during renneting are similar for all types of particle, since it allows similar behaviour at the surfaces of shear. The premature aggregation of the larger particles can be explained by interpenetration of partly-renneted bound micelles, with binding of these entities arising from the interaction of the partly renneted surface of the micelle with the smaller casein aggregates on the fat surfaces.

Caseins and Sodium Caseinate at Fat-Water Interfaces

Effective models of casein micelles might in principle be
constructed by binding caseins to a fat/water interface in an
emulsion or to a polystyrene latex matrix. This is in fact not the
case. It is questionable whether simple systems such as these will
explain the behaviour of either casein micelles or homogenized milk,
and we have sought an explanation for this anomaly.

Individual caseins bound to oil/water interfaces under
favourable conditions can interchange relatively rapidly with those
in solution. It is of course known that the different caseins have
different affinities for the interface (26). An emulsion, prepared
from oil and pure α_{s1}-casein, which is resuspended in a solution of
pure β-casein, loses α_{s1}-casein and gains β-casein within a few
hours (27). α_{s1}-casein, however, does not readily displace β-casein
from a β-casein/oil emulsion. In emulsions prepared using sodium
caseinate it would be expected that similar exchanges should occur,
but analyses of supernatants in emulsions prepared from butter oil
and sodium caseinate showed somewhat less enrichment of the aqueous
phase in α_s-casein and β-casein than was expected, although a degree
of preferential adsorption did occur (Robson, unpublished results).
This is presumably the result of the presence of more structured
protein particles in the solutions of sodium caseinate, which
therefore behaves in a more complex way than would be predicted from
a knowledge of the behaviour of the individual proteins.

The influence of the aggregation state of the caseins is
emphasized by attempts to prepare emulsions using κ-casein as the
surface active agent. Surprisingly, in view of its clearly defined
role as the stabilizing protein in native casein micelles and in
homogenized milks, κ-casein as isolated from milk was not able to
produce stable emulsions (27). Both α_{s1}- and β-caseins produced more
stable emulsions. However, as isolated, the κ-casein exists in the
form of disulphide-linked oligomers (28), and these remain
undissociated during the fractionation of the casein complex. These
κ-casein oligomers present a strongly hydrophilic exterior to their
surroundings, and κ-casein in this state is a poor surfactant.
Treatment of the κ-casein with 2-mercaptoethanol breaks the
disulphide links and it was found that such treatment, followed by

dialysis to remove the mercaptoethanol, greatly improved the
emulsifying capacity of the protein, and much more stable emulsions
with a smaller particle size were produced. As expected, these
emulsions were stable in the presence of calcium ions. The state of
the κ-casein in sodium caseinate may also explain the unexpected
results with respect to casein partition during and after emulsion
formation. The inability of oligomeric κ-casein to form emulsions
also implies that the stabilizing micelles in homogenized milk must
have been altered before they can bind to the fat surface.

The different caseins, α_{s1}-, β- and κ-casein, have diffferent
net charges (29), and the first two bind Ca^{2+} much more strongly
than the third (30). However, emulsions and polystyrene latices
coated with the individual caseins showed rather similar
electrophoretic mobilities irrespective of the particular casein
used. The mobilities of both coated and uncoated latices are
reduced by increasing salt concentration, and by the concentration
of Ca^{2+}, but the decline in the mobility appears to be determined
mainly by the ionic strength rather than by specific effects of the
binding of Ca^{2+} to the proteins, since only small differences were
found to exist between the α_{s1}- and β-caseins on the one hand and
κ-casein on the other (31). Moreover, these simple systems show
appreciable differences from native casein micelles in their
response to Ca^{2+}. In casein micelles, the binding sites for Ca^{2+}
appear to be some distance from the surface of the hairy layer (13)
and the same argument can be presumably used for the individual
caseins, and show that the calcium binding sites in the synthetic
particles are within the surface of shear. On the other hand, the
binding of Ca^{2+} may cause conformational changes in the interfacial
layer.

The stabilities of the emulsions incorporating individual
caseins show resemblances to those of the same caseins in solution,
particularly in respect of their sensitivity to the presence of
calcium ions. The emulsions based on κ-casein show no tendency to
aggregate when calcium is added: on the other hand, both the
β-casein and α_s-casein emulsions are reversibly precipitated by
Ca^{2+}, showing that the original properties of the proteins are not
extensively modified by their binding to the fat surfaces (27). In
solution, the behaviour of both α_s- and β- caseins is believed to

depend on the reduction of the original negative charge of the proteins by the binding of Ca^{2+} (32). However, the lack of significant specific effect of Ca^{2+} on the electrophoretic mobilities of the complexes may indicate that a different mechanism obtains, probably the simple shielding of charge by increased ionic strength.

The most striking evidence of the differences between the native structures of caseins and their conformation at oil/water interfaces is evidenced by κ-casein. It has already been shown that in its native state on the surface of the micelle the protein is highly susceptible to proteolysis by rennet. It was found that, when an emulsion stabilized by κ-casein was treated with rennet, the electrophoretic mobilities of the particles were decreased, as occurs with casein micelles. In such an emulsion, there is of course free κ-casein in the aqueous phase. When the emulsion was washed to remove the free protein, and was treated with rennet, the decrease in electrophoretic mobility occurred only slowly (27). The kinetics of the interaction of the enzyme with the rennet-sensitive bond in κ-casein have been altered, presumably as a result of the conformation of κ-casein in a monolayer on an oil/water interface being such that the sensitive bond is no longer readily accessible to the active site of the enzyme. The bond is known to be situated at the junction of the hydrophilic and hydrophobic parts of the κ-casein, and it is assumed that the latter is involved in the binding of the protein to the interface. If κ-casein on the surface is thus rendered inaccessible to the enzyme, then the change in electrophoretic mobility which occurs during renneting of the original emulsion must result from either (i) normal enzyme action on the κ-casein in solution, and rapid exchange between the para-κ-casein formed in this way and the intact κ-casein on the interface, or (ii) the κ-casein in the original emulsion being more than a monolayer. Adsorption isotherms are lacking to allow these two possibilities to be distinguished.

Conclusion

Most of the work which has been described has been within a context of general research into milk, and is generally in response to specific problems in this area, especially with respect to the behaviour of homogenized milk systems and caseinate emulsions. The studies described above are mostly at an early stage. Further research is required and is in caseins when they are on an oil/water interface. The studies on κ-casein emulsions and their susceptibility to specific enzymic attack suggest a fruitful area of research which may be more generally applied. Indeed, since κ-casein is very specifically attacked by chymosin or rennet, it is possible only to discuss the location or conformation of one bond. More general enzymic attack is desirable, and may offer a better opportunity of defining clearly the particular parts of the proteins which are actually involved with binding to the interface. This in turn may allow better functional definition of the properties of the particles in the emulsion. Such research is actively underway at the time of writing, but results are not yet available.

The functional properties of the particles in homogenized milk also require better definition. It is only partly evident how instability develops in the larger particles, but the problem of the conformation of the stabilizing entities in the particles remains. Definition of an adequate model of the complex fat/casein particles in the milk is still to be achieved, and until this is done there seems to be little chance of understanding the complex behaviour of this material.

The emulsions formed using α_{s1}- and β-caseins appear to be understandable in terms of similarity to the behaviour of the proteins themselves. However, it appears that more attention must be paid to the behaviour of caseins in mixed systems, especially in terms of the factors which govern the exchange of caseins on the interfaces. Such interchange can influence the behaviour of the particles by altering the nature of the surface layer (e.g. β-casein precipitates less readily than α_{s1}-casein when subjected to Ca^{2+}). It may be possible to control these exchange reactions to produce emulsions of desirable defined properties.

Literature Cited

1. Phillips, M.C. Chemistry and Industry 1977, 170-6.
2. Oortwijn, H.; Walstra, P.; Mulder, H. Neth. Milk Dairy J. 1977, 31, 134-7.
3. Holt,C.; Hasnain, S.S.; Hukins, D.W.L. Biochim. Biophys. Acta 1982, 719, 299-303.
4. Schmidt, D.G. In Developments in Dairy Chemistry; Fox, P.F., Ed.; Applied Science: London, 1982; Vol. 1, Ch.2.
5. Holt, C. Proc. Int. Conf. Coll. Surf. Sci.; Wolfram, E., Ed.; Akademiai Kiado: Budapest, 1975; pp. 641-4.
6. Mitchell, J.; Irons, L.; Palmer, G.J. Biochim. Biophys. Acta 1970, 200, 138-50.
7. Tornberg, E., Lundh, G. J. Food Sci. 1978, 43, 1553-1558.
8. Donnelly, W.J.; McNeill, G.P.; Buchheim, W.; McGann, T.C.A. Biochim. Biophys. Acta 1984, 789, 136-43.
9. Waugh, D.F.; Creamer, L.K.; Slattery, C.W.; Dresdner, G.W. Biochemistry 1970, 9, 786-95.
10. Payens, T.A.J. J. Dairy Res. 1979, 46, 291-306.
11. Walstra, P. J. Dairy Res. 1979, 46, 317-23.
12. Walstra, P.; Bloomfield, V.A.; Wei, J.G.; Jenness, R. Biochim. Biophys. Acta 1981, 669, 258-60.
13. Holt, C.; Dalgleish, D.G. J.Coll.Int.Sci. 1986, in press.
14. Woychik, J. H.; Kalan, E.B.; Noelken, M.E. Biochemistry 1966, 5 2276-82.
15. Horne, D.S. Biopolymers, 1984, 23, 989-93.
16. Horne, D.S. Colloid Polymer Sci. 1986 in press
17. Griffin, M.C.A.; Roberts, G.C.K. Biochem. J. 1985, 228, 273-6.
18. Raap. J.; Kerling, K.E.T.; Vreeman, H.J.; Visser, S. Arch. Biochem. Biophys. 1983, 221, 117-24.
19. Dalgleish, D.G. J. Dairy Res. 1984, 51, 425-38.
20. Oortwijn, H; Walstra, P. Neth. Milk Dairy J. 1979, 33, 134-154.
21. Walstra, P. Neth. Milk Dairy J. 1980, 34, 181-190.
22. Dalgleish, D.G.; Robson, E.W. J. Dairy Res. 1985, 52, 539-46.
23. Dalgleish, D.G.; Robson, E.W. J. Chem. Soc. in press.
24. Dalgleish, D.G. J. Dairy Res. 1979, 46, 653-61.

25. Robson, E.W.; Dalgleish, D.G. J. Dairy Res. 1984, 51, 417–424.

26. Dickinson, E.; Robson, E.W.; Stainsby, G. J. Chem. Soc. Far. Trans. 1 1983, 79, 2937–52.

27. Dickinson, E.; Whyman, R.H.; Dalgleish, D.G. J. Chem. Soc., in press.

28. Yaguchi, M; Davies, D.T.; Kim, Y.K. J. Dairy Sci. 1968, 51, 473–7

29. Swaisgood, H.E. In Developments in Dairy Chemistry; Fox, P.F., Ed; Applied Science: London, 1982; Vol. 1 Ch.1. 28.

30. Dickson, I.R.; Perkins, D.J. Biochem. J. 1971, 124, 235–40.

31. Dalgleish, D.G.; Dickinson, E.; Whyman, R.H. J. Coll. Int. Sci. 1985, 108, 174–9.

32. Horne, D.S.; Dalgleish, D.G. Int. J. Biol. Macromol. 1980, 2, 154–160.

RECEIVED January 28, 1987

Chapter 42

Interfacial Properties of Milk Casein Proteins

P. Paquin, M. Britten, M.-F. Laliberté, and M. Boulet

Groupe de Recherche en Sciences et Technologie du Lait (STELA), Département de Sciences et Technologie des Aliments, Université Laval, Québec G1K 7P4, Canada

Casein-casein and casein-monoglyceride interactions have been investigated at air-water interface. Casein micelles were dissociated by dialyzing against EDTA. Monolayer techniques were used to characterize the interfacial properties of the resultant Fractions. Fraction I contained highly cohesive complexes that did not unfold at the interface and had an average diameter of 9.1 nm. These particles are thought to represent submicelles, previously identified in micelle formation. Fraction II showed interfacial properties that are characteristic of spread casein monomers, and contained mainly α_s-casein. The results are discussed in relation to casein interactions and micellar formation. Mixed monolayers of sodium caseinate/glyceride monostearate (NaCas/GMS) were also examined at different composition ratios. The results show that for low surface pressures (0-20 mNm^{-1}), there is a condensation ascribable to hydrophobic interactions in the mixed film. At high surface pressures, the hydrophobic interaction is modified and the protein is expelled from the monolayer into the subphase. These results are discussed in relation to emulsion stability.

Milk is one of the oldest constituent of human nutrition. Raw milk is, however, frequently unsafe for human consumption. The basic function of the dairy and milk industry is the transformation of this perishable product into a hygienic one, for widespread consumption. However, the evolution of the milk industry requires new markets for milk and its derivatives. We should now consider milk as a source of ingredients to fulfil the needs of food, pharmaceutical and even cosmetic industries. To take advantage of the rich potential of milk, we must first optimize the techniques for isolation of its constituents. Secondly, we must characterize these constituents in an attempt to evaluate their potential for utilization. The physical chemistry of milk constituents is our main interest, and, as a first step. We have concentrated our efforts on milk caseins.

Caseins are the principal group of proteins in milk. They are present in three different forms (α, β, κ), whose amino acid composition and sequence have been determined (1). These proteins show a micellar organization, which form a colloidal dispersion in milk. The size of particles ranges from 20 to 600 nm (2). Casein micelles are mainly responsible for the high sensitivity of milk to physical, chemical and enzymatic treatments (3). Many models have been proposed to explain the structure of the micelle (4). However, the nature of the interactions between caseins which lead to micelle formation remains unclear. One recently proposed model presents the micelle as an aggregate of subunits (5). On the basis of this model, calcium phosphate maintains the integrity of the micelle, but the mecanism is unknown.

A better knowledge of casein micelle structure is a prerequiste for further characterization of casein derivatives. One way to study the structure of the micelle is by inducing controlled dissociation. The analysis of resultant Fractions provides information on the initial state of aggregation. Removal of calcium phosphate has been used to dissociate the micelle (6). The resultant complexes have been separated by chromatography, and their composition, and average size evaluated (7,8,9). However, the nature of interactions leading to their formation is still obscure. In the present work, we removed calcium to induce dissociation of the micelle, but afterwards, we paid attention to the interactive properties of the isolated complexes.

A second aspect of casein interactions has been studied by our group. We have been concerned with the use of caseins to improve the stability of dairy product emulsions. In raw milk, the fat dispersion is stabilized by a natural membrane which surrounds the fat globules. The role of this membrane is to maintain a low interfacial tension at the fat—serum interface. To avoid creaming, it is a common industriel practice to homogenize the product. By this process, the size of globules is reduced to decrease the gravity effect on phase separation. However, this process is accompanied by an increase of the fat—serum interfacial area. A new membrane will form at this interface to maintain the stability of the dispersion. Surfactants are often added prior to homogeneization to control the composition and the properties of the new membranes. Caseins are among the surfactants available to produce stable dispersions. Interfacial properties of pure casein species have been investigated and shown their ability to form membranes at interfaces (10,11,12,13,14). However, a mixture of surfactants usually improves the membrane properties. In the work presented here, we have investigated the interaction between caseins and glyceride monostearate at interfaces. This sort of interaction has already been reported by Durham (15) and Friberg (16,17). They concluded that caseins interact with monoglycerides, but they did not further investigate this phenomenon.

In the last few years, our research group evaluated the possibility of using monolayer techniques to caracterize caseins interactions. These techniques have allowed us to investigate the interfacial properties of milk caseins which are believed to be responsible for the interactions leading to micellar organization. Secondly, preparation of mixed films by these techniques provides relevant information about interactions between caseins and monoglycerides related to emulsion stability.

The present paper is a review of the work we have carried out to investigate colloidal aspects (micellar structure) and emulsion stability (fat globule membrane) of milk through measurement of interfacial properties. These results lead to a better understanding of milk or dairy products behaviour when various treatments are applied to them.

Experimental Methods

Casein micelles were dissociated using EDTA as the calcium sequestering agent. After dissociation, two factions were isolated by gel chromatography. Dissociation conditions and chromatography methods were as reported by Britten et al. (18). Pressure-Area (π-A) isotherms of casein micelles and their Fractions were obtained by using a Wilhelmy surface balance (19) consisting of a trough (14x60x1.2 cm) and two barriers coated with teflon and filled with double deionized water. The force was measured by an electro balance (Cahn. Cerritos, Mexico). Protein samples were carefully deposited according to the method of Trunit (20).

Caseinate/glyceride monostearate mixture were also deposited at air-liquid interface. The same surface balance was used to monitor the π-A isotherms. Five mixtures using different caseinate/monoglyceride ratios were investigated. Further details on films preparation are reported by Laliberté et al. (Can. Inst. Food Sci. Technol. J. submitted for publication).

Results and Discussion

Micellar Structure. The diagram in Figure 1 shows the π-A isotherms of casein micelles and its two fractions obtained after dissociation. The close-packing area (A_o) of casein micelles, found at minimum compressibility of the film is 0.43 $m^2 mg^{-1}$ (Fig. 1a). This value is very low compared to the values measured for pure casein species (≈ 1.0 $m^2 mg^{-1}$) (14). The isotherms for Frantion I and II are shown in Figures 1b and 1c. Fraction I showed a close-packing area of 0.08 $m^2 mg^{-1}$. This extremely low value is unusual for proteins spread at interfaces (21) and indicates that Fraction I contains complexes resistant to unfolding. The complexes appear to be highly cohesive since the interfacial energy of the air-water interface does not open the structure to any great extent. Fraction II (Fig. 1c) shows a close packing area of 1.05 $m^2 mg^{-1}$ which is in agreement with the average value of pure casein species (14,22). We have concluded that this fraction contained either non-associated caseins, or weakly associated caseins which dissociate when deposited at air-water interface.

The rate of change of the monolayer area ($-dA/d\pi$) and the compressibility coefficient (κ) were derivated from isotherms of Figure 1. These calculations confirmed the differences observed between casein micelles, Fraction I, and Fraction II. We have tried to measure ellipsometric properties of these different films, at the air-water interface (Salesse et al. unpublished results). Results obtained to date support previous ones.

In relation to the structure of casein micelles, our results indicate that the micelles are composed of two different fractions. Using the additivity law (23), we found that Fraction I and Fraction

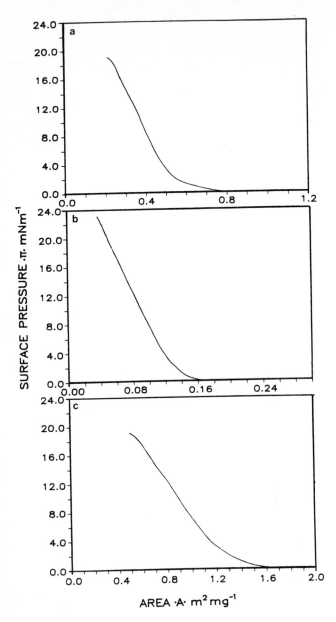

Figure 1. π–A isotherms of casein micelles and fractions. Casein micelles (a); fraction I (b); fraction II (c). (EDTA: Calcium ratio for dissociation = 5.0.) (Reproduced with permission from Ref. 18. Copyright 1986 M. Britten.)

II were present in the casein micelle film in a weight ratio of
0.64:0.36. From the close-packing area of Fraction I, and the specific volume for caseins reported by Schultz and Bloomfield (7), we
have calculated the average diameter of complexes in this fraction.
The diameter is 9.1 nm, which is in agreement with results published
by Pepper and Farrell on the diameter of the submicelles (24). The
submicellar structure of caseins has also been observed in electron
microscopy studies. It has been reported that such submicelles could
adsorb at fat-serum interface without becoming disrupted (17,25).
This is in agreement with the results obtained here at the air-water
interface.

We have used the information from interfacial studies of casein
fractions to propose a model for micelle structure (Figure 2). This
model is a compromise between the core-coat model of Waugh (26) and
the submicellar model of Schmidt (5). The micelle is composed of two
protein entities held in a coherent structure by calcium phosphate
bridges. Fraction I, which represents the submicelles, plays a stabilizing role in the micelle. The high concentration of κ-casein in
this Fraction (18) is responsible for its amphiphilic nature. It is
believed that submicelles are exposed to serum phase, as proposed by
Slattery and Evard (27). These submicelles surround the core of the
micelle in the same manner as surfactant molecules surround the fat
globule. Fraction II forms the core of the micelle. It is mainly
composed of α_s-casein which forms a framework in the presence of calcium phosphate. This structural role of α_s-casein has already been
suggested by Lin et al. (28) and Heertje et al. (29). β-casein, also
present in Fraction II is believed to interact with the framework
via hydrophobic interactions. Our model fulfils the various structural aspects present in the literature for casein micelles. Nonetheless, additionnal study is required to clarify to exact type of interaction involving α_s-casein and calcium phosphate.

Fat Membrane Structure, Figure 3 shows the isotherms for different
ratios of NaCas/GMS at pH 6.8. The curves present two distinct regions. The first one occurs below the collapse pressure of pure caseinate (0 to 20 mNm^{-1}). In that region, we observed an increase in
area as the fraction of caseinate in the film increased. We used the
additivity law to calculate the interaction between the two components at the interface. We have observed that mixed films do not follow ideality. In fact, there is a condensation of the film (30).
This condensation has been attributed to hydrophobic interactions
between caseinate and monoglyceride. In the high pressure region
(20 to 55 mNm^{-1}) the isotherms of NaCas/GMS mixtures almost overlap-
ped the pure GMS isotherm. The amount of protein remaining in the
film in this pressure range is very low but has been calculated for
different surface pressures. We found that for a surface pressure
above 30 mNm^{-1}, there is about 1.0 μg of caseinate remaining in the
film. This value is independent of the initial quantity deposited in
the mixed film.

It has been shown that the interfacial tension at a fat globule
membrane is in the vincinity of 0-2 mNm^{-1} (31). It implies that the
interfacial pressure is in the order of 40 mNm^{-1}. At that pressure,
we demonstrated that the casein fraction is almost completely expelled from the interface into the srum. The hydrophobic portion of

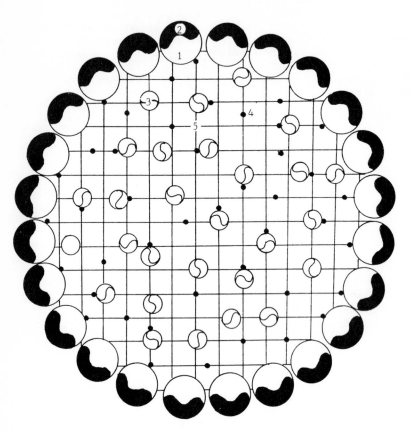

Figure 2. Proposed Model of the Casein Micelle.
(1), submicelles; (2), κ-casein; (3), α_s-caseins aggregates;
(4), colloidal calcium phosphate; (5), α_s framework.

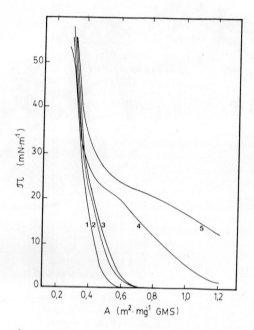

Figure 3. π–A Isotherms of NaCas/GMS Mixtures.
NaCas:GMS respective fraction of area initially occupied:
1,0:100; 2,20:80; 3,40:60; 4,60:40; 5,80:20.

proteins remaining in the membrane is limited. The organization of a membrane containing caseinate and glyceride monostearate is shown on Figure 4. Unfortunately, it is not possible from our results to know to what extent the caseins expelled from the interface are still associated with the membrane.

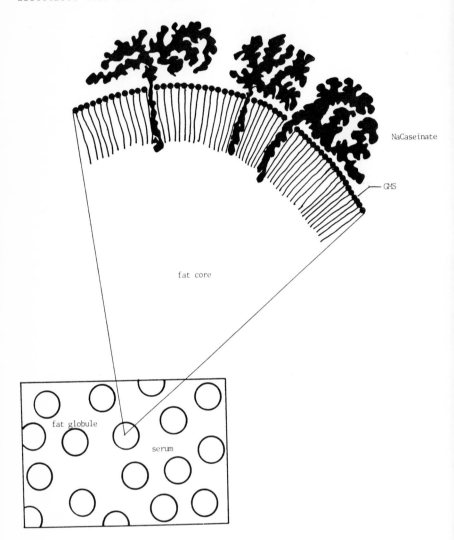

Figure 4. Proposed Model of NaCas:GMS mixed membrane in an oil/water emulsion.

The mixture of NaCas/GMS showed an interaction that could be interesting in an attempt to improve the emulsion stability. However, this interaction decreases rapidly when the surface pressure reaches the conditions found in a fat globule membrane. We believe that a mixed film that possesses stronger interactions would be more suitable. The utilization of other surfactants with NaCas should be investigated. Such studies are in progress.

Conclusion

The work presented in this paper shows the usefulness of the monolayer technique to determine the interfacial properties of caseins. Characterization of casein Fractions at the air-water interface contributed to a better understanding of the nature of casein interactions in micellar organization. Preparation of NaCas/GMS mixed films provided relevant information about the manner in which proteins might be involved in the structure of fat globule membrane.

We believe that interfacial properties of milk colloids are often responsible for milk response to various treatments. As a next step, our group will investigate the role of these properties in different processing problems such as heat stability and fouling in heat exchangers.

Acknowledgments

This work has been supported by a grant from the National Science and Engineering Research Council of Canada. We wish to express our gratitude to Nancy Kariel for her contribution to the writing of this article.

Literature Cited

1. Cheftel, J.-C.; Cuq, J.-L.; Lorient, D. Proteines Alimentaires; Tec. & Doc. Lavoisier: Paris, France, 1985, Chapter 3.
2. Schmidt, D.G.; Walstra, P.; Buchheim, W. Neth. Milk and Dairy J. 1973, 27, 128-42.
3. Walstra, P.; Jenness, R. Dairy Chemistry and Physics; John Wiley & Sons Ed.; New-York, 1984, Chapter 12.
4. Weeb, B.H.; Johnson, A.H. Fundamentals of Dairy Chemistry; AVI Ed.; Wesport, 1985.
5. Schmidt, D.G. in Developments in Dairy Chemistry-1; Fox, P.F. Ed.; Applied Science Publishers: London, 1982; p 61-86.
6. Bloomfield, V.A.; Morr, C.V. Neth. Milk Dairy J. 1973, 27, 103-20.
7. Schultz, B.C.; Bloomfield, V.A. Arch. Biochem. Biophys. 1976, 173, 18-26.
8. Creamer, L.K.; Berry, G.P. J. Dairy Res. 1975, 42, 169-83.
9. Ono, T.; Odatiri, S.; Takagi, T. J. Dairy Res. 1983, 50, 37-44.
10. Phillips, M., Food Technol., 1981, 1, 50-57.
11. Graham, D.E.; Phillips, M.C. J. Colloid Interface Sci. 1979, 70, 403-14.
12. Graham, D.E.; Phillips, M.C. J. Colloid Interface Sci. 1979, 70, 415-426.
13. Graham, D.E.; Phillips, M.C. J. Colloid Interface Sci. 1979, 70, 427-39.

14. Mitchell, J.; Irons, L.; Palmer, G.J. Biochim. Biophys. Acta.
    1970, 200, 138–50.
15. Durham, K. Chem & Ind. 1963, 631.
16. Berger, K.G. in Food Emulsion; Friberg, S. Ed.; Marcel Dekker:
    New-York, 1976; p 142–196.
17. Graf, E.; Bauer, H. in Food Emulsion; Friberg, S. Ed.; Marcel
    Dekker: New-York, 1976; p 296–377.
18. Britten, M.; Boulet, M.; Paquin, P. J. Dairy Res. 1986, 53,
    573–84.
19. Rothfield, L.I.; Fried, V.A. in Methods in Membrane Biology;
    Korn, E.D. Ed.; Plenum Press: New-York, 1975; Vol. 4, p 277.
20. Trurnit, H.J. J. Colloid Interface Sci. 1960, 15, 1–13.
21. MacRitchie, F. in Advances in Protein Chemistry; Afinsen, C.B.;
    Edsall, J.T., Richards, F.M. Ed.; Academic Press: New-York,
    1978; Vol. 32, p 283.
22. Boyd, J.V.; Mitchell, J.R.; Irons, L.; Musselwhite, P.R.; Sher-
    man, P. Colloid Interface Sci. 1973, 45, 478–86.
23. Tancrède, P.; Parent, L.; Paquin, P.; Leblanc, R.M. J. Colloid
    Interface Sci. 1981; 83, 606–13.
24. Pepper, L.; Farrell, H.M. J. Dairy Sci. 1982; 65, 2259–66.
25. Mulder, H.; Walstra, P. The Milk Fat Globule; Emulsion Science
    as applied to Milk Products and comparable Foods; Pudoc, Wage-
    ningen & CAB, Farnham Royal, 1974; p 101.
26. Waugh, D.F. in Milk Proteins: Chemistry and Molecular Biology;
    McKenzie, H.A. Ed.; Academic Press: New-York, 1971; Vol. II,
    p 3–85.
27. Slattery, C.W.; Evard, R. Biochim. Biophys. Acta. 1973; 317,
    529–38.
28. Lin, S.H.C.; Leong, S.L.; Dewan, R.K.: Bloomfield, V.A.; Morr,
    C.V. Biochemistry. 1972; 11, 1818–21.
29. Heertje, I.; Visser, J.; Smits, P. Food Micro. Struc. 1985; 4,
    267–77.
30. Gaines, G.L. Insoluble Monolayers. Interscience Publishers,
    John Wiley & Sons: New-York, 1966.

RECEIVED February 18, 1987

Author Index

Affiliation Index

Subject Index

N

O

Production by Cara Aldridge Young
Indexing by Deborah H. Steiner
Jacket design by Carla L. Clemens

Elements typeset by Hot Type Ltd., Washington, DC
Printed and bound by Maple Press Co., York, PA

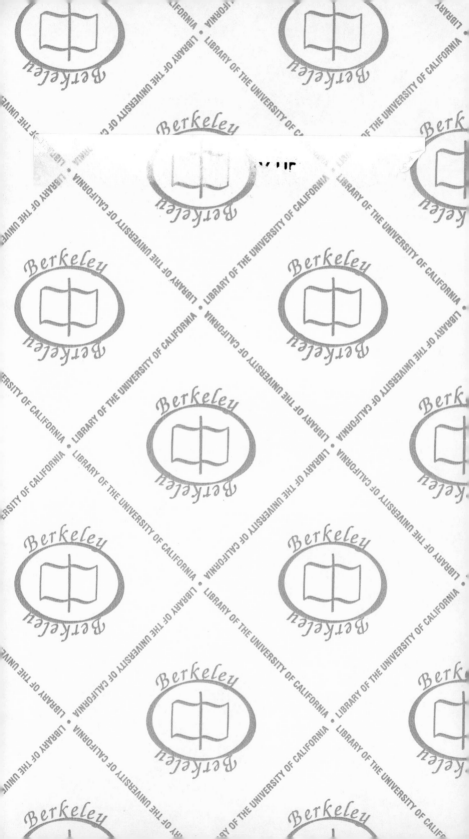